"May he walk in beauty."
Traditional first words heard by a Navajo child.

BURNHAM'S CELESTIAL HANDBOOK

An Observer's Guide to the Universe
Beyond the Solar System

ROBERT BURNHAM, JR.

Staff Member, Lowell Observatory, 1958–1979

IN THREE VOLUMES

Volume Two, Chamaeleon–Orion

REVISED AND ENLARGED EDITION

DOVER PUBLICATIONS, INC.
NEW YORK

The author takes great pleasure in offering his special thanks and appreciation to Herbert A. Luft, whose unflagging interest and support has helped immeasurably to make the *Celestial Handbook* a reality.

FRONTISPIECE: Orion Nebula

For information about our audio products, write us at:
Newbridge Book Clubs, 3000 Cindel Drive, Delran, NJ 08370

Published in Canada by General Publishing Company, Ltd., 30 Lesmill Road, Don Mills, Toronto, Ontario.
Published in the United Kingdom by Constable and Company, Ltd.

This Dover edition, first published in 1978, is an expanded and updated republication of the work originally published by Celestial Handbook Publications, Flagstaff, Arizona, in 1966.

INTERNATIONAL STANDARD BOOK NUMBERS:
paperbound edition: 0-486-23568-8
clothbound edition: 0-486-24064-9
Library of Congress Catalog Card Number: 77-082888

Manufactured in the United States of America
Dover Publications, Inc.
31 East 2nd Street
Mineola, N.Y. 11501

PREFACE

This is the second volume of the *Celestial Handbook*, covering the constellations Chamaeleon through Orion.

The format and arrangement of material is the same as in the preceding volume. A list of terms, symbols, and abbreviations used appears in the first volume, beginning on page 98. As before, all positions are given for the Epoch 1950.0, and the star charts are all oriented with north at the top and east at the left.

Comparison magnitudes have now been added to most of the identification charts, with the exception of a few fields for which no accurate photometry seems to have yet been done. Users of the *Handbook* may, of course, add star magnitudes to any of the charts when measurements have been made and published by the A.A.V.S.O. or other observers.

A number of observers, after examining the manuscript for Volume I, have recommended that detailed charts for each constellation be included along with the tables of objects and descriptive data. This would possibly add to the usefulness of the book for actual use at the telescope, but would also be essentially a huge duplication of effort for the present author, since several excellent star atlases already exist, and may be obtained at moderate cost. This *Handbook* was designed to be used with either *Norton's Atlas* or the Skalnate Pleso *Atlas of the Heavens*, both of which should be in the possession of any serious observer.

R.B.

Flagstaff, Arizona
February, 1977

CHAMAELEON

LIST OF DOUBLE AND MULTIPLE STARS

NAME	DIST	PA	YR	MAGS	NOTES	RA & DEC
h4109	26.1	128	40	7 - 8½	relfix, spect A2 & A	08240s7616
I 288	0.6	267	47	8 - 8	PA inc, spect A0	09068s7433
h4214	9.2	193	40	8 - 8½	relfix, spect G5	09202s7726
h5444	41.9	235	19	7 - 9	spect B8	10322s8140
δ'	0.6	76	46	6 - 6½	(I 294) PA inc, cpm; spect gG8	10448s8012
I 212	0.7	183	47	7½- 7½	PA inc, spect F5	10587s8117
h4440	22.4	162	19	7½- 12	spect K0	11255s7814
I 893	1.1	118	41	7½- 9½	relfix, spect A0	11548s8013
ε	0.9	188	41	5½- 6	(h4486) PA inc, dist dec, cpm; spect B9	11571s7757
h4566	30.5	229	18	6½- 13	spect A0	13040s7012
h4590	22.4	134	31	6 - 9½	cpm; spect F5; primary may be variable (S Cham)	13289s7719

LIST OF VARIABLE STARS

NAME	MagVar	PER	NOTES	RA & DEC
R	7.5--14..	334	LPV. Spect M6e--M7e	08230s7612
S	6.0--7 ?		Uncertain, possibly not variable; spect F5. = double star h4590	13289s7719
T	9.5--12..	Irr	RW Aurigae type	11548s7905
Z	11.4--15.3	70:	U Geminorum type	08087s7624

LIST OF STAR CLUSTERS, NEBULAE, AND GALAXIES

NGC	OTH	TYPE	SUMMARY DESCRIPTION	RA & DEC
3195		◎	Diam 40" x 30"; pB,S,1E	10101s8037

CIRCINUS

LIST OF DOUBLE AND MULTIPLE STARS

NAME	DIST	PA	YR	MAGS	NOTES	RA & DEC
h4632	6.4	14	33	6 - 10	relfix, spect gG8	13548s6533
α	15.7	232	51	3½- 9	(△166) PA slow dec, spect F0,K5	14384s6446
Hd 239	36.3	64	00	5½- 10	spect A7	14413s6240
△ 169	68.5	106	38	6 - 7½	spect B2, K0	14415s5523
I 235	0.5	111	27	7½- 8½	spect A0	14436s6819
h4699	0.1	302	53	7½- 7½	(⌀ 298) PA inc,	14453s5912
	36.9	125	17	- 9½	spect G5, A3, A	
Rst 5008	2.8	117	42	7½- 12	spect B8	14460s6208
-63° 3436	0.5	143	13	6½- 6½	spect G3 + B8; not resolved in recent years.	14485s6336
h4707	0.8	69	26	8 - 8½	binary, about 290 yrs; spect G0	14501s6613
Gls 213	5.2	334	43	7½- 9	relfix, spect B9	14567s6747
h4735	7.1	31	33	7½-10½	relfix, spect F8	15088s6012
I 329	0.8	336	47	6½- 8	slight PA inc, spect B3	15099s6109
h4746	3.0	45	10	8 - 9	spect B0	15115s5853
	12.7	349	17	- 11½		
	18.1	312	17	- 11½		
I 370	5.4	116	34	5½- 12	spect 09	15148s6019
	44.5	244	00	- 11		
Cor 187	1.0	282	42	8½- 9½	spect A0	15187s6011
	11.5	233	19	- 8½	(Rst 5014)	
γ	0.9	49	49	5 - 5	(h4757) PA dec, cpm; spect B5,F8	15194s5909
Cp 16	2.3	27	36	7 - 8	spect A3; PA inc slightly	15256s5811

LIST OF VARIABLE STARS

NAME	MagVar	PER	NOTES	RA & DEC
R	9.0--11..	220	Semi-reg; spect M	15239s5733
T	9.3--10.5	3.298	Ecl.bin; spect B9	13397s6513
X	6.5--17..	---	Nova 1926	14386s6500
VW	9.5--11..	200	Semi-reg; spect M2e	14178s6609
AI	11-----16	---	Nova 1914	14450s6839
AR	10---15	---	Nova 1906	14443s5948

LIST OF STAR CLUSTERS, NEBULAE, AND GALAXIES

NGC	OTH	TYPE	SUMMARY DESCRIPTION	RA & DEC
5315		◎	Mag 12, diam 5"; virtually stellar appearance	13502s6616
5715	△333	⠬	Diam 6'; L,pRi,cCM; about 30 stars mags 11..... class E	14398s5720
5823		⠬	Diam 9'; cL,Ri,lCM; 80 stars mags 13..... class F	15019s5524

DESCRIPTIVE NOTES

ALPHA Mag 3.18; spectrum F0 Vp; position 14384s 6446. Alpha Circini is approximately 65 light years distant and has an actual luminosity of about 17 suns (absolute magnitude +1.7). The annual proper motion is 0.31" in PA 217°; the radial velocity is about 4 miles per second in recession.

The 9th magnitude companion at 15.7" is a K5 star, sharing the proper motion of the primary, and maintaining the same separation since the first measurements were made in 1837. The PA is decreasing slowly, from 244° to 232° between 1837 and 1951. Projected separation = 320 AU.

COLUMBA

LIST OF DOUBLE AND MULTIPLE STARS

NAME	DIST	PA	YR	MAGS	NOTES	RA & DEC
h3728	10.0	260	51	6½- 9½	relfix, spect G0	05069s4117
h3735	7.2	153	52	7½- 8	relfix, spect F2	05117s3157
h3740	23.9	287	33	6½- 8	relfix, spect G5	05134s3642
Hu1389	1.0	116	43	8 - 9	spect G0	05143s3100
I 61	0.8	88	50	8 - 9	PA dec, slight dist inc, spect F5	05168s4106
§166	2.2	7	33	6 - 11	spect B5p	05195s3424
h3757	15.1	315	33	7½-11½	PA inc, spect F0	05213s3147
h3760	7.4	222	35	8 - 8½	relfix, spect F5	05241s3524
	26.0	282	35	- 10½		
I 346	20.3	172	28	6 - 14	spect A2	05255s4059
△ 22	7.5	169	52	7½- 8	relfix, spect A5; cpm pair	05297s4221
Hu1393	0.6	335	46	7 - 7½	PA dec, spect F5	05336s3318
§ 167	0.3	204	60	8½- 9	(Daw 118) spect K0	05348s3445
h3781	15.9	135	19	7½- 9½	spect F5	05370s4119
α	13.5	359	50	3 - 11	spect B8 (*)	05378s3406
Cor 33	3.4	202	30	9 - 9	spect A5	05482s3613
	7.9	325	30	-13		
I 64	16.5	250	60	5½-11½	PA inc, dist dec; spect K0	05508s3738
I 16	1.3	126	37	7 - 11	relfix, spect K0	05511s3832
Cp 5	1.7	217	60	8½- 9	(Hu1398) PA inc; cpm; spect F2	05556s4146
h3823	0.3	285	27	9 - 9½	(Hu1399) binary, 72 yrs; PA dec, spect K5. All cpm	05585s3102
	1.7	42	60	- 9		
h3825	32.6	338	20	7- 10½	spect F5	06001s2725
h3831	2.5	129	42	8½- 8½	PA dec, spect F5	06027s4109
	15.1	186	38	- 14		
I 750	0.7	349	51	7½- 10	PA dec, spect F2	06096s2755
h3849	39.6	53	50	6½- 8	relfix, spect K0 & G5	06192s3928
h3857	12.9	256	60	5½- 9½	relfix, spect G5	06223s3641
	64.8	72	60	- 7		
h3858	3.8	311	59	7½- 8½	slight PA dec, spect A3	06238s3500
h3860	8.6	227	47	7 - 9	relfix, spect A3	06242s4057
I 4	0.9	303	60	7½- 7½	relfix, spect B8	06291s4025
β754	0.9	43	47	7½- 7½	PA inc, spect F0	06329s3358

LIST OF DOUBLE AND MULTIPLE STARS (Cont'd)

NAME	DIST	PA	YR	MAGS	NOTES	RA & DEC
β755	1.3	258	59	6 - 7	(h3875) slight	06337s3644
	21.4	301	32	- 11	dist inc, spect B9	
Cp 6	1.1	240	59	8 - 8	relfix, spect B8	06337s3846
φ 19	0.4	350	53	7 - 7½	binary, 26½ yrs; spect F5	06342s3603
I 1118	1.2	93	43	7½- 9½	PA dec, spect F5	06342s3508
Rst 4816	0.2	34	51	8 - 8	PA inc, spect G0	06344s3605

LIST OF VARIABLE STARS

NAME	MagVar	PER	NOTES	RA & DEC
R	8.0--14..	327	LPV. Spect M3e--M4e	05486s2913
S	8.9--14..	326	LPV. Spect M6e	05451s3143
T	6.7--12.6	225	LPV. Spect M4e--M6	05175s3345
U	9.9--10.5	1.246	Ecl.bin; spect A	06130s3303
V	9.3--14..	300	LPV.	06083s3044
W	8.5--10..	327	LPV.	06262s4004
RS	8.6--9.0	14.66	Cepheid	05133s2848
RV	8.8--9.8	106	Semi-reg.	05338s3051
RY	9.5--10.8	.4789	Cl.Var; spect A7	05136s4141

LIST OF STAR CLUSTERS, NEBULAE, AND GALAXIES

NGC	OTH	TYPE	SUMMARY DESCRIPTION	RA & DEC
1792		⊖	Sc; 10.7; 3.0' x 1.0' vB,L,mE, glbM	05035s3804
1800		⊖	S0/Sa; 12.9; 0.8' x 0.4' pB,pmE, gpmbM	05046s3201
1808	△549	⊖	SB; 11.2; 4.0' x 1.0' B,L,E, BN, F outer arms	05059s3734
1851	△508	⊕	Mag 7, diam 5', vB,vL,R, vBM,vRi; stars faint	05124s4005
2090	△594	⊖	Sc; 12.4; 2.5' x 1.0' B,pL,lE,gbM	05452s3415
2188		⊖	SB; 12.6; 3.0' x 0.6' pF,pL,vmE, glbM	06083s3405

DESCRIPTIVE   NOTES

ALPHA
Name- PHACT or PHAET. Mag 2.64; spectrum
B8 Ve. Position 05378s3406. The computed
distance is about 140 light years; the actual luminosity
about 145 times that of the Sun (absolute magnitude -0.6).
The star has an annual proper motion of 0.025"; the radial
velocity is 21 miles per second in recession.

The 11th magnitude companion at 13.5" is evidently
not physically connected with the primary; the separation
is increasing slowly from 11.6" in 1900.

BETA
Name- WEZN. Mag 3.12; spectrum K1 III. The
position is 05492s3547. The distance of this
star is approximately the same as Alpha Columbae, about
140 light years. The computed luminosity is then about 90
times that of the Sun (absolute magnitude 0.0). The annual
proper motion is 0.40" in PA 7°; the radial velocity is
54 miles per second in recession.

MU
Mag 5.16; spectrum 09.5 V; sometimes given
as B0. Position 05441s3219. Mu Columbae is
a famous "Runaway Star", one of three known early type
stars which appear to be moving out at high speeds from
the nebulous region of Orion. All three stars seem to be
at about the same distance as the Orion Association, and
have apparently been ejected from that region during the
last few million years, possibly by some process connected
with the explosion of supernovae. The other two stars are
53 Arietis and AE Aurigae, both identified by abnormally
high space velocities. AE Aurigae is the most interesting
of the three; its rapid passage through the heavens is
presently carrying it through the large diffuse nebulosity
IC 405, and the structure of the nebulosity is evidently
being greatly altered by the star's intense radiation.
Mu Columbae itself has an annual proper motion of about
0.025" in a direction slightly east of due south; the true
space velocity is about 74 miles per second.

For a diagram of the plotted paths of all three stars
refer to AE Aurigae (page 288).

COMA BERENICES

LIST OF DOUBLE AND MULTIPLE STARS

NAME	DIST	PA	YR	MAGS	NOTES	RA & DEC
2	3.6	237	62	6 - 7½	(ζ) (Σ1596) relfix, spect A8,F2	12017n2144
Σ1615	26.7	88	31	6 - 8	relfix, cpm pair; spect K0	12116n3304
OΣ245	8.6	280	58	5½- 10	cpm; spect A4; slight PA inc.	12150n2913
β27	3.5	105	55	7 - 11	cpm; spect K0	12176n1408
Σ1633	9.0	245	58	7 - 7	neat pair; cpm; relfix, spect both dF2.	12181n2720
11	9.1	44	58	5 -13	relfix, cpm; (Ho 52) spect G8	12182n1804
12	66.1	167	25	5 - 8	In Coma Cluster;	12200n2607
	35.0	54	35	- 11	AB cpm; spect G0, F8; C = optical	
Σ1639	1.3	329	62	6½- 7½	In Coma Cluster; binary, about 600 yrs; PA dec, spect F0	12219n2552
Σ1643	2.2	18	63	8½- 8½	PA dec, spect K2	12247n2719
17	145	251	28	5½- 6½	(ΣI 21) In Coma Cluster, cpm field-glass pair; spect A0, A3	12264n2611
17b	1.8	157	00	6½-13½	(β1080)	
Σ1652	6.0	180	22	9 - 9	relfix, spect G0	12300n2123
24	20.3	271	58	5 - 6½	relfix, cpm, nice colors, yellow & bluish. spect K2, A7 (Σ1657)	12326n1839
Σ1663	0.6	85	68	8 - 8½	PA dec, spect F2	12347n2128
Σ1678	34.1	181	58	6½- 7½	Probably optical; PA dec, spect A0	12429n1438
30	42.5	13	12	6- 11½	(h522) optical; spect A2	12469n2749
Σ1685	16.0	202	49	7 - 7½	relfix, cpm pair; spect F2, A2	12494n1927
32+33	195	49	22	6- 6½	(ΣI 23) field-glass pair; spect gM0, F8	12497n1721

LIST OF DOUBLE AND MULTIPLE STARS   (Cont'd)

NAME	DIST	PA	YR	MAGS	NOTES	RA & DEC
Σ1686	5.6	187	41	8 - 8	relfix, spect F5	12505n1518
35	1.0	144	62	5 - 7½	(Σ1687) binary;	12508n2131
	28.7	126	58	- 9	about 675 yrs; PA inc; spect G8, F6. ABC all cpm.	
Σ1696	3.2	204	63	8 - 8	relfix, spect F8	12550n3038
Σ1699	1.6	7	62	8 - 8	perhaps slight PA inc, spect G5	12563n2745
37	5.2	351	58	5 - 13	(β1081) cpm; spect G9	12579n3103
Σ1709	2.4	250	55	7 - 10	relfix, spect F2	13001n2346
h2638	6.3	219	58	6 - 11	cpm; spect A3	13038n2918
h2638b	0.4	221	01	11½ - 11½	(β1083)	
Σ1722	2.8	338	62	8 - 8½	perhaps slight PA dec, spect K0	13060n1546
α	0.3	14	62	5 - 5	(42 Comae) (Σ1728) binary, 25.8 yrs; spect F5 (*)	13076n1748
A2225	3.0	75	63	7½ - 12	PA inc, spect F2	13098n1624
Ho 55	0.7	166	05	7 - 11	spect K0	13113n3005
	73.4	154	00	- 11		
β800	6.2	108	62	7 - 10	dist inc; cpm; spect K3, K6	13143n1717
A2166	0.1	202	63	8 - 8	PA dec, spect F5	13178n1802
Ho 259	9.7	242	08	7 - 12	spect F8	13202n2623
OΣ266	2.0	350	62	7½ - 8	PA inc, spect F5	13260n1558
A567	1.5	262	58	6 - 12	relfix, cpm; spect K0	13304n2436
Σ1756	14.5	177	13	8½ - 9	relfix, spect G5	13310n2316
Σ1760	8.6	65	50	8 - 8	relfix, spect F2	13320n2632

COMA BERENICES

LIST OF VARIABLE STARS

NAME	MagVar	PER	NOTES	RA & DEC
R	7.5--14..	362	LPV. Spect M5e	12017n1904
RZ	9.0---9.8	.3385	Ecl.bin; W Ursae Majoris type; spect K0	12326n2337
SY	9.0--10..	176	Semi-reg; spect M4	12078n1947
TW	9.5--10.1	Irr	spect K5	12162n2222
UU	5.4 ± 0.02	2.195	α Canum type; spect A3p	12285n2451

LIST OF STAR CLUSTERS, NEBULAE, AND GALAXIES

NGC	OTH	TYPE	SUMMARY DESCRIPTION	RA & DEC
4032		⊘	I; 13.0; 1.1' x 1.0' pF,pL,lE,gbM	11580n2021
4064		⊘	Sb; 12.8; ?.5' x 1.0' B,E,gbM	12016n1843
4136	321[2]	⊘	Sc; 12.1; 3.2' x 2.9' F,vL,vgbM	12067n3012
4147	19[1]	⊕	Mag 11, diam 4'; class IX; vB,pL,R,eRi; stars faint	12076n1849
4150	73[1]	⊘	E2; 12.6; 1.3' x 0.8' B,S,lE, pgmbM	12080n3041
4152	83[2]	⊘	Sc; 12.7; 1.3' x 1.0' pB,pL,lE; pgmbM	12081n1619
4158	405[2]	⊘	Sa; 12.9; 0.8' x 0.7' F,pS,lE,bM	12086n2027
4162	353[2]	⊘	Sc; 12.6; 1.9' x 1.0' B,L,lE,bM	12094n2424
4192	M98	⊘	Sb; 11.0; 8.2' x 2.0' B,vL,vmE,vmbM (*)	12113n1511
4203	175[1]	⊘	E3p/S0; 11.7; 2.0' x 1.8' vB,S,R,psmbM	12125n3329
4212	108[2]	⊘	Sc; 12.0; 2.4' x 1.5' B,L,E,sbM	12131n1411
4237	11[2]	⊘	Sb/Sc; 12.6; 1.4' x 0.9' pB,pL,lE,vgbM	12147n1536
4245	74[1]	⊘	SBb; 12.3; 1.5' x 0.9' cB,pL,vlE,smbM	12152n2953
4251	89[1]	⊘	Sa; 11.3; 2.0' x 0.8' lens shape; vB,S,E, BN	12157n2827

LIST OF STAR CLUSTERS, NEBULAE, AND GALAXIES   (Cont'd)

NGC	OTH	TYPE	SUMMARY DESCRIPTION	RA & DEC
4254	M99		Sc; 10.4; 4.5' x 4.0' ! B,L,R,gbM; (*)	12163n1442
4262	110[2]		S0/SB; 12.6; 1.1' x 1.0' B,S,R	12170n1509
4274	75[1]		Sb; 11.5; 5.0' x 1.2' (*) vB,vL,E, mbMN. Saturn-like inner ring, faint outer halo	12174n2953
4278	90[1]		E1; 11.4; 1.2' x 1.0' vB,pL,R,mbM; 4283 nf 3.2'	12177n2934
4283	323[2]		E0; 12.9; 0.5' x 0.5' B,S,R,bM. nf 4278	12179n2935
4293	5[5]		Sa/pec; 11.7; 4.8' x 1.8' F,vL,E,1bM; spiral pattern dim, heavy dust lanes	12187n1840
4298	111[2]		Sc; 11.9; 2.7' x 1.1' F,L,E,vgbM, 4302 foll 2'	12190n1453
4302	112[2]		Sc?; 12.9; 4.5' x 0.5' L,vmE; flat streak, edge-on spiral with equat. dust lane	12192n1453
4314	76[1]		SBa; 11.6; 3.0' x 2.7' cB,L,E, sbM, faint outer arms	12200n3010
4321	M100		Sc; 10.4; 5.2' x 5.0' ! pF,vL,R,vgpsbM. Fine spiral (*)	12204n1606
4340	85[2]		SBa; 13.0; 2.2' x 1.4' pB,S,R,psbM. θ structure	12210n1700
4350	86[2]		E7; 11.9; 1.9' x 0.5' lenticular, cB,vS,mE,sbM	12214n1658
4377	12[1]		E1; 12.9; 0.9' x 0.7' B,S,1E, smbM	12227n1502
4379	87[2]		E1; 13.0; 0.7' x 0.6' pS,R,psbMN	12228n1553
4382	M85		Ep/S0; 10.5; 3.0' x 2.0' vB,pL,1E,bM. (*) 4394 foll 7.8'	12228n1828
4394	55[2]		SBb; 12.0; 3.0' x 3.0' pB,1E,bM. Pair with M85	12234n1829
4383			E2 or S0; 12.9; 0.8' x 0.5' vS,B,1E	12230n1645

LIST OF STAR CLUSTERS, NEBULAE, AND GALAXIES (Contd)

NGC	OTH	TYPE	SUMMARY DESCRIPTION	RA & DEC
4414	77[1]	⊖	Sc; 11.0; 3.1' x 1.5' vB,L,E,vsmbM	12240n3130
4419	113[2]	⊖	SB?; 12.2; 2.3' x 0.7' B,pmE,sbM	12244n1519
4448	91[1]	⊖	Sb; 11.7; 2.9' x 1.0' B,L,mE,sbM	12258n2854
4450	56[2]	⊖	Sb; 11.1; 3.8' x 3.0' B,L,1E,vmbMN	12259n1721
4455	355[2]	⊖	S ; 13.0; 2.4' x 0.5' F,L,mE,gbM. edge-on.	12262n2306
4459	161[1]	⊖	S0; 11.7; 1.5' x 1.0' pB,pL,1E,bM; internal dust ring around nucleus	12265n1415
4474	117[2]	⊖	E6; 12.8; 1.4' x 0.6' pF,mE; spindle shaped	12274n1421
4477	115[2]	⊖	SBa; 11.6; 2.5' x 2.0' pB,cL,1E	12276n1355
4494	83[1]	⊖	E1; 10.9; 1.5' x 1.4' vB,pL,R,vsmbMN	12289n2603
4501	M88	⊖	Sb; 10.5; 5.7' x 2.5' B,vL,vmE, multiple-arm spiral (*)	12295n1442
4540	94[2]	⊖	I or Sd; 12.9; 1.2' x 0.9' F,pS,bM	12323n1550
4548	120[2]	⊖	SBb; 10.9; 3.9' x 3.4' B,L,1E,bM; faint outer arms in S-pattern	12329n1446
4559	92[1]	⊖	Sc; 10.5; 10.0' x 3.0' vB,vL,mE,gbM. multi-arm spiral, coarse structure	12335n2814
4561	407[2]	⊖	Sc; 12.5; 1.1' x 1.0' pB,pL,v1E, 1bM. coarse structure	12336n1936
4565	24[5]	⊖	Sb; 10.5; 15.0' x 1.1' ! B,eL,eE,mbMN. Superb edge-on spiral (*)	12339n2616
4571	602[3]	⊖	Sc; 12.2; 2.8' x 2.4' vF,L,E,vgbM,SN. spiral arm pattern faint. 9^m star nf 2.7'	12343n1428

LIST OF STAR CLUSTERS, NEBULAE, AND GALAXIES   (Cont'd)

NGC	OTH	TYPE	SUMMARY DESCRIPTION	RA & DEC
4595	622^2	⊖	Sc; 13.1; 1.1' x 0.8' pF,pL,R,gbM	12373n1534
4635		⊖	Sc; 13.0; 1.5' x 0.9' pF,L,vglbM	12402n2012
4651	12^2	⊖	Sc; 11.4; 3.0' x 2.5' cB,L,gbM	12412n1640
4670	328^3	⊖	Ep/S0; 12.7; 0.8' x 0.6' pF,cS,1E,bM	12428n2723
4689	128^2	⊖	Sb; 11.7; 2.8' x 2.0' pB,vL,E,vglbM	12452n1401
4710	95^2	⊖	S0; 12.0; 3.4' x 0.5' cB,pL,vmE,smbMN. Probably edge-on S0; equatorial dust lane.	12471n1526
4712		⊖	Sb; 13.2; 1.9' x 0.8' vF,pL,E	12472n2544
4725	84^1	⊖	SBb; 10.5; 7.5' x 4.8' vB,vL,mE,vsmbM, BN (*)	12481n2546
4747	344^2	⊖	Sp; 12.8; 3.0' x 0.5' F,pL,E, distorted form, extending filament on NE	12494n2601
4793	93^1	⊖	Sc; 12.4; 1.9' x 0.8' pB,pS,1E; 8^m star $1\frac{1}{2}$' n	12523n2913
4826	M64	⊖	Sa; 8.6; 7.5' x 3.5' ! vB,vL,mE,bM,SBN. Central dust cloud. "Black-Eye Galaxy" (*)	12543n2157
4874		⊖	S0; 13.5; 1.0' x 1.0' F,vS,R. 7' pair with 4889; members of Coma Galaxy cluster	12572n2814
4889	321^2	⊖	E4; 13.2; 1.0' x 0.6' pB,pmE,bM. Brightest member of Coma Galaxy Cluster (*)	12577n2815
4961	398^2	⊖	Sc; 13.2; 1.0' x 0.7' F,S,1E	13034n2800
5012	85^1	⊖	Sb; 12.6; 2.2' x 1.2' pF,cL,E	13093n2311
5016	356^2	⊖	Sb; 12.8; 1.1' x 0.8' pB,S,1E	13097n2421

LIST OF STAR CLUSTERS, NEBULAE, AND GALAXIES   (Cont'd)

NGC	OTH	TYPE	SUMMARY DESCRIPTION	RA & DEC
5024	M53	$\oplus$	B,R,vC,eRi, mag 8, diam 10', class V; stars mags 12..... (*)	13105n1826
5053	7^6	$\oplus$	vF,pL,R,vgbM; mag 10½, diam 8'. 1° sf NGC 5024. (See note on M53)	13139n1757
5074	309^3	⊘	S?/pec; 13.2; 0.5' x 0.5' F,vS. On Coma-Canes border	13162n3144
5116	368^3	⊘	Sb; 12.9; 1.6' x 0.6' F,pS,pmE,glbM	13205n2714
5172		⊘	Sb; 12.5; 1.9' x 1.0' F,pL,lE,gbM	13269n1719

DESCRIPTIVE NOTES

ALPHA   (42 Comae) Mag 4.23; spectrum F5 V. Position 13076n1748. Double star, discovered by F.G. W.Struve in 1827. This is a close and rather difficult binary system, but of special interest from the rare cir- cumstance that the orbit is seen almost exactly edge-on; the two stars thus appear to move back and forth in virtu- ally a straight line, with a nearly constant PA. The period is 25.85 years, with the apparent separation varying from practically zero to a maximum of about 0.9". The semi- major axis of the orbit is 0.67"; the eccentricity is 0.5. Owing to the orientation of the orbit with respect to the Solar System, the time of apparent closest approach (1949) does not coincide with true periastron passage (1963).

Although T.W.Webb speaks of the two stars as "making an occultation about every 13 years" there is no real evi- dence that either star actually eclipses the other. From computations by H.Haffner (1948) and F.Pavel (1949) it now appears that the orbit is inclined just one-tenth of a degree from the edge-on position, implying that no real eclipse can occur. The two stars are nearly identical in size, type, and brightness; the magnitudes are both 5.1 and both spectra are F5 V. Each star is about 3 times as luminous as the Sun, and the mean separation of the pair is close to 10 AU, about the separation of Saturn and the Sun.

The distance of the system is approximately 65 light years, the annual proper motion is 0.45" in PA 287°, and the radial velocity is 10½ miles per second in approach.

THE COMA STAR CLUSTER

One of the best known galactic star clusters, the Coma Berenices Star Cluster is a conspicuous scattered group of naked-eye stars, centered about midway between Alpha Canum and Beta Leonis, near 12220n2600. The cluster is not listed in the NGC or in Messier's catalog, but is sometimes referred to by its number in the catalog list of Melotte - "Mel 111". Covering an area about 5° in diameter, the cluster contains, as its brightest members, the fifth magnitude stars 12, 13, 14, 16, and 21 Comae. About thirty fainter stars have been identified as probable members, and a few others may await discovery. The bright stars 15, 18, and 7 Comae do not appear to be physical members of the group, but the star 31 Comae, some 5° from the cluster center, is probably a member. The information in the table below has been compiled from the Yale "Catalogue of bright Stars" (1964).

| Star | Mag. | Spect. | Yearly proper motion | | Radial |
			RA	Dec	Velocity
12	4.83	G0 III + A3V	-0.009"	-0.015"	+1 km/sc
13	5.18	A3 V	-0.016	-0.016	+1
14	4.95	F0 p	-0.015	-0.014	-3
16	5.00	A4 V	-0.007	-0.016	+2
21	5.46	A p	-0.012	-0.015	0
22	6.29	A4 V	-0.018	-0.010	+1

As a cluster, the Coma group is at its best in a pair of good binoculars, but is completely lost in the much narrower field of the telescope, which can show only a small portion at any one moment. G.P.Serviss spoke of noting a

THE COMA BERENICES STAR CLUSTER. The members of this nearby cluster are scattered over the entire field of this 13-
inch telescope photograph.   Lowell Observatory

DESCRIPTIVE NOTES (Cont'd)

"curious twinkling, as if gossamers spangled with dewdrops were entangled there. One might think the old woman of the nursery rhyme who went to sweep the cobwebs out of the sky had skipped this corner, or else that its delicate beauty had preserved it even from her housewifely instinct.."
T.W.Webb remarked that such a group obviously requires only distance to become a nebula to the naked eye.

The cluster is named in honor of Berenice II of Egypt, who was the Queen of Ptolemy III (246 - 221 BC). In one of the most appealing star legends we are told of the Queen's vow to sacrifice her famed "amber tresses" in the temple of Aphrodite at Zephyrium, following the king's safe return from battle. After the offering mysteriously vanished from the temple, the court astronomer Conon convinced the royal couple that the lost tresses had been transformed by the gods into a constellation, and enshrined forever among the stars. The Roman poet Catullus (about 60 BC) refers to this legend when he speaks of

"the consecrated offering of Berenice's golden hair, which the divine Venus placed, a new constellation among the ancient ones, preceding the slow Bootes, who sinks late and reluctantly into the deep ocean..."

In the coins of Berenice, we find some of the most exquisite coin portraits which have come down to us from the ancient world, particularly in the large gold octa-drachms and decadrachms dating from the early days of the reign of Ptolemy III. "During this reign.....Egypt had command of the sea", states Professor Barclay V.Head in his monumental work *Historia Numorum* (1911) "and her empire embraced many of the maritime districts of Asia Minor, even extending across the Aegean into Thrace. Hence the appear-ance of Egyptian influence at mints like Ephesus and

DESCRIPTIVE NOTES   (Cont'd)

Ptolemais-Lebedus. In the absence of specific local and other marks, the Ptolemaic coins issued in these regions can seldom be attributed with certainty...." Prof. Head suggested, however, that the much-admired gold octadrachm of Berenice was probably minted at Ephesus on the west coast of Asia Minor, famous as the site of the great Temple of Diana, one of the ancient Seven Wonders. The respected German authority K.Regling dated this same coin, on the basis of a study of die-styles, to 258 BC, the year of the marriage of the young Berenice of Cyrene to the future Ptolemy III. The ancient city of *Berenice*, named for the queen, is still in existence as the modern Benghazi in Libya, capital and chief port of the province of Cyrenaica.

  Before the days of Berenice, however, the star group seems to have been regarded as a part of the constellation Leo, marking the tuft of hair at the end of the Lion's tail. The Arabian names, *Al Halbah* and *Al Dafirah* refer to this identification and have been translated "the Coarse Hair" or "The Tuft". Eratosthenes, however, in the 3rd Century BC, refers to the group as "Ariadne's Hair" though in another passage he does connect it with the Egyptian Berenice. In various star maps of the late Middle Ages the cluster is identified as a rose-wreath or ivy-wreath, and occasionally as a Sheaf of Wheat held in the hands of Ceres or Virgo. These identifications seem to date back to a time when Coma was considered a part of the constellation Virgo. According to R.H.Allen, the astronomer Tycho Brahe "set the question at rest in 1602 by cataloguing it separately, adopting the early title as we have it now."

  At a distance of about 250 light years, the Coma group is one of the nearest of all the star clusters; probably only the Ursa Major Group and the Hyades in Taurus are closer. Cluster members may be identified by the annual proper motion of the group, about 0.02" in P.A. 218°, and by the fact that the radial velocity of the members averages nearly zero. Any star with a high radial velocity is thus immediately disqualified. In one of the first really comprehensive studies of the group, R.J.Trumpler (1938) identified 37 stars as true cluster members on the basis of both proper motion data and radial velocity measurements. The most luminous members, as 14 and 16 Comae, have about

50 times the luminosity of the Sun. The faintest stars accepted as members are about 1/3 the brightness of the Sun. Spectral types range from A3 to about G9. As a standard of comparison, our Sun at the distance of the Coma Cluster would appear as a star of magnitude 9.2.

There are eight known spectroscopic binary stars in the Coma Cluster, with orbital elements presently available for several of the brighter ones. The bright star 12 Comae is the most interesting of these. It is a spectroscopic binary with a period of 396.49 days, but there is also a distant visual companion of the 8th magnitude at 66" in PA 167°, almost certainly a true cluster member. The spectrum of 12 Comae is composite, G0 III + A3 V; the spectrum of the visual companion is about F8. In addition, about 0.5° to the southeast is the binary Σ1639, with a computed period of some 600 years and a semi-major axis of about 1". The orbit has an eccentricity of about 0.9 and the spectrum of the brighter star is about F0.

The Coma Cluster contains no giant stars, though the H-R Diagram (below) shows that the brightest members are just beginning to evolve toward the giant stage. The total mass of the group is probably under 100 solar masses. One of the peculiar features of the cluster is the apparent

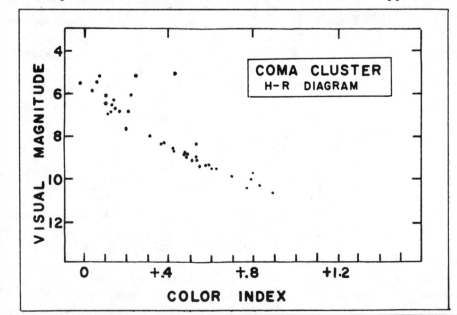

lack of fainter stars; the main sequence seems to terminate abruptly at apparent magnitude 10½, which corresponds to an absolute magnitude of about 6. Fainter and redder dwarf stars, actually the commonest stellar types, seem to be missing completely in the Coma group. If we compare this cluster with the famous Pleiades, it seems relatively poor in stars and only about a quarter as thickly populated, although the volumes of the two clusters are nearly equal. The sparseness of the Coma group suggests the possibility that the cluster may be gradually dispersing due to the small total mass. The computed star density is about one star per 10 cubic parsecs, a value very near the theoretical lower limit for stable clusters. In terms of age, the Coma group appears to be older than the Pleiades, but younger than Praesepe (M44) in Cancer or the Hyades Cluster in Taurus. (For a discussion of cluster age-dating, refer to M13 in Hercules)

M53   (NGC 5024) Position 13105n1826. A rich globular star cluster which forms a pair of 1° separation with the more unusual cluster NGC 5053. M53 itself lies 1° northeast of the binary Alpha Comae, and was first observed by J.E.Bode in February 1775. He referred to it as "a new nebula, appearing through the telescope as round and pretty lively"; the use of the last term suggests some hint of resolution. Messier's independent discovery occurred just two years later, in February of 1777; he found it "round and conspicuous" but "without stars" and compared it afterwards with the comet of 1779. Bright and well condensed, the cluster is an easy object for the small telescope, appearing as a round nebulous spot in a three-inch glass, but resolving into a wonderful swarm of tiny star images with larger instruments. Partial resolution usually requires a 6-inch telescope. In the greatest telescopes M53 is more than worthy of Webb's description: "a brilliant mass of minute stars, blazing in the center." John Herschel spoke of the radiating curves and streams of stars which adorn the outer edges, an appearance which is seen in many other bright globulars as well. Sir William Herschel, with his great reflector, spoke of M53 as "one of the most beautiful sights I remember to have seen in the heavens. The cluster appears under the form of a solid ball

GLOBULAR STAR CLUSTERS IN COMA BERENICES. The bright M53 is separated by 1° from its fainter neighbor NGC 5053.

Mt.Wilson and Palomar Observatories

consisting of small stars quite compressed into a blaze of light with a great number of loose ones surrounding it and distinctly visible in the general mass. Similar in appearance to M10". Admiral Smyth saw M53 as "a mass of minute stars 11-15 mag. and from thence to gleams of star-dust, with stragglers....and pretty diffuse edges.." K.G.Jones (1968) rates it as "certainly one of the most beautiful" of the globulars. "It consists of a brilliant nucleus about 2' in diameter, surrounded by a dusky halo of light that gives it a glittering gem-like appearance. The background is dark but glimpses of numerous, faint stars can be obtained in clear conditions."

The total photographic magnitude of M53 is about 8.7, the extreme diameter about 14', and the integrated spectral type is F4. The cluster lies at a distance of about 65,000 light years; the resulting total luminosity is about 200 thousand times the Sun. Radial velocity studies show an approach velocity of about 70 miles per second. Some 45 variable stars are known in M53.

One degree to the southeast is the peculiar cluster NGC 5053, an unusual object which from its appearance could be classified as either a very loose globular or a very rich galactic cluster. It was discovered by Sir William Herschel in 1784, and may be detected in an 8-inch glass as a faint hazy spot of about magnitude 10½. It contains no dense nucleus of stars, and the faint members are widely separated even at the cluster center. Resolution can be achieved only with rather large telescopes.

The classification of this system as a globular is supported by the color-magnitude diagram and the presence of short-period pulsating variables of the RR Lyrae type. The population is about 3400 stars down to the 21st magnitude, decidedly sparse for a globular cluster. With an actual diameter of close to 100 light years, the resulting density is evidently not much more than 0.3 star per cubic parsec. The distance of the cluster seems to be comparable to that of M53; in fact the studies of the variables in the two groups suggest that NGC 5053 is probably slightly nearer to us than its brighter neighbor. The best present figure is about 55,000 light years, which gives the cluster a total luminosity of some 16,000 suns. This is one of the lowest luminosities derived for any globular cluster; the

NGC 5053. The unusual globular star cluster in Coma Berenices, photographed with the 48-inch Schmidt telescope at Palomar.

DESCRIPTIVE NOTES (Cont'd)

great Omega Centauri, for comparison, shines with nearly
the light of a million suns. (Refer also to NGC 5139 in
Centaurus and M13 in Hercules)

M64   (NGC 4826) Position 12543n2157. The Black-Eye
           Galaxy, a large oval spiral of the 8th magnitude
and measuring about 7½' X 3½' in size, easily located about
1° ENE from the star 35 Comae. M64 was discovered by J.E.
Bode on April 4, 1779, and was recorded merely as looking
like a "small nebulous star". Messier found it in March
1780 and thought it to be about half the brightness of M53.
Observers today will find both objects visible in good
field glasses. Admiral Smyth called M64 "a conspicuous
nebula, magnificent both in size and brightness, being
elongated in a line N.p. & S.f. and blazing to a nucleus".
In the Shapley Ames Catalogue it is rated among the dozen
brightest spirals in the entire heavens; this suggests that
it is not a member of the great Virgo Galaxy Cluster which
is centered some 9° away, or at least that it is considerably closer than most of the other galaxies in this part of
the sky.
     The structure of M64 is somewhat unusual, and the
galaxy has been classified as type Sa by some authorities,
type Sb by others. The spiral arms show a beautifully
smooth and uniform texture with no trace of resolution into
star clouds or knots of nebulosity. Separating and defining
the arms are thin, dusky bands which appear similarly
smooth and soft-textured. In the region of the central
nucleus, however, a huge dust cloud suddenly makes its
appearance, bordering the entire north and east side of the
oval central mass. The dust cloud shows much fine detail
on photographs made with great telescopes, and breaks up
into an intricate region of mixed dark and luminous material
as it rims the near side of the central hub. D'Arrest
thought the center partly resolvable, while Lord Rosse
interpreted the mottling as a "close cluster of well-defined
little stars". The visibility of the dark mass is naturally
a controversial point among observers, but it is definitely
within the capabilities of a good 6 or 8-inch glass; J.H.
Mallas has detected it in a 4-inch refractor. The observation of such details requires very dark clear skies, and
the observer must keep his eyes well dark-adapted. In the

study of such things, the inter-galactic traveller may also temporarily ignore the usual rule against using high-power oculars on nebulae and galaxies; a somewhat higher magnification in this case may help to darken the field and increase the visibility of the "black eye" of M64.

The exact distance of M64 is not precisely determined but appears to be in the range of 20 - 25 million light years; the red shift of the system is about 225 miles per second, about a third the average value for members of the Virgo Galaxy Cluster. The apparent size of about 7.5' then corresponds to about 48,000 light years, and the total luminosity is about 13 billion times the sun (absolute magnitude -20.5). E.Holmberg, in his *Catalogue of External Galaxies*, however, adopts a much greater distance, of 13.5 megaparsecs, or close to 44 million light years; this would double the value given above for the linear diameter, and quadruple the total derived luminosity. At the present time, the smaller distance appears to fit better the currently accepted value (1976) for the Hubble Constant. The light of the system is somewhat yellower than that of many spirals, and the integrated spectral class is about G7. Whatever the exact distance, there is general agreement that M64 is one of the more massive and more luminous galaxies, and is certainly among the most noteworthy objects of its type for small telescopes. As of 1976, no supernovae have yet been recorded in this galaxy.

M85   (NGC 4382) Position 12228n1828. One of the bright members of the Virgo Galaxy Cluster, the great aggregation of external systems which is centered some 5° to the south. The chief facts concerning this very remarkable cluster are given in the constellation section on Virgo; the most notable members lying north of the border in Coma are: M85, M88, M98, M99, M100, and NGC 4565.

M85 is, as T.W.Webb says, a "fair specimen of many nebulae in this region". It was discovered by Mechain in 1781 and was described by Messier in the same year as "a nebula without star, above and near to the ear of Virgo.... this nebula is very faint." M85 appears to most cameras and to the eye as a normal elliptical galaxy about 3' X 2' in size, magnitude about 10½, elongated nearly north to south. There is a strong concentration of luminosity toward

SPIRAL GALAXY M64. The famous "Black-Eye Galaxy". This photograph was made with the 61-inch reflector of the U.S. Naval Observatory at Flagstaff, Arizona.

OFFICIAL U.S. NAVY PHOTOGRAPH

the center, but no telescope has shown resolution into
stars. Plates made at Palomar, however, show faint elonga-
tions or tufts of material at the north and south tips of
the system, either the vague beginning of a spiral pattern
or the last surviving remnant of one. From this appearance
the classification has been changed to "SO" in many modern
catalogues.

The corrected red shift of M85 is about 450 miles per
second; according to be best current value for the Hubble
relationship, this implies a distance of about 44 million
light years. E.Holmberg (1964) has derived a slightly
closer distance of about 41 million light years, and finds
a total mass of about 100 billion solar masses; the total
luminosity is close to absolute magnitude -20.5. The 3'
apparent size corresponds to about 40,000 light years at
the accepted distance. A supernova was recorded in this
galaxy in 1960.

The faint barred spiral NGC 4394 lies in the same
field, 7.8' to the east.

M88   (NGC 4501) Position 12295n1442. A bright,
nicely symmetrical galaxy of the multiple-arm
type, oriented about 30° from the edge-on position, and
measuring about 5.5' X 2.5'. This is one of Messier's discoveries, found March 18, 1781, and described with the
usual terse phrase "nebula without star". Admiral Smyth
found it to be a "long elliptical nebula...pale white.....
and with its attendant stars forms a pretty pageant....the
N.part is brighter than the S." Lord Rosse in 1850 found
it to be of spiral form, while modern photographs reveal
many closely packed whorls curving about the bright and
elongated central hub; the central nucleus is almost star-
like. The general appearance resembles M63 in Canes, but is
apparently a more distant object.

M88 is one of the notable spirals of the Virgo-Coma
Galaxy Cluster, and lies near the northern end of a bright
chain of objects which form a sweeping 3° curve toward the
southwest. This chain of galaxies is evidently the center
or nucleus of the entire vast cluster, but the majority of
the members lie to the south of the Coma border, in Virgo.
For identification charts of this area, refer to the Virgo
section of this Handbook. The Skalnate-Pleso Atlas also

DESCRIPTIVE NOTES   (Cont'd)

charts nearly all the galaxies listed on pages 663- 667,
and the voyager in this fabulous region may explore island
universes by the dozens. The intergalactic traveler, however, must learn to observe with the mind as well as the
eye, since the galaxies reveal their full splendor only to
the more sensitive eye of the camera.

M88 has a much larger red shift than the two other
galaxies described previously, M64 and M85. Taken at face
value, the figure of about 1280 miles per second (corrected for the solar motion) would seem to place this system
far beyond many of the bright spirals of the Virgo group.
Yet, from other studies, M88 is believed to be at about
the same distance as M85; E.Holmberg adopts a distance of
about 12.5 megaparsecs, or about 41 million light years;
the resulting true diameter is about 60,000 light years
and the computed absolute magnitude about -21. Up to 1976,
no supernovae have been observed in this galaxy.

K.G.Jones (1968) considers M88 one of the best of
the Virgo Group galaxies for the small telescope, and finds
it seemingly brighter than its reported magnitude of 10½.
"It will stand magnifying power well, and with averted
vision, a considerable amount of faint detail can be made
out from time to time."

A little more than 1° to the ESE is the position of
one of the mysterious missing Messier objects, M91, at
1235Un1402. The discoverer, in March 1781, described it as
a "nebula without star...its light even fainter than M90".
Although Shapley and others have theorized that M91 may
have been a true comet that got away, it seems more likely
that it was merely a duplicate observation of one of the
other objects in the region, with perhaps an erroneous
position. The nearby galaxy NGC 4571 has been suggested as
a probable candidate; O.Gingerich on the other hand, found
that a duplicate observation of M58 (about 2° to the south
in Virgo) was the most probable explanation. Questions as
these may never be settled with absolute finality.

M98   (NGC 4192) Position 12113n1511. A large and
              much-elongated galaxy measuring about 8' x 2'
and located ½° west of the 5th magnitude star 6 Comae. This
is another of Mechain's discoveries, found in 1781, and
confirmed by Messier in the same year; the noted comet-

hunter described it as "a nebula without star, of an extremely faint light, above the northern wing of Virgo..."
William Herschel in 1783 found it to be "a large extended
fine nebula. Its situation shows it to be M.Messier's 98th;
but from the description it appears that that gentleman has
not seen the whole of it, for its feeble branches extend
above a quarter of a degree....my field of view will not
quite take in the whole nebula". Smyth thought it "large,
but rather pale....on keeping a fixed gaze it brightens up
toward the centre."

          Photographs show M98 to be a nearly edge-on spiral,
probably of type Sb, the long dimension oriented from NNW
to SSE. The central region has a total magnitude of about
11, with a small nearly stellar nucleus; the faint and fine-
structured spiral arms sweep out for vast distances at the
north and south ends of the galaxy. Curiously, M98 is one
of the few galaxies in the Virgo-Coma region that does not
show a red shift; according to an extensive list published
by M.L.Humason, N.U.Mayall, and A.R.Sandage (1956) the
corrected radial velocity is about 125 miles per second in
approach! It is difficult to know how to interpret such an
anomalous result. Ordinarily a blue shifted galaxy would
be an excellent candidate for membership in the Local Group
of Galaxies. Studies of photographs and photometry of the
system, however, indicate that the distance must be close
to 35 million light years, somewhat closer than most of the
members of the Virgo group; if a true member, it would be
necessary to conclude that an occasional galaxy may have a
random velocity so large as to over-compensate for the
recession of the group as a whole. This circumstance would
seem to make it unwise to base galaxian distances solely on
the observed red shift.

          If the derived distance of about 35 million light
years is correct, M98 has an actual diameter of about 80
thousand light years and an absolute magnitude of about -21
which makes it more or less equal in luminosity to M88. The
integrated spectral type is G0. E.Holmberg finds a total
mass of about 130 billion solar masses for this galaxy. Up
to 1976 no supernovae have been recorded in the system.

          The bright round spiral M99 lies about 1.3° distant
toward the ESE; another faint spiral, NGC 4237, is about
1° away, toward the NE.

GALAXIES IN COMA BERENICES. Top: The fine open spiral M99.
Below: The more compact spiral NGC 4274.
　　　Palomar Observatory 200-inch telescope photographs.

SPIRAL GALAXY M100. The largest spiral of the Virgo
Galaxy Cluster, photographed with the 200-inch telescope.

Palomar Observatory

DESCRIPTIVE NOTES (Cont'd)

M99 (NGC 4254) Position 12163n1442. A bright
round spiral of type Sc, called by R.H.
Allen the "Pinwheel Nebula" though this name is more often
applied to the great M33 in Triangulum. M99 is located just
50' southeast of the star 6 Comae, and about 1.3° distant
from M98. It was discovered by Mechain in 1781 and confirmed by Messier the same year; he found it a "nebula without
star, of a very pale light, nevertheless a little clearer
than the preceding, M98." D'Arrest described it as "large,
round, with vividly sparkling light; nucleus more or less
resolvable." M99 seems to have been the second galaxy to
be recognized as a spiral; Lord Rosse in 1848 found it
"spiral with a bright star above; a thin portion of the
nebula reaches across this star and some distance past it."
     M99 is a nearly circular, face-on spiral of magnitude
10½, some 4' in apparent size. The spiral pattern is very
well defined, although somewhat asymmetric, with an unusually far-extending arm on the west side; the arms are beautifully marked by a series of bright star clouds and nebulous
regions. This system has sometimes been called a "three-
branch spiral" although there are only two major arms. The
supposed third arm actually consists of a number of short
segments which radiate out from one of the main arms on the
northeast side. Thin dust lanes may be traced deep into the
bright central mass, and there is a small, almost stellar
nucleus. Unlike its neighbor M98, this galaxy has one of
the largest red shifts of any member of the Virgo Cloud,
about 1490 miles per second. Supernovae have appeared in
M99 in 1967 and 1972.
     The probable distance of the system is about 45 to
50 million light years, though the red shift alone would
imply a much larger distance. The true diameter would seem
to be about 50,000 light years, and the computed absolute
magnitude about -21. E.Holmberg finds a total mass of about
50 billion solar masses; the integrated spectral type is
G2.

M100 (NGC 4321) Position 12204n1606. The largest
spiral of the Virgo-Coma Galaxy Cluster,
located near the center of the large triangle formed by
M85, M88, and M98. This is another of the galaxies found by

DESCRIPTIVE NOTES (Cont'd)

Mechain in 1781, and observed by Messier only a few weeks
later; Messier found M98, M99, and M100 "very difficult to
recognize because of their feeble light; one can observe
them only in good weather and near meridian passage." Sir
William Herschel found M100 to be about 10' in apparent
size, and thought that the bright central mass consisted
of "a small bright cluster of supposed stars." To Admiral
Smyth the galaxy appeared "pearly white; this is a large
but pale object of little character, though it brightens
from its attenuated edges toward the centre and is therefore thought to be globular". The spiral form appears to
have been first detected by Lord Rosse in 1850.

In the small telescope, M100 appears as a round glow
about 5' in diameter with a total magnitude of about 10½.
On photographs taken with great telescopes this spot of
structureless haze is wonderfully transformed into one of
the most impressive spirals in the whole region. M100 is an
Sc system, oriented not quite face-on; a large number of
secondary arms and segments fill the spaces between the two
chief spiral arms, and a complex system of dust lanes
carries the spiral pattern all the way into the actual
nucleus of the galaxy. Bright nebulous regions ornament
the spiral arms like so many pearls on a string, and some
of the star clouds appear to be partially resolved on 200-
inch telescope plates, showing individual supergiant stars.
Supernovae have appeared in this galaxy in 1901, 1914, and
1959.

According to A.Sandage (1962) the major spiral arms
of M100 have an average thickness of nearly 3000 light
years, about twice the estimated thickness of the arms in
our own Galaxy. M100 is another of those systems in which
the observed red shift (about 960 miles per second) seems
to give too great a distance; from observations of the
giant stars of the spiral arms, M100 appears to be about
40 million light years distant. This makes the true linear
diameter about 110,000 light years, closely comparable to
the great Andromeda Galaxy M31. The total luminosity is
close to 20 billion times the Sun (absolute magnitude about
-21) and the computed total mass is about 160 billion solar
masses. Apparently owing to the number of high-temperature
giants in the spiral arms, M100 is somewhat bluer than M85
or M98; the integrated spectral type is F5.

NGC 4565. A perfect example of an edge-on spiral galaxy.
Lowell Observatory photograph made with the 13-inch wide-
angle telescope.

DESCRIPTIVE   NOTES   (Cont'd)

NGC 4565   (HV 24) Position 12339n2616. The largest
of the edgewise spiral galaxies, and undoubtedly the most famous object of its type. It is located
1.7° east of the star 17 Comae, and less than 3° from the
North Galactic Pole. The galaxy is possibly an outlying
member of the Virgo Galaxy Cluster, though it lies some 13°
north of the main concentration.

NGC 4565 is an interesting object for the small telescope, and appears as a bright narrow streak in a good 6-
inch glass. With a 10-inch and dark skies it is a perfect
little needle of light which can be traced out to nearly
its full photographic diameter of 15'. At the central hub
the thickness is about 1.4', and the total magnitude is
about $10\frac{1}{2}$. The very small bright nucleus is mentioned in
the NGC catalog as a "central star". Lord Rosse (1855)
described the system as "a beautiful object, very well seen
in the finding eyepiece; the whole nebula is much broader
at nucleus than elsewhere, narrowing off suddenly, and the
nucleus projects forward into the dark space..." Rosse's
drawing was made with his 6-foot reflector and shows the
dark absorbing lane running the full length of the system.
Such drawings, of course, convey no hint of the true com-
plexity of bright and dark material which is revealed by
modern photography. These dark bands of absorbing matter
appear to be a standard feature of spiral galaxies, and
are prominent in other similar systems such as NGC 891 in
Andromeda and NGC 4594 in Virgo. A similar absorption lane
in our own Galaxy is undoubtedly the cause of the "Great
Rift" in the Milky Way.

Radial velocity measurements of NGC 4565 show a red
shift (corrected) of about 750 miles per second, comparable
to typical members of the Virgo Group. The large apparent
size, however, seems to indicate that the galaxy is much
closer than the Virgo Cluster, and the degree of resolution
seems to point to the same conclusion. If the distance is
about 20 million light years, the apparent size of 15' then
corresponds to about 90,000 light years, which seems quite
reasonable. A distance much greater than about 30 million
light years seems unlikely, as the true diameter then becomes much greater than any other spiral known. The total
light received from the galaxy is equal to about 3 billion
suns. (See also NGC 891 in Andromeda and NGC 4594 in Virgo)

GALAXY NGC 4565. The most famous of the edge-on spirals, photographed with the 200-inch reflector at Palomar.

CENTRAL PORTION OF THE COMA GALAXY CLUSTER, photographed with the 200-inch telescope. This print is centered on the bright elliptical galaxy NGC 4889.

Palomar Observatory

NGC 4889   The brightest member of a very remote and very rich cluster of galaxies, located in the northeast corner of the constellation at 12577n2815, about 2.3° west of Beta Comae. This very distant group should not be confused with the well-known Virgo Cluster which lies some 10 times closer, and which includes many of the bright galaxies in Coma.

NGC 4889 and 4874 are the only two members of the Coma Cluster of Galaxies which are likely to be detected in most amateur telescopes, though owners of larger reflectors may find it interesting to record some of the fainter members with long-exposure photography. NGC 4889 is a giant elliptical system of the 13th magnitude, while the slightly fainter NGC 4874, about 7¼' to the west, is usually classified as type S0. The majority of the other members are 15th magnitude and fainter.

A survey by F.Zwicky (1957) identified 804 galaxies brighter than magnitude 16.5 (pg) within 160' of the cluster center, and 29,951 galaxies brighter than 19.0 within 6°. A large proportion of these fainter objects may be very distant background systems, but the true cluster population is probably at least 1000 galaxies. This vast swarm of island universes lies at an estimated distance of nearly 400 million light years. The full diameter of the group is at least 3° and the richest central portion is about 1° across. At the adopted distance, these dimensions correspond to about 20 million and 7 million light years, respectively. The central "core" of the Coma Cluster is thus not much more than twice the size of the Local Group of Galaxies, but is incomparably richer; the total mass must exceed that of the Local Group by a factor of at least 50. A very rich cluster of galaxies in Corona Borealis has a similar degree of concentration, and many other such aggregations are now known as a result of the Palomar Sky Survey made with the 48-inch Schmidt camera. All these groups are at chillingly vast distances; some are among the most remote objects ever identified.

The red-shift of the Coma Galaxy Cluster is close to 4250 miles per second; the value for NGC 4889 being a little less than the mean, while that of NGC 4874 is a little greater. (Refer also to the Galaxy Cluster in Corona Borealis)

SPIRAL GALAXY NGC 4725. The supernova which appeared in this galaxy in 1940 is indicated by the marker in the lower print.

Mt.Wilson and Palomar Observatories

CORONA AUSTRALIS

LIST OF DOUBLE AND MULTIPLE STARS

NAME	DIST	PA	YR	MAGS	NOTES	RA & DEC
I 230	7.1	138	52	7½- 8½	relfix, spect K0	17571s3742
	62.4	117	51	- 10		
λ 345	16.0	141	13	7½- 12	spect K0	17594s4328
h5011	29.6	350	19	8 - 9	relfix, spect A0	18029s4146
h5014	1.8	215	43	6 - 6	binary, about 190 yrs; spect A5, A5. NGC 6541 in field	18032s4326
h5023	8.7	276	51	8 - 8	relfix, spect A2	18073s4026
I 1018	1.4	283	43	7½- 11	PA inc, spect G0	18095s3910
I 1020	0.4	294	44	8 - 8½	PA dec, cpm pair, spect B8	18129s4029
χ	21.4	359	36	6 - 6½	(△222) relfix, easy cpm pair, spect A0, B8	18299s3846
I 1372	8.2	18	59	7½-11½	spect M0	18329s3811
I 250	1.0	120	51	7½- 9	cpm; PA dec, spect A0	18377s4213
I 1379	0.3	182	36	8 - 8	spect B8	18395s3921
λ	29.2	214	53	5 - 9	(Cor 227) relfix	18404s3822
	40.0	57	00	- 10	spect A1, K0	
h5066	10.1	85	33	7 - 10	relfix, spect B5	18475s4108
Hd 291	36.0	342	59	6½- 10	relfix, spect K2; B = 0.6" pair	18486s4147
h5074	15.9	246	20	6½- 12	spect A0	18558s3937
Brs 14	12.7	281	51	6½- 7	nice cpm pair, relfix, spect both B8	18577s3708
γ	2.7	54	43	5 - 5	(h5084) binary, about 120 yrs; PA dec, spect F8, F8	19030s3708
Cor 231	5.6	198	31	8½- 10	spect K5	19031s3812
I 1396	5.2	91	39	7½- 11	perhaps slight PA dec, spect K0	19097s4050

CORONA AUSTRALIS

LIST OF VARIABLE STARS

NAME	MagVar	PER	NOTES	RA & DEC
ε	4.7---4.9	.5914	Ecl.Bin; spect F5; W Ursae Majoris type	18553s3711
R	9.7--12..	Irr	Peculiar, erratic, in nebula NGC 6729 (*)	18583s3702
U	9.0--13.4	147	LPV. Spect M2e	18377s3753
V	9.8--14..	Irr	R Coronae Borealis type Spect R0	18441s3813
X	9.8--11.7	324	Semi-reg; spect M7	18063s4525
RR	8.8--13..	280	LPV. Spect M2e	18024s3815
RU	8.5--11..	Irr		19139s3942
RW	9.3--10.3	1.684	Ecl.bin; spect A0	17559s3752
RX	9.6--14..	286	LPV. Spect M3e	18228s4419
RZ	8.9--12..	460	LPV. Spect M6e	19050s4226
TY	8.8--12.6	Irr	Spect B2, erratic; in nebula NGC 6726 (*)	18583s3657
TZ	9.6--10.2	.6867	Ecl.bin; spect A0	18152s4324
UX	9.9--16..	347	LPV. Spect Me	18541s3752
YY	9.6--15..	126	LPV. Spect Me	18379s3706
AM	7.6--11.7	188	Semi-reg; spect M3	18378s3732
V 394	7.5--14..	---	Nova 1949	17570s3900

LIST OF STAR CLUSTERS, NEBULAE, AND GALAXIES

NGC	OTH	TYPE	SUMMARY DESCRIPTION	RA & DEC
6496		⊕	Mag 10, diam 3', class XII; pL,lE,gbM. stars faint	17555s4415
6541	△473	⊕	Mag 6, diam 6', class III B,R,eC,rrr, stars mags 13	18044s4344
6726	m393	▢	Complex nebulous region surrounding variable	18583s3657
6727	m394			
6729	m395		stars TY & R Cor. (*)	18584s3702
---	I.1297	◎	Mag 11½, diam 2", appears stellar	19140s3942

DESCRIPTIVE NOTES

NGC 6726-6727-6729   Position 18583s3657. A complex
field of mixed bright and dark
nebulosity, remarkable for its content of erratic nebular
variable stars of the T Tauri type, and located in a region
of heavy obscuration about 7½° south of Zeta Sagittarii.
The brightest portion is the double nebulosity NGC 6726 and
6727, two roughly circular patches in contact forming a
"figure 8" pattern measuring about 2' x 1.3', oriented NE-
SW, and adorned around the edges by many wispy nebulous
filaments. The SW section is illuminated by an A-type star
of magnitude 7.2, but the NE section surrounds the erratic
variable TY CorA which has varied from 8.8 to about 12.5.
The spectral type, about B2, is one of the earliest known
for any nebular variable; E.Hubble and W.J.Luyten, however,
have classed it somewhat later, B9 or early A. Observations
are rendered difficult by the light of the nebulosity.
        About 4.7' to the SE lies the most interesting por-
tion of the nebulosity, the small comet-like wisp NGC 6729,
about 1.3' in length, containing as a nucleus the erratic
variable star R CorA. Star and nebula form a combination
reminiscent of Hubble's Nebula NGC 2261 in Monoceros and
its illuminating star R Mon. Variations in the light of
the nebula generally follow the fluctuations of the star,
but there are also peculiar changes in the shapes of various nebular details, often too rapid to be physically real,
and evidently attributable to changing light and shadow
effects. R CorA has a spectral class of about F5 and the
extreme range is over 3 magnitudes. The variations, first
detected by J.Schmidt in 1866, are generally unpredictable
but the star sometimes shows changes of up to 2 magnitudes
within a few days. About 1.3' to the SE, in the cometary
"tail" of NGC 6729, lies the fainter variable T CorA; it
shows similar variations with a total range of about 11.7
to 13.5. The spectral type is close to F0. A fourth variable in the field is S CorA, located at virtually the same
declination as R, but 43 seconds preceding in RA. It has a
range of 10.8 to about 12.5. The spectral type is near dG
but with many emission lines; the spectral features are
very similar to those of the famous nebular variable T
Tauri. Stars of the type are main sequence stars and dwarfs
rather than giants, and are believed to be among the youngest known stars. (Refer also to T Tauri and R Monocerotis)

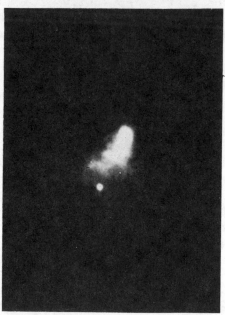

NEBULOUS REGION IN CORONA AUSTRALIS. Top: The field of NGC 6726 & 6727 photographed with the 40-inch reflector at Lowell Observatory. Below: Variations in NGC 6729, recorded with the 100-inch reflector at Mt.Wilson Observatory.

CORONA BOREALIS

LIST OF DOUBLE AND MULTIPLE STARS

NAME	DIST	PA	YR	MAGS	NOTES	RA & DEC
Σ1932	1.0	57	61	5½- 6	binary, about 190 yrs, PA inc, spect dF8	15162n2701
Σ1935	8.5	289	55	8½- 9	relfix, spect G5	15182n3053
Σ1936	20.3	232	16	8½- 9	relfix, spect F2	15190n2713
η	0.6	169	68	5½ - 6	(Σ1937) binary,	15211n3028
	57.7	12	20	-12½	41½ yrs; PA inc, spect G2,G2 (*)	
Σ1941	1.6	219	58	8½- 8½	PA dec, spect F5	15236n2648
Σ1950	3.1	92	62	6½- 8	relfix, spect K2	15278n2541
Σ1963	5.0	296	62	7½- 8	slight dist inc,	15359n3016
	31.0	116	08	- 12	spect F8	
Σ1964	1.0	90	66	7½- 12	(Hu 1167) spect	15363n3624
	15.2	86	33	- 7½	F5. C = 1.6" pair	
ζ	6.3	305	57	5 - 6	(Σ1965) relfix, fine pair, spect B6, B7	15375n3648
γ	0.2	265	62	4 - 5½	(Σ1967) binary, 91 yrs (*)	15406n2627
Σ1973	30.6	322	25	7 - 8	relfix, spect F5	15445n3636
OΣ302	28.6	52	25	7 - 9	relfix, spect A2	15530n3430
Ho 399	3.2	119	54	7½- 10	relfix, spect A0	15534n2941
ε	1.8	3	58	4- 12½	(AGC 7) cpm, PA slow inc, spect K3	15555n2701
OΣ304	10.7	175	26	6½- 10	relfix, spect A0	15591n3919
Σ2011	2.4	68	63	7- 9½	relfix, spect A3	16056n2908
τ	2.2	186	58	5½- 13	PA inc, cpm, spect K0 (β1087)	16072n3637
OΣ305	5.4	263	41	6 - 10	relfix, spect gK2	16098n3328
Σ2022	2.3	145	63	6 - 9½	PA inc, spect F2	16107n2648
Σ2029	6.3	187	42	7½- 9½	relfix, spect F2	16118n2851
σ	6.2	229	62	5 - 6	(Σ2032) binary,	16128n3359
	71.0	85	33	- 10½	PA & dist inc (*)	
Hh511	34.7	19	14	6 - 8½	relfix, spect A3 (23 Herculis, but now in Corona)	16210n3227
Σ2044	8.5	342	63	7½- 8	relfix, spect K0	16224n3709

CORONA BOREALIS

LIST OF VARIABLE STARS

NAME	MagVar	PER	NOTES	RA & DEC
α	2.2---2.3	17.360	Ecl.bin; spect A0+dG6 GEMMA (*)	15326n2653
β	3.67±0.02	18.487	Magnetic spectrum var; Spect F0 pec (*)	15258n2917
γ	3.8--3.86	.030	δ Scuti type, spect A0; also visual binary (*)	15406n2627
R	5.6--14.8	Irr	Remarkable "fade-out" variable; spect F7 (*)	15465n2819
S	6.0--14..	360	LPV. Spect M6e--M8e	15194n3133
T	2.3---10	---	"The Blaze Star", famous recurrent nova, maxima in 1866 & 1946 (*)	15574n2604
U	7.6---8.8	3.452	Ecl.bin; spect B5 + A2	15161n3150
V	6.9--12.5	358	LPV. Spect N2	15477n3943
W	7.9--14.3	238	LPV. Spect M2e--M5e	16136n3754
X	8.5--14.2	241	LPV. Spect M5e--M7e	15470n3624
Y	9.8--11.7	300:	Semi-reg; Spect M8	15449n3829
Z	8.8--15.5	251	LPV. Spect M4e	15541n2923
RR	7.1---8.6	61	Semi-reg; Spect M5	15396n3843
RS	7.6---9.6	331	Semi-reg; spect M7	15567n3609
RY	9.2--10.4	96	Semi-reg; spect M1	16211n3058
SS	8.7--10..	Irr	Spect M5	15396n3647
ST	9.0--10..	Irr	Spect M5	15484n3829
SW	7.6---8.3	Irr	Spect M0	15389n3852
SX	8.5--9...	Irr	Spect M5	16135n3327

LIST OF STAR CLUSTERS, NEBULAE, AND GALAXIES

NGC	OTH	TYPE	SUMMARY DESCRIPTION	RA & DEC
---	---	⊘⊘	Rich, faint cluster of very distant galaxies. (*)	15205n2750

CORONA BOREALIS

DESCRIPTIVE NOTES

ALPHA   Name- GEMMA or ALPHECCA. Mag 2.23 (slightly variable). Spectrum A0 V + dG6. Position 15326n2653. This is the brightest star in the Northern Crown, approximately 75 light years distant and about 45 times the brightness of our Sun. The annual proper motion is 0.15"; the radial velocity is about 1 mile per second in recession.

Gemma is a spectroscopic binary with a period of 17.359907 days. The components have a moderately eccentric orbit (0.38) and are separated by a mean distance of some 17 million miles. As an eclipsing system the star is not of great interest to the visual observer, since the depth of the eclipse light curve is only about 1/10 magnitude. Primary eclipse is annular, the small G-star passing in transit across the face of the bright primary. Facts of interest about the two stars are given in the table.

	Spect.	Diam.	Mass	Lum.	Abs.Mag.
A	A0 V	2.9	2.5	45	+0.7
B	dG6	0.9	0.9	0.5	+5.6

BETA   Name- NUSAKAN. Mag 3.66; spectrum F0 p. Position 15258n2917. The distance of Beta Coronae is approximately 100 light years; the resulting luminosity is about 25 times that of the Sun. The annual proper motion is 0.19"; the radial velocity is 10½ miles per second in approach. The star shows very nearly the same space motion as the Hyades Cluster in Taurus, and is probably an outlying member of that group.

Beta Coronae is a spectroscopic binary star with a period of 10.496 years; the presence of a third body of low mass has been suspected. The spectrum of the primary is unusual for the strength of the lines of the rare - earths elements, and resembles the spectrum of the "magnetic variable" Alpha Canum Venaticorum. And, like Alpha Canum, Beta Coronae has a remarkably intense magnetic field which varies periodically, with a reversal of polarity, in a cycle of 18½ days. The light changes- if any- are not yet accurately measured, but evidently cannot exceed a few hundredths of a magnitude. (Refer also to Alpha in Canes Venatici)

DESCRIPTIVE NOTES (Cont'd)

GAMMA   Mag 3.85; spectrum AO IV. Position 15406n
2627. The distance is about 140 light years,
and the actual luminosity about 40 times that of the Sun.
The annual proper motion is 0.11"; the radial velocity is
6 miles per second in approach.

Gamma Coronae is a close visual double with a period
of 91 years, discovered by F.G.W.Struve in 1826. Since the
orbit is inclined only 6° from the edge-on position, the
PA changes very little because of orbital motion. P.Baize
(1953) finds that periastron occurred in 1931; the semi-
major axis of the system is about 0.74" and the eccentricity is 0.42. Individual magnitudes are 4.2 and 5.6; the
true separation of the stars averages about 30 AU. The
primary is a δ Scuti type variable with a period of about
43 minutes, and a spectroscopic binary of uncertain period.

ETA   Mag 5.02; spectrum G2 V. Position 15211n
3028. Eta Coronae is a well-observed bin-
ary system which has completed more than three revolutions
since discovery by F.G.W.Struve in 1826. It is a rather

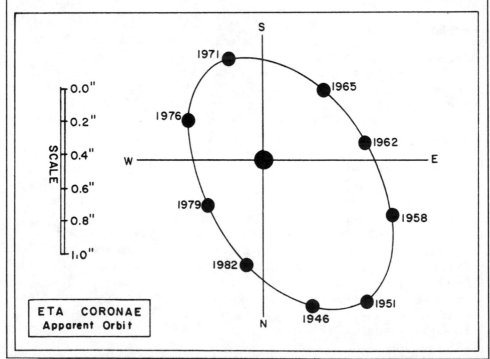

ETA CORONAE
Apparent Orbit

DESCRIPTIVE NOTES (Cont'd)

close pair for small telescopes, but may be divided with fairly high powers on a good 6-inch glass throughout much of its orbit. The period is 41.56 years with the widest separation of about 1.0" occurring in 1951, 1993, etc. The individual magnitudes are 5.7 and 6.0; both stars are very nearly twins of our Sun in size, mass, and luminosity. An orbit computation by A.Danjon (1938) gives the system an eccentricity of 0.28 and a semi-major axis of 0.84" or about 13 AU. A third component, at 58", is not a physical member of the system.

The computed distance of Eta Coronae is about 50 light years; the annual proper motion is 0.24" in PA 146°; the radial velocity is 4 miles per second in approach.

SIGMA   Mag 5.36; spectrum F8 + dG1. The position is 16128n3359. This is an attractive binary star for the small telescope, the separation increasing steadily from 1.3" at discovery by F.G.W.Struve in 1827. The exact period is uncertain, and values ranging from 340 years up to about 1600 years have been computed. The shorter periods are not compatible with the newest observations; a recent orbit by K.Strand gives the period as about 1160 years, with periastron occurring about 1828 when the two stars were some 30 AU apart. Maximum separa-

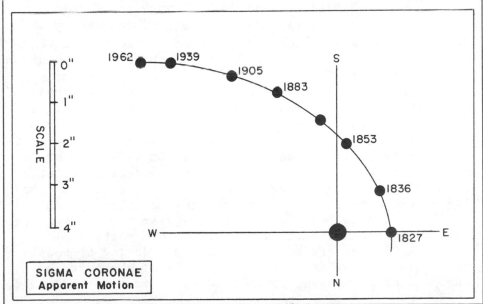

SIGMA CORONAE
Apparent Motion

DESCRIPTIVE NOTES (Cont'd)

tion of about 13" will occur about 500 years from now. Strand's orbit has an eccentricity of 0.80 and a semi-major axis of 7.3"; the mean separation of the stars may be about 150 AU. T.W.Webb says of this pair "a great divergence as to colors", evidently from the attempt to detect a color contrast where virtually none exists! The two stars are of very similar type, and the total light of the system is about 3 times that of the Sun. The primary is a spectroscopic binary with a period of 7.974 days and a total mass of about 1.7 suns. In addition, Sigma Coronae has a very distant proper motion companion which lies 12' away toward the southwest. This is LTT 14836, a red dwarf of the 13th magnitude, about 400 times fainter than the Sun. The projected separation from the AB pair is 15,000 AU. In addition to this star, there is also an optical companion of the 10th magnitude which was 44" distant in 1836; the separation is increasing from the proper motion of the bright pair.

The computed distance of Sigma Coronae is close to 70 light years; the annual proper motion is 0.30" in PA 252°, and the radial velocities (somewhat different owing to the orbital motion of the stars) are $6\frac{1}{2}$ and $10\frac{1}{2}$ miles per second, both in approach.

R Position 15465n2819. A very remarkable irregular variable star, discovered by the English observer E.Pigott in 1795. It is the typical example of a rare class of variables which remain at maximum throughout most of their cycle. For several years at a time this star shines with a nearly constant brightness of near 6th magnitude. Then it will suddenly begin to fade, and within a few weeks will have fallen to almost any magnitude between 7 and 15, but normally about $12\frac{1}{2}$. The minimum usually lasts for several months. At times it may last for a year or two, and on at least one occasion the star remained below normal for 10 years. During a minimum the light is not constant, but fluctuates erratically. This star is often called the "ideal irregular variable" since the changes seem to follow no predictable pattern.

R Coronae has been well observed during the past 150 years, and selected sections of the light curve are shown on pages 703 and 704. The deep minima of 1917, 1938,

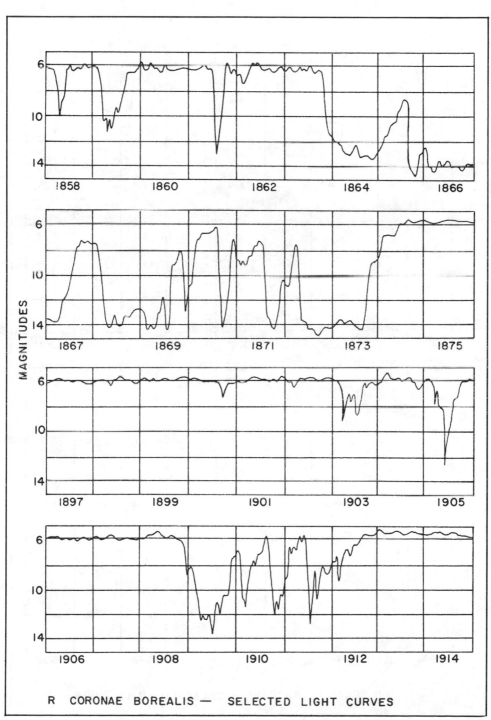

MAGNITUDES

R CORONAE BOREALIS — SELECTED LIGHT CURVES

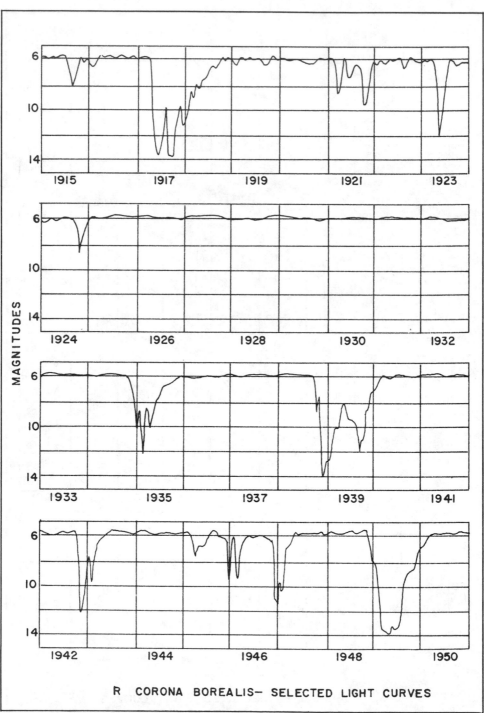

MAGNITUDES

R CORONA BOREALIS— SELECTED LIGHT CURVES

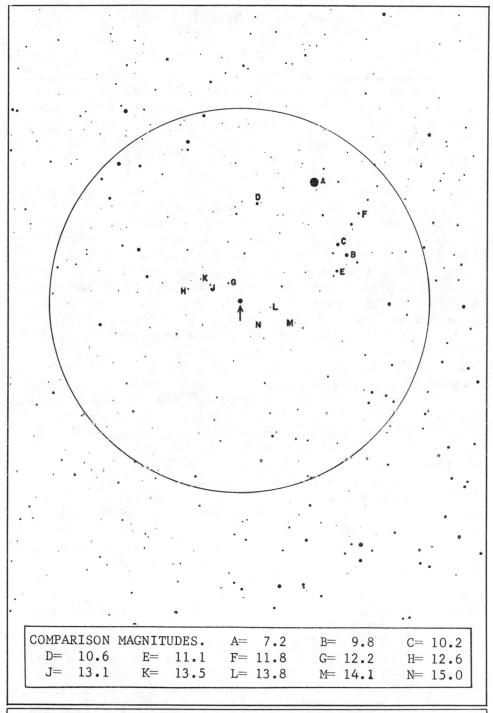

COMPARISON MAGNITUDES. A= 7.2   B= 9.8   C= 10.2
D= 10.6   E= 11.1   F= 11.8   G= 12.2   H= 12.6
J= 13.1   K= 13.5   L= 13.8   M= 14.1   N= 15.0

R CORONAE BOREALIS. Identification field, made from a 13-inch telescope plate at Lowell Observatory. Circle diameter = 1° with north at the top. Limiting magnitude about 14.

and 1949 may be studied as standard examples of the curious activity of this star. The light curves of the three
minima are like "reverse novae"; the fall is quite rapid
and the return to normal is much slower, with occasional
relapses. At other times there are "short minima" like
those of 1903 and 1924, and "intermediate minima" like the
ones of 1905 and 1923. Perhaps most interesting are the
rare periods of strong oscillation, such as the one lasting from 1908 to 1912. The ten-year minimum of 1863-1873
is another example; the star seemed unable to return to
normal after its decline in late 1863, and oscillated
strongly for a number of years. On one occasion, in 1870,
it rose to nearly 6th magnitude, but then fell immediately
to 14th. In contrast, the magnitude remained virtually
constant for the ten-year period 1924-1934. Activities of
R Coronae in recent years have been more typical; it
faded in the summer of 1962, returned to 6th magnitude at
the end of the year, and declined again early in 1963. In
autumn of that year the star was near 14th; it brightened
to 10th in 1964, but during most of 1965 was again at 14th.
In December of 1965 it began to brighten and eventually
reached 7th magnitude toward the end of summer 1966. A
rapid fading to 13th occurred in October, but by the end
of the year the magnitude was back to about $7\frac{1}{2}$.

R Coronae has a peculiar spectrum, classified by
various authorities as type F, G, or occasionally M. The
most recent observations suggest that the spectrum is basically that of a supergiant of near F7, but peculiar in
showing stong absorption bands caused by carbon in the
star's atmosphere. According to spectroscopic analysis
the composition of the atmosphere appears to be about 67%
carbon and only 33% hydrogen and other elements. These
observations have suggested a theory that the light variations are due to the emission of clouds of carbon, the
material condensing into a "soot cloud" around the star
and effectively obscuring the light. As the cloud clears
or is reabsorbed into the star, the light returns to normal. This theory also considers R Coronae to be an ancient
star which has consumed most of its hydrogen, and is now
operating on the helium-to-carbon cycle. A study of the
spectrum appears to support the suggestion by showing that
the star is indeed hydrogen-poor; what we see may be, in

R CORONAE BOREALIS. The remarkable variable is shown in
its normal state (top) in May 1940, and during a deep
minimum (below) in June 1964. The photographs were made
with the 13-inch telescope at Lowell Observatory.

fact, chiefly a helium "core", the outer hydrogen having been consumed or ejected into space.

Very little is known about the distances and actual luminosities of the R Coronae type stars. If the spectrum of a normal F7 supergiant may be used as a standard the absolute magnitude may be in the range of -4.4 to -5.0; the corresponding distance is then about 4,000 to 5,000 light years. These results imply that the star is at a very great distance above the plane of the Galaxy, at least 2500 light years. R Coronae is evidently a Population II star and a member of the great spherical halo which encloses the Galaxy; this finding agrees with other evidence that R Coronae is an ancient star.

There appear to be only two direct trigonometrical parallaxes recorded in the literature for this star; an attempt made with the 60-inch at Mt.Wilson some years ago yielded 0.04"; while a more recent measurement obtained at Allegheny gave 0.012". The resulting distances (82 and 270 light years) are not only discordant, but would also seem to be unacceptably small; the reason for these peculiar discrepancies is not yet understood, but is does not seem likely that the star can be much closer than a few thousand light years. Rho Cassiopeiae is another star which present similar problems.

The annual proper motion of R Coronae is 0.02"; the radial velocity is about 15 miles per second in recession. Other stars of this rare class are: RY Sagittarii, S Apodis, SU Tauri, XX Camelopardi, and RS Telescopii.

T   Position 15574n2604. This is the "Blaze Star", the best known example of a recurrent nova. This remarkable star is now of the 10th magnitude, but on two occasions in the past century it has suddenly risen to 2nd magnitude and then faded away again. There are 6 other stars known which have undergone more than one nova outburst, but T Coronae Borealis is by far the brightest one on record. The only other which reaches naked-eye visibility is RS Ophiuchi.

The first observed outburst of T Coronae occurred on the night of May 12, 1866, and the star when noticed was about the same brightness as Alpha Coronae, magnitude 2.2. The maximum magnitude of about 2.0 was reached the same

night and the star began to fade rapidly, falling below
naked-eye visiblity in only 8 days. By June 7 it had dimmed to its former state of low luminosity. About 100 days
after maximum the star brightened again to magnitude 8½
and remained at that brightness about 90 days before beginning its final decline.

With the spectroscope, Huggins detected bright lines
of hydrogen in the light of the nova, and stated that these
appeared to be superimposed upon a normal spectrum which
resembled that of the Sun. This was the first spectroscopic examination of a nova.

Because of its unusually short maximum and comparatively small amplitude, some astronomers hesitated to accept T Coronae as a normal nova. Its light increase at the
maximum amounted to only 2500 times, whereas most typical
novae increase more than 10,000 times as they rise to maximum. A second peculiarity was the shortness of the outburst; the star faded to the pre-nova level in less than a
month, while a typical nova usually requires a number of
years to reach that stage. Much later it was pointed out
that a few other novae- U Scorpii in 1863 and RS Ophiuchi
in 1898- had shown a similar behavior. When U Scorpii exploded for the second time in 1906, and RS Ophiuchi also
staged a repeat performance in 1933, it seemed likely that

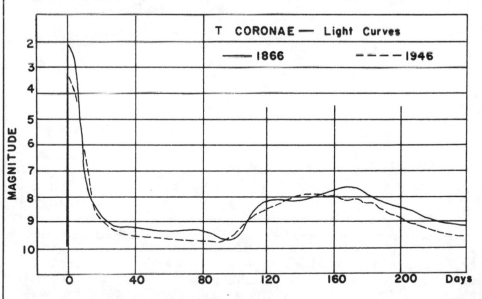

T CORONAE — Light Curves

——— 1866   − − − − 1946

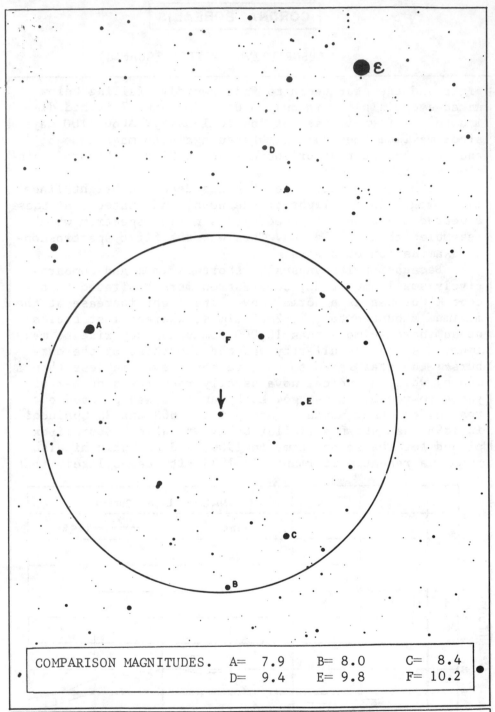

COMPARISON MAGNITUDES.   A= 7.9   B= 8.0   C= 8.4
                         D= 9.4   E= 9.8   F= 10.2

T CORONAE BOREALIS. Identification field, from a 13-inch
telescope plate obtained at Lowell Observatory. Circle
diameter = 1° with north at the top; limiting magnitude
about 14. Bright star at top is Epsilon Coronae, mag 4.15.

DESCRIPTIVE NOTES (Cont'd)

T Coronae would also prove to be a recurrent nova. On the night of February 9, 1946, this theoretical prediction was spectacularly verified as T Coronae flared up again, reaching the 3rd magnitude, and soon subsided, almost exactly as it had done some 80 years before. For a short time it was a conspicuous object to the naked eye.

It is likely that the star was already past the maximum when discovered at magnitude 3.2. By the following evening it had fallen to 3.6, and the decline thereafter proceeded rapidly at about 0.5 magnitude per day, compared with about 0.63 magnitude per day in 1866. The light curve was a virtual duplicate of the earlier explosion, and the secondary maximum, 100 days after the main outburst, also repeated itself faithfully. It is interesting to note that T Coronae is the only fast nova, and the only recurrent nova, in which a strong secondary maximum appears. In RS Ophiuchi a weak secondary maximum may be detected, but the phenomenon is most pronounced in the ordinary novae DQ Herculis (1934) and T Aurigae (1891).

The outstanding feature of T Coronae at the 1946 outburst was the enormous expansion velocity, which has not been equalled by any other nova, the supernovae alone excepted. The outer layers of the star showed an expansion rate of about 2700 miles per second. This was the highest velocity ever measured in our Galaxy with the exception of the two supernova remnants "Cassiopeia A" and Tycho's Star of 1572. This is a difficult feature to explain, in view of the fact that T Coronae and the other repeating novae do not show the large magnitude range of the classical or ordinary novae, and are therefore sometimes regarded as relatively feeble objects. This conclusion, however, now appears to be much in error. Recent studies of T Coronae indicate a distance of about 2600 light years; this gives an absolute magnitude of -8.4 at maximum and an actual luminosity of some 200,000 suns. This is quite comparable to the luminosities deduced for many of the normal novae, and is decidedly superior to some. The smaller range seems to result from the greater brightness of the star at minimum, some 40 or 50 times the brightness of the Sun. Most of the normal novae appear to have minimal luminosities equal to or fainter than that of the Sun. A study of RS Ophiuchi leads to a rather similar result. The recurrent

DESCRIPTIVE NOTES (Cont'd)

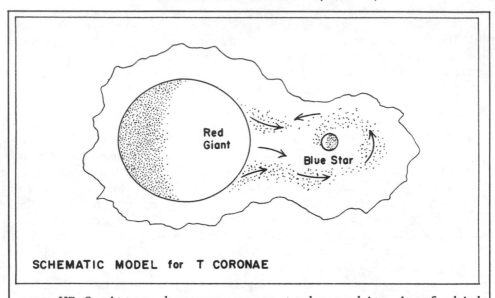

Red Giant

Blue Star

SCHEMATIC MODEL for T CORONAE

nova WZ Sagittae, however, seems to be a white dwarf which
attains a luminosity of a mere 25 suns at maximum. So it
appears that the stars classed as recurrent novae are not
necessarily members of a true physical group. These facts
complicate the problem of determining the place of these
stars in the picture of stellar evolution. But, before
considering such questions, another fact must be presented. The spectrum of T Coronae is peculiar and shows a dual
or composite nature. There is little doubt that the star
is a close double. At minimum the spectrum is type M, but
during an outburst the M-spectrum is overpowered by the
nova-spectrum. A similar situation exists in such variable
stars as Z Andromedae and R Aquarii; the so-called "symbiotic variables". In the case of T Coronae the radial velocity is found to vary in a cycle of 227.6 days, supplying
additional evidence for binary motion.

The red star seems to be normally the brighter of the
pair. The visual brightness of the blue component is possibly one or two magnitudes fainter; thus the true light
range of the nova component may be about 10 magnitudes. In
a study of the system, R.P.Kraft (1963) finds that the
masses of the two stars must be at least 2.1 and 2.9 solar
masses for the blue and red stars, respectively. The red
star, of class gM3, has a diameter of about 120 million
miles; the computed orbit places the blue star virtually

DESCRIPTIVE NOTES (Cont'd)

in contact with the outer edges of the red giant. In tech-
nical language, the red star fills the "inner Lagrangian
point" of the system, and material must be passing across
to go into orbit around the blue star or to fall into it.
It is tempting, though perhaps somewhat premature, to spec-
ulate that in this phenomenon we have the whole key to the
mystery of nova outbursts.

Erratic fluctuations in the light of the system have
been measured by various observers, and are attributed to
minor outbursts of the blue star. As far as is known, the
red star is stable. These fluctuations appear to have increased in violence just before the outburst of 1946, and
reached a total amplitude of two magnitudes. An unusually
low minimum was reported just before the explosion. If this
pattern is characteristic of T Coronae a similar behavior
in the future may provide advance notice of the next maximum. The observed maxima have been so short that it is
quite possible some have been missed. The star deserves
continuous attention.

THEORIES OF RECURRENT NOVAE. From the binary nature of T
Coronae, RS Ophiuchi, SS Cygni and other similar freaks of
the heavens, some astronomers have been led to postulate
a new theory of nova formation, attributing the outbursts
of the SS Cygni stars and the recurrent novae to the same
basic process. The theory assumes that these stars are all
close binaries in which one component is a hot dwarf which
is nearing the white dwarf state and is at least partially
degenerate. The other component is a larger and cooler star
of more normal type. The outbursts of the dwarf star are
triggered in some way by an interchange of material; the
simplest picture assumes merely that the instability of
the hot dwarf is pushed past the critical point by accretion of material from the companion. Current studies of SS
Cygni and U Geminorum seem to indicate that the real circumstances are probably somewhat more complicated and may
very well defy interpretation for some time yet. But if
duplicity is not an essential feature in all nova-type
stars, it is at least an important factor in a number of
the best-studied examples. Another fascinating possibility
is that the recurrent novae form an intermediate stage between the SS Cygni stars and the true full-scale novae. If
so, it may be that all novae are recurrent, though the

intervals between explosions may be hundreds or thousands of years. (For a review of novae in general refer to Nova Aquilae 1918. See also DQ Herculis, T Aurigae, GK Persei, and CP Puppis. The SS Cygni stars are described under SS Cygni and U Geminorum.)

LIST OF RECURRENT NOVAE. As of 1967, the following seven stars constitute the complete list of known recurrent novae. For data refer to the respective constellations:

T Coronae Borealis (1866, 1946). RS Ophiuchi (1898, 1933, 1958, 1967). U Scorpii (1863, 1906, 1936, 1979. T Pyxidis (1890, 1902, 1920, 1944, 1967). WZ Sagittae (1913, 1946, 1978). VY Aquarii (1907, 1962). V1017 Sagittarii (1901, 1919, 1973).

CORONA BOREALIS GALAXY CLUSTER A very rich cluster of distant galaxies, called by some astronomers a "super-galaxy", one of the most remarkable of all such aggregations. It is located in the southwest corner of the constellation at 15205n2750. In the cluster are more than 400 galaxies, all concentrated in an area of the sky half a degree wide, about the apparent width of the Moon. Most of the galaxies of this group appear to be of the elliptical type, and the brightest members have apparent magnitudes of about $16\frac{1}{2}$, placing them far beyond the reach of the usual amateur telescope. The cluster is extremely remote; recent investigations have shown that it must be at least four times more distant than was believed when E.Hubble wrote his "Realm of the Nebulae" in 1936. The best present estimate of the distance is between 1 and 1.3 thousand million light years while the red shift in the spectra of these remote systems implies a recession velocity of about 13,000 miles per second, or about 1/14 the velocity of light.

The Corona Borealis Galaxy Cluster is very similar to the rich aggregation in Virgo, but is at a much greater distance and consequently much fainter and more compact. Members of the Corona Cluster average about 6.1 magnitudes fainter than members of the Virgo group, implying a distance factor of about $16\frac{1}{2}$; this is in good agreement with the difference in radial velocities, which amounts to about 17 times. (Refer also to the Coma Galaxy Cluster, and the Virgo Galaxy Cluster)

CENTRAL PORTION OF THE CORONA BOREALIS CLUSTER OF GALAXIES. The brightest members are about 16th magnitude. Photograph made with the 100-inch reflector at Mt. Wilson Observatory.

CORVUS

LIST OF DOUBLE AND MULTIPLE STARS

NAME	DIST	PA	YR	MAGS	NOTES	RA & DEC
h4481	3.6	194	54	8 - 8	relfix, spect F5	11547s2216
β1079	11.7	148	30	6 - 13	spect K0	11581s2134
β412	2.0	160	52	8 - 8½	relfix, spect F2	12058s1818
Σ1604	10.0	91	62	6½- 9	AB cpm; spect G0,	12068s1135
	18.4	73	62	- 9	AC optical, dist dec.	
S634	5.4	291	62	7 - 8	PA slow inc, spect G5	12078s1631
B222	4.3	95	35	7½- 14	cpm; spect K0	12121s2430
β920	1.4	292	62	7 - 8	PA inc, spect F5	12132s2305
β921	3.4	219	43	7½- 11	relfix, spect B9.	12153s2344
β921b	0.3	47	51	11- 11	slow PA inc.	
β605	0.4	180	62	6½- 8½	PA inc, spect dG2, in field with Zeta	12176s2154
ζ	11.2	66	46	5 - 13	(β1245) optical, dist & PA inc, spect B8	12180s2156
Hwe 26	5.2	150	29	8½- 8½	spect A3	12189s2357
β606	0.3	78	62	7 - 9	PA & dist dec, spect F5	12234s1440
δ	24.2	214	58	3 - 8½	cpm pair, relfix, spect A0, dK2 (*)	12273s1614
β28	1.2	300	67	6½- 10	binary, about 180 yrs; PA inc, spect dF8	12275s1307
Lv 5	1.0	356	59	7 - 10	PA dec, spect G0	12315s1755
h1218	0.1	118	60	7½- 7½	(Fin 368) spect	12331s1633
	11.8	259	33	- 11	F2, all cpm	
Σ1669	5.4	309	58	6 - 6	neat pair, slow PA inc, spect dF6 & dF1	12387s1244
Hu738	9.1	259	59	6½- 11	PA & dist inc, spect K0	12412s1144
S643	23.4	294	15	7½- 9	relfix, cpm pair; spect A0, A2	12513s1746

CORVUS

LIST OF VARIABLE STARS

NAME	MagVar	PER	NOTES	RA & DEC
R	6.7--14.4	317	LPV. Spect M5e--M7e	12170s1859
U	9.0--14..	283	LPV.	12324s1811
V	9.8--14..	194	LPV. Spect M2e	12379s1721
X	9.5--11..	112	Semi-reg; spect M6	12461s1914
RV	8.4---9.2	.7473	Ecl.bin; spect F0+G0	12351s1918
ST	9.9--13..	225	LPV.	11572s1258
SU	9.5--11..	351	LPV. Spect Me	12189s2141
SV	6.8---7.6	60:	Semi-reg; spect gM5	12472s1448
SX	9.0--9.25	.3166	Ecl.bin; W Ursa Maj type Spect F8	12376s1832

LIST OF STAR CLUSTERS, NEBULAE, AND GALAXIES

NGC	OTH	TYPE	SUMMARY DESCRIPTION	RA & DEC
4024	295[2]	⊖	E2/S0; 12.9; 1.1' x 0.4' F,vS,bM	11560s1805
4027	296[2]	⊖	Sc/pec; 11.6; 2.0' x 1.7' F,pL,R (see note on 4038)	11570s1859
4033	508[2]	⊖	E5; 12.8; 0.9' x 0.5' pB,S,1E,bM	11580s1734
4038	28[4]	⊖	S?/pec; 11.0; 2.5' x 2.5' pB,cL,R. Two extending filaments. The "Ringtail Galaxy" (*)	11593s1835
4050	509[2]	⊖	SBb; 12.5; 2.2' x 1.4' F,cL,R,bM	12004s1606
4094		⊖	Sb; 13.0; 3.5' x 1.0' eF,pL,pmE, vgbM	12033s1416
4361	65[1]	◎	Mag 10½; diam 80", with 13^m central star. L,pF, R disc	12219s1829
4462	764[3]	⊖	Sb; 13.0; 2.7' x 0.8' pB,pS,E,sbM	12267s2254
4594	43[1]	⊖	"Sombrero Galaxy"; on Virgo-Corvus border. See Virgo.	12373s1121

LIST OF STAR CLUSTERS, NEBULAE, AND GALAXIES   (Cont'd)

NGC	OTH	TYPE	SUMMARY DESCRIPTION	RA & DEC
4756	281[3]	⊖	E2; 13.3; 0.5' x 0.4' vF,pS,R	12503s1508
4763	489[3]	⊖	Sa; 13.2; 0.9' x 0.6' vF,S,1bM	12506s1643
4782	135[1]	⊖	E0; 12.9; 0.5' x 0.5' pF,pS,R,mbM; contact pair with 4783	12520s1219
4783	136[1]	⊖	E0; 12.9; 0.5' x 0.5' pF,pS,R,mbM; pair with 4782, separation 0.7'	12520s1218
4802		⊖	E3; 13.2; 0.9' x 0.6' F,S,1E; 11^m star on East edge	12532s1147

DESCRIPTIVE   NOTES

BETA   Mag 2.66; spectrum G5 III. Position 12318s2307. The computed distance is about 110 light years and the actual luminosity about 85 times that of the Sun. Beta Corvi has an annual proper motion of 0.06"; the radial velocity is 4½ miles per second in approach.

GAMMA   Name- GIENAH. Mag 2.58; spectrum B8 III. The position is 12132s1716. The star is some 450 light years distant, giving a computed luminosity of some 1200 suns (absolute magnitude -3.1). The annual proper motion is 0.16"; the radial velocity is about 2½ miles per second in approach.

DELTA   Name- ALGORAB. Mag 2.95; spectrum A0 V or B9.5 V. Position 12273s1614. Delta Corvi is an easy double star for the small telescope, first observed by J. Herschel and J.South in 1823.   There has been no change in either the separation or PA of the pair in the last 140

DESCRIPTIVE NOTES (Cont'd)

years, although the physical association is proven by the common proper motion of 0.26" per year in PA 235°. Most observing lists have called the colors "yellowish and pale lilac"; occasionally "white and orange". The small star, magnitude 8.4, has a spectral type of dK2 and an actual luminosity of about half that of the Sun. The projected separation of the pair is about 900 AU.

Delta Corvi is approximately 125 light years distant, which gives the primary a luminosity of about 75 suns. Radial velocity measurements of the system show a speed of about 5½ miles per second in recession.

EPSILON   Mag 2.99; spectrum K3 III. Position 12075s2221. The statistics on Epsilon Corvi are: Distance about 140 light years, actual luminosity about 100 times the Sun, absolute magnitude -0.2, annual proper motion 0.07", radial velocity about 3 miles per second in recession.

NGC 4038   Position 11593s1835. The "Ring-Tail Galaxy", a spiral (?) of rather peculiar structure, sometimes regarded as a gravitationally interacting pair of galaxies or an actual collisional system. It is located about 3.7° WSW from Gamma Corvi. On short exposure photographs the object resembles a reversed question-mark, an irregular loop about 2' in length. The curved northern rim of the loop is the brightest portion and consists of many bright and dark patches - probably star clouds and masses of dust. There is no central nucleus and no readily apparent spiral pattern. The extension to the south, sometimes identified by a separate NGC number (4039) is fainter but has much the same appearance, resembling the mottled central mass of an irregular galaxy or a barred spiral.

Long-exposure photographs completely transform the appearance of the object, revealing details which have excited much speculation. The main mass and the southern extension appear as two large luminous elliptical blobs, fused together at their eastern tips, and from the area where they join are seen two long curving filaments reaching out for enormous distances. The northern filament can be traced out about 5' and the southern one for more than 10' in a great sweeping curve which first reaches south

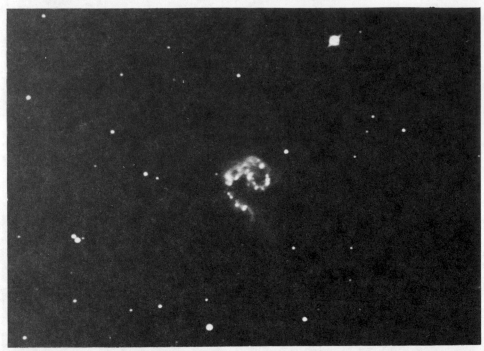

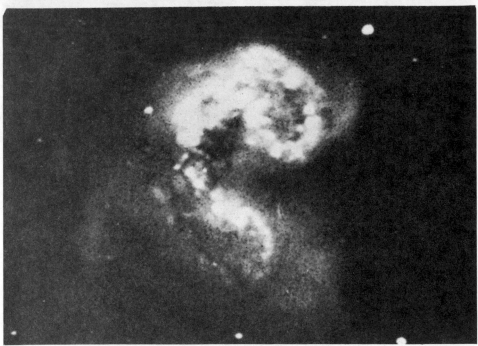

THE RING-TAIL GALAXY (NGC 4038) in CORVUS. This peculiar
system is shown here (top) on a plate made with the Lowell
42-inch reflector, and (below) with the McDonald 82-inch
reflector.

720

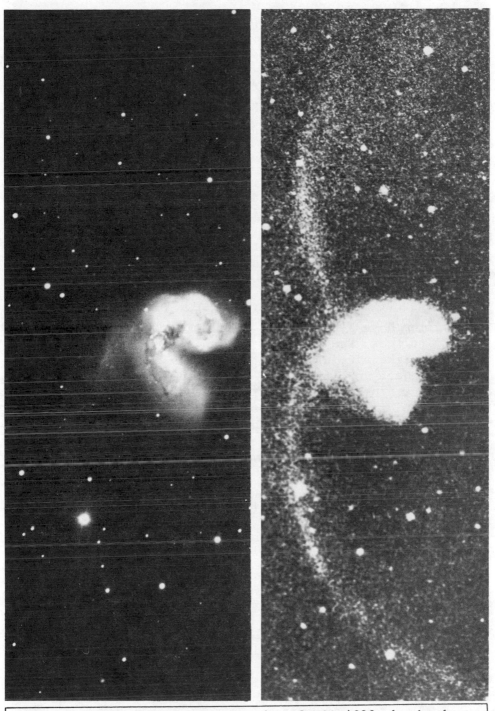

THE RING-TAIL GALAXY. Photographs of NGC 4038 obtained at Palomar Observatory with the 48-inch Schmidt camera. The long extending filaments are shown in the print at right in which the contrast has been artificially increased.

but then curves to point nearly due west. On the assumption
that NGC 4038 is actually a collisional system, these odd
structures are often referred to as "tidal filaments", but
it is possible that electric and magnetic fields also play
a part in their formation. A strange galaxy in Cancer, NGC
2623, has a very similar appearance (photo on page 343).
Like many other peculiar galaxies, the Ring-Tail is a
source of radio radiation, first detected in 1957.

The red shift of the system is about 910 miles per
second, and the computed distance is close to 90 million
light years. This gives the main body an actual diameter
of some 100,000 light years; the total luminosity appears
to be about 20 billion times that of the Sun.

Radial velocity measurements at various points across
the system have been made by E.M. and G.R.Burbidge (1966);
they find that "the velocity field departs strongly from
anything like a regular rotation of two galaxies, or of a
single object". Still, it is uncertain that this system
actually represents a collision of two galaxies; a process
of fissioning has also been suggested. V.Ambartsumian has
shown that the number of double galaxies is too great to
be explained by a capture or collision process; he concludes that the majority were actually formed as doubles.
The tidal filaments of NGC 4038, however, probably imply
that some strong interaction is occurring here between two
systems, if not an actual collision. H.Shapley (1943) has
called attention to the fact that another peculiar galaxy
(NGC 4027) lies only 0.7° distant, to the SW; the two
systems are "almost certainly physically associated, or
have been". This probable companion system resembles a one-
arm spiral with its single tail curving to the north and
east, in the general direction of the Ring-Tail. The projected separation of the two systems is about a million
light years, which implies that the interaction- if any-
must have occurred at some distant time when the separation
was much less. These two objects possibly form one of those
intriguing double systems which are linked by luminous
bridges and filaments, though in this case the presumed
"connecting filament" fades out- perhaps becoming too faint
to be recorded- after covering less than a third of the
intervening distance. Systems of this type may also be related to "eruptive galaxies" as NGC 5128 and M82.

CRATER

LIST OF DOUBLE AND MULTIPLE STARS

NAME	DIST	PA	YR	MAGS	NOTES	RA & DEC
A1774	3.7	271	52	5½- 10	cpm; spect gG6	11007s1102
Σ1509	32.9	16	25	7 - 9	relfix, spect K0	11040s1309
ψ	0.2	358	62	6½- 7	(β220) dist & PA dec, spect A0	11100s1814
Hu 461	1.6	70	57	8½- 9½	slow PA inc, spect G0	11128s1713
β600	1.0	210	59	6 - 12	AB cpm; PA dec,	11144s0652
	57.2	98	59	- 10	spect Ap; AC dist dec, optical	
Hu 130	1.1	122	63	8 - 8½	PA dec, spect F5	11164s1130
Σ1530	7.7	313	55	7½- 8	relfix, spect F2	11172s0637
β26	2.8	68	48	7 - 10	relfix, spect K0	11212s1009
γ	5.2	96	55	4 - 9	(15 Crat) (h040) cpm, relfix, spect A7	11224s1725
Hu 462	0.5	76	68	8 - 8½	binary, 48 yrs; spect K0	11248s1522
Jc 16	8.2	80	54	5½- 9	cpm; PA inc, spect dF1	11272s2411
Hd 130	10.0	344	32	7½- 13	spect A2	11305s2310
h4456	15.2	122	59	7½- 11	dist dec, spect	11343s2410
	39.5	328	10	- 13	K2	
ι	1.4	226	58	5½- 11	(24 Crat) (Kui 58) spect dF5	11361s1256
β1078	7.8	52	34	6½- 12	spect A0	11373s1411
h4479	7.3	93	59	8 - 9½	cpm; spect G5	11508s2419
A2381	4.4	356	60	7½- 12	relfix, spect K0	11510s1550
	68.7	321	23	- 11		
h4481	3.6	194	54	8 - 8	relfix, spect F5	11547s2216

LIST OF VARIABLE STARS

NAME	MagVar	PER	NOTES	RA & DEC
R	8.0---9.5	160:	Semi-reg; spect M4	10581s1803
S	9.0--11..	155	Semi-reg; spect M6e--M7e	11502s0719
T	8.0--10..	70:	Semi-reg; spect M4	11214s1938
U	9.5--13..	169	LPV. Spect M0e	11103s0701
V	9.7--10.6	.7020	Ecl.bin; spect A6	11217s1624
Y	8.5--13..	158	LPV. Spect M3e--M5e	11180s2439
RR	8.0--9..	Irr	Spect M5	11291s1206
RT	9.5--12..	343	LPV.	10593s0723
RU	7.5--8.5	Irr	Spect M3	11487s1055
SU	8.63 ±0.01	.055	Delta Scuti type, spect F2	11303s1145
SV	6.32 ±0.02	5.905	Alpha Canum type, spect composite: A3 + A8	11144s0652

LIST OF STAR CLUSTERS, NEBULAE, AND GALAXIES

NGC	OTH	TYPE	SUMMARY DESCRIPTION	RA & DEC
3511	39[5]	⊘	Sc; 11.9; 4.2' x 1.5' vF,vL,mE	11008s2250
3513	40[5]	⊘	SBc; 12.0; 2.0' x 1.6' vF,vL,vmE, S-shaped spiral	11011s2258
----	I.2627	⊘	SBb; 12.8; 2.0' x 2.0' eF,L,R, S-shaped spiral	11075s2328
3571	819[2]	⊘	Sa; 13.1; 2.6' x 1.0' pF,pL,mE, bM	11089s1801
3637	551[2]	⊘	SBa; 12.8; 1.0' x 1.0' F,S,R, psbM; 7^m star 2.9' s prec.	11181s0958
3672	131[1]	⊘	Sb; 11.8; 3.5' x 1.4' pB,L,E, gbM. Highly tilted multiple-arm spiral	11225s0932

LIST OF STAR CLUSTERS, NEBULAE, AND GALAXIES (Cont'd)

NGC	OTH	TYPE	SUMMARY DESCRIPTION	RA & DEC
3732	552[2]	⊖	E0; 12.9; 0.5' x 0.5' F,S,R, psbM	11317s0934
3865		⊖	Sb; 13.0; 1.6' x 1.0' F,pL,1E	11427s0856
3887	120[1]	⊖	Sc; 11.6; 2.8' x 2.0' pB,L,1E, vgpmbM	11446s1635
3892	553[2]	⊖	Sa/S0; 12.9; 1.7' x 1.5' pB,pL,R,gbM	11455s1041
3955	623[2]	⊖	Sb/pec; 12.8; 1.9' x 0.5' cF,S,E	11515s2254
3957	294[2]	⊖	E8; 12.9; 2.3' x 0.4' F,S,E, spindle-shaped, edge-on lenticular; possibly S0 system.	11516s1917
3956	290[3]	⊖	Sc; 12.6; 2.8' x 0.7' cF,pL,pmE, nearly edge- on spiral	11516s2018
3962	67[1]	⊖	E2; 12.0; 0.8' x 0.7' cB,pL,1E,gbM	11522s1342
3981	274[3]	⊖	Sb; 12.7; 3.5' x 1.0' vF,pL,E; nearly edge-on	11537s1937

LIST OF DOUBLE AND MULTIPLE STARS

NAME	DIST	PA	YR	MAGS	NOTES	RA & DEC
Cor 133	8.3	21	54	8 - 8 -11½	spect F5	11567s6230
△ 117	23.0	149	18	8½ -8½	relfix, spect B8,	12002s6148
	25.0	18	18	- 10½	F5	
η	44.0	299	18	4½- 10	(h4501) spect F0	12043s6420
I 902	2.7	168	44	8 - 10	spect G5	12128s5855
h4516	17.4	96	18	7½- 9½	spect A5	12210s6343
Cor 137	10.0	342	19	8½- 9½	spect A2	12211s6150
Brs 8	5.3	335	43	8 - 8	spect G0; in star cluster 4337	12221s5750
a	4.4	115	59	1½- 2	ACRUX, splendid	12238s6249
	90.1	202	13	- 5	object, spect B1, B3 (*)	
R 200	2.5	137	41	8½- 9½	relfix, spect B5, in cluster H5	12245s6031
h4524	31.1	338	17	7½- 9½	spect A0, in cluster 4439	12254s5946
Cp 12	1.9	221	60	8 - 8½	PA dec, spect G0	12255s6129
Cp 12b	0.3	311	60	8½- 8½	(Rst 4499) PA dec	
Hd 217	27.2	275	03	6 - 13	(L 5169) spect K0	12259s5607
	49.0	290	02	- 12		
I 36	21.8	325	02	7 - 12	spect B5	12261s6136
γ	111	31	19	1½- 6½	optical, dist inc, spect M3, A2 (*)	12284s5650
Hld 116	1.9	193	56	7 - 9	relfix, spect A2	12353s5540
h4543	2.5	163	39	6½- 11	spect K0	12406s5837
	31.1	96	13	- 9		
ϕ 65	0.2	238	59	7½- 7½	spect A0	12417s5701
Cor 140	4.9	97	34	7½- 9	spect F8	12423s6156
ι	26.9	22	42	4½- 7½	(h4547) PA & dist dec, spect K1	12427s6042
h4548	52.6	169	13	5 - 8½	cpm; spect B3	12435s5613
β	44.3	322	01	1½- 11	BETA CRUCIS. Spect B0 (I 362) (*)	12448s5925
μ	34.9 •	17	52	4½- 5½	(△126) relfix, easy pair, spect B3, B5; cpm	12516s5654

CRUX

LIST OF VARIABLE STARS

NAME	MagVar	PER	NOTES	RA & DEC
β	1.25--1.3	0.2365	Spect B0, Beta Canis Majoris type (*)	12448s5925
R	6.3---7.1	5.826	Cepheid; spect F6--G7	12209s6121
S	6.1---6.8	4.690	Cepheid; spect F5--G5	12514s5810
T	6.2-- 6.8	6.733	Cepheid; spect G1--G5	12186s6200
U	9.1--13..	351	LPV. Spect M4e--M6e	12296s5718
V	8.8--12..	376	LPV. Spect Ne	12536s5738
W	8.5---9.3	198.53	Giant eclipsing binary, spect G0p	12093s5830
X	8.2---8.9	6.220	Cepheid; spect G0--K0	12434s5851
Y	9.5--14..	214	LPV.	12004s6327
Z	9.5--12.0	341	Semi-reg; spect N0	12085s6411
ST	9.4--12..	440	LPV. Spect M6e	12230s5943
VW	9.6 -10.4	5.265	Cepheid; spect K0	12304s6314
ZZ	9.6--10.1	1.862	Ecl.bin; lyrid type, spect B8	12031s6313
AB	8.9---9.5	3.413	Ecl.bin; spect B0	12149s5753
AE	9.0---9.7	3.478	Ecl.bin; spect B9	11560s6053
AG	7.8---8.7	3.837	Cepheid, spect F5	12385s5931
AI	9.3--10.0	1.418	Ecl.bin; spect B	12035s6059
AO	7.5--9..	Irr	spect M0	12150n6320
AP	10.7--16	---	Nova 1935	12285s6410
BH	7.2--10..	421	LPV, spect S	12136s5601

LIST OF STAR CLUSTERS, NEBULAE, AND GALAXIES

NGC	OTH	TYPE	SUMMARY DESCRIPTION	RA & DEC
4052			diam 10', pRi,1C, about 50 stars; class E	12006s6254
4103	△291		pL,Ri; diam 9', about 25 stars mags 10...14, class D	12041s6058
4337			Diam 4'; pRi,1C, about 25 stars mags 12...14, class F	12212s5750

LIST OF STAR CLUSTERS, NEBULAE, AND GALAXIES   (Cont'd)

NGC	OTH	TYPE	SUMMARY DESCRIPTION	RA & DEC
4349	△292		vB,vL,lC, diam 15', about 100 stars mags 12...14; class G	12214s6137
----	H5		pL, diam 7'; about 30 stars, class D; includes double star R 200.	12252s6029
4439	△300		S; diam 3', about 15 stars mags 11.... class D; with double star h4524	12254s5949
4463			Crux-Musca border, refer to Musca	
----	H7		L,vRi, diam 18', about 200 stars, members faint, class C	12359s6020
4609	△272		pL,pC,E; diam 4', about 20 stars mags 10... class E	12394s6242
4755	△301		! vL,Ri; diam 10', about 50 stars mags 6...10; the Kappa Crucis or "Jewel Box Cluster". Class G (*)	12506s6005

DESCRIPTIVE NOTES

ALPHA   Name- ACRUX, although the star is often called simply "Alpha Crucis". Position 12238s6249.
This is the 14th brightest star in the sky; magnitude 0.87; spectrum B1 IV and B3. Acrux is the star at the foot of the Southern Cross. Midnight culmination occurs about March 28.
     Acrux is a brilliant double star, one of the very finest of the heavens, and first noted as such, according to R.H.Allen, by "some Jesuit missionaries sent by King Louis XIV to Siam in 1685". Both components are highly luminous B-type "helium stars" of magnitudes 1.39 and 1.86, with actual luminosities of about 3000 and 1900 suns; the computed absolute magnitudes are -3.9 and -3.4. Each star seems to be a spectroscopic binary, but the published periods of 59.3 and 56.0 days have been questioned recently at Radcliffe Observatory by A.D.Thackeray (1974) who finds the brighter star to have a period of 75.769 days, while the evidence for duplicity in the other star still remains very

DESCRIPTIVE NOTES (Cont'd)

uncertain. The visual pair, separated by about 4½", form
a long-period binary with extremely slow orbital motion.
The only change detected since discovery is a very small
decrease in the separation, which was 5.4" in 1826; the PA
may also have decreased slightly in the last century, but
not by more than a few degrees. The present projected separation of the pair is about 500 AU, or a little more than
six times the distance across the Solar System.

Acrux lies at a computed distance of about 370 light
years. The annual proper motion is 0.04"; the radial velocities of the two stars differ somewhat owing to their orbital motion around a common center of gravity; the measurements give 6½ and ½ miles per second, both in approach.
The star is a member of a large moving group of which Beta,
Delta, and Zeta Crucis are other members. A number of stars
in Scorpius and Centaurus show about the same proper motion and apparently form a large moving association. (For
details concerning this group, refer to "Scorpio-Centaurus
Association" under Scorpius)

BETA Mag 1.28 (slightly variable). Spectrum B0.5
or B0 IV. Position 12448s5925. This is the
20th brightest star in the sky, marking the end of the
eastern arm of the Southern Cross. Midnight culmination
occurs about April 2.

Beta Crucis is approximately 490 light years distant
and about 5800 times the luminosity of the Sun. (Absolute
magnitude about -4.6). The annual proper motion is 0.05";
the radial velocity is 12 miles per second in recession.

The star is a pulsating variable of the Beta Canis
Majoris type, with a primary period of 5h 40.5m; the
light range, however, is only about 0.06 magnitude. There
is also an 11th magnitude companion at 44", discovered by
R.T.Innes in 1901, probably not a true physical companion.
The projected separation of the two stars is some 6900 AU.

GAMMA Mag 1.67; spectrum M3 II; position 12284s
5650. This is the 28th brightest star in the
sky, marking the top of the Southern Cross. Gamma Crucis
is a giant with an absolute magnitude of about -2.5, similar in general type to Antares but not so large or brilliant. The distance is about 220 light years and the actual

luminosity about 900 times that of the Sun. The star has
an annual proper motion of 0.27" in PA 175°; the radial
velocity is 13 miles per second in recession.
        The star GC 17055, magnitude 6.4, spectrum A2, is
located 111" distant (1919). This star, however, does not
share the proper motion of Gamma itself and the separation
is now increasing from the southward motion of the primary
star.

DELTA   Mag 2.82; spectrum B2 IV. Position 12125s5828.
               This star marks the west arm of the Southern
Cross. The computed distance is about 570 light years,
leading to a true luminosity of about 1900 suns. Delta
Crucis shows an annual proper motion of 0.04"; the radial
velocity is 15½ miles per second in recession.

NGC 4755 Position 12506s6005. The Kappa Crucis star
             cluster, a brilliant and beautiful galactic
cluster ranking among the finest and most spectacular objects of the southern Milky Way. The popular name "Jewel
Box" was derived from Sir John Herschel's statement that
this cluster produced upon him the impression of a superb
piece of jewelry. The 50 or so brightest members are all
compressed into a space about 10' in diameter, or approximately 25 light years. The brightest stars are supergiants
and include some of the most luminous stars known in our
Galaxy. According to a study at Radcliffe Observatory in
1962, the 10 brightest cluster members have the following
apparent magnitudes and spectra:

1.	Mag 5.75; spect B9	6.	Mag 9.09; spect B2
2.	" 5.94 B3	7.	" 9.79 B0
3.	" 6.80 B2	8.	" 10.01 B1
4.	" 7.58 M2	9.	" 10.58 B1
5.	" 8.35 B1	10.	" 11.42 B8

With the exception of #4, all the brighter members are B-
stars, with absolute magnitudes ranging from -7.5 to about
-1.8. These figures have been corrected for a light loss
of 1.3 magnitudes through space absorption. Star #1 thus
has a true luminosity of about 80,000 suns, and star #2
about 75,000 suns. Star #4 is a red supergiant, located
near the center of the cluster. The computed absolute mag-

THE MILKY WAY IN CRUX. A striking portion of the Galaxy, in the vicinity of the Southern Cross, showing the dark "Coal Sack" Nebula.
Boyden Station, Harvard Observatory

THE JEWEL BOX. Star Cluster NGC 4755, photographed with
the 60-inch reflector at the Boyden Station of Harvard
College Observatory. This brilliant galactic cluster is
one of the show objects of the Southern Milky Way.

nitude of the star is about -5.7 and the true luminosity
is about 16,000 suns, more or less the equal of Betelgeuse
(Alpha Orionis). Another supergiant, of type A1, some 10'
from the edge of the main cluster, is very probably a member. This star (HD 111613) is magnitude 5.7, giving a true
luminosity of about 83,000 suns.

These luminosities are based on a derived distance of
about 7700 light years, from recent studies at Radcliffe;
published distances are very discrepant, owing to heavy
obscuration in the region. The rich central mass of the
cluster appears to be about 25 light years in diameter with
outliers scattered over a 50-light-year area. The radial
velocity is about 11 miles per second in approach.

NGC 4755 is one of the youngest star clusters known
and appears to be comparable in type and age to the fine
Double Cluster in Perseus. The presence of highly luminous
supergiants indicates an age of probably not more than a
few million years for either cluster. The Perseus group
also resembles NGC 4755 in the possession of red supergiants, evidently objects of large mass which have already
evolved away from the main sequence and are now nearing the
point of ending the hydrogen-burning reaction which powers
the majority of the stars. The bright blue stars of NGC 4755
will presumably begin their own evolutionary expansion in
the near astronomical future.

The cluster lies in a rich and remarkable region of
the heavens, well worth exploring with low power telescopes
and instruments of the "rich-field" type. Immediately to
the south, the star clouds are obscured by the vast dark
blot called the "Coal Sack", certainly the most famous of
the naked-eye dark nebulae. (It is one of the quaint odd-
ities of astronomical nomenclature that the Jewel Box should
lie next to the Coal Sack.) Measuring about 7° x 5° across,
the Coal Sack appears as a nearly starless spot just east
of Alpha Crucis, and was noticed by Portuguese navigators
in the 16th century. This is possibly the nearest of the
dark nebulosities, believed to be no more than 500 or 600
light years distant, and some 60 or 70 light years in diameter. Many curious dark lanes and channels ornament the
surrounding region. (See photographs, pages 731 and 734)

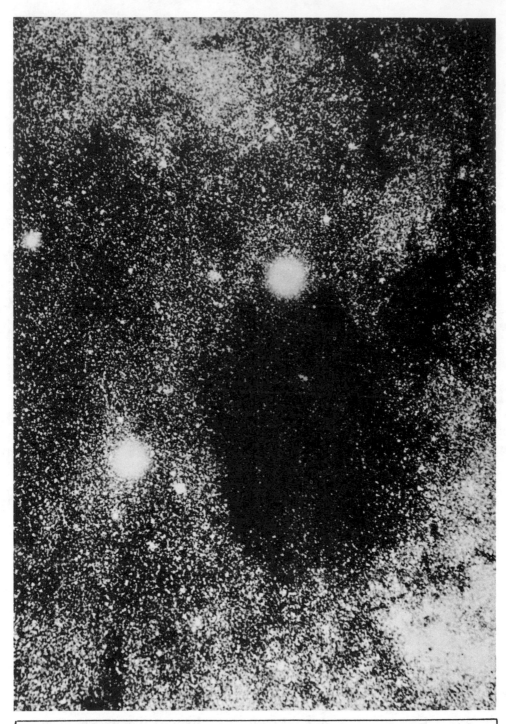

THE COAL SACK. One of the curious features of the Southern Milky Way, this great dark cloud appears as a black
blot against the star clouds of Crux.
                              Harvard College Observatory

CYGNUS

LIST OF DOUBLE AND MULTIPLE STARS

NAME	DIST	PA	YR	MAGS	NOTES	RA & DEC
Σ2479	0.7	1	53	7 - 10	PA dec, spect A3	19073n5515
	6.8	31	53	- 9½		
Σ2486	8.2	212	62	6 - 6½	neat pair in fine	19108n4945
	47.1	13	24	- 13	field; cpm; slow	
					PA & dist dec,	
					spect both G4.	
Σ2496	2.3	78	58	7 - 11	relfix, cpm pair;	19140n4959
					spect G6	
Σ2511	12.4	42	12	7½- 11	cpm; spect K5	19192n5015
	79.3	116	12	- 10		
β1129	0.2	324	60	6½- 6½	PA dec, spect A5	19204n5217
OΣ373	1.7	231	59	7½- 10	relfix, spect A0	19226n4620
OΣΣ182	73.4	300	56	6½- 7½	wide pair, slow	19255n5002
	34.7	150	56	- 11	dist inc, spect F5	
Σ2534	7.0	64	49	8 - 8	relfix, spect A0	19259n3626
Hu1303	0.7	305	56	7½- 9	PA dec, spect A3	19268n3705
	4.5	70	19	- 14		
Es 654	7.6	192	08	7- 12½	spect G5	19271n5441
β	34.3	54	59	3 - 5½	splendid fixed	19287n2751
					pair, superb color	
					contrast (*)	
Λ 712	0.1	289	59	7 - 7½	PA dec, spect A0	19292n5632
OΣ374	18.7	298	09	7- 10½	spect K0	19297n5005
A 713	0.4	256	60	7 - 7½	PA inc, spect A3	19298n4722
A 369	4.2	6	29	7½- 14	spect B9	19349n3013
θ	4.2	53	34	5 - 13	(13 Cygni) (β1131)	19351n5006
	48.4	182	23	- 11	AB cpm, PA inc,	
					spect F4, AC =	
					optical, dist inc.	
h1423	20.8	128	24	6½- 11	spect B5	19352n2913
	28.0	340	24	- 13		
	35.1	357	24	- 12		
Σ2557	11.2	104	50	7½- 10	relfix, spect A0	19376n2938
	21.5	303	16	- 11	(β54)	
Ho 111	1.1	4	58	6½- 11	cpm, spect A3	19379n3352
β145	0.8	267	43	7 - 9½	relfix, spect G5	19393n3036
	9.0	29	25	- 10		
	26.9	157	25	- 12		
Kui 94	0.2	130	60	6½- 8	PA dec, spect A3	19402n4008
16	39.0	134	55	5 - 5	nice & easy pair,	19406n5024
					cpm; spect G2,G5	

LIST OF DOUBLE AND MULTIPLE STARS (Cont'd)

NAME	DIST	PA	YR	MAGS	NOTES	RA & DEC
OΣ383	0.9	19	56	7 - 8	PA dec, spect A0	19412n4036
OΣ384	1.0	195	59	7 - 7½	relfix, spect B8	19420n3812
Ho453	15.4	53	34	6½- 13	relfix, spect B8	19420n3402
	33.7	136	24	- 12		
δ	2.2	247	61	3 - 6½	(Σ2579) binary, PA dec, spect B9. Difficult (*)	19434n4500
Σ2576	1.5	14	65	8 - 8	binary, 240 yrs; spect K5, cpm with 17 Cygni	19436n3330
Σ2578	15.1	125	56	6½- 7½	relfix, spect B9, B9	19438n3558
	45.9	358	08	- 12		
OΣΣ191	38.7	27	56	6½- 8½	Spect K0, A2	19440n3453
OΣ385	1.3	54	49	7½- 10	relfix, spect B8	19441n4026
A 1404	0.2	72	58	7½- 8	PA dec, spect B9	19442n3946
	19.6	273	18	- 13		
	21.8	79	18	- 12		
17	26.0	70	53	5 - 8	(Σ2580) relfix, spect F5, dK6; cpm with Σ2576. Rich region !	19445n3337
OΣΣ192	2.7	217	43	6½- 12½	(Ho 114) PA dec, ABC cpm; spect K2, AD optical, dist dec.	19447n3246
	9.7	214	43	- 13½		
	30.7	197	33	- 9		
h601	24.7	241	34	5½- 12½	spect B8	19457n3817
OΣ386	0.9	74	59	7½- 8	relfix, spect A2	19465n3702
OΣ387	0.5	204	61	7 - 8	binary, 157 yrs; PA dec, spect F2	19468n3511
Σ2588	9.6	160	32	7 - 7½	relfix, spect B8	19474n4415
Σ2588b	0.3	48	59	8 - 8½	(A718)	
Es 84	11.6	161	55	7 - 12	spect G3	19477n3835
	23.0	98	55	- 12		
19	54.6	106	23	6 - 10	optical (h603) spect gM2	19488n3835
OΣ389	9.7	307	07	6½- 12	(Es 357) spect A5	19497n3101
	12.5	183	46	- 9		
OΣ390	9.7	22	53	7 - 9	relfix, spect B5	19531n3004
	16.4	175	32	- 11		
J781	2.8	306	46	9½- 9½		19533n3003

LIST OF DOUBLE AND MULTIPLE STARS (Cont'd)

NAME	DIST	PA	YR	MAGS	NOTES	RA & DEC
Ho 581	0.1	283	62	8 - 8½	binary, 25.7 yrs; PA inc, spect K0	19533n4144
Hu 687	0.1	262	62	8½- 8½	binary, 112 yrs; spect A0	19536n5041
η	7.2	208	56	5 - 12	(21 Cygni) (β980)	19544n3457
	46.0	328	24	- 11	AB cpm, spect K0;	
	49.5	169	24	- 11	C+D optical	
ψ	3.1	177	62	5 - 7½	(Σ2605) PA dec, spect A3, A7	19544n5218
AX	18.5	51	14	9 - 10	(OΣ391) spect N, A2; primary N-type variable, 9---11^m	19556n4408
Σ2607	0.1	264	62	7 - 9	(OΣ392) PA dec,	19562n4207
	3.1	290	58	- 9	spect A2, AC= relfix.	
OΣ393	18.9	228	53	7½- 8½	dist dec, spect K0 & A0; contrast	19563n4415
Σ2606	1.0	137	56	7½- 8½	slight PA inc, spect A5	19566n3308
Σ2609	2.1	23	54	7 - 8	slight PA dec, spect B5	19568n3758
Σ2611	5.3	27	54	8 - 8	relfix, spect K0	19574n4714
h1464	13.4	32	02	8 - 12½	relfix, spect K0	19577n5032
β1133	0.9	338	56	7- 9½	relfix, spect A0	19577n3141
β1258	1.4	149	24	8 - 12	PA dec, spect A0	19582n2947
OΣ394	11.5	295	33	7 - 10	relfix, spect K0	19583n3617
h1468	13.0	279	28	8 - 9	relfix, spect A0	19597n4010
26	41.7	148	49	5½- 8½	relfix, spect K1	20000n4958
26b	9.0	74	33	8½ - 11		
Σ2624	1.6	172	66	7 - 7½	slight PA & dist	20016n3553
	42.6	328	56	- 9½	dec, spect 0	
	29.0	172	00	- 11½		
h1470	28.8	337	25	7 - 9	relfix, spect M	20018n3811
Ho 454	5.6	57	05	7 - 12	spect A0	20018n5019
β440	6.9	65	43	7 - 12	(β429) multiple	20041n3539
	11.3	24	56	- 11	group, spect 0;	
	11.4	300	56	- 9½	AC dist inc.	
	28.1	107	56	- 11½		
	36.1	28	56	- 8		

LIST OF DOUBLE AND MULTIPLE STARS   (Cont'd)

NAME	DIST	PA	YR	MAGS	NOTES	RA & DEC
0Σ398	0.9	82	45	7½- 10	relfix, spect B0	20055n3534
	5.4	133	29	- 14	(A280)	
Σ2633	11.7	103	05	8 - 11	spect B9	20060n3226
A382	1.6	94	53	7 - 10½	PA inc, spect K0	20063n4214
Σ2639	6.2	301	25	7½- 8½	relfix, spect B2	20070n3521
0Σ400	0.2	264	62	7 - 8	binary, 84 yrs;	20086n4348
					PA dec, spect G4	
A1418	2.9	324	31	7½-11½	spect K0	20098n3844
A282	0.2	220	58	7½- 8	PA slow inc,	20101n3420
	23.0	17	29	- 12	spect A2	
	42.5	15	04	- 12		
0Σ401	13.8	60	29	7- 10½	relfix, spect G5	20104n3818
AC 17	4.0	80	58	6- 11½	relfix, cpm;	20111n5119
					spect gK1	
Es 502	12.1	222	33	7½- 10	spect A0	20116n4902
β660	9.6	319	58	7- 13	relfix, spect K2	20120n4314
O¹	107	173	26	4 - 7	(31 Cygni) wide	20120n4635
	338	323	26	- 5	group. A= ecl.bin	
	36.6	331	03	-13	V695; 5^m = 30 Cyg;	
					spect K1,B5,A3 (*)	
Σ2658	5.5	113	52	7 - 9	PA dec, spect F5;	20122n5258
	49.3	208	31	- 10	AC dist inc	
	34.0	313	13	- 13		
Es 244	40.6	130	26	7½-11½	spect A2	20125n3515
Es 244b	5.0	12	26	11½-12	C = 4" pair	
Σ2657	0.8	172	68	7 - 7	(0Σ403) spect	20126n4157
	11.5	33	62	- 9½	B8, all cpm	
A283	3.0	290	58	6- 13½	cpm; spect gG6	20134n3335
A387	4.8	151	28	7½-13½	spect A0	20134n4109
Ho 455	33.6	82	25	7 - 11	spect M	20140n5359
	32.3	257	25	- 11		
	38.1	75	25	- 10		
Ho455b	3.7	182	25	11½-11½	PA dec	
	5.7	210	25	- 12		
0Σ404	29.6	114	52	7 - 9½	spect K5	20143n5221
Ho 588	51.1	297	49	6½- 8½	relfix, spect A0	20149n3121
Ho 588b	8.5	16	25	8½-12½		
Σ2663	5.4	324	52	8 - 8½	relfix, spect A0	20150n3932
Σ2666	2.5	246	62	6½- 8½	relfix, spect 08	20164n4035
	34.2	208	08	- 9		

LIST OF DOUBLE AND MULTIPLE STARS (Cont'd)

NAME	DIST	PA	YR	MAGS	NOTES	RA & DEC
A1425	0.2	280	60	8 - 8	PA dec, spect B1	20170n3807
	8.8	310	17	-13		
Ho125	3.1	194	41	7- 11½	relfix, spect K2	20171n3851
Σ2671	3.5	338	58	6 - 7½	relfix, spect A2,	20172n5514
	83.9	55	11	- 12½	F3	
β1206	2.0	1	38	7½-10½	spect F2	20172n3636
OΣ406	0.5	120	61	7 - 8	binary, 96 yrs;	20182n4512
					PA inc, spect F5	
Σ2668	0.2	97	62	7 - 9	(A1427) binary,	20184n3915
	3.3	285	62	- 9	90 yrs; spect A1;	
					all cpm.	
Es 800	28.2	315	09	8½- 9	C = 2" pair	20186n5106
	40.3	104	09	- 10		
β1207	5.8	217	19	7½-13½	spect O	20188n4342
β663	7.2	314	58	6½- 14	AC PA dec, spect	20192n5326
	6.5	69	58	-12½	K5	
A46	0.2	93	53	8½- 8½		20194n4331
	2.0	265	53	-10½		
γ	142	196	23	2 - 10	(β665) Spect F8;	20204n4006
					B = 2" pair (*)	
OΣΣ206	43.0	256	49	7 - 8½	relfix, spect B9	20211n3903
	23.4	262	05	- 11		
OΣΣ207	1.2	12	57	6½- 11	(Ho128) AB cpm,	20212n4249
	93.2	63	20	- 7½	PA dec, spect K0	
Σ2681	6.7	40	12	7½-10½	relfix, spect A0	20215n5315
	39.6	201	49	- 8		
	42.9	170	13	- 11		
D22	3.1	155	59	8 - 9	PA inc, spect G5	20237n3956
	15.6	97	26	-13½		
A730	0.2	349	62	7 - 7	binary, 84 yrs;	20240n5926
					PA dec, spect A0	
Σ2687	26.6	118	50	6½- 8½	relfix, cpm;	20252n5628
					spect A0	
Ho594	18.2	309	25	7- 12½	spect K2	20272n3540
Wei 35	4.0	213	49	8 - 8½	relfix, spect F5	20274n3721
	87.1	100	49	- 9		
Wei 35c	12.0	202	33	9 - 11		
h1525	8.7	232	49	8½- 8½	relfix, spect A2	20281n4012
Kui 97	0.7	137	62	6 - 8½	PA dec, spect B9	20283n5555
ω¹	17.9	341	35	5½- 13	(45 Cygni) (β669)	20285n4847
	56.4	87	18	- 10	spect B2	

CYGNUS

LIST OF DOUBLE AND MULTIPLE STARS (Cont'd)

NAME	DIST	PA	YR	MAGS	NOTES	RA & DEC
44	2.0	159	58	6½- 11	relfix, spect F5	20291n3646
	16.5	76	58	- 13		
S 755	60.1	279	25	6½- 10	relfix, spect A2	20293n4903
OΣ408	1.6	190	54	7 - 10	relfix, spect B8	20321n3430
Σ2700	23.7	285	51	6½- 8½	relfix, spect K0	20327n3220
Σ2702	3.3	205	55	8½- 8½	relfix, spect A0	20336n3500
Σ2705	3.1	261	54	7 - 8	relfix, spect K0	20358n3311
Σ2707	22.8	31	21	7 - 9	relfix, spect A0	20362n4746
	55.4	195	23	- 9		
Σ2708	34.2	327	26	7 - 9	optical, dist inc,	20368n3828
	16.9	14	11	- 13	spect G0	
Σ2717	1.8	258	56	7- 9½	PA dec, spect G0	20368n6035
	42.7	52	56	- 9½	AC dist dec.	
A746	2.1	147	44	7½- 13	spect A0	20373n4731
Σ2711	2.4	223	55	8 - 9	relfix, spect A0	20375n3020
OΣ410	0.8	9	61	6½- 7	PA dec, spect B8;	20377n4024
	69.1	70	39	- 8½	all cpm	
A748	1.4	24	27	7½- 12	spect A0	20381n4710
Σ2714	6.8	338	06	8½- 12	relfix, spect F8	20382n2936
Ho 137	0.8	314	62	6½- 11	PA inc, spect A3	20385n2938
49	2.8	46	61	6 - 8	(Σ2716) relfix;	20390n3208
					spect G8, A	
51	3.0	102	34	6 - 12	(β675) relfix;	20407n5010
	25.7	182	34	- 11½	spect B2	
	32.7	329	24	- 11½		
OΣ411	21.0	331	40	7½- 10	Dist & PA inc,	20407n4539
					spect G0	
Arg 39	10.3	157	62	8½- 8½	PA inc, spect K5	20409n4905
Ho 140	7.0	312	06	7 - 13	spect G5	20420n4608
52	6.6	67	55	4 - 9	(Σ2726) slight	20436n3032
					PA inc, spect K0;	
					NGC 6960 in field	
OΣ412	25.5	282	01	7 - 13	spect F5; B= 5"	20442n5029
					pair	
ε	54.9	272	59	3 - 12	(β676) optical,	20442n3347
					dist inc (*)	
Kui 101	0.5	105	51	6½- 8½	spect A2	20450n4620
T	9.9	121	34	5½- 12	(β677) cpm; A =	20452n3411
	14.0	201	34	- 11	variable, spect K3,	
					AC optical, distinc	

LIST OF DOUBLE AND MULTIPLE STARS (Cont'd)

NAME	DIST	PA	YR	MAGS	NOTES	RA & DEC
OΣ414	10.0	95	49	7 - 8	relfix, spect B9	20454n4213
λ	0.7	28	61	5 - 6½	(54 Cygni) PA dec,	20455n3618
	84.6	105	55	- 9	binary, about 400	
					yrs; spect B5	
β268	0.5	206	55	7½- 8½	PA dec, spect A0	20459n4153
V367	0.1	50	53	7½- 7½	(A1434) PA dec,	20461n3906
	2.3	255	60	-13½	Ecl.bin; spect A7	
Es 93	12.2	302	35	6 - 11	optical, PA & dist	20464n5214
					inc, spect dK0	
Σ2731	4.1	86	35	7½-10½	relfix, spect A0	20471n3936
	16.4	171	22	-12½	(Ho 596)	
	18.5	249	22	-13		
Σ2732	4.0	74	62	7 - 9	relfix, spect B9	20472n5143
					cpm pair	
β250	19.2	7	34	7 - 12	spect B3	20482n4628
β67	1.6	294	24	7 - 10	PA inc, spect B7;	20485n3043
					near center of the	
					Veil Nebula	
β155	0.9	36	57	7½- 8	PA inc, spect F0	20496n5114
	16.2	23	15	- 12		
OΣ416	8.7	124	62	8 - 8	PA dec, spect A3	20502n4334
OΣ422	2.8	333	45	7½- 9	relfix, spect B9	20524n4455
OΣ420	5.8	3	30	7 - 11	relfix, spect B6e	20525n4031
OΣ418	1.2	287	61	7 - 7	PA dec, dist inc,	20528n3231
					spect G0	
OΣ419	1.8	34	35	7 - 10	relfix, spect A0	20528n3653
OΣ423	2.9	80	54	7 - 9½	relfix, spect B9	20535n4219
β1137	7.2	348	34	6- 13½	cpm; spect F0	20548n5032
Σ2741	1.8	27	62	6 - 7½	slight PA dec,	20569n5016
					cpm; spect B8	
Hu 764	0.4	198	51	7½- 8½	PA inc, spect B9	20574n3614
59	20.3	352	51	4½- 9	(Σ2743) relfix,	20581n4720
	26.7	141	21	- 12	cpm; spect B1,	
	38.3	220	13	-11½	Primary variable	
β1210	2.3	119	24	7½- 12	(OΣ425) Spect A0	20584n4829
	14.9	27	24	- 10½		
60	2.6	164	62	6 - 10	(OΣ426) slight	20594n4557
					PA dec, spect B1e	
Σ2746	1.1	312	62	8 - 8½	PA inc, spect F0	20599n3904
Ho 600	1.9	80	24	7 - 12	spect A2	21000n4359

LIST OF DOUBLE AND MULTIPLE STARS   (Cont'd)

NAME	DIST	PA	YR	MAGS	NOTES	RA & DEC
Σ2748	19.0	301	33	6½- 9	relfix, spect K2	21004n3919
	25.9	250	05	-12		
Σ2747	4.7	263	54	8 - 8	slow PA inc,	21004n3728
					spect G5	
β1138	0.2	180	59	7 - 7	(Ho 282) spect B8	21010n4539
β445	4.8	110	25	7½- 12	spect K0	21014n2854
	48.4	19	14	- 12		
Σ2757	1.8	268	37	8 - 9½	relfix, spect A0	21030n5212
β158	11.1	310	43	7½- 12	slight PA dec,	21040n4736
					planetary nebula	
					7026 in field	
61	28.4	144	68	5½- 6	(Σ2758) PA inc,	21044n3828
					famous binary,	
					spect K5, K7 (*)	
Σ2760	0.7	291	60	7½- 8	PA inc, dist dec,	21047n3356
	60.0	152	25	- 9½	spect A2, optical	
63	15.7	153	34	4 - 13	(Es 32) spect K4	21049n4727
A881	4.2	218	37	7½- 12	spect B9	21054n4428
Es817	11.4	352	09	7½- 12	spect A0	21055n4732
	24.0	351	09	- 12		
Σ2762	3.3	308	62	6 - 8	(V389) variable.	21065n3000
	57.8	226	31	- 9	cpm; spect A0,	
					AC = optical	
β159	1.3	315	55	6 - 9	relfix, spect B5	21087n4729
	135	189	21	- 7½		
Ho 283	21.5	211	60	7- 12½	relfix, spect B1	21090n3606
OΣ432	1.4	119	57	6½- 7	PA dec, spect G5	21124n4056
τ	0.9	229	61	4 - 6½	(65 Cygni) binary,	21128n3749
	29.5	228	14	- 13	50 yrs; PA dec,	
					spect F0 (*)	
Ho285	8.9	26	05	7 - 12	spect B8	21136n3702
Ho153	0.9	124	59	8 - 9	PA inc, spect A5	21156n3332
Σ2785	2.7	235	37	8 - 10	relfix, spect F0	21156n3932
υ	15.1	220	58	4½- 10	(OΣ433) (66 Cyg)	21158n3441
	21.5	181	58	- 10	cpm; spect B2e	
OΣ434	24.7	122	46	6½- 9½	relfix, spect A0	21170n3932
Es 98	26.7	311	28	6½- 9	relfix, spect A0	21178n5206
	30.1	87	28	- 9		
Es 98b	52.1	108	12	9 - 10		
Σ2789	6.7	115	62	7 - 7	dist inc, spect G5	21184n5246

LIST OF DOUBLE AND MULTIPLE STARS   (Cont'd)

NAME	DIST	PA	YR	MAGS	NOTES	RA & DEC
0Σ437	2.2	27	65	6½- 7	PA dec, spect G8,	21187n3214
	81.7	141	62	-11	G5	
A765	0.4	32	58	7- 8½	PA dec, spect B2	21220n4657
	26.5	25	28	- 14	C = 6½" pair	
A1892	0.7	347	59	7 - 9	relfix, spect A0	21221n5505
69	33.0	30	23	6 - 10½	(S790) spect B0	21237n3627
	54.0	98	22	- 9		
Σ2803	24.8	287	14	7½- 9	Spect A0	21282n5242
Ho161	2.9	6	43	7 - 11	PA inc, spect F5	21294n3950
Σ2802	3.8	9	54	8 - 8	spect A5	21297n3335
h1669	15.6	240	02	7½- 12	spect B9	21344n5016
β167	2.1	89	16	7 - 11	cpm; spect G8	21340n2949
A1445	1.5	281	27	6½-12½	spect A3	21378n3917
Es825	9.6	253	09	7½ 11½	spect K2	21381n4854
	55.1	286	09	- 8½		
75	2.7	324	58	5 - 10½	(AC 20) relfix.	21382n4303
	57.9	254	24	- 9½	AB cpm, spect gM0;	
					AC = optical	
Ho164	3.8	68	63	8 - 8	PA inc, spect K0	21389n3451
	25.3	238	23	- 12		
RU	11.1	224	00	7 - 11½	(Es 35) primary	21390n5406
	18.6	29	00	- 10	semi-reg variable,	
					spect M7	
77	0.1	337	66	6 - 6	binary, 24.4 yrs;	21404n4051
					PA dec, spect A0	
Σ2820	16.3	232	21	8 - 10½	Spect A0	21405n4212
β688	0.3	210	59	7½- 7½	Spect F0	21406n4049
	29.3	34	14	- 12		
79	1.5	159	48	5½- 11	(Kui 109) cpm. PA	21413n3804
	151	60	25	- 7	dec, spect A0.	
					Very red variable	
					RV Cygni 17' south	
μ	1.7	289	67	4½- 6	(Σ2822) binary,	21419n2831
	48.6	277	24	-11½	PA inc (*)	
Σ2832	13.1	213	46	8 - 8½	relfix, spect A0	21474n5016
	45.9	319	12	-9½		
0Σ456	1.5	35	60	8 - 8	PA inc, spect F2	21537n5217
	25.3	187	12	- 10		
	35.2	283	12	- 9		
Hu 774	0.2	256	53	7½- 7½	PA inc, spect A0	21578n4853

LIST OF VARIABLE STARS

NAME	MagVar	PER	NOTES	RA & DEC
χ	3.6--14.2	407	LPV. Spect M6ep (*)	19486n3247
31	3.8---4.2	3803	(V695) (Omicron-1 Cygni) Ecl.bin. (*)	20120n4635
32	4.0---4.2	1148	Ecl.bin; (*)	20139n4735
59	4.5---4.9	Irr	Spect B1. Also visual double (Σ2743)	20581n4720
P	3.0--5...	---	"Permanent nova" peculiar spectrum (*)	20159n3753
Q	3.0-- 15	---	Nova 1876	21398n4237
R	6.6--14.1	426	LPV. Spect S3e--S6e	19355n5005
S	9.2--15..	323	LPV. Spect S5e	20044n5750
T	4.8---5.3	Irr	Spect K3. Also visual double	20452n3411
U	6.7--11..	465	Spect Npe. Strong color	20181n4744
V	7.8--13.8	420	LPV. Spect Npe. Red!	20397n4758
W	5.0---7.6	131	Semi-reg; spect M4e- M6	21341n4509
X	6.0---7.1	16.39	Cepheid; spect F7--G8	20414n3524
Y	7.0---7.6	2.996	Ecl.bin; spect B0 (*)	20501n3428
Z	7.6--14..	264	LPV. Spect M5e	20000n4954
RR	9.9--10.8	Irr	Spect M3	20443n4441
RS	6.6---9.4	417	Semi-reg; spect Npe	20116n3835
RT	6.5--12.6	190	LPV. Spect M2e--M4e	19422n4839
RU	7.5--10..	234	Semi-reg; spect M7e	21390n5406
RV	7.1---9.3	300:	Semi-reg or Irr; spect N5; very red star B592	21412n3747
RW	8.5--10..	600:	Semi-reg; spect M3	20270n3949
RX	8.0--	?	Spect B3; possibly not variable.	20093n4740
RY	8.5--10..	Irr	Spect N	20085n3548
RZ	9.8--14.1	276	Semi-reg; spect M7	20502n4710
SS	8.2--12..	Irr	Nova-like variable, "cataclysmic star" (*)	21407n4321
ST	9.4--14.5	336	LPV. Spect M6e	20312n5447
SU	6.4---7.1	3.846	Cepheid; spect F0--G1	19428n2909
SW	9.3--11.8	4.573	Ecl.bin; spect A2+K0	20054n4609
SX	8.2--15.2	412	LPV. Spect M7e	20136n3055
SZ	9.5--10.3	15.107	Cepheid; spect F8--G8	20313n4626
TT	7.8---9.1	118	Semi-reg; spect N3e	19390n3230
TU	8.7--14.9	219	LPV. Spect M3e	19448n4857
TV	9.0--10..	Irr	Spect M0	20317n4624

LIST OF VARIABLE STARS (Cont'd)

NAME	MagVar	PER	NOTES	RA & DEC
TW	8.9--15.0	341	LPV. Spect M9e	21038n2912
TX	9.7--10.9	14.708	Cepheid; spect F5--G6	20583n4224
TY	9.0--15.0	350	LPV. Spect M6e	19318n2813
TZ	9.6--11.7	Irr	Spect M6	19148n5004
UV	8.7---9.2	136	Semi-reg; spect M6	19296n4332
UX	8.0--14.6	561	LPV. Spect M4e--M6e	20530n3013
VY	9.5--10.4	7.857	Cepheid; spect F6--G1	21024n3946
VZ	8.5---9.2	4.865	Cepheid; spect F5--G5	21497n4254
WW	9.9--13.7	3.318	Ecl.bin; spect B8+G	20023n4127
WX	8.8--13.2	411	LPV. Spect N3e	20167n3717
WY	8.0--14.9	304	LPV. Spect M6e	21467n4401
WZ	9.9--10.7	.5845	Ecl.bin; spect dF0	20512n3838
XY	9.5--14..	298	LPV. Spect Se	19469n4130
XZ	8.6---9.5	.4666	Cl.Var; spect A2--F0	19314n5617
AA	8.4--11.4	213	Semi-reg; spect S7	20026n3640
AB	7.4---8.5	520	Semi-reg; spect M4e	21344n3152
AC	8.6---9.4	142	Semi-reg; spect M7	20113n4917
AF	6.4---8.4	94	Semi-reg; spect M6e	19287n4603
AH	9.8--11.3	112	Semi-reg; spect M6	19589n4002
AI	8.2--10.8	197	Semi-reg; spect M6	20298n3221
AU	8.5--14.4	435	LPV. Spect M6e	20166n3414
AZ	9.0--11..	Irr	Spect M2	20563n4616
BF	9.3 -13.3	754	Peculiar (*)	19219n2934
BG	9.1--12.4	292	LPV. Spect M7e	19369n2824
BI	8.4--10.0	Irr	Spect M4	20195n3646
BR	9.4--10.5	1.333	Ecl.bin; spect B9	19394n4640
BU	9.9--14.3	158	LPV. Spect M0	20041n5013
CD	8.9--10.5	17.07	Cepheid; spect F8--K0	20025n3358
CE	8.5--10.1	---	RW Aurigae type ? spect K5	21171n4648
CH	6.8-- 8.0	97	Semi-reg; spect M6	19232n5008
CN	7.3--14.0	199	LPV. Spect M5e	20169n5938
CO	9.0--10..	Irr	Spect K5	20588n4434
CU	9.5--14.1	216	LPV. Spect M6e	19513n5512
DD	9.3--12..	148	LPV. Spect M0e	19297n3436
DL	9.6--10.2	4.830	Ecl.bin; spect B3	21379n4819
DR	8.3--14.5	314	LPV. Spect M5e	20418n3758
DT	6.1---6.5	2.499	Cepheid; spect F5--F7	21044n3059
DU	9.8--14.2	216	LPV.	21107n3632
FF	8.2--14.2	324	LPV. Spect M4e	20370n3743
GN	9.1--11.4	141	Semi-reg; spect M7	20026n4018

LIST OF VARIABLE STARS (Cont'd)

NAME	MagVar	PER	NOTES	RA & DEC
G0	8.3--8.9	.718	Ecl.bin; spect B9+A0	20354n3515
KR	9.2--10.0	.845	Ecl.bin; spect A2	20071n3024
MR	8.5--9.3	1.677	Ecl.bin; spect A0	21570n4745
MW	9.6--10.4	5.955	Cepheid, spect F8--G1	20104n3243
MY	8.7---9.4	2.003	Ecl.bin; spect A5	20181n3347
V366	9.8--10.4	1.096	Ecl.bin; spect A5	20431n5355
V367	7.2---7.9	18.60	Ecl.bin; lyrid, spect A7pe; also visual double	20461n3906
V369	9.8--14.1	105	LPV. Spect M4e	19408n5433
V380	5.5--5.6	12.43	Ecl.bin; spect B1	19489n4028
V382	9.1--10.1	1.885	Ecl.bin; spect 07+08	20169n3611
V388	9.7--10.3	.859	Ecl.bin; spect A3, lyrid	20272n3113
V389	5.5---5.7	1.129 1.193	Class uncertain; spect A0; Two periods! Also visual double Σ2762	21065n3000
V395	7.8---8.4	Irr	Spect F8	20070n4355
V402	9.4--10..	4.365	Cepheid; spect G0	20073n3700
V405	8.9--10..	Irr	Spect M6	20198n4017
V425	9.5--10.1	Irr	Spect F5	20062n3559
V427	9.2--11..	143	Semi-reg; spect M3	20062n3728
V441	8.7--11..	375		20252n3623
V444	8.3---8.6	4.212	Wolf-Rayet Ecl.bin (*)	20177n3834
V448	8.1---8.6	6.520	Ecl.bin, lyrid, spect B1	20043n3514
V449	6.8-- 8.0	Irr	Spect M3	19515n3349
V450	7.0--17..	---	Nova 1942	20567n3545
V453	8.3---8.6	3.890	Ecl.bin; spect B1 + B1	20047n3536
V460	6.1---7.0	Irr	Spect N1	21399n3517
V465	7.3-- 17	---	Nova 1948	19508n3626
V470	8.7---8.8	1.873	Ecl.bin; spect B2+B2	20176n4044
V476	2.0---17	---	Nova 1920 (*)	19571n5329
V477	8.2---8.9	2.347	Ecl.bin; spect A3+F5	20035n3149
V478	9.1---9.5	2.881	Ecl.bin; spect B0+B0	20178n3811
V480	9.4--10..	50:	Semi-reg; spect M0	19506n3429
V485	7.9---8.8	Irr	Spect M5	19593n3347
V532	9.4--10.0	3.284	Cepheid; Spect F5	21187n4515
V539	9.1--10.4	160	Semi-reg; spect M6	21376n4457
V548	8.9---9.7	1.805	Ecl.bin; spect A0	19558n5440
V568	6.6---6.9	Irr	Spect B2e	20404n3516
V697	7.9---9.8	Irr	Spect M2	19480n5239
V778	9.4--11..		LPV. Spect N	20361n5955

LIST OF VARIABLE STARS (Cont'd)

NAME	MagVar	PER	NOTES	RA & DEC
V788	9.3--10.2	23.925	Ecl.Bin; Spect F8	20256n2956
V819	6.1± 0.01	.3775	Beta Canis type? Spect B0	19516n4740
V828	9.7--10.1	2.104	Ecl.bin; lyrid, spect B8	20366n5419
V829	9.6--10.8		LPV? Spect M4	20436n3218
V832	4.5---4.9	Irr	(59 Cygni) Spect B1	20581n4720
V836	8.6---9.3	.6534	Ecl.bin; lyrid, spect A0	21194n3531
V840	9.0-10...	275:	Semi-reg	19201n2834
V885	9.9--10.3	1.695	Ecl.bin; lyrid	19309n2955
V891	9.3---9.9	1.906	Ecl.bin; spect A0	19316n2910
V909	9.3---9.7	1.403	Ecl.bin; spect A0	19339n2810
V918	9.5--10.2		LPV? Spect M7	19357n3006
V927	9.8--10.5	49:	Semi-reg; spect M6	19361n4840
V973	6.3-- 7.1		LPV? Spect M1	19431n4036
V977	9.1-- 9.8		LPV. Spect M5	19438n3106
V1016	11...17.5	Irr	Z Andromeda type	19553n3942
V1027	9.4--10.4		LPV or Semi-reg; spect K8	20005n2957
V1034	9.6--10.6	.9769	Ecl.bin; lyrid, spect A0	20036n3050
V1042	8.0 ±0.1	Irr	Wolf-Rayet star	20100n3603
V1059	8.0--9..	372:	LPV or semi-reg, spect M	21027n4215
V1061	8.6--9.2	2.347	Ecl.bin; spect F8	21056n5151
V1070	6.7---7.7		Semi-reg? Spect M7	21209n4043
V1073	8.2-- 8.7	.7859	Ecl.bin; W UrsaMaj type, Spect F0	21229n3329
V1125	7.7--9...		LPV?	19299n3145
V1143	5.9-- 6.1	7.641	Ecl.bin; Spect F5+F5	19376n5451
V1154	8.7-- 9.4	4.925	Cepheid; Spect G2	19464n4300
V1183	9.0--10..		LPV? Spect M6	20115n3736
V1276	6.44±0.01	.088	δ Scuti type, spect F1	19408n2913
V1330	7.5--18..	---	Nova 1970	20508n3548
V1334	5.8--5.95	3.334	Cepheid? Spect F2	21174n3801
V1339	5.9--7.1	35:	Semi-reg, spect M3	21402n4532
V1357	8.8---8.9	5.601	Ecl.bin; spect O9	19565n3504
V1362	8.0---8.2	7:	Ecl.bin? spect B5	20018n3617
V1372	7.36±0.03	18:	Alpha Canum type, spect A4p	20123n5330
V1425	7.9---8.4	1.252	Ecl.bin; spect B9+A0	21095n5509
V1500	1.8--21+	---	Nova Cygni 1975 (*)	21099n4757

LIST OF STAR CLUSTERS, NEBULAE, AND GALAXIES

NGC	OTH	TYPE	SUMMARY DESCRIPTION	RA & DEC
6811		⋰	L,pRi, mag 9, diam 15'; about 50 stars mags 11... 14; class D	19367n4627
6819		⋮	vL,vRi, mag 10, diam 6'; 150 stars mags 11...15, class D	19396n4006
6826	73^4	◎	vB,pL, mag 8.8; diam 25"; 11^m 0-type central star	19434n5024
6833		◎	Mag 14, diam 2"; F,vS; virtually stellar	19485n4850
6857	144^3	◎	S,F, diam 40"; central star 14^m	20000n3323
6866	59^7	⋰	L,vRi,cC, mag 8, diam 8'; about 50 stars mags 10... class D	20021n4351
6884		◎	Mag 12½; diam 7", nearly stellar, eF central star	20088n4619
6881		◎	vF,vS, mag 14, diam 5"; 15^m central star	20090n3716
----	I.1311	⋰	S, diam 5', about 30 F stars; class E	20091n4102
6888	72^4	▢	F,vL,vmE; diam 18' x 12'; crescent-shaped neby with 7^m Wolf-Rayet star	20107n3816
6894	13^4	◎	F,S,v1E, mag 14, diam 44"; annular, with 17^m central star. Ring-shaped neby.	20144n3025
6910	56^8	⋰	pB,pS, mag 6½, diam 8'; about 40 stars mags 10.... class D	20213n4037
6913	M29	⋰	S,1C, Mag 7, diam 7'; about 20 stars mags 8... class D (*)	20222n3821
6914		▢	F,L,Irr; diam 6', 9^m star inv.	20234n4210
6960	15^5	▢	!! eL,F, great Cygnus Loop	20436n3032
6992	14^5		or "Veil Nebula". Fine filamentary structure (*)	20543n3130
----	I.5067	▢	vvF,vL,Irr; diam 80', star 56 Cygni inv. "Pelican Nebula". Near NGC 7000.	20469n4411

LIST OF STAR CLUSTERS, NEBULAE, AND GALAXIES

NGC	OTH	TYPE	SUMMARY DESCRIPTION	RA & DEC
----	I.5076	□	F,pL,R, diam 7', surrounds 6^m B8e star.	20542n4713
6992			Refer to NGC 6960	
6997	58^8	∴	L, scattered cluster in west part of NGC 7000	20547n4427
7000	37^5	□	vF,eeL,Irr; diam 100' "North American Neb" (*)	20570n4408
7008	192^1	◎	cB,L,E, mag 12, diam 85" X 70"; 13^m central star	20591n5421
7026		◎	pB,mE, mag 12, diam 25" x 6"; 15^m central star	21046n4739
7027		◎	eB,S, mag 9, diam 18" x 11"; bluish-green; "has richest spectrum of all the planetaries"	21051n4202
----	I.1369	□	S,F, diam 2', with small cluster inv.	21105n4733
7048		◎	Mag 11; diam 60" x 50" pF,pL,iR, 18^m central star	21126n4604
7062	51^7	∴	pS,pRi,pC, 30 stars mags 12... class E; diam 5'	21215n4610
7082	52^7	∴	vL,pRi,cC, stars mags 10.. 13	21270n4650
7086	32^6	∴	cL,vRi,pC, mag 9, diam 8'; about 50 stars mags 11... 16, class E	21298n5122
7092	M39	∴	vL,P, mag 5, diam 30', 25 stars mags 7... class E (*)	21304n4813
----	I.5117	◎	vS,F, mag 13, diam 2"; appearance nearly stellar	21306n4423
7127		∴	S,P,1C, mag 10, diam 2'; about 10 F stars, class D	21422n5424
7128	40^7	∴	S,pRi, mag 9, diam 2'; about 30 stars mags 10... class D	21424n5329
----	I.5146	□	L,F,1E, complex neby with bright & dark masses; diam 12' x 10'. 10^m B-star inv. Lies in east end of dark streamer	21513n4702

"These are the last days of the waning year;
 High in the west now stands Deneb,
 Great Star of the Cross....."

DESCRIPTIVE NOTES

ALPHA Name- DENEB. An older name, "Arided", is now nearly obsolete. Deneb is the 19th brightest star in the sky; magnitude 1.26, spectrum A2 Ia. The star is located at the top of the Northern Cross at 20397n4506; the star also marks the Swan's tail in the mythological outline of Cygnus. Midnight culmination (opposition date) is about August 1. Deneb is the faintest of the three 1st magnitude stars which outline the well known "Summer Triangle", consisting of Vega, Deneb, and Altair.

The name of the star is from *Al Dhanab al Dajajah*, which in Medieval Arabic signifies "The Hen's Tail"; on antique star globes it also appears as *Denebadigege* and *Deneb Adige*. Other Denebs are found in the heavens in such cases where a mythological beast requires a proper name for his tail; *Denebola* for Beta Leonis and *Deneb Kaitos* for Beta Ceti are two well known instances. Another Arabic name for Alpha Cygni, according to R.H.Allen, was *Aridif* from *Al Ridf*, the Hindmost, another allusion to its location in the Tail of the Swan. The identification of Cygnus with some sort of bird, though not necessarily a swan, was quite common in the ancient world; in various writings it is mentioned as a flying eagle, a hen, a partridge, or a pigeon; a star-bird of some sort called *Urukhga* appears in Babylonian writings, and may be the original of the *Rukh* or *Roc* of Arabian legend. On some of Bayer's charts of the early 17th century it is called "Gallina". But probably more inspiring to the star pilgrims of the western world is the identification of Deneb as the *Star of the Cross*, particularly at the Christmas season when the great figure of the Northern Cross assumes an increasingly vertical position in the western sky. James Russell Lowell in his *New Year's Eve, 1844*, speaks of

"*The Lyre whose strings give music audible
To holy ears, and countless splendors more,
Crowned by the blazing Cross high-hung o'er all;*"

Deneb is one of the greatest supergiant stars known, and is probably equalled only by Rigel among all the 1st magnitude stars. The actual luminosity is computed to be about 60,000 times that of the Sun, and the absolute magnitude about -7.1. Deneb probably has a mass of about 25 solar masses. At a distance estimated to be nearly 1600 light years the star is one of the most remote of all the bright stars. Our sun at the same distance would appear as a telescopic star of magnitude 13.3. The annual proper motion is about 0.003"; the radial velocity about 3 miles per second in approach. Deneb is an early "Sirian" or hydrogen-type star, with a surface temperature of 9700°K; the true diameter may be about 60 times that of the sun.

Deneb shows a variable radial velocity with a semi-regular period of about 11.7 days; there is also evidence for a slight variability of about 0.05 magnitude. Other A and F-type supergiants show similar variations, which are attributed to pulsation or large-scale turbulence in the star's atmosphere.

As one of the most luminous of supergiants, Deneb has recently been used as a test case in an attempt to detect spectral changes resulting from a star's evolution. From studies by C.Hayashi and R.C.Cameron (1962) it now appears that a very massive star may require only about 20 thousand years to evolve from type A to the red-giant type M, with a consequent change in temperature of about 0.3° per year. This steady decrease in temperature should produce definite changes in the spectrum which might be detected in an interval of well under a century. M.H. and W.Liller (1964) have studied spectra of Deneb obtained from 1887 to 1964 and have detected one definite change - a strengthening of the so-called "H-line" of ionized calcium, relative to the nearby "Hε" line of hydrogen. The greater part of this change, however, evidently occurred in the early years of this century, suggesting that the explanation lies in some sort of temporary atmospheric activity or long-period cycle. At present there appears to be no definite change attributable to stellar evolution in the spectrum of Deneb or of any other star.

Deneb is often considered to be the chief source of illumination of the huge "North American Nebula" NGC 7000 which lies 3° to the east. (Refer to page 811)

About 1.8° to the SE lies the star 56 Cygni (mag 5.1, spectrum A4m), a suspected nova-like variable, based on a single observation by James L.Kuhns of Savannah, Georgia. On the morning of July 12, 1973, he reported an object of magnitude 1.5 whose position appeared to coincide with 56 Cygni; by the next evening the star was observed to be in its normal 5th magnitude state.

STAR CLOUDS IN CYGNUS. Deneb and the North American Nebula are at upper left. Photographed with a 50mm Takumar lens in red light by David Healy of Manhasset, New York.

BETA Name- ALBIREO. Mag 3.09; spectrum K3 II; located at the foot of the Northern Cross at 19287n2751. The name appears to have originated in a mis-translation of the term *ab ireo* in the 1515 edition of Ptolemy's star catalogue called the *Almagest*. The original Arabian name, according to R.H.Allen, was *Al Minhar al Dajajah,* or the "Hen's Beak". On Riccioli's maps it appears as *Menkar Eldigiagieh,* or occasionally as *Hierzim.*

Albireo is one of the most beautiful double stars in the sky, considered by many observers to be the finest in the heavens for the small telescope. The brighter star is a golden yellow or "topaz", magnitude 3.09, spectrum K3; the "sapphire" companion is magnitude 5.11, spectrum B8 V. The separation is 34.3", an easy object for the low power telescope. Even a pair of good binoculars, if steadily held will split the pair. Albireo is noted for its superb color contrast, best seen with the eyepiece slightly displaced from the sharpest focus. Miss Agnes Clerke (1905) called the tints "golden and azure", giving perhaps "the most lovely effect of color in the heavens". For the average amateur telescope there is probably no pair so attractive, though the color effect seems to diminish in either very small or very large telescopes, or with too high a magnification. No more than 30X is required on a good 6-inch to show this superb pair as two contrasting jewels suspended against a background of glittering star-dust. The surrounding region is wonderfully rich, and for wide-angle telescopes the star clouds to the NE are probably unequalled in splendor in the entire heavens.

Albireo is believed to be a physical pair, although no evidence of orbital motion has been detected since the first observations of F.G.W.Struve in 1832. The computed distance is 410 light years, the resulting actual luminosities are 760 and 120 suns. The annual proper motion is about 0.01"; the radial velocities are 14½ and 11 miles per second for the two stars, both in approach.

Albireo A has a composite spectrum and evidently consists of two stars too close for telescopic resolution; the spectrum of the unseen companion is either late B or early A-type. The 5th magnitude visual component is also very interesting to the spectroscopist, showing a hydrogen

THE CYGNUS STAR CLOUD. This is one of the most spectacular regions of the Milky Way, just north and east of Beta Cygni. Gamma Cygni is at upper left.

Lowell Observatory photograph

NEBULOSITY NEAR GAMMA CYGNI. A curious region of bright and dark nebulosity, about a degree east of Gamma Cygni. Palomar Observatory 48-inch Schmidt photograph.

emission spectrum. The projected separation of the visual
pair is about 4400 AU, or just over 400 billion miles.
This is of course a minimum value; the true separation
may obviously be much greater if one star is somewhat more
distant than the other. It is worth contemplating, in any
case, the fact that at least 55 solar systems could be
lined up, edge-to-edge, across the space that separates the
components of this famous double!

GAMMA   Name- SADR. Mag 2.23; spectrum F8 Ib; position
              20204n4006. The computed distance of this
star is about 750 light years, giving a true luminosity of
some 5800 suns. (Absolute magnitude about -4.6). Gamma
Cygni shows an annual proper motion of a mere 0.001", and
a radial velocity of 4½ miles per second in approach. The
star marks the intersection of the two arms of the Northern Cross. There are fine Milky Way fields throughout the
region, especially in the direction of Beta Cygni. The
area around Gamma itself contains vast stretches of faint
diffuse nebulosity whose intricate details may be studied
only on the photographic plate. The curious field shown
on page 756 lies about a degree east of Gamma.
        Between Beta and Gamma Cygni lies the great Cygnus
Star Cloud, a superb region for study with a small telescope or even a pair of binoculars. The stars are richer
here than in any similar area of the sky, and present a
glorious sight in the rich-field telescope. Webb says "Its
low-power fields are overpowering in magnificence." Sweep
the area with a low power wide-angle eyepiece, and notice
the huge numbers of stars, groups, and clusters, and the
occasional dark gaps caused by clouds of non-luminous
material.
        It is in Cygnus that the belt of dark dust clouds
known as the "Great Rift" begins; it runs from Cygnus to
Centaurus and apparently divides the Milky Way into two
parallel streams. The obscuring clouds of the Great Rift
lie at an average distance of 4 or 5 thousand light years
and in the Sagittarius region they prevent any direct observation of the nuclear regions of our Galaxy. The star
clouds of Cygnus, however, evidently mark out a portion
of one of the spiral arms of our Galaxy, lying some 7000
light years away.

At 142" from Gamma lies a 10th magnitude companion, probably not physically related to the primary. This small star was found to be a close double by S.W.Burnham in 1878; the separation is about 2" and the individual magnitudes are 10 and 11.

DELTA   Mag 2.87; spectrum B9 or A0 III. Position 19434n4500. The computed distance is about 270 light years, and the actual luminosity about 400 times that of the Sun. (Absolute magnitude -1.7). Delta Cygni shows an annual proper motion of 0.06"; the radial velocity is 12½ miles per second in approach.

The star is a rather difficult binary, discovered by F.G.W.Struve in 1830. It may be regarded as a fairly severe test for a 3-inch or 4-inch telescope. The small companion lies virtually on the first diffraction ring of the primary and is thus often obscured except at times of very steady seeing. Owing possibly to this circumstance the companion has often been rated as only 8th magnitude, whereas current measurements show that the true value is near 6½.   There appears to be no published spectral class for the small star in the literature, but observers have

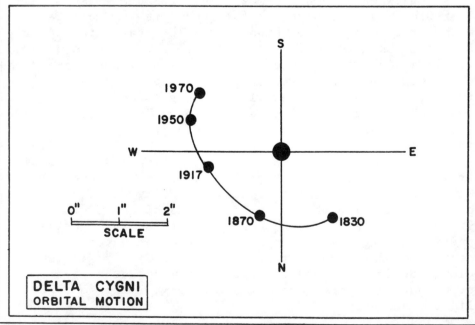

DELTA CYGNI
ORBITAL MOTION

generally agreed that the color appears bluish. J.H.Mallas (1967) finds the components "blue-green and blue".

Orbital elements for the system are still uncertain, and periods of between 300 years and 540 years have been computed. The separation has been increasing somewhat from a minimum of 1.5" in 1875, and is probably now approaching the maximum. An orbit by Jackson (1920) gave the period as 321 years, but more recent computations by E.Rabe (1961) give about 537 years, with periastron in 1890. According to Rabe's results, the semi-major axis of the system is 2.56" and the eccentricity is 0.30; the projected mean separation of the pair is some 220 AU.

EPSILON   Name- GIENAH. Mag 2.46; spectrum K0 III. Position 20442n3347. The star is at a distance of about 75 light years; the actual luminosity is about 40 times that of the Sun. Epsilon Cygni has an annual proper motion of 0.48" in PA 48°; the radial velocity is about 6 miles per second in approach.

Epsilon is probably a spectroscopic binary, but the period is undetermined. A 12th magnitude field star is sometimes listed as a companion, but the two stars form only an optical pair. The separation was 38" in 1878 but is gradually increasing from the proper motion of the bright star, and was 55" in 1960. However, Epsilon Cygni does have a much fainter physical companion which shares the proper motion of the primary. Designated LTT 16072, it is a red dwarf of the 15th magnitude, 78" distant, nearly due west.

Epsilon marks the eastern arm of the Northern Cross. The irregular variable T Cygni lies in the field, about ½° to the northeast.

ZETA   Mag 3.20; spectrum G8 II. Zeta Cygni is located at 21108n3001. The computed distance is about 390 light years; the actual luminosity is about 600 times that of the Sun. (Absolute magnitude -2.2) The annual proper motion is 0.06"; the radial velocity is 10½ miles per second in recession.

One degree to the west lies the visual double Σ2762; the primary of this pair is the unusual variable star V389.

MU   Mag 4.45; spectrum F6 V and dF3. Position is
21419n2831. Mu Cygni is a close visual binary
with a computed period of 4½ or 5 centuries. It was discovered by Sir William Herschel about 1780, when the separation was 6". In the next century the two stars gradually
grew closer together and reached a minimum separation of
0.55" in 1937. The distance is now again increasing, and
in 1961 was 1.8" in PA 285°. The individual magnitudes are
4.7 and 6.1, and both components appear to be normal main
sequence stars of very similar type. The total luminosity
of the system is about 5 times that of the Sun.
     The best recent orbital computation is probably
that made by T.Jastrzebski (1960) and based upon 180 years
of recorded observations of the pair. He obtains a period
of 444 years, a semi-major axis of 4.24", and an eccentricity of about 0.62. Periastron was in 1965. The mean true
separation of the components is about 85 AU.
     Mu Cygni is approximately 65 light years distant,
and shows an annual proper motion of 0.34" in PA 132°; the
radial velocity is about 11 miles per second in recession.
     In addition to the orbiting pair there is an 11th
magnitude optical companion at 48½" whose distance from
the primary is increasing, and a 6th magnitude field-glass
companion at 204" in PA 51°, also not a true physical member of the system. This latter star is also known as
Es 521, and has its own 14th magnitude companion at 14" in
PA 270°.

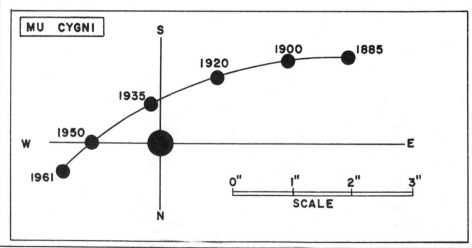

TAU   Mag 3.74; spectrum F0 IV. Position 21128n3749.
Tau Cygni is a close visual binary star, discovered by A.G.Clark in 1874. An orbit by Van Biesbroeck
gives the period as 49.8 years with retrograde motion, and
periastron occurring in 1939. Although visual observers
have estimated a difference of about 4 magnitudes between
the components, modern measurements show that the true
magnitudes are about 3.8 and 6.4. According to the orbit
computations (plotted below) the semimajor axis is 0.85"
and the eccentricity is about 0.24. The true separation of
the pair is about 20 AU, comparable to Uranus and the Sun.
Tau Cygni is approximately 75 light years distant,
giving an actual luminosity of about 15 suns. The primary
shows radial velocity changes in the short period of 0.1425
days. At first attributed to rapid binary motion, the variations are now thought to originate in some type of pulsation in the atmosphere of a single star. The radial velocity is about 13 miles per second in approach, and the
annual proper motion is 0.46" in PA 15°.

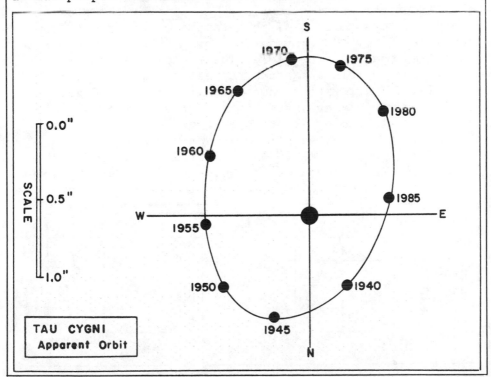

TAU CYGNI
Apparent Orbit

The 13th magnitude star at 29" (1914) is not a physical member of the system, and the separation is increasing from the proper motion of Tau itself. However, at the much larger distance of 88" in PA 185° lies a 12th magnitude star which is a true member of the system; it is an M-type red dwarf of about 1/170 the solar luminosity. The orbital period around the double primary is estimated to be about 45,000 years; the projected separation at present is just over 2000 AU.

CHI   Variable. Position 19486n3247. This is the second of the long-period variable stars to be discovered, found by Gottfried Kirch in 1686. The star is located about 60% of the way along a line drawn from Gamma Cygni to Beta Cygni, or some 2½° SSW from Eta Cygni. The star is the brightest and most easily observed of the long-period variables, with the exception of the famous Mira (Omicron Ceti) and is often visible to the naked eye at maximum, reaching a brightness of 4th or 5th magnitude; the maximum recorded brightness is about 3.5. At minimum the star nearly always drops to below 12th magnitude, and cannot be identified among the many faint stars in the region without a detailed chart of the field. The whole area is sprinkled with myriads of distant stars, appearing like a background of luminous dust. Except when near minimum, Chi Cygni stands out vividly against this background like a glowing red beacon. The color is stronger than that of Mira, but does not, of course, equal the N-type stars.

Miss Agnes Clerke (1905) referred to Chi Cygni as "a variable in the neck of the Swan, which Bayer, ignorant of its changing character, set down in his maps as of the

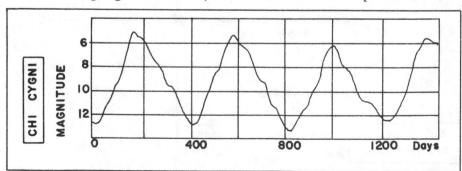

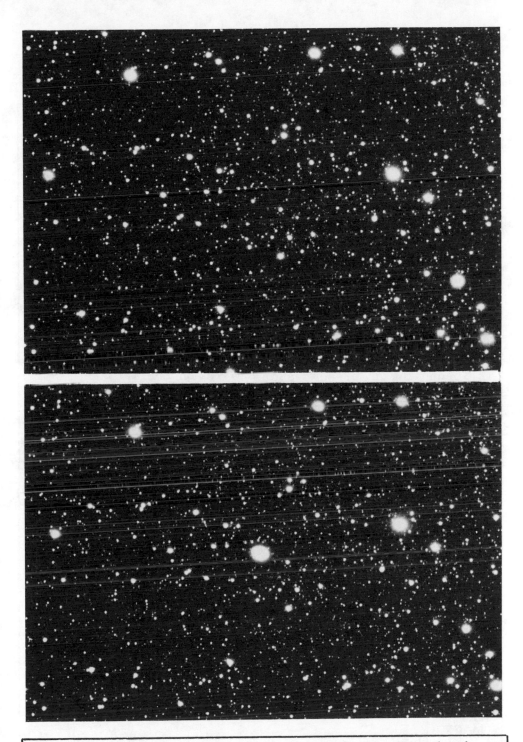

CHI CYGNI. The star is shown here on May 24, 1931 (top) and August 30, 1940 (below). Lowell Observatory photographs made with the 13-inch telescope.

fifth magnitude. It still retains the name he gave it of
"χ Cygni". Missed by Gottfried Kirch in July 1686, it re-
appeared October 19, and subsequently disclosed to his
vigilant watch fluctuations even wider than those of the
'wonderful' star in Cetus. It descends below the thirteenth
and rises nearly to the fourth magnitude, sometimes indeed
stopping short when barely visible to the naked eye, but
more commonly remaining lucid for a couple of months. Nor
is its course much better regulated as regards time.
'Errors' up to forty days often attach to its phases, and
the attempt to correct them by the introduction of cyclical
terms has proved only partially successful. The period,
estimated at 402 days by Kirch, now averages 406. Olbers
noticed that it had been steadily lengthening down to 1818
and is lengthening still......As usual in such cases, the
ascent to maximum is much more rapid than the descent from
it, occupying at present about 171 days."

     Chi Cygni has an abnormally large range in bright-
ness, averaging about 3000 times, and occasionally exceed-

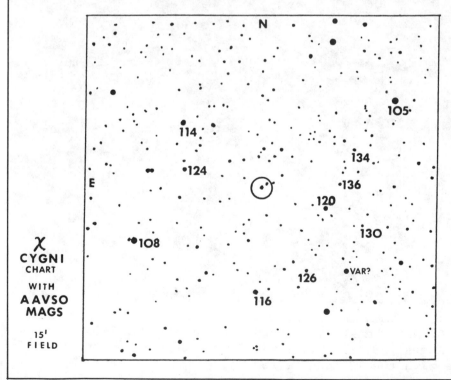

χ
CYGNI
CHART

WITH

AAVSO
MAGS

15'
FIELD

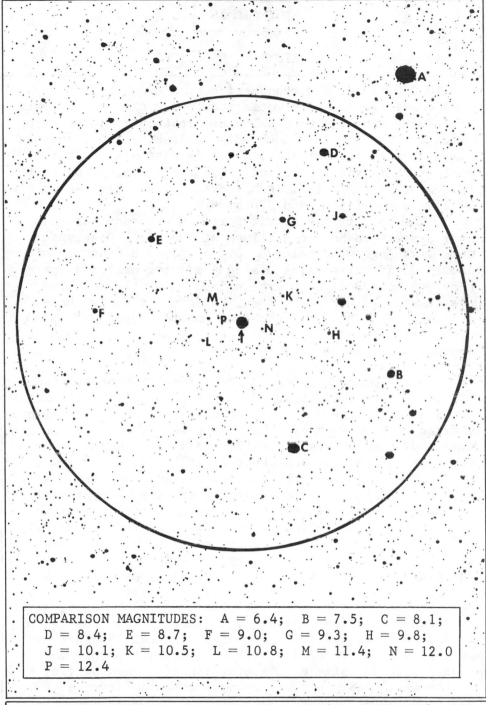

COMPARISON MAGNITUDES: A = 6.4; B = 7.5; C = 8.1;
D = 8.4; E = 8.7; F = 9.0; G = 9.3; H = 9.8;
J = 10.1; K = 10.5; L = 10.8; M = 11.4; N = 12.0
P = 12.4

CHI CYGNI. Identification chart, made from a 13-inch telescope plate obtained at Lowell Observatory. Circle diameter = 1° with north at the top; limiting magnitude about
14.

DESCRIPTIVE NOTES (Cont'd)

ing 10,000 times. The mean period at present is 407 days,
with considerable differences from one cycle to the next.
A true red giant, the actual size probably compares with
that of Mira, and is undoubtedly at least several hundred
times the diameter of our Sun. It is also one of the
coolest stars yet identified, with a temperature of less
than 1900°K at minimum. The spectral class is given as M6e
by some authorities, and S7e by others.

The only trigonometric parallax recorded for this
star appears to be one obtained at Van Vleck Observatory
some years ago; the value of 0.014" gives a distance of
about 235 light years. A somewhat greater distance, perhaps 300 to 400 light years is implied by the computed
absolute magnitude at maximum, believed to be in the −1
to −2 range. Chi Cygni shows an annual proper motion of
0.05"; the radial velocity is about 1 mile per second in
approach. (Refer also to Omicron Ceti)

31 (Omicron-1 Cygni) Position 20120n4635. Mag
3.76; spectrum gK1 or K2 II. This is the primary star of a fine wide color-contrast group which forms a
very attractive sight for the low power telescope. The 5th
magnitude star 30 Cygni lies 338" distant, and a closer
companion of the 7th magnitude is 107" away. Both stars are
noticeably bluish (spectra A3 and B5) and contrast strongly
with the bright golden tint of the primary. The group is
probably only an optical one. 31 Cygni itself shows no
measurable proper motion or parallax; the radial velocity
is about 4½ miles per second in approach. For 30 Cygni the
catalog values are: Annual proper motion = 0.01"; radial
velocity 12 miles per second in approach; parallax = .002".
The primary star is an eclipsing variable (V695)
with the long period of 10.42 years or 3802.84 days, and a
magnitude range (photographic) of 4.9 to 5.3. The K-star
is an orange giant with a diameter of about 150 suns; it
is evidently surrounded by a huge gaseous "corona" more
than double the size of the star itself. The spectral type
of the small star is given by various authorities as B3,
B5, or B8; the Arizona-Tonantzintla Catalogue (1965) has
B3 V. The B-star may be about 5 times the size of the Sun.
From the spectra, the total light of both stars is computed
to be about 500 times the light of the Sun.

DESCRIPTIVE NOTES (Cont'd)

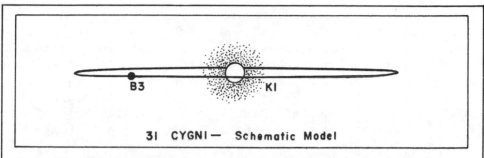

31 CYGNI — Schematic Model

The two stars are 1.2 billion miles apart, and the orbit is oriented nearly edge-on, so that the smaller blue star is totally eclipsed by the giant once in each revolution. This phenomenon begins with an "atmospheric eclipse" in which the radiation of the small star must come through gradually deeper layers of the giant's atmosphere. Total eclipse lasts for 63 days; the atmospheric eclipses each last about 2½ months. Dates of beginning of totality are: January 10, 1962, June 9, 1972, etc.

32   (Omicron-2 Cygni) Position 20139n4735. Magnitude 3.97; spectrum K3 Ib + A3. This is a giant binary system and an eclipsing variable of small amplitude. The primary is a K-giant probably about 200 times the size of the Sun; it is a low temperature yellow-orange star surrounded by a deep and extensive atmosphere. Orbiting the star at a distance of some 550 million miles is the smaller bluish companion, apparently a normal A-type star about 5 times the size of the Sun. The period of the pair is 3.15 years or 1148 days. From the derived luminosities, the distance of the system appears to be in the range of 500 to 600 light years. 32 Cygni shows an annual proper motion of only 0.007", and a radial velocity of about 8½ miles per second in approach. The total absolute magnitude may be about -3.5.

Total eclipse occurs when the smaller star passes behind the giant, and is preceded and followed by the usual "atmospheric eclipses" lasting a little more than a month each. Totality lasts 11 days. The first such eclipse of 32 Cygni was observed at the University of Michigan Observatory on November 1, 1949, when it was discovered that the light of the A-star had vanished from the spectrum. It began to reappear about 10 days later, and the now familiar spectral changes made it possible to picture the "rising" of the eclipsed star above the giant's atmosphere, a process which took more than a month to complete. This system is very similar to the star 31 Cygni, which lies only 1° away, to the south, and also to the well known Zeta Aurigae. Comparison of the light curves and spectra would suggest that 32 Cygni has the more extensive atmosphere, and that the eclipses are nearly grazing while those of Zeta Aurigae are nearly central. The photographic range of 32 Cygni is 5.3 to 5.6.

Just 45' to the east and slightly north lies the very red N-type variable U Cygni, its ruddy glow forming a striking contrast with a 7.8 mag bluish star 65" distant.

6I   Position 21044n3828. A very famous and note-worthy double star, particularly suited for small instruments, and historically one of the most interesting objects in the heavens. The components are in very slow orbital revolution, requiring some 7 centuries to complete. Computed periods of 720 and 691 years have been quoted in many current texts; the most reliable modern orbit, however, is probably that derived by E.de Caro and G. Veca (1948) which gives a period of 653 years. The components were closest (11") in 1650, and will reach their maximum separation of 34" around the year 2100 AD. Other orbit elements are: Semi-major axis = 24.3"; eccentricity = 0.40; and periastron in the year 1676. Facts of interest about the two stars are given in the table:

	Mag.	Spect.	Diam.	Mass.	Lum.	Dens.
A	5.3	K5 V	0.48	0.6	0.065	5.4
B	5.9	K7 V	0.43	0.5	0.038	6.2

Both components show a fine orange tint in the small telescope, and both are dwarf stars, smaller and fainter than the Sun, with absolute magnitudes of +7.8 and +8.4. The average separation is about 84 AU.

6I Cygni became famous as early as 1792, when Piazzi detected the abnormally large proper motion of 5.22" per year (PA = 52°) and christened it "The Flying Star". The first measurements of the star as a double, however, seem to have been made by F.G.W.Struve in 1830. In 1838 the star was chosen by F.W.Bessel for the first successful attempt to measure the actual distance of a star by trigonometrical

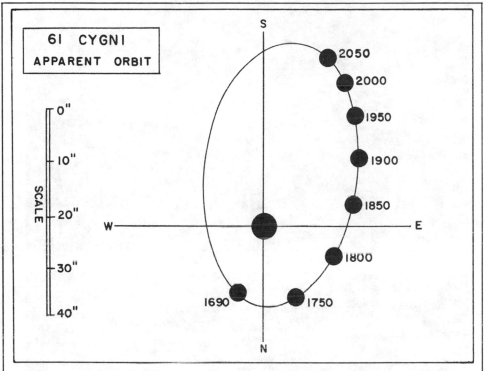

61 CYGNI
APPARENT ORBIT

methods. Bessel was able to detect and measure the tiny
parallax of about 0.29", and calculated the distance to be
10.3 light years. The presently accepted value is 11.1
light years. Among all the naked-eye stars 61 Cygni is the
fourth nearest to the Earth; probably only Sirius, Alpha
Centauri, and Epsilon Eridani are closer. The large proper motion is thus partly a result of the relatively small
distance, combined with a rather high space velocity. The
radial velocity of the system is 38 miles per second in
approach.

In addition to the two visible components of this
famous binary, there is an unseen third body, detected by
small systematic variations in the orbits of the visible
pair. This object, now called 61 Cygni C, is of greatest
interest since it is only some 8 times as massive as our
greatest planet Jupiter, and there is some question as to
whether it should be regarded as a star or a planet. It
revolves with one of the visible stars, apparently the A-
component, in a period of 4.8 years. Its actual size is
not known, but theoretical calculations suggest that it

61 CYGNI. The noted binary is shown here in 1916 and 1948; illustrating the proper motion over an interval of 32 years. These photographs were made with the 42-inch reflector at Lowell Observatory.

DESCRIPTIVE NOTES (Cont'd)

may be less than a tenth the diameter of our Sun. In any case, an object of such small mass would be expected to be non-luminous.

The existence of the mysterious dark companion was first detected by K.A.Strand at Dearborn Observatory in 1942. In an article in *"Sky & Telescope"* for December 1956 further information on this interesting object is given as follows:

"Dr.Strand has now completed fourteen additional years of observation and analysis that confirm the existence of the dark companion. It moves with a period of 4.8 years... its mass is 0.008 that of the Sun, or only 8 times the mass of Jupiter. Dr.Strand pointed out that the unseen 61 Cygni C could be regarded either as an exceptionally small star or as a planetary companion."

It is interesting to note that similar bodies seem to exist in several other star systems. The fourth nearest of the stars, Lalande 21185 in Ursa Major, is a famous case; the unseen companion has a period of about 8 years and a mass of about 1/100 that of the Sun. Perhaps the most re-markable of such objects is the unseen companion to the second nearest star, Barnard's Star in Ophiuchus. According-ing to P.van de Kamp (1963) the invisible body has a mass of about 1½ times the mass of Jupiter, and a period of 24 years. There seems to be no doubt that such a body should be called a planet.

These unseen companions are all less massive than any known star. The smallest stellar masses actually observed are those of the binary L726-8 (UV Ceti system) where each component has a mass of 40 times that of Jupiter, or a total of 0.08 the solar mass for the system. For Ross 614b and Krueger 60b the figures are 0.08 and 0.14 the solar mass, respectively. Up to 1966 these were the least mass-ive stars known. but it is likely that such objects as Proxima Centauri and Van Biesbroeck's Star in Aquila have even smaller masses. In the most extreme cases, where the masses are 0.02 or less, the "star" would be unable to sustain any nuclear reactions in its interior, and would never become self-luminous. Thus there is some theoretical justification for considering objects of still smaller masses as planets, rather than true stars. (Refer also to Barnard's Star in Ophiuchus, and Lalande 21185)

DESCRIPTIVE NOTES (Cont'd)

P   (34 Cygni) Magnitude 4.88; spectrum B1 eq.
                 Position 20159n3753. Sometimes called "Nova
Cygni #1. This is a remarkable variable which might be
called a "permanent nova". It was first seen in the year
1600 as a 3rd magnitude star where none had been noticed
before. Possibly the earliest well-documented observation
was that made by the Dutch astronomer and mathematician
Willem Blaeuw of Amsterdam, in August 1600. An astronomical
globe made by Blaeuw about 1640, and now in the Museum of
National Literature in Prague, contains an inscription referring to the star, as follows:
         "The new star in Cygnus that I first observed on
August 8, 1600, was initially of the 3rd magnitude. I determined its position.... by measuring its distance from
Vega and Albireo. It remains in this position but now is
no brighter than 5th magnitude."
         The star remained bright for about 6 years, then
faded gradually, reaching 6th magnitude in 1620, and drop-
ping below naked-eye visibility in 1626. From 1626 to 1654
it remained invisible, but brightened again about 1655 and
eventually rose to magnitude 3.5 where it remained until
1659. Three years later it had vanished again, but rose
once more in 1665 and after numerous fluctuations became
steady at 5th magnitude in 1715. It has remained at that
brightness ever since.
         In spite of its popular title, there seems to be
no real reason to class P Cygni among the true novae; modern studies suggest instead that it is one of the high-
luminosity ejection variables as typified by the enigmatic
Eta Carinae. C.P.Gaposchkin refers to it as "a bright star
which underwent something akin to shell activity when it
brightened to 3rd magnitude in 1600". The evidence for a
surrounding expanding shell is still evident in the strange
spectrum of the star; the emission lines are bordered on
their violet edges by sharp absorption lines, and the spectral peculiarities seem to imply that the outer regions of
the shell are expanding more rapidly than the inner. The
acceleration is most probably caused by radiation pressure
and the star seems to be losing about 1/100,000 of its mass
each year in this way.
         The star is evidently very remote, but the true
distance is not well known. A distance of about 3000 light

years has been tentatively accepted for some years, but
has recently been challenged on the basis of studies of
galactic structure. It seems that the star may be a member of the Cygnus star cloud which marks one of the spiral
arms of the galaxy, and the distance might then be as much
as 7000 light years, raising the apparent absolute magnitude to -6.8. After a further correction for absorption
the true absolute magnitude would be something like -8.9,
which would make the star one of the most luminous objects
known, even in its present state. The peculiar "nova" Eta
Carinae is one of the few objects which attained a comparable luminosity, and which may be a very similar type of
star.

The measured annual proper motion of P Cygni is
0.01"; the radial velocity is about 5½ miles per second in
approach. (Refer also to Eta Carinae)

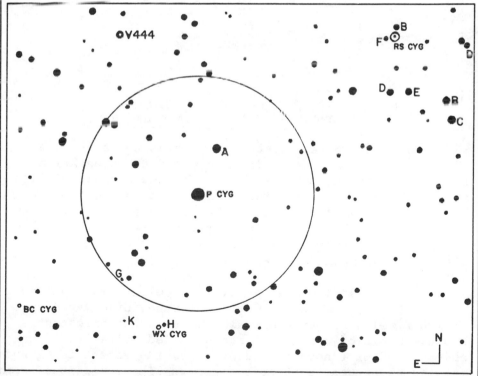

P CYGNI FIELD. Circle diameter = 1°. Comparison magnitudes are: A= 7.1; B= 7.2; C= 7.4; D= 8.2; E= 8.6; F=
9.0; G= 9.3; H= 9.6; J= 10.1; K= 10.8.

DESCRIPTIVE NOTES (Cont'd)

Y   Position 2050ln3428. Spectrum 09.5 or B0 IV.
                 Y Cygni is a massive binary system, claimed by
a few authorities to be the equal of A0 Cassiopeiae and UW
Canis Majoris; it is, at any rate a gigantic system by any
other standards. The star is located approximately 1.3° NE
of Epsilon Cygni, and consists of two giant 0 or B-stars
orbiting each other in a period of 2.99633 days, mutually
eclipsing each other at every revolution. The light variations were first detected by S.C.Chandler at Harvard in
1886.
    Each star is some 5 million miles in diameter and poss-
ibly about 5000 times more luminous than the Sun; current
studies suggest an absolute magnitude of about -4.5 for
each star and a diameter of about 5.9 times that of the

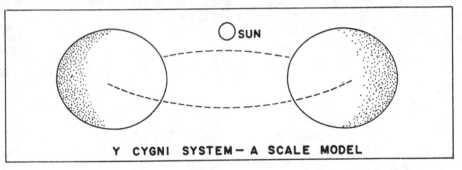

Y CYGNI SYSTEM— A SCALE MODEL

Sun. From the orbital elements the computed masses are
17.3 and 17.1 times the solar mass, and the separation of
the components (center-to-center) is just over 12 million
miles. The orbit has the rather small eccentricity of 0.13.
From the derived luminosities the distance of Y Cygni may
be something like 9000 light years. R.O.Redman (1930) has
found some evidence for a third body in the system. (See
also A0 Cassiopeiae and UW Canis Majoris)

SS   Position 21407n4321. A peculiar dwarf variable
                 star of the U Geminorum type, discovered by
Miss L.D.Wells at Harvard in 1896. It is the brightest
known example of a class of stars called "cataclysmic variables" or "dwarf novae", and presents the entertaining and
intriguing spectacle of frequent nova-like explosions which
occur very suddenly, several times a year. The star is normally of the 12th magnitude, but at intervals which vary

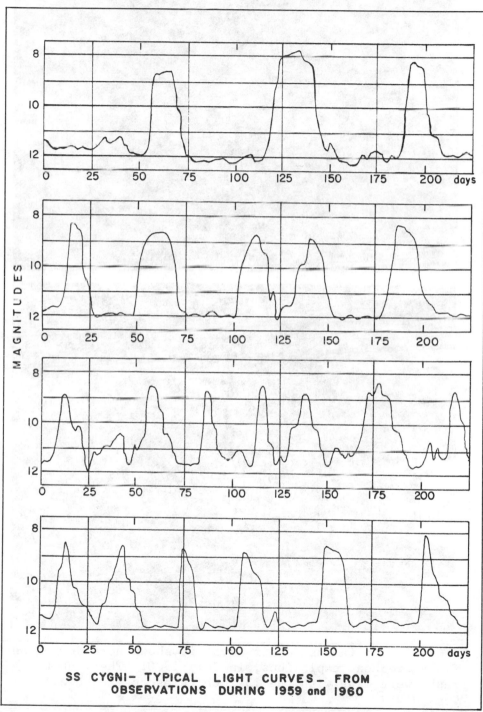

SS CYGNI— TYPICAL LIGHT CURVES— FROM
OBSERVATIONS DURING 1959 and 1960

SS CYGNI. The famous "dwarf nova" is shown here during one of its frequent explosions, in July 1960. These photographs were made with the 13-inch telescope at Lowell Observatory.

between 20 and 90 days it suddenly brightens to the 8th
magnitude. The light increase thus amounts to about 40
times and is usually accomplished in a period of scarcely
more than a day or two. The average interval between the
outbursts is approximately 51 days. For observers with
small telescopes, SS Cygni is the classic example of this
erratic variety of star, and certainly the most rewarding
object of its type. The total number of recorded observations of the star is now close to 100,000, and it is believed that not a single maximum has been missed since
discovery in 1896.

SS Cygni exhibits two types of maxima: a wide maximum of about 18 days duration, and a narrow maximum of some
8 days duration. The two types usually occur alternately,
though at times two maxima of the same type will occur in
succession. Both types are shown on the accompanying light
curves. A third type of maximum, called "anomalous", occurs
at rarer intervals; it is wide and rather symmetrical in
outline, with a comparatively slow rise to maximum.

SS Cygni has a peculiar spectrum, resembling class G
at minimum, but with a continuous background which gains
greatly in intensity during an outburst. From spectroscopic observations made at Mt.Wilson in 1956, it is now known
that SS Cygni is actually a binary star. One component is
a yellowish dwarf of the solar type which has been classed
as dG5, the other is a hot bluish star which is definitely
underluminous and may be related to the white dwarfs. It
has been classed as "sdBe". The two stars complete their
orbital revolution in the astonishingly short period of
0.276244 days, or slightly over $6\frac{1}{2}$ hours. This is one of
the shortest orbital periods known, and implies that the
components must be of abnormally small size and revolve
nearly in contact.

Current studies suggest that each component may have
about half the mass of the Sun, and that the diameters may
be about 0.9 and 0.1 the solar diameter, for the yellow
and blue stars respectively. The surface-to-surface separation is believed to be 100,000 miles or less. At the
present time these figures are somewhat uncertain. During
an outburst the faint spectrum of the yellow star is more
or less obliterated by the strong, nearly continuous spectrum of the hot star, which at this time resembles class

777

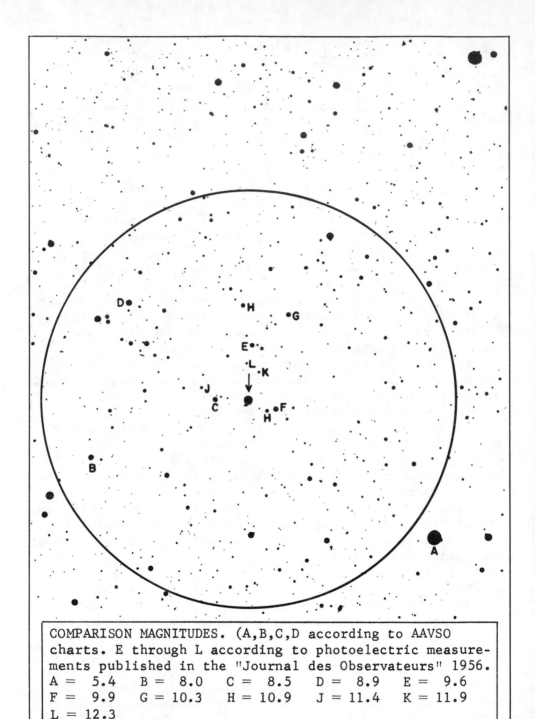

COMPARISON MAGNITUDES. (A,B,C,D according to AAVSO
charts. E through L according to photoelectric measure-
ments published in the "Journal des Observateurs" 1956.
A = 5.4   B = 8.0   C = 8.5   D = 8.9   E = 9.6
F = 9.9   G = 10.3   H = 10.9   J = 11.4   K = 11.9
L = 12.3

SS CYGNI- Identification chart made from a 13-inch telescope plate at Lowell Observatory. Circle diameter = 1° ;
north is at the top. Limiting magnitude about 14. Bright
star "A" is 75 Cygni, magnitude 5.4.

DESCRIPTIVE NOTES (Cont'd)

AO. The blue star seems to be erratically variable even at
minimum, since rapid fluctuations have been recorded by
several observers. Some of these variations show a wave-
like period of 2 or 3 hours, while others last less than
10 minutes. The average amplitude, in both cases, is on
the order of 0.2 magnitude.

The bluish star is the component usually thought to
be responsible for the nova-like behavior of SS Cygni.
However, the correctness of this view has recently been
questioned following a study by W.Krzeminski (1965) of the
very similar star U Geminorum. In this system the two stars
form an eclipsing binary system which makes it possible to
gain more information concerning the physical characteristics - sizes, separation, etc. - than in the case of SS
Cygni. Krzeminski finds evidence that the outbursts of U
Geminorum probably originate in the redder component - in
some manner which is still not at all clear - rather than
in the blue dwarf. If further research supports this view,
a new picture of these stars will begin to emerge, a picture in which the red component is possibly "triggered" to
violent flares by some effect connected with the nearby
hot companion.

W.J.Luyten (1965) has measured an annual proper motion of 0.12" for SS Cygni, the largest motion so far recorded for any star of this type. Evidently the star is at
no great distance. A parallax determination by K.Strand
agrees well with an earlier determination by P.P.Parenago
and B.V.Kukarkin, and leads to a distance of 90 or 100
light years. Resulting absolute magnitudes for the system
are:

    At minimum: +9.5   (Luminosity = 0.0145 X Sun)
    At maximum: +5.6   (Luminosity = 0.5236 X Sun)

The blue component thus appears to have an absolute magnitude fainter than 10, which places it virtually in the
realm of the white dwarfs. Even at maximum the luminosity
is only about half that of our Sun. Similar results have
been obtained for another well-known star of the type, U
Geminorum, and it is now certain that these stars are a
few thousand times fainter than the true novae and the recurrent novae. But this does not rule out the possibility
that the three types of exploding stars form some sort of
developmental sequence, or at least that the same general

DESCRIPTIVE NOTES (Cont'd)

process is operating in all cases. The central question, of course, remains: In which direction would such an evolution proceed? Is SS Cygni on its way to becoming a full scale nova? Or is it now dying out after having been one at some time in the past? Or are the SS Cygni outbursts much less violent than those of the classical novae merely because the stars are much less massive? If so, it may be that the various types of stellar explosions are produced by a similar mechanism, but do not necessarily form an evolutionary sequence.

Several investigators have pointed out that the SS Cygni stars display a "period-luminosity relationship" in which the stars with longer average periods show the greater range in magnitude during an outburst. This relation has been extended to cover the recurrent novae, though with uncertain success, since only 7 stars of the type are now known, and only a few have been thoroughly studied. But from the available data, it seems that the relationship holds true reasonably well. This implies that these two classes of exploding stars are members of the same physical group, since the total amount of energy released in a given interval of time seems to be a constant. Whether the full-scale novae follow the same rule is an unsolved question.

A fact of great interest is the binary nature of SS Cygni, a characteristic which seems to be shared by all other stars of the type. U Geminorum is apparently a very similar system; SS Aurigae and AE Aquarii are other well studied examples. In such stars as Z Andromedae and R Aquarii, a hot bluish dwarf appears to be mated with a red giant. These "symbiotic stars" are represented also among the recurrent novae such as T Coronae and RS Ophiuchi, both known to be close binary systems. The great similarity in all these systems is unlikely to be coincidental, and it appears almost certain that the outbursts are in some way related to the presence of a close companion. This was first suggested for the variations of AE Aquarii, not strictly a true SS Cygni type star, but clearly a member of the same general physical group. (Refer also to U Geminorum, SS Aurigae, SU Ursa Majoris, AE Aquarii, T Coronae Borealis, RS Ophiuchi, and WZ Sagittae. For an account of novae in general, see Nova Aquilae 1918)

BF   Position 19219n2934. A peculiar variable star belonging to the rare class called "symbiotic stars", of which R Aquarii seems to be the prototype. The variations in brightness often occur in a fairly regular cycle of about 750 days, though on occasion the star has remained nearly constant for as long as 6 years (1929-1935). The range during any one cycle is about a magnitude, but the mean brightness varies from one cycle to the next so that the total amplitude is about 4 magnitudes.

The star is easily located in the field of the bright star 2 Cygni, which is magnitude 4.9. In the chart below, the portions of the circle represent the usual 1° field, with north at the top. The total magnitude range of BF itself is about 9.3 to 13.

Like all the symbiotic stars, the spectrum is composite, the brighter star showing a B-type spectrum with emission lines. The companion is a red giant of type M4. Most of the variations are attributed to the B-star, whose spectrum normally dominates, but it is likely that the red star varies also. The star grows bluer as it brightens.

The spectrum shows many complex irregularities which are difficult to interpret. The bright lines change in intensity and displacement in a sudden and unpredictable manner. Both stars appear to be imbedded in a diffuse nebulosity whose spectral features vary greatly in intensity. The computed density of the cloud is much higher than that of a typical planetary nebula, and is comparable to that of a nova shell; its radius may be about 10,000 times that

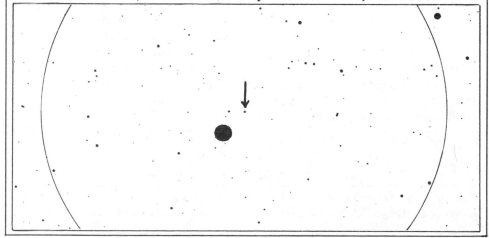

of the Sun. The cloud has presumably originated in outbursts of the B-star, suggesting the possibility of some
connection with the novae. (Refer also to R Aquarii and
Z Andromedae)

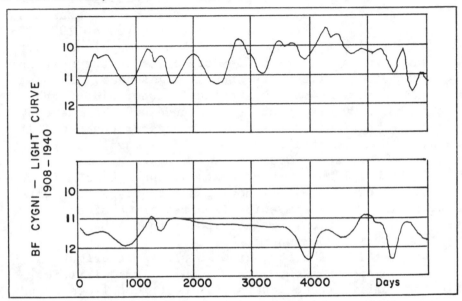

BF CYGNI — LIGHT CURVE 1908 – 1940

V444   Position 20177n3834. An eclipsing binary
system of special interest, one of the few
binaries known in which one component is a Wolf-Rayet star.
It was the first such case to be discovered, found in 1937
by O.C.Wilson and identified as an eclipsing system by S.
Gaposchkin in 1940. The period of revolution is 4.21238
days and the eclipses are partial with the rather small
amplitude of 0.3 magnitude. As in many giant binaries of
this type, there is some uncertainty about the actual sizes
and masses of the components, due to a distortion of the
radial velocity measurements by moving gas streams encircling the stars. From a summary of current evidence the
following table has been compiled:

	Spect.	Diam.	Mass	Lum.	Abs.Mag.
A	06	10	32	6900	-4.8
B	WN5	2.3	18	1450	-3.1

The two stars are about 17 million miles apart, the total
absolute magnitude of the system is -5.0, and the computed

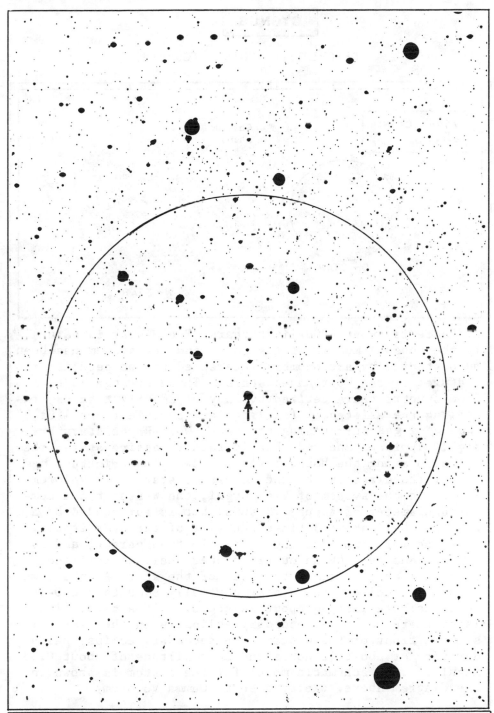

V444. Identification field, from a Lowell Observatory 13-inch telescope plate. Circle diameter = 1° with north at the top. The bright star at lower right is the famous nova-like variable P Cygni.

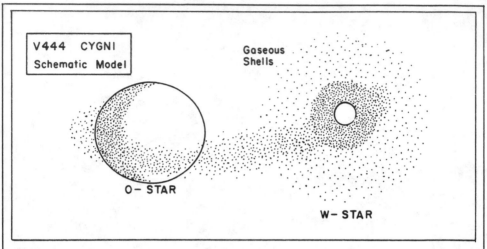

distance is about 4900 light years. The W-star is naturally the unusual member of the pair, and the system has been
much studied by astronomers interested in the peculiar
features of the Wolf-Rayet stars. These strange objects
show spectra characterized by many broad bright bands, revealing the existence of a large and turbulent shell of
expanding gas surrounding the star. Wolf-Rayet stars are
highly luminous and undoubtedly rather massive stars; they
are also among the hottest stars known. The brightest member of the class is the 2nd magnitude star Gamma Velorum.
        In the system of V444 Cygni, the W-star has a com-
puted diameter of about 2.3 suns, but is surrounded by a
gaseous shell about 3 times the size of the star itself.
This shell is luminous and partially transparent, and is
in turn enclosed by an outer thinner shell of rarified
material, about 8 times the size of the star. At the time
of primary eclipse the W-star partially occults the O-star
in an eclipse lasting about 24 hours. Half a revolution
later there occurs a secondary eclipse when the O-star is
seen in front of the W-star. The secondary eclipse has a
duration of about 12 hours and an amplitude of about 0.15
magnitude. A schematic model for the system is shown in
the diagram above. (Refer also to Gamma Velorum)

V 476   Nova Cygni 1920, sometimes called Nova
            Cygni #3. Position 19571n5329. A bright
nova, first seen on August 20, 1920, when it had attained
magnitude 3.5; it reached maximum brightness of 1.8 some

four days later. Subsequently, its image was found on a plate taken at Copenhagen on August 16. No earlier photograph shows any star brighter than 15th magnitude in the place of the nova. This indicates a rise of at least 13 magnitudes, corresponding to a light increase of 160,000 times within a few days.

The computed distance of Nova Cygni is about 4000 light years, the maximum luminosity about 250,000 times that of the Sun, and the absolute magnitude at maximum about -8.7. The light curve classes V476 Cygni as a "fast nova". After reaching peak brightness it began to decline at the rate of 0.29 magnitude per day, and had faded to below naked-eye visibility by the middle of September. Spectroscopic measurements revealed the existence of an expanding gas cloud around the star, growing at the rate of about 390 miles per second. This cloud was ultimately detected visually in 1944, and appeared as a nebulous disc about 4.3" in width. The star itself was then an object of magnitude 15½, with an O-type spectrum showing some bright lines. A further fading appears to have taken place in the last 15 years; the apparent magnitude in 1962 was no brighter than 17. (For a general review of nova facts and theories, refer to Nova Aquilae 1918)

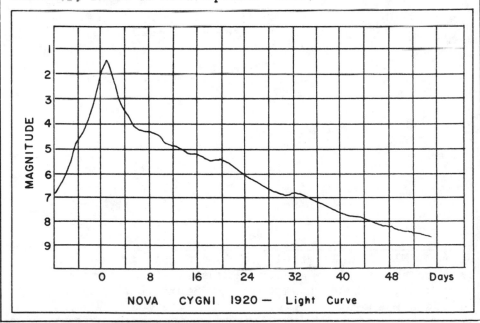

NOVA CYGNI 1920 — Light Curve

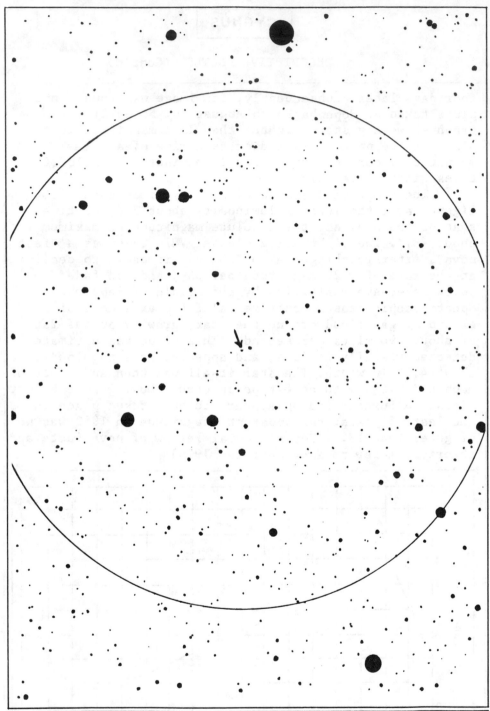

NOVA CYGNI 1920. The field of V476 Cygni, traced from a 13-inch telescope plate at Lowell Observatory. Circle diameter = $\frac{1}{2}°$, with north at the top. Stars to about 16th magnitude are shown.

CYGNUS

DESCRIPTIVE NOTES   (Cont'd)

V1500   Nova Cygni 1975. Position 21099n4757.   The bright nova which flared up in the Cygnus Milky Way in August 1975, reaching a magnitude of 1.8 and becoming the brightest nova seen from the Earth since Nova Puppis in 1942.   The position was about 5° ENE from Deneb and just 1° ENE from 63 Cygni, less than a degree from the Galactic Equator. Eventually becoming slightly brighter than Polaris, the new star was first seen by hundreds of persons all around the world on August 29, when it had attained 3rd magnitude. According to a summary of the observations reported in *Sky and Telescope* in October 1975, the earliest known visual sighting of the new star was probably made by Mr. Kentaro Osada in Japan at about 12^h U.T. on August 29; the star was then magnitude 3.0 and brightening rapidly. During the next six hours it increased by nearly a magnitude. On the evening of August 29 it was close to 2nd magnitude, and was independently discovered by scores of observers throughout the U.S., including N.G. Thomas of Lowell Observatory who noted the presence of the strange star in Cygnus while camping out in Arizona's Havasu Canyon. The following evening the star attained its maximum brilliancy of magnitude 1.8.

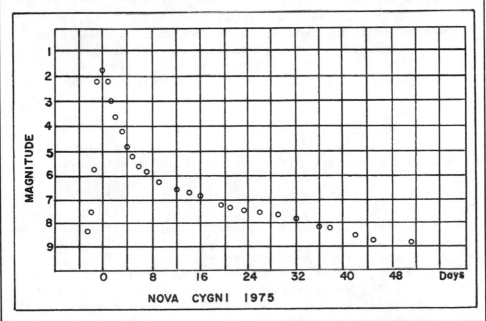

NOVA CYGNI 1975

DESCRIPTIVE NOTES   (Cont'd)

Pre-discovery photographs of the region showed
that the star had been fainter than magnitude 9.6 (photovisual) on the evening of August 27; twenty-four hours
later it had reached 7.5, and within another eight hours
it had become a naked-eye object of about 3rd magnitude and
was still brightening hour by hour.

When older photographic plates of the region were
examined, it became evident that Nova Cygni was an exploding star of exceptionally great magnitude range. No pre-
outburst image at all appears on the Lowell 13-inch telescope plates which reach to about magnitude 17.3. On 48-
inch Schmidt camera plates at Palomar, no image brighter
than 21st magnitude appears near the position. The range of
at least 19 magnitudes corresponds to a difference of 40
million times in brightness, the greatest range ever observed for a nova, exceeding even Nova Puppis 1942 which
had an 18 magnitude range. The rise of at least 19 magnitudes, in fact, suggests a supernova, though all other
evidence indicates that the star was an ordinary "classical" nova. The light curve, spectrum, and expansion rates
were all reasonably normal. According to G.de Vaucouleurs
(1975) a study of all the data suggests a maximum absolute
magnitude of about -9.5, equal to about 500,000 times the
luminosity of the Sun, certainly one of the most luminous
normal novae on record. Allowing for an obscuration of
about 1 magnitude, the derived distance is about 5000 light
years.   I.S.McLean of the University of Glasgow (1976)
found a somewhat higher absolute magnitude of about -10.1;
this would decrease the computed distance to about 3700
light years.

The expansion velocity of the nova-shell, measured
just before maximum, was about 1200 miles per second, quite
comparable to other bright novae. The strong intensity of
the hydrogen-alpha radiation at this time produced a rosy
or orange tint reported by many observers of the star. This
has been seen also in many other novae; the color was
quite striking in the 1967 outburst of the recurrent nova
RS Ophiuchi.

Following peak brilliancy, Nova Cygni began to
fade rapidly; by the evening of September 3 it was down to
4th magnitude, and two days later had fallen to 5th. The
fading thereafter continued steadily; in December 1976 the

NOVA CYGNI 1975, photographed at Lowell Observatory with the 13-inch telescope. The three photographs were made on August 26, 1940; August 30, 1975; and September 26, 1975.

DESCRIPTIVE NOTES   (Cont'd)

star was at magnitude 12.0, while a measurement made in early January 1977 gave 12.4. The fading is expected to continue for a number of years.

The unusually great amplitude of Nova Cygni 1975 is the outstanding feature of this star, and has excited much speculation. Most novae are believed to be extremely close double stars in which at least one of the components is a degenerate or semi-degenerate dwarf; this star is gradually accreting matter from the close companion. For many years this matter, chiefly hydrogen, may remain inert on the surface of the white dwarf, since the temperature and pressure are not sufficient to "ignite" any nuclear reactions. Eventually, however, the temperature at the bottom of the hydrogen layer reaches the nuclear ignition point, and the star then "goes nova", blasting the entire outer shell into space. Obviously, this chain of events can occur repeatedly, which suggests that all novae may be recurrent, though the interval between outbursts may be many centuries. As of 1976, seven recurrent novae are actually known, with periods averaging several decades; in the eruptive dwarf variable stars such as SS Cygni and U Geminorum, a similar process seems to be occurring on a much smaller scale.

The abnormally great ranges of Nova Cygni 1975 and Nova Puppis 1942 may indicate that these stars are "virgin novae" which have not previously exploded. The very smooth and rapid decline also suggests the possibility that Nova Cygni, unlike many other novae, was a single star; as it is the mutual interaction of the nova-component and its companion which is thought to produce the oscillations and "flickerings" seen in many nova light curves. If the nova-star had no companion, then the accreted material was presumably collected from interstellar gas and dust clouds. From the known distribution of the interstellar material, it seems that only about 1% of nova outbursts could be triggered in this way. However, in January 1977, Nova Cygni was showing oscillations of up to 0.5 magnitude in a period of about 3 hours, thus apparently rendering all such speculations obsolete. (For a survey of novae in general, refer to Nova Aquilae 1918. See also Nova Persei 1901, Nova Pictoris 1925, Nova Herculis 1934, and Nova Puppis 1942. Supernovae are treated under "Tycho's Star" in Cassiopeia.)

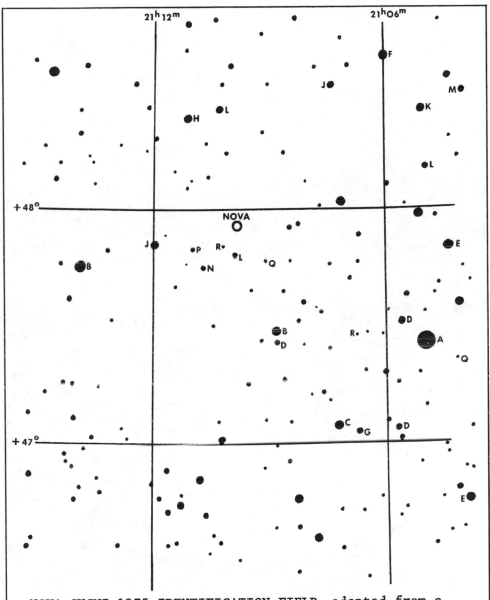

NOVA CYGNI 1975 IDENTIFICATION FIELD, adapted from a
preliminary AAVSO Chart. Bright star "A" is 63 Cygni,
Mag 4.56, spectrum K4. Comparison star magnitudes (AAVSO)
are: B= 6.5; C= 7.0; D= 7.3; E= 7.6; F= 7.8; G= 7.9;
H= 8.0; J= 8.6; K= 8.9; L= 9.0; M= 9.1; N= 9.7; P= 9.8;
Q= 10.3; R= 10.5. (See detailed chart on page 792)

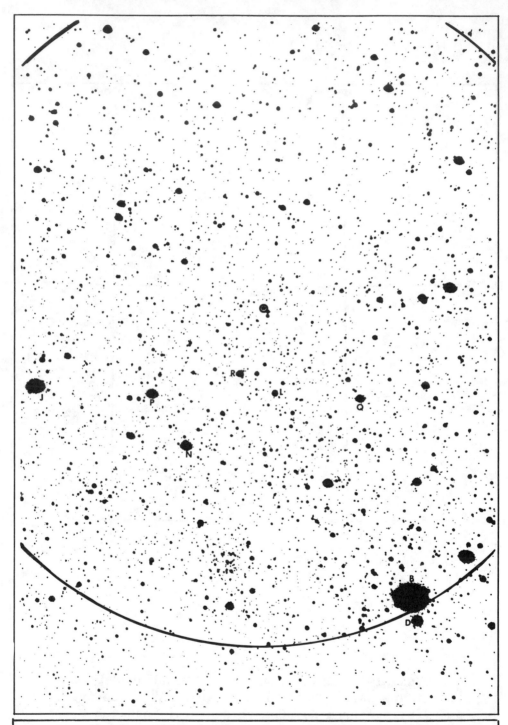

NOVA CYGNI 1975 IDENTIFICATION CHART, from a Lowell Observatory 13-inch telescope plate. The circle is 1° in diameter with north at the top. Limiting magnitude about 16.

CYGNUS X-1   Position 19565n3504. One of the strongest sources of X-ray energy in the sky, and thought to be probably the most convincing candidate for a "black hole". It was among the first interstellar X-ray sources detected, during the experiments made from an Aerobee rocket launching in June 1962; by 1965 it was known that the Cygnus source was variable. Cygnus X-1 was thus a prime target of NASA's *UHURU* satellite, launched in December 1970, and designed specifically to detect and measure X-ray sources in the sky. Within a few months it was shown that the source displayed very large changes in X-ray intensity on a time-scale of less than 50 milliseconds. The very rapid variability suggested that the source was something of unusually small dimensions, probably a gravitationally collapsed object, but the lack of any strong radio emission seemed to imply that Cygnus X-1 was not a supernova remnant. Radio emission from this mysterious object has, however, been detected in recent years; it appears to have "turned on" in March 1971 while at the same time the observed X-ray emission decreased by a factor of three.

With the position accurately measured, the coincidence of the source with a visible 9th magnitude star was soon demonstrated. The star, HDE 226868 or BD+34°3815, lies slightly less than 0.5° ENE from Eta Cygni; it may be located easily in small telescopes, and is marked with a superimposed circle on the photograph on page 795. The star is a very hot and luminous supergiant of spectral type about B0 Ib; a precise spectral type of O9.7 Ib has also been published. Further studies of the star soon showed that HDE 226868 is a single-line spectroscopic binary with a period of 5.599823 days and an orbital inclination of about 27°; the radial velocity varies from about 42 miles per second in approach to 46 miles per second in recession. This visible star has a surface temperature close to 30,000°K and a computed absolute magnitude of about -6. The expected mass of such a star is in the range of 20- 30 suns; from the derived luminosity the distance of Cygnus X-1 appears to be 2.0 to 2.5 kiloparsecs, or 6500 to 8000 light years.

Studies of the system in 1973 have established the existence of a strongly heated stream of gas, passing from the B-star to the unseen component; the X-ray energy originates in this tremendously hot stream as it falls into the

unseen *something* which orbits the visible star. From the
known orbital elements, the mass of the unseen body would
seem to be at least 10 solar masses, and is more likely in
the range of 15 to 20. A normal star of this mass should be
detectable in the combined spectrum of the pair, but a
"normal star", in any case, appears to be ruled out by the
very rapid variations of the X-ray source, which imply that
the source is something of very small dimensions, almost
certainly less than a hundred miles in diameter. As the
derived mass is well above that considered possible for
any white dwarf or neutron star, the most likely interpretation at present (1976) is that the unseen *something* is a
completely collapsed body or "black hole". The existence
of such bizarre objects has been debated on and off since
1798, when Pierre Laplace showed that, from the principles
of gravitational theory, it was possible to imagine a body
of such mass and density that the escape velocity would
exceed the velocity of light. Such a body would logically
be invisible, since light itself could not escape the very
strong gravitational field. From a formula given in 1916
by K.Schwarzschild, it can be shown that a mass of one sun,
to achieve this "black hole" state, would need to be compressed down to a sphere about 3.6 miles in diameter. For
the Earth, the Schwarzschild radius is slightly under one
centimeter. According to current ideas, a "black hole"
should be the end-product of a sufficiently massive star
which has exhausted its nuclear energy sources, allowing
gravitational contraction to "take over". Stars of up to
about 1.25 solar mass merely become conventional white
dwarfs; stars of up to about 3.2 solar masses probably go
through a supernova explosion and leave a neutron star as
a remnant; so it is among the stars of masses greater than
about 3.2 suns that we expect to find candidates for the
"black hole" state. In addition to Cygnus X-1, the possible existence of a black hole has been suggested in such
odd binary systems as Beta Lyrae, Epsilon Aurigae, and SV
Centauri. (Refer to page 413) Even the existence of "mini-
black holes" has been postulated, and it has been suggested
that the famous Tunguska Meteorite of June 30, 1908 might
have been such a body, a particle smaller than a dust grain
but weighing billions of tons. Many such entertaining
speculations will be found in current books on the subject.

FIELD OF CYGNUS X-1. The visible object HDE 226868 is marked near left center. Bright star at right is Eta Cygni. Lowell Observatory photograph made with the 13-inch telescope in blue light.

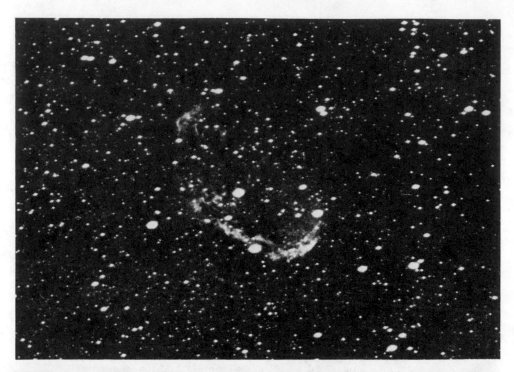

NEBULAE IN CYGNUS. Top: The curious "Crescent Nebula" NGC 6888. Below: A nebulous field 1° ESE from Gamma Cygni. Lowell Observatory photographs in red light with the 13-inch telescope.

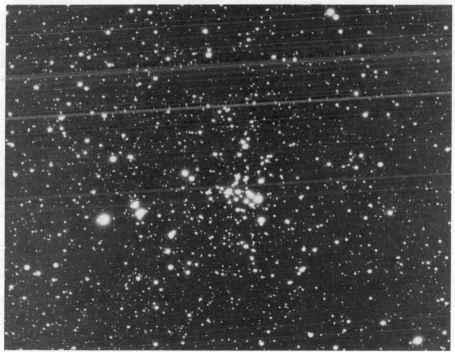

STAR CLUSTERS IN CYGNUS. Top: NGC 6819, about 7° west of Gamma Cygni. Below: NGC 6866, about 4.7° NW from Gamma. Lowell Observatory photographs with the 13-inch camera.

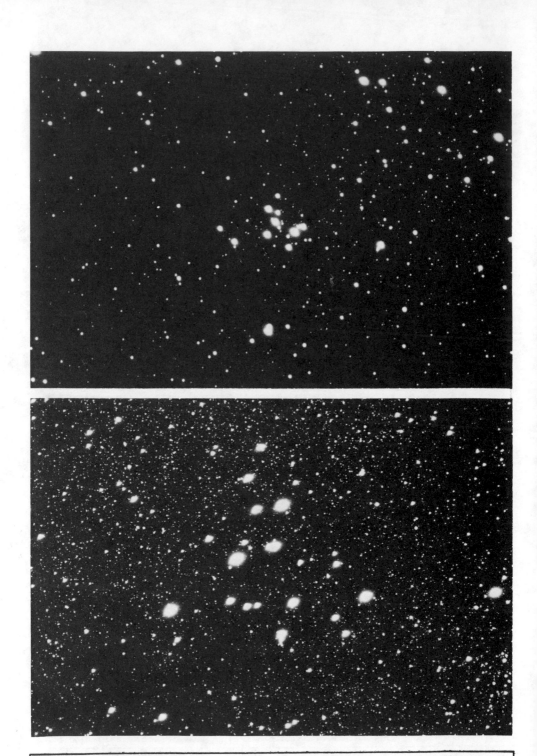

STAR CLUSTERS IN CYGNUS. Top: The compact cluster M29 which
lies near Gamma Cygni. Below: The large scattered group M39
is located about 9° ENE from Deneb. Lowell Observatory
photographs with the 13-inch telescope.

M29   (NGC 6913) Position 20222n3821. A small and visually rather indistinguished star cluster located in a rich and crowded area of the Cygnus Milky Way about 1.7° SSE from Gamma Cygni. Discovered by Messier in July 1764, it appears in small telescopes as a trapezoid-shaped knot of a dozen or so 8th-9th magnitude stars; the group measures about 5' with a few outliers increasing the total size to possibly 7' or 8'. M29 lies in a heavily obscured region of the Galaxy where interstellar absorption produces an estimated 3 magnitudes of dimming. In studies of polarization in the cluster, W.A.Hiltner (1954) found that the density of dust within the cluster is nearly 1000 times the mean value for the Galaxy; evidently M29 might be a rather striking cluster if it could be seen "in the clear". The brightest members are all B-type stars; the stellar population resembling M36 in Auriga, which, however, is much nearer to the Solar System. Current studies suggest a distance of about 7200 light years for M29, agreeing very well with the early estimates of R.J.Trumpler in 1930. This gives M29 an actual diameter of about 15 light years and a total luminosity of about 50,000 suns.

M39   (NGC 7092) Position 21304n4813. A large but very loose-structured galactic star cluster located some 9° ENE from Deneb, easily found in binoculars but too large and sparse a group to make a good telescopic object, though T.W.Webb called it a "grand open cluster" and Smyth found it a "splashy field of stars". Discovery is often credited to Le Gentil in 1750, but P.Doig (1925) quotes a statement made by J.E.Gore that this cluster was noted by Aristotle as a cometary appearing object about 325 BC.

E.G.Ebbighausen (1940) determined the annual proper motion of the group to be about 0.024" in PA 222°; the radial velocity is about 8½ miles per second in approach. From these criteria, some 30 stars are now accepted as true cluster members. The 12 brightest stars are all A or B-type, scattered over a field more than ½° in diameter. In 1953 studies by H.L.Johnson and H.F.Weaver gave a distance of about 800 light years, which gives the cluster a true diameter of about 7 light years. From the H-R diagram of M39 it is found that virtually all the members are main

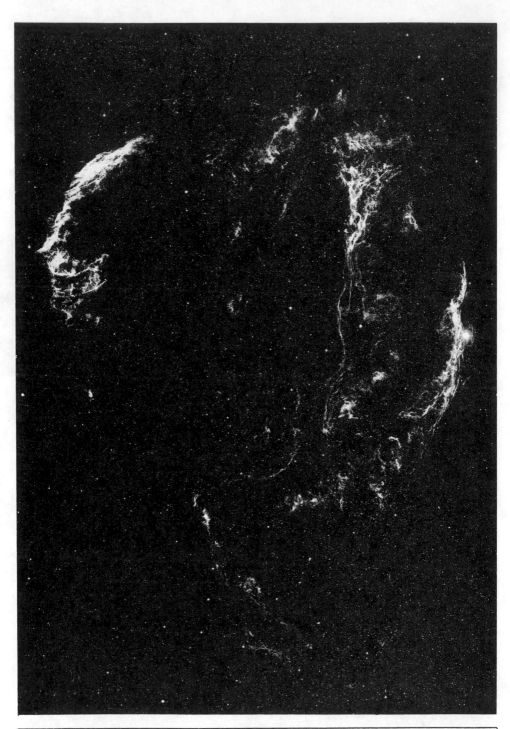

THE VEIL NEBULA. The complete arc of this famous nebula is shown in this photograph made with the 48-inch Schmidt at Palomar. NGC 6992 is the bright mass at the left; NGC 6960 and the bright star 52 Cygni are at the right edge.

sequence objects, but that the few brightest stars appear
to be just at the point of beginning their evolution toward the giant stage. A comparison of many cluster diagrams
shows that M39 is probably somewhat younger than the Coma
Berenices cluster but older than the Pleiades.

NGC 6960   Two wonderful lace-like gaseous nebulae,
NGC 6992   forming together the "Bridal Veil Nebula" in
                 Cygnus, one of the most striking objects in
the heavens on long-exposure photographs taken with great
telescopes. The nebula was discovered visually by William
Herschel in 1784 with his 18-inch reflecting telescope.
NGC 6992 is often called the "Filamentary Nebula" and NGC
6960 the "Network Nebula"; however these names are often
used indiscriminately and are sometimes used in referring
to either portion or to both objects together.
      NGC 6992 is located at 20543n3130 and is visible in
a 6-inch or 8-inch glass with low power, looking like a
miniature Milky Way in itself in the field. It appears as
a faint curved arc like a ghostly white rainbow, over one
degree in length. A very dark clear night and a wide-field
eyepiece of low power are absolutely essential for the detection of this difficult object. About $2\frac{1}{2}°$ to the WSW is
NGC 6960, appearing as another faint narrow arc of nebulosity crossing the field of the 4th magnitude star 52 Cygni
which is located at 20436n3032. This nebulosity is a difficult and indistinct object in a 6-inch glass. A third main
nebulous mass lies about 2° directly west of NGC 6992; it
is the faintest of the three segments and remains unnumber-
ed. It cannot be detected visually in small telescopes.
      It is a curious fact, however, that the brightest
portion of the Veil Nebula, NGC 6992, may be seen faintly
but unmistakably in 7X50 binoculars. This has been accomplished by numerous observers, and is due to the very low
magnification and great light concentrating power of such
an instrument. Some of the filamentary structure may be
seen visually with large telescopes, but the finest details may be seen only on the photographic plate.
      The peculiar filamentary structure, resembling frost
patterns or fine lace, makes this nebula one of the most
noted of all such objects, and therefore of special interest to the amateur, even though the direct view through

WESTERN PORTION OF THE VEIL NEBULA. NGC 6960 and the star 52 Cygni, photographed with the 60-inch reflector at Mt. Wilson Observatory.

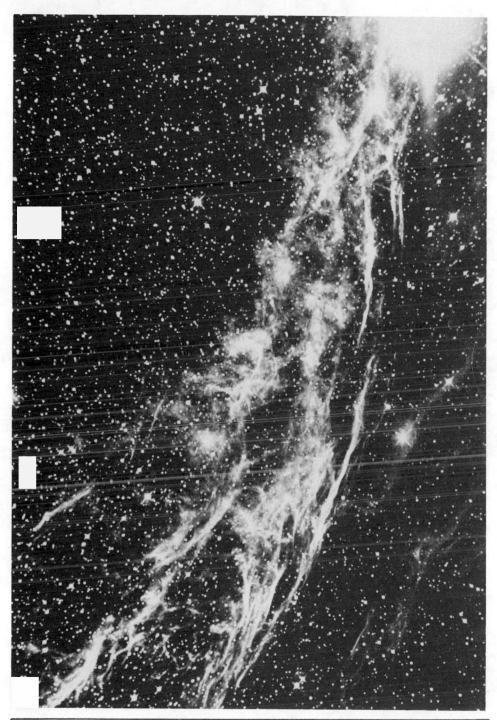

DETAILS IN THE VEIL NEBULA. A region in NGC 6960, just south of 52 Cygni, showing the filamentary structure. Mt. Wilson Observatory 60-inch reflector photograph.

FILAMENTARY STRUCTURE IN THE VEIL NEBULA. A portion of NGC 6992, photographed with the 61-inch astrometric reflector at the Flagstaff Station of the U.S. Naval Observatory.

the small telescope will show nothing of its true splendor.
Kelvin McKready, in his *"Beginner's Star Book"* described
the nebula as seeming like *"the long and shelving undula-
tions of a thin cataract of light, as it slips from star to
star in its shining fall through space"*. Some fantastically
beautiful photographs of the Veil have been obtained in
color by W.C.Miller at Palomar Observatory; these show that
the hues of various filaments range from blue to white to
red, according to the temperature. One of these photographs
appears in the *Life Nature Library* book *"The Universe"* with
the caption *"The Veil Nebula, a diaphanous filigree of
star-spangled red, white and blue"*. These colors, however,
cannot be seen visually in even the largest telescopes because of the extremely low intensity of the light.

The visible portions of the Veil appear to be segments of an expanding wreath-like gas cloud some 2.6° in
diameter, often called the Great Cygnus Loop. The distance
of the nebula is still somewhat uncertain, but an average
of several estimates gives about 1500 light years as the
probable value. The nebulosity on this basis is about 70
light years in diameter, and is expanding at about 0.06"
per year. This suggests that the gas cloud had its origin
in a great explosion, probably a supernova outburst, in
the distant past. At the present rate of expansion the
cloud would have required over 150,000 years to attain its
present diameter. However, it seems certain that the rate
of expansion has decreased to a fraction of its original
value. A supernova cloud begins its growth at a rate of
more than 1000 miles per second, but the largest radial
velocity measured in the Veil is less than 45 miles per
second. The deceleration may have been caused chiefly by
resistance from gas and dust clouds in the region. There
is no doubt that the Veil is expanding into a dusty region
and sweeping up the interstellar material as it does so.
The sky within the loop is noticeably clearer than the
area outside, as evidenced by the difference in the number
of faint stars visible in the two regions. This effect is
plainly seen on the photograph on page 802.

Allowing for the deceleration of the nebulosity, the
explosion of the supernova is estimated to have occurred
possibly 30 to 40 thousand years ago. The Veil Nebula,

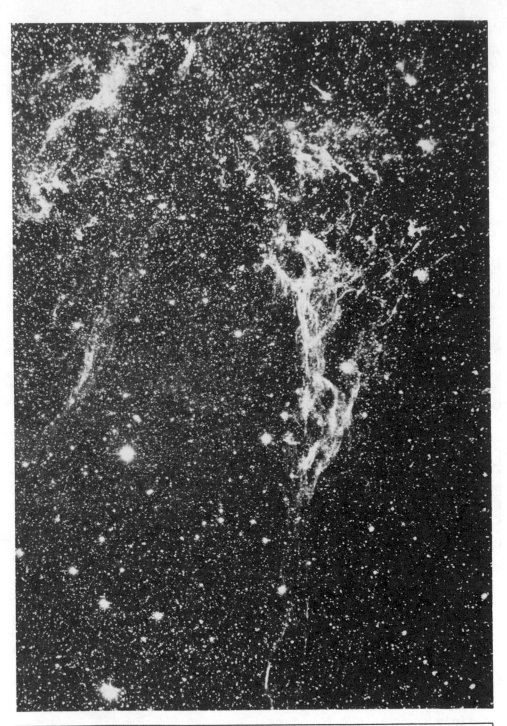

NORTH CENTRAL PORTION OF THE VEIL NEBULA. Details in the faint nebulosity between NGC 6960 and 6992, photographed with the Crossley Reflector at Lick Observatory.

DESCRIPTIVE NOTES (Cont'd)

like all other known supernova remnants, is a source of
radio radiation, but a much weaker source than the famous
Crab Nebula in Taurus whose age is only 900 years. Radio
emission from the Veil was first detected by D.Walsh and
R.H.Brown at Jodrell Bank in England.

The source of illumination of the nebula is still
something of a mystery. Spectrum analysis shows that the
light is fluorescence, but the exciting star has not been
identified; neither is there any star near the center of
the loop displaying post-nova characteristics. The western
segment of the Veil appears to be drifting past the 4th
magnitude star 52 Cygni, but the star is known to be a
foreground object and has no connection with the nebula.
R.Minkowski has suggested that the post-nova may be a very
close companion to one of the stars within the loop, most
probably the double star GC 290/1 or β67 which lies near
the apparent center of the ring. The primary of this pair
has a spectral type of about B7, and cannot be the source
of light of the Veil, but there is a faint companion of
the 10th magnitude about 1.6" distant, discovered by S.W.
Burnham in 1875. The spectrum of the companion has not
been obtained, so that its identification as the ancient
source of the Veil is still conjectural. If it is not the
former nova, some other unseen companion may be.

The fine filamentary structure of the Veil Nebula is
its outstanding feature, and has not yet been fully explained. The thickness of a typical filament ranges from
500 or 700 AU up to about 2000 AU; the apparent widths
being from 1" to 5". It is possible that some of the filaments are actually thin sheets of gas seen edge-on; a similar sheet if seen from the front would possibly be unde-
tectable because of low surface brightness. The expansion
of the cloud into the interstellar medium probably plays a
part in the formation of filaments, which may be largely
shock-front phenomena. Electromagnetic forces must also
be considered. The Russian astronomer V.G.Fessenkov has
obtained photographs with the 50-centimeter Maksutov camera at Alma Ata in Central Asia, seemingly resolving some
of the filaments into chains of stars. The existence of
the majority of these chains does not seem to be verified
by the fine photographs taken with larger telescopes in
America. On the whole, the evidence seems doubtful, and

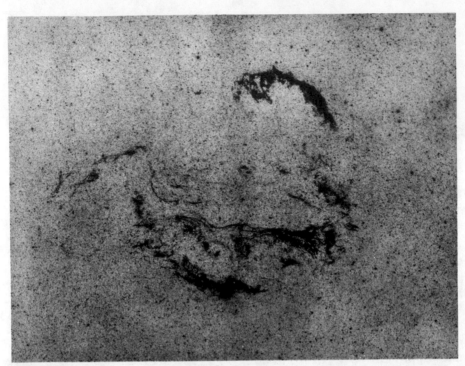

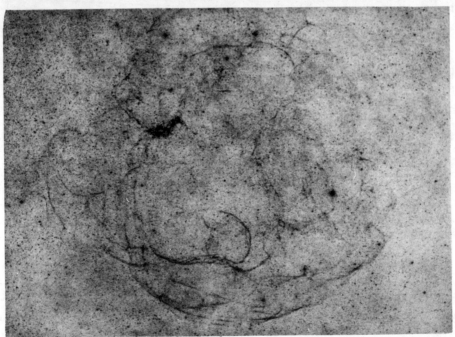

COMPARISON OF TWO FILAMENTARY NEBULAE. Top: The Veil Nebula in Cygnus, NGC 6960--6992. Below: The faint nebula S147 in Taurus. Palomar Observatory 48-inch Schmidt telescope photographs.

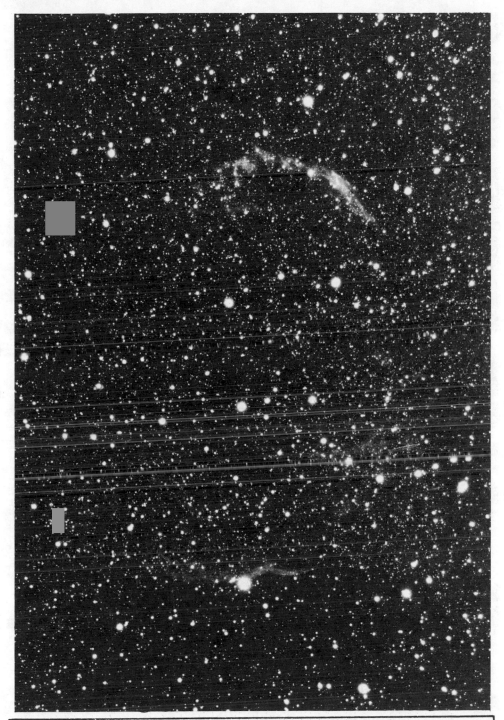

THE VEIL NEBULA. NGC 6992 is at top; NGC 6960 below. This photograph gives some impression of the appearance in the 10 or 12-inch telescope.

Lowell Observatory photograph

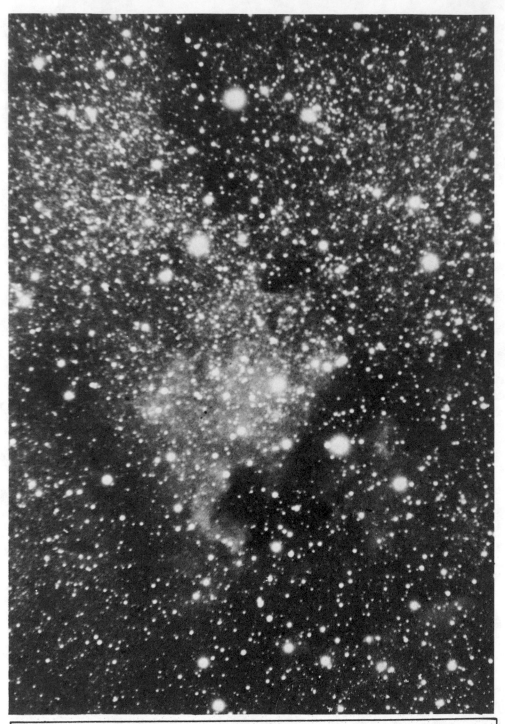

NGC 7000. The North American Nebula in Cygnus, photographed in red light with a 200mm Takumar lens by David Healy of Manhasset, New York.

the estimated mass of the Veil does not seem to be nearly
sufficient for the formation of large groups of stars.

A very similar, but much fainter filamentary nebula
in Taurus is known as S147; this too is believed to be a
supernova remnant, comparable in size and age to the Veil.
It consists entirely of an intricate network of fine
thread-like filaments forming a ring about 2° x 3° in size.
A third object is identified with the radio source Vela X.
The similarity in these objects invites speculation about
the future history of the Crab Nebula; perhaps in 50,000
years it will have developed the fine filamentary structure
now so admired in the Veil. (Refer also to NGC 1952 in
Taurus. For a survey of supernovae, refer to Tycho's Star
in Cassiopeia)

NGC 7000   Position 20570n4408. The North American
Nebula, a vast cloud of mixed nebulosity,
dust and stars, named by Dr.Max Wolf, and considered one
of the outstanding features of the northern Milky Way. It
lies some 3° east of Deneb and may be detected with the
naked eye as a region of increased brightness in the Cygnus
Milky Way. Binoculars show an irregular glow more than 1½°
in diameter with the "North American" shape becoming un
mistakable on a clear night. Perhaps the best view of the
unusual outline is obtained with a 3 or 4-inch rich-field
telescope and a wide-angle eyepiece. The dark Gulf of
Mexico stands out clearly and the Pacific coast can be
traced from Canada to Mexico. In larger instruments the
outline is invariably lost because of reduction in size
of the field and lack of contrast.

The main mass of the nebula has a mottled appearance, which in the region of the "Atlantic coast" develops
into an amazing pattern of shreds and streamers of nebulosity, both bright and dark. Still farther out in the "Atlantic" this same pattern reappears even more strikingly
in the detached nebulous island IC 5067, often called the
"Pelican Nebula". The entire face of the pelican appears
on red-sensitive plates as a virtual cascade of fine wisps
and streamers, flowing out toward the southeast. The north
portion of the nebula is traversed by several beautifully
defined narrow tendrils of dark nebulosity, sharply outlined against the glowing background. About 1½° to the

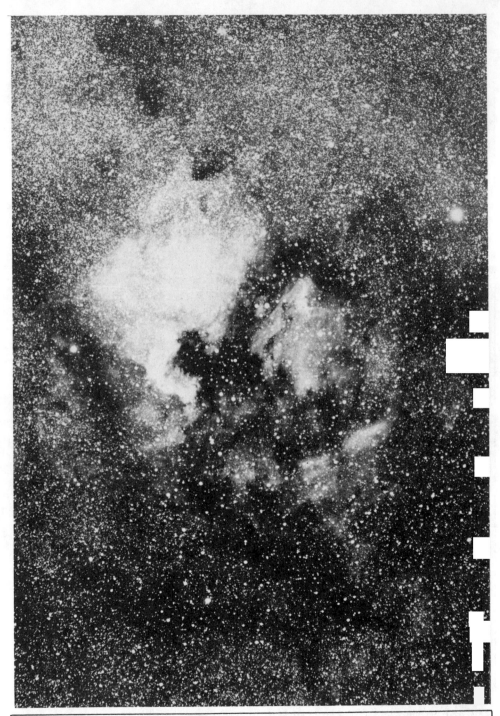

NGC 7000. The North American Nebula, one of the striking features of the northern Milky Way. This photograph was made with a 4-inch lens by Alan McClure.

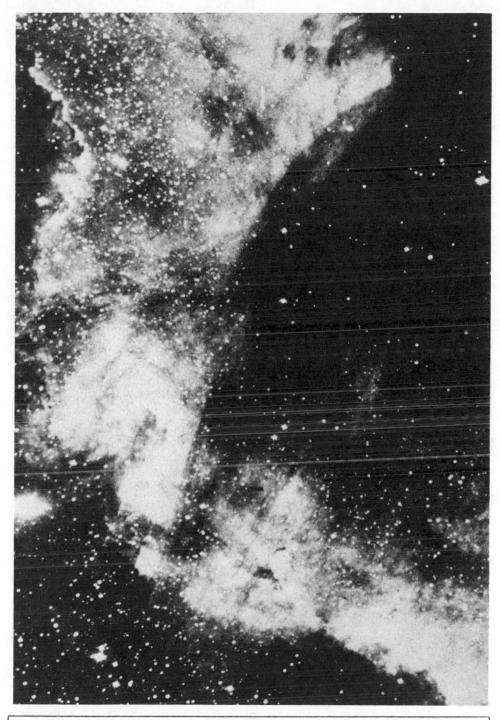

DETAILS IN NGC 7000.   The southern portion of the North American Nebula, showing the "Mexico" and "Central America region."

100-inch telescope, Mt.Wilson Observatory

south, in the detached region IC 5068, is a remarkable
field where numerous streamers and filaments intersect at
right angles to produce an intricate cross-weave pattern
which surely challenges explanation. In the North American Nebula itself the most remarkable region is found in
"Mexico and Central America" where bright and dark masses
meet to produce many a striking effect. Among specific details we must mention the splendid bright rim nebula which
lies on the "Pacific coast" like a silver-lined cloud, and
the gauzy semi-transparent veil of dark nebulosity which
streams southward from the Gulf region to pass across the
southern portion of "Mexico". Many of these details are
even more striking on the beautiful color photographs obtained at Palomar Observatory; the prevailing color being
a rich red which reaches its greatest intensity in the
bright-rim feature just noted.

    The 1st magnitude star Deneb has traditionally
been considered to be the chief source of the illumination
of the Nebula, though it is likely that many stars in the
region are contributors to the effect. The loose galactic
star cluster NGC 6997 lies within the nebulosity, but is
not definitely known to be at the same distance. The Nebula itself is estimated to be about 1600 light years distant, which agrees well with the computed distance of Deneb.
However, the true separation of star and nebula cannot be
much less than 70 light years. The nebula itself is some
45 light years in diameter.

CYGNUS A   Position 19578n4035. This is one of the
best known radio sources, first detected
by radio telescopes long before it was actually identified
optically. Although it is the second strongest radio source
in the sky, no bright visual object appears at that position. The mystery was partly solved at Palomar Observatory
in 1951, when 200-inch telescope plates revealed an unusual
galaxy now believed to be between 500 and 700 million light
years distant. The object appears as a pair of fuzzy 18th
magnitude images about 2" apart. The spectrum shows gaseous emission lines, and the object was at first thought to
show us a direct collision between two giant galaxies.
Such a collision would not affect the billions of stars in
the two galaxies, since they are separated by relatively

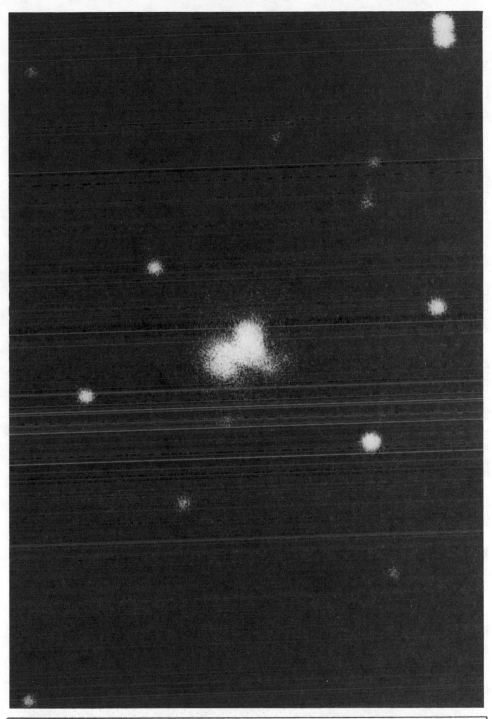

CYGNUS A. This unusual galaxy is one of the greatest known sources of radio radiation in the Universe.
Palomar Observatory 200-inch telescope

enormous distances, but the gaseous and nebulous matter would be heated to incredible temperatures. The passage of the two galaxies through each other would require about a million years.

This interpretation of Cygnus A as a colliding galaxy system now appears rather doubtful. It has been suggested that the object is not a double galaxy at all, but possibly a single one with a dark central band, resembling the odd NGC 5128 in Centaurus. Also, the radio waves do not come from the visible nuclei, but from two large and invisible "lobes" about 100" apart, on either side of the optical object. These lobes are probably connected with a strong magnetic field, but it is not known what processes are operating in this galaxy to produce such strange phenomena and create one of the greatest known radio transmitters in the universe. As of 1963, it appears impossible that the observed amount of energy can be generated by colliding galaxies, and astronomers have begun to look elsewhere for an explanation. Several other strong radio sources have now been identified with optically visible objects resembling Cygnus A; these appear to be abnormal galaxies which are undergoing violent outbursts of material from their nuclear regions. Among the nearby examples are such unusual systems as NGC 5128 in Centaurus, M82 in Ursa Major, M87 in Virgo, and NGC 1275 in Perseus. The outpourings of material in these systems seems to involve masses of several million suns. Explosions on this scale may result from the formation of a "hyperstar" or contracting gas cloud of several million solar masses in the central region of a galaxy. Such a hypermassive object could never condense into a stable superstar, it seems, but would go into a state of gravitational collapse, become violently unstable, and eventually blow itself apart. The idea of hyperstars brings to mind some of the seemingly fantastic speculations concerning the so-called "quasars", strong radio sources which look stellar in appearance but show enormous red-shifts. According to one school of thought, it is possible that the peculiar galaxies, radio galaxies, and quasars are all basically the same type of object, powered by the same mechanism. Too little is known at present to permit much more to be said. (For an account of quasars, refer to 3C273 in Virgo)

LIST OF DOUBLE AND MULTIPLE STARS

NAME	DIST	PA	YR	MAGS	NOTES	RA & DEC
Σ2665	3.3	17	43	6½- 9	relfix, spect A0	20170n1413
Σ2665b	0.3	253	45	9½- 10	(A1672) PA dec	
Hu 1197	1.1	303	21	7½- 14	spect A0	20196n1326
Σ2673	2.5	331	55	8 - 9½	relfix, spect F2;	20204n1311
					76" from Σ2674	
Σ2674	15.6	1	35	8 - 10½		20204n1311
Σ2679	23.5	78	55	7½- 8½	slow dist inc,	20221n1925
	39.1	152	10	- 12	spect A2	
β664	8.9	288	26	7- 12½	spect B9	20221n0521
Σ2680	16.2	289	55	8½- 8½	relfix, spect A0	20225n1442
Ho 131	4.1	324	38	7½- 11	AB cpm, spect C5	20260n1836
	95.8	81	21	- 10½		
1	0.9	346	58	6 - 8	(β63) relfix	20279n1044
	16.8	349	15	- 13	spect A0pe	
β987	2.4	128	41	7- 11½	relfix, spect B9	20280n1915
	22.1	71	24	- 11	AB cpm	
	106	288	15	- 7½		
A1675	0.2	316	62	7½- 7½	binary, 48 yrs;	20288n1538
					PA dec, spect A2	
Σ2690	16.7	256	58	7 - 7½	dist inc, spect A0	20288n1105
	23.5	106	58	- 12	(h269) (0Σ407)	
Σ2690b	0.2	261	67	8 - 8	PA inc (Da 1)	
Σ2696	0.6	300	67	8 - 8½	relfix, spect A2	20310n0516
β1208	3.0	333	23	7½- 12	spect B5	20321n0642
Σ2703	25.3	290	56	8 - 8	relfix, spect A5;	20345n1433
	73.7	235	52	- 8	AC dist inc	
Σ2701	2.1	222	65	8 - 8½	relfix, spect G5	20346n1152
β	0.6	346	74	4 - 5	(β151) binary,	20352n1425
	18.7	126	61	-13	PA inc, 26.6 yrs;	
	42.4	323	61	-11	Spect F5, AC dist	
					dec, AD dist inc	
κ	28.8	286	47	4½- 11	(0Σ533) optical,	20367n0955
	214	101	03	- 9	dist inc, AC cpm;	
					AC spect G5, dK1	
β288	6.8	159	34	7 -12½	optical, spect B4;	20368n1540
	43.8	279	13	-11	slow PA dec	
0Σ409	16.9	84	26	7- 10½	relfix, spect K0	20378n0316
	65.3	334	15	- 8		
Σ2713	4.9	63	39	9 - 9	relfix, spect B9	20385n1024
Σ2715	12.0	3	25	7½- 10	relfix, spect F8	20394n1220

LIST OF DOUBLE AND MULTIPLE STARS   (Cont'd)

NAME	DIST	PA	YR	MAGS	NOTES	RA & DEC
Σ2718	8.5	86	54	7½- 7½	relfix, spect F5	20402n1233
Σ2720	3.8	182	54	8½- 8½	relfix, spect F8	20412n1646
Σ2721	2.6	28	52	8 - 10	Spect G5, perhaps slight PA dec	20412n1942
Σ2722	7.5	307	62	8 - 8½	relfix, spect G5	20413n1933
Σ2723	0.9	119	66	6½- 8	PA inc, spect A0; A= 0.1" pair	20425n1208
Σ2725	5.7	9	61	7½- 8	PA & dist inc; Spect K0; in field with Gamma Delph.	20439n1543
γ	10.1	268	57	4½- 5	(Σ2727) fine pair with slight PA dec Spect K2, F8 (*)	20444n1557
13	1.6	194	58	5½- 9	(β365) slight PA inc, spect A1	20453n0549
Σ2730	3.3	335	61	8 - 8	relfix, spect K0	20486n0612
Ho 597	10.0	221	43	7½- 12	spect A3	20514n1924
Σ2735	1.8	285	61	6½- 7½	cpm; spect sgG6	20532n0420
16	38.2	21	11	6 - 12	optical, spect A4	20532n1223
J157	3.5	172	54	9½- 9½		20534n0823
Σ2736	5.1	218	39	7½- 8½	relfix, spect F2	20544n1248
Σ2738	14.9	254	51	7 - 8	relfix, spect A0	20562n1615
Σ2739	3.2	252	36	8½- 9	relfix, spect F0	20575n1952
h1608	19.8	258	10	7 - 11	spect F2	21025n1215
Σ2750	16.2	280	31	8½- 9½	relfix, spect K0	21026n1231
Σ2750b	0.6	122	62	9½- 10	(A1690) relfix	

DELPHINUS

LIST OF VARIABLE STARS

NAME	MagVar	PER	NOTES	RA & DEC
δ	4.5--4.56	.1351	Dwarf cepheid; Delta Scuti type, Spect A7	20411n1454
R	7.6--13.7	284	LPV. Spect M5e--M6e	20125n0856
S	8.3--13.2	277	LPV. Spect M6e	20408n1654
T	8.5--15.2	332	LPV. Spect M3e--M6e	20430n1613
U	5.7--7.6	Irr	Spect M5	20432n1754
V	8.1--16..	534	LPV. Spect M4e--M6e (*)	20455n1909
W	9.5--12.3	4.806	Ecl.bin; spect A0, G5	20354n1806
X	8.2--14.6	281	LPV. Spect M4e	20526n1727
Y	8.8--14.7	470	LPV. Spect M8e	20393n1142
Z	8.3--15.3	304	LPV. Spect S5e--S7e	20304n1717
RR	9.9--11.5	4.600	Ecl.bin; spect A	20412n1346
RS	8.2--9.0	60	Semi-reg; spect M8	20268n1606
RU	9.9--14..	259	LPV. Spect M3	20151n1001
RX	8.7--15..	186	LPV.	20276n1236
TW	9.5--10.8	77:	Semi-reg; spect M2	20329n0801
TX	9.2--9.8	6.167	Cepheid, spect G0--G5	20477n0328
TY	9.7--10.9	1.191	Ecl.bin. Spect B9	21020n1300
VW	9.5--14..	219	LPV.	20203n1727
AG	9.6--14..	238	LPV. Spect M4	20320n1824
BR	8.7--11.5	337	LPV. Spect M8e	20441n0412
CT	8.0--8.8	Irr	Spect M7	20270n0943
CZ	8.0--9.2	123	Semi-reg; spect M5	20312n0921
DM	8.6--8.9	.8447	Ecl.bin; Spect A3	20372n1415
DX	9.4--10.1	.4726	Cl.Var. Spect A9--F6	20451n1216
EI	9.3--10.6	20:	Semi-reg; spect M2	20238n0502
EU	6.0--6.9	60	Semi-reg; spect M5	20356n1805
HR	3.5--12.0	---	Nova Delphini 1967 (*)	20401n1859

LIST OF STAR CLUSTERS, NEBULAE, AND GALAXIES

NGC	OTH	TYPE	SUMMARY DESCRIPTION	RA & DEC
6891		◎	Mag 10, diam 15" x 7" Central star mag 11	20128n1235
6905	16[4]	◎	B,S,R, mag 12, diam 44" x 38"; 14[m] 0-type central star	20202n1957
6934	103[1]	⊕	Mag 9, diam 2'; B,S,R, rrr; Class VIII	20317n0714
7006	52[1]	⊕	Mag 11.5; diam 1', class I; B,S,R, gbM, stars eF. Very remote globular (*)	20591n1600

DESCRIPTIVE NOTES

BETA　　　Double star. Magnitude 3.78; spectrum F5 III or IV. Position 20352n1425. Beta Delphini is a close and rapid binary, generally a difficult object, but discovered, surprisingly, with an aperture of only 6 inches, by the indefatigable S.W.Burnham in August 1873. The two stars are magnitudes 4.0 and 4.9 and revolve in their orbit in a period of 26.65 years with periastron in late 1957. Greatest apparent separation of the pair occurs at PA 356° as in 1949; the stars are then 0.65" apart. Closest approach is near 90° as in 1959, when the components narrow down to about 0.2". Orbital elements according to Finsen (1938) are given here: Semi-major axis = 0.48" = about 20 AU at the adopted parallactic distance of 125 light years; eccentricity = 0.35; inclination = 62°. The total light of the system for the same distance is about 36 times the luminosity of the Sun. Beta Delphini shows an annual proper motion of 0.11"; the radial velocity is 14 miles per second in approach.

　　　The curious names "Sualocin" and "Rotanev" for Alpha and Beta Delphini, spell in reverse the name of Nicholas Venator (the Latinized version of Niccolo Cacciatore) the valued assistant of Piazzi at Palermo Observatory. The names, according to R.H.Allen, first appeared for these stars in the Palermo Catalogue of 1814.

DELPHINUS. The entire pattern of the Dolphin appears in this print, centered on β Delphini. The interloper is Comet West 1975n, photographed with the Lowell Observatory 5-inch camera on April 7, 1976.

NORTHERN PORTION OF DELPHINUS. The four bright stars below center are γ, α, δ and β. Near the top of the print, the brightest star is Nova Delphini 1967, near maximum in late October 1967. (See page 828) Lowell Observatory photograph

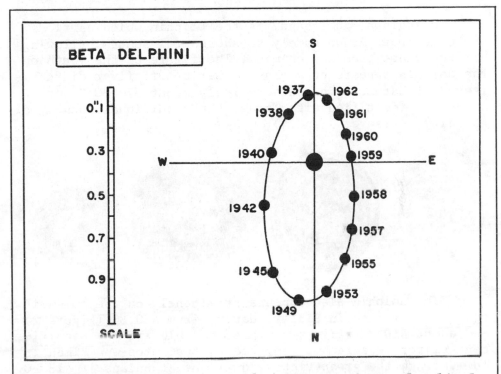

BETA DELPHINI

Alpha and Beta are the brightest members of a little
lozenge-shaped asterism some 2.7° in length, lying about
14° NE of Altair. The group includes Alpha (mag 3.77, B9),
Beta (3.78, F5), Gamma (3.91, K2+F8), and Delta (4.53, Am).
A few degrees to the south, the star Epsilon (3.98, B6)
completes the mythological outline of Delphinus, the Dolphin. How much of this group may be a true physical association remains debatable. Beta, Epsilon, Eta and Zeta all
show about the same radial velocity (about 12 miles per
second in approach) while Epsilon and Zeta show a nearly
identical proper motion, but Beta has a larger motion than
the other three. Gamma also seems disqualified on the basis
of a much larger motion than the others; whereas Alpha has
only about 1/3 the radial velocity of Beta and Epsilon. A
table of data for the brighter stars in the area is given
on the following page according to figures published in
the Yale Catalogue of Bright Stars. The reader may make his
own analysis as to the probability of any true physical
group existing among these stars. The computed distances
also seem discordant.

Delphinus, in legend, is the Dolphin which carried the Greek poet Arion safely to shore at Tarentum, allowing him to escape from his enemies. The figure of the youth on the dolphin appears on a classic series of silver didrachms issued at Tarentum in southern Italy about 370 BC, where Arion was identified with Taras, the traditional founder of the city. (Figures 1 and 4)

The dolphin also appears prominently on the splendid coins of Syracuse in Sicily, dating from 480 BC (Figure 2) and about 410 BC (Figure 3). Most notable are the magnificent silver decadrachms designed by Euainetos and Kimon to commemorate the great victory over the Athenians in 413 BC. This issue has been rightfully considered the supreme example of the art of coinage, and possibly the most beautiful coin of all time:

Delphinus is mentioned by Chaucer as *Delphyn*; it is *Dauphin* in France and *Delfino* in Italy; to the Latins it was *Delphis* or *Delphin*. Ovid referred to the constellation as *Amphitrite*, the Goddess of the Sea; the Greeks called it *Vector Arionis*, while Riccioli has it labeled *Hermippus* on his star charts. According to R.H.Allen the Hindus knew it

DESCRIPTIVE NOTES (Cont'd)

as *Shi-shu-mara* or *Zizumara*, the Porpoise; to the Arabians however it was *Al Ka'ud*, the Riding Camel. In early Hebrew tradition it was sometimes identified with the *Great Fish* of Jonah, while early Nestorian Christians identified it with the Cross of Christ. Delphinus is also popularly called *"Job's Coffin"* though the origin of this name does not appear to be known today.

| STAR | Mag. | Spectrum | Yearly Proper Motion | | Radial |
			RA	DEC	Velocity
α	3.77	B9 V	+0.062"	-0.003"	-6 km/sec
β	3.78	F5 IV	+0.106	-0.034	-23 "
γ	3.91	K2,F8	-0.032	-0.197	-8 "
δ	4.53	Am	-0.025	-0.048	+9 "
ε	3.98	B6 III	+0.007	-0.022	-19 "
ζ	4.62	A3 V	+0.007	-0.002	-22 "
η	5.22	A2 V	+0.067	+0.026	-18 "
κ	5.02	G5 IV	+0.313	+0.015	-52 "
θ	5.82	K3 Ib	-0.006	0.004	-14 "
ι	5.42	A2 V	+0.035	-0.007	-4v

GAMMA Mag 3.91; spect K2 IV and F8 V. Position 20444n1557. The most attractive double star in Delphinus, located at the NE corner of the diamond-shaped pattern formed by Alpha, Beta, Gamma and Delta. Gamma is a physical pair of 10" separation, magnitudes 4.3 and 5.1, easy in any good small telescope and first noted by F.G.W.Struve in 1830. The components show a common proper motion of 0.20" per year in PA 189° but the relative alignment of the two stars has changed by only 5° in the last 130 years. Color estimates for the pair have ranged from South's "white and yellowish" to Gore's "reddish-yellow and greyish lilac". Modern observers generally find both stars yellowish, the fainter star often appearing slightly tinged with green.

Trigonometric and spectroscopic parallaxes agree in giving a distance of about 100 light years; the resulting absolute magnitudes of the stars are +1.8 and +2.6, equal to luminosities of 16 and 8 suns. Gamma Delphini shows an approach radial velocity of 4.5 miles per second, and the projected separation of the two stars is just over 300 AU.

In the same field with Gamma Delphini, about 15' to the south and slightly west, lies the closer but fainter double star Σ2725. The components of this pair have shown a PA increase of about 10° since 1830, about twice the orbital motion observed for Gamma itself. The two doubles appear to be at approximately the same distance from us, but are evidently not physically related since Σ2725 has a smaller proper motion in a different direction: 0.14" per year in PA 50°.

V   Position 20455n1909. A long period pulsating red variable star of the Mira class, discovered by Fleming at Harvard in 1891. It is located about 3° north of the bright double star Gamma. V Delphini is remarkable for its unusually large range of about 9 magnitudes, over 4000 times in light intensity between maximum and minimum. The light curve shows large differences in brightness from one maximum to the next; sometimes the star nearly reaches 8th magnitude while at another cycle it may scarcely exceed 13th. Observers must therefore expect to find the star totally invisible on occasion; at a low minimum it may be beyond the range of any but the greatest telescopes. The minimum of August 1900 was 17.1 as observed at Yerkes with the 40-inch refractor; this was probably the faintest minimum ever recorded visually for any of the regularly observed long period variables. The spectral type of the star varies from M4 to about M6, and shows the usual bright lines of hydrogen which are characteristic of stars of the class. V Delphini has an estimated absolute magnitude of about -2 at maximum, indicating a distance of nearly 5000 light years. As in the case of most of the variables of abnormally large range, the period is noticeably longer than average, about 534 days. (Refer also to Omicron Ceti)

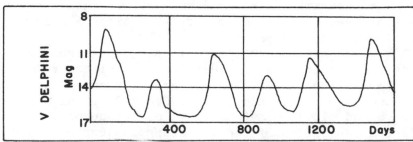

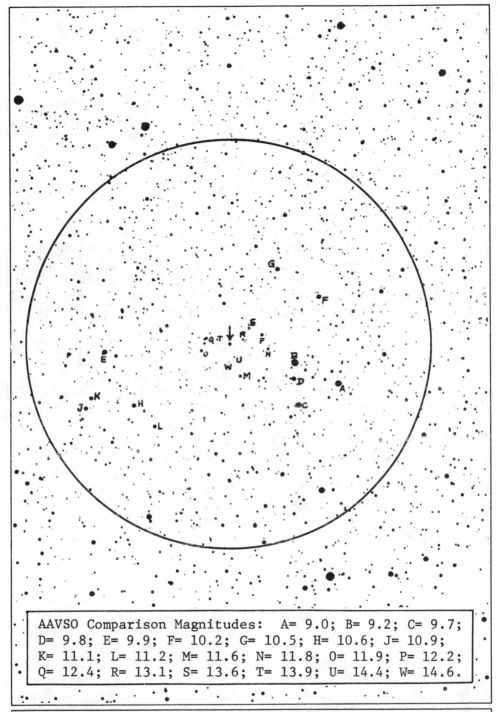

AAVSO Comparison Magnitudes: A= 9.0; B= 9.2; C= 9.7;
D= 9.8; E= 9.9; F= 10.2; G= 10.5; H= 10.6; J= 10.9;
K= 11.1; L= 11.2; M= 11.6; N= 11.8; O= 11.9; P= 12.2;
Q= 12.4; R= 13.1; S= 13.6; T= 13.9; U= 14.4; W= 14.6.

V DELPHINI. Identification field made from a plate obtained with the 13-inch telescope at Lowell Observatory. North
is at the top; circle diameter = 1°. Limiting magnitude
about 15.

DESCRIPTIVE NOTES   (Cont'd)

HR Nova Delphini 1967. Position 20401n1859. A very
unusual nova, remarkable for its slow development
and a maximum lasting more than a year; consequently one of
the most widely observed modern novae. The star was first
noticed by the experienced English amateur and comet discoverer G.E.D.Alcock on July 8,1967, as an object of magnitude 5.6; pre-discovery photographs of the region showed
that the nova had been magnitude 6.7 on June 25 and about
6.0 on July 3. A check of older patrol plates revealed a
star of magnitude 11.9 at the exact position; this object
had apparently varied slightly for a number of years and
the spectral type was O or B. No previous maximum was found
in a check of plates made from 1890 to 1966, so the star is
probably not a recurrent nova. The rise to 5th magnitude
required about 30 days.

During the first six months of its career, the
nova remained near maximum light with slight variations. In
August 1967 it brightened about half a magnitude to 5.0 and
in mid-September it rose still further to 4.6, followed by
a slight fading to 5.1 in mid-October. After another slight
increase to 4.8 in early November the star brightened rapidly to its ultimate peak magnitude of 3.5 in mid-December.
Following this outburst the star began a pattern of oscillations with well defined maxima occurring at intervals of
about 10 days; in March and April 1968 the activity became
more erratic with changes of more than half a magnitude
occurring from one night to the next. The fluctuations died
away in June 1968 and the nova, then at 6th magnitude,
began to fade steadily but slowly. In March 1969 it was
still magnitude 8.3; by 1975 it was close to 11.5.

Although the range of only 7 magnitudes seems
unusually small, the normal spectral features of a nova
were present, including evidence for 4 sets of expanding
gaseous shells with velocities of 325, 380, 420 and 485
miles per second. The distance is not accurately known so
the true luminosity of the star is uncertain, but it is
generally thought that the very slow novae are considerably
less luminous than the very fast ones such as Nova Aquilae
1918 which show a single sharp maximum and a rapid decline.
Perhaps the star most closely resembling HR Delphini was
DO Aquilae (1925) which had a long flat maximum lasting for
some 200 days.

NOVA DELPHINI 1967. The star is shown in its normal 12th magnitude state in August 1937, at at maximum in October 1967. Lowell Observatory 13-inch telescope photographs.

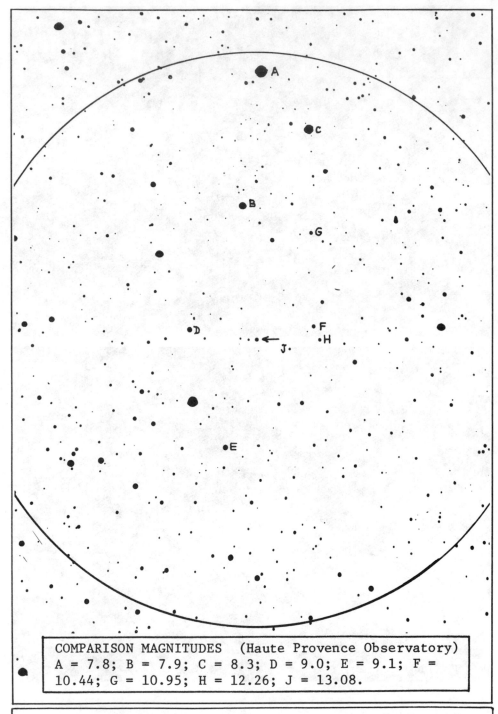

COMPARISON MAGNITUDES (Haute Provence Observatory)
A = 7.8; B = 7.9; C = 8.3; D = 9.0; E = 9.1; F = 10.44; G = 10.95; H = 12.26; J = 13.08.

NOVA DELPHINI Identification chart, from a 13-inch telescope plate made at Lowell Observatory. Circle diameter =1°, North is at the top; limiting magnitude about 15.

DESCRIPTIVE NOTES   (Cont'd)

NGC 7006   Position 20591n1600. An exceedingly remote globular star cluster, probably the farthest object of its type with the exception of the well known NGC 2419 in Lynx, which may actually be an extra-galactic object. NGC 7006 appears to be about 150,000 light years from the center of our Galaxy and some 185,000 light years from the Solar System. This is almost comparable to the distances of the Magellanic Clouds and suggests that these two clusters might possibly be regarded as extra-galactic objects.

Despite its vast distance, NGC 7006 is visible in a 6-inch telescope as a fuzzy spot of magnitude 11.5, about 1' in apparent size. The diameter increases to about 2.2' on long exposure photographs, giving a true size of about 120 light years. This is one of the most difficult of all globulars to resolve, since even the brightest members are of the 16th magnitude. Allowing for a certain loss of light through absorption by interstellar dust, the true luminosity of the cluster appears to be about 130,000 times that of the Sun, and the absolute magnitude is near -8. The integrated spectral type is F1. This cluster shows a very large radial velocity of 215 miles per second in approach.

Harlow Shapley, in his book "Star Clusters" (1930) pointed out the interesting fact that the extreme dimensions of our Galaxy are indicated by the extent of its globular cluster system. Shapley's observations of eleven RR Lyrae type stars in NGC 7006 showed it to be about five times as remote as M3 or M5, certainly the most distant globular known at the time. On the other side of the Galaxy the most distant system appears to be NGC 2298 in Puppis; the separation of these two globulars must be slightly over a quarter of a million light years. The size of the Galaxy is thus more than doubled when these clusters are used as boundary markers! The validity of this approach, however, now appears somewhat doubtful owing to the recent discovery of a number of other very distant and extremely faint globulars on the Palomar 48-inch Sky Survey plates. Some of these are evidently true inter-galactic objects, so it now appears possible to regard NGC 7006 and 2419 as either the most distant members of the Galaxy or the nearest of the intergalactic clusters. (Refer also to NGC 2419 in Lynx and M13 in Hercules)

NGC 7006 in DELPHINUS. One of the most remote globular
star clusters, photographed with the 48-inch Schmidt
Camera at Palomar Observatory.

DORADO

LIST OF DOUBLE AND MULTIPLE STARS

NAME	DIST	PA	YR	MAGS	NOTES	RA & DEC
Fin 87	0.6	133	52	6 - 9	PA dec, spect F5	04174s5259
Rmk 4	5.9	242	55	7 - 7½	PA slow inc, spect G0	04232s5711
Slr 6	1.0	99	42	7 - 9	relfix, spect F0	04243s5313
h3683	2.3	91	46	7 - 7½	binary, about 550 yrs; dist inc, PA dec, spect G5	04394s5902
h3686	7.3	220	36	8½- 8½	relfix, spect A0	04410s6119
Cp 4	2.8	40	42	7 - 9	slight PA inc, spect F0	04470s6134
I 342	2.1	148	51	8 - 8½	PA dec, spect G5	04485s5359
h3755	21.7	280	16	7½- 12	spect K0	0517/s6200
I 276	1.2	172	59	6½- 7	PA dec, spect F0	05273s6840
L1922	9.2	68	30	6½- 12	(Hd 192) Spect F0	05297s6358
I 1152	0.5	213	27	9 - 10	(h3796) Multiple	05390s6908
	0.9	168	27	- 14	group near nebula	
	1.3	110	27	- 13½	NGC 2070 in Large	
	2.6	116	27	- 12½	Magellanic Cloud	
	7.4	346	27	- 13½		
	11.6	185	27	- 13½		
	11.1	297	27	- 13		
	11.6	291	27	- 13		
	12.2	52	27	- 13½		
	20.3	4	27	- 11½		
	22.2	47	27	- 12½		
	26.1	136	27	- 12½		
I 745	0.8	251	42	7½- 10½	Spect F8	05427s6710
Slr 15	0.2	286	59	7½- 8	PA & dist dec, spect A0	05528s6150
△26	20.9	117	17	7 - 8½	slow PA inc, spect F5, G	06121s6531

DORADO

LIST OF VARIABLE STARS

NAME	MagVar	PER	NOTES	RA & DEC
β	3.8---4.7	9.842	Cepheid, spect F6- G5 (*)	05332s6231
R	6.0--7...	340	Semi-reg; spect M7	04362s6211
S	8.4---9.5	Irr	Spect A0e; super-lumin-ous star in cluster NGC 1910, in Large Magellan-ic Cloud (*)	05186s6918
T	9.3--13..	168	LPV. Spect M3e	04451s5953
U	8.0--14..	394	LPV. Spect M7e	05099s6423
SU	9.0--12..		LPV.	04465s5546
WZ	5.1---5.3	40:	Semi-reg; spect gM4	05070s6328

LIST OF STAR CLUSTERS, NEBULAE, AND GALAXIES

NGC	OTH	TYPE	SUMMARY DESCRIPTION	RA & DEC
1515	△348	⊘	Sb; 12.1; 5.0' x 1.1' B,L,vmE, bM	04027s5414
1533		⊘	S0; 12.3; 2.1' x 1.9' vB,vL,R,bN	04088s5615
1546		⊘	S0; 12.5; 1.8' x 0.7' pB,pL,gmbM,BN	04136s5611
1549		⊘	E1; 11.0; 2.8' x 2.5' B,pS,R,BN; 13' pair with 1553	04147s5542
1553		⊘	S0; 10.2; 3.1' x 2.3' vB,pS,R,bM	04152s5554
1566	△338	⊘	Sb; 10.5; 5.0' x 4.0' vB,L,vmbM, SBN	04189s5504
1596		⊘	E7/S0; 12.3; 2.9' x 0.7' B,pL,mE, SBN	04266s5507
1574		⊘	S0; 12.2; 1.5' x 1.5' pB,S,R, pgbM	04210s5705
1617	△339	⊘	SBa; 11.7; 3.5' x 1.6' B,L,mE, smbM, BN	04306s5442
1672	△296	⊘	SBb; 11.4; 4.0' x 3.5' B,L,smbM, vBN	04449s5920

LIST OF STAR CLUSTERS, NEBULAE, AND GALAXIES (Cont'd)

NGC	OTH	TYPE	SUMMARY DESCRIPTION	RA & DEC
1688		⊘	SB; 12.7; 2.0' x 1.6' pB,pL,1E, mbM	04476s5953
1714		◎	Mag 10; diam 8"; vB,S,E	04520s6701
1743	△114	◎	Diam 15"; B,pL,R, bM; on edge of Magellanic Cloud	04544s6917
1796		⊘	SBb; 12.9; 1.5' x 0.8' pF,pS,pmE, glbM	05021s6112
1910	△129	⠿	L,Ri, irregular cluster in Magellanic Cloud; contains variable S Doradus (*)	05185s6916
----	LMC	⊘	Large Magellanic Cloud (Nubecula Major) Irregular galaxy, total mag about 1.0; diameter 6° (*)	05200s6900
1936		☐	Giant stellar association in Large Magellanic Cloud; stars mags 11.... with bright nebulosity	05224s6801
1947		⊘	S0; 12.2; 1.8' x 1.5' pB,L,R, with dark lane	05260s6349
2070	△142	☐	↓↓↓ vB,vL,Irr; diam 20'; "Tarantula Nebula"; in Large Magellanic Cloud (*)	05399s6904
2082		⊘	SBb; 12.8; 1.4' x 1.3' F,pL,R	05416s6420
2369		⊘	SBa; 13.1; 4.1' x 1.0' pF,pS,1E, glbM	07160s6216

DORADO

DESCRIPTIVE NOTES

BETA Variable. Position 05332s6231. Beta Doradus is a bright cepheid variable star, one of the half dozen brightest stars of the class in the entire heavens, but almost unknown among North American observers owing to its position in the far southern sky. It is located about 8° north of the Large Magellanic Cloud and about 12° SW from Canopus. The star is visible to the naked eye during every part of its cycle; the magnitude range is 3.8 to 4.7 visually, and 4.5 to 5.7 photographically. Beta Doradus has a precise period of 9.84235 days, and the spectral class varies from F6 Ia at maximum to about G5 at minimum. These variations greatly resemble those of the pulsating star Delta Cephei, the standard star of the class, to which the reader is referred for a brief review of the cepheid variables. Convenient comparison stars for Beta Doradus are:

Delta Doradus=	4.34	Lambda Doradus=	5.13
Zeta Doradus =	4.71	Gamma Pictoris=	4.50
Eta-2 Doradus=	4.88		

The cepheid period-luminosity relation (See page 591) implies an absolute photographic magnitude of about -3.5 for the star at midrange, and a peak visual luminosity of nearly 7000 suns; the resulting distance is close to 1700 light years. Beta Doradus shows a very small annual proper motion of less than 0.01"; the radial velocity is about 4 miles per second in recession. S.Gaposchkin (1958) reports a secondary fluctuation in the light curve of the star, an effect which causes some cycles to be as much as 0.2 magnitude higher than others. (See also Delta Cephei)

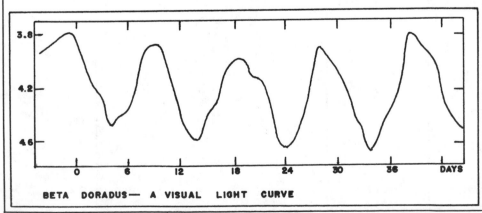

BETA DORADUS— A VISUAL LIGHT CURVE

**THE LARGE
MAGELLANIC CLOUD**
(NUBECULA MAJOR) One of the nearest of the external galax-ies, a famous irregular sys-tem which appears to the naked eye like a detached Milky Way star cloud, and bright enough to be visible even in full moonlight. It is located at RA 5h 20m, Dec -69°. The Large Cloud and its fainter companion- the Small Cloud in Tucana-were discovered by Portuguese seamen in the 15th century and later named in honor of Ferdinand Magellan. Owing to their position in the far southern sky, the Clouds are not accessible to observers in the U.S.; much of our knowledge concerning them has resulted from observations made with the 74-inch reflectors at Mt.Stromlo in Australia and at Radcliffe in South Africa.

The Magellanic Clouds are irregular galaxies, members of the Local Group, and close enough to be regarded as possibly satellites of the Milky Way. According to recent studies (1965) the Large Cloud is at a distance of about 190,000 light years, and the Small Cloud about 200,000. The two Clouds are some 22° apart in the sky, corresponding to an actual separation of about 80,000 light years, center to center. These are the nearest of the external galaxies, with the single exception of a nearby system discovered by radio studies in 1975, lying about 70,000 light years away some 8° NE from Betelgeuse in Orion. This dwarf galaxy, now the nearest system known to our Milky Way, has been given the appropriate- and irreverent- name of *Snickers*.

The Large Cloud is a gigantic system, rivalling an average spiral galaxy in size and luminosity. The central core or bar measures over 20,000 light years in length and the outer haze of stars and clusters increases the total size to at least 50,000 light years. Through the telescope the Cloud is revealed as a marvelous aggregation of stars, nebulae, groups, and clusters, including some supergiant stars which exceed anything known in our own Galaxy, and vast gas and dust clouds where new stars are even now in the process of formation. This remarkable galaxy is an astronomical treasure-house, a great celestial laboratory for the study of the growth and evolution of the stars. The Clouds are at least 10 times closer than the Andromeda Galaxy M31, and a 20-inch telescope used on the Clouds is the equal of the 200-inch used on the Andromeda System.

THE MAGELLANIC CLOUDS. The Large Cloud (below) and its smaller companion, photographed with a 3-inch lens at the Southern Station of Harvard Observatory in South Africa.

DESCRIPTIVE NOTES (Cont'd)

The true shape and classification of the Large Magellanic
Cloud is an interesting question. Although it is often
classed simply as "irregular" the elongate core strongly
resembles the central "bar" of a barred spiral galaxy. There
is some evidence for the beginning of a spiral pattern at
the ends of the bar, and wide-angle photographs record a
faint curved streamer extending some 20° to the NW, seemingly an incipient spiral arm. This unusual feature points
out in a direction away from the center of our own Galaxy,
and measures something like 50,000 light years in length.
The discovery of this faint spiral pattern in the Cloud is
extremely interesting to astronomers studying the problems
of the dynamics and evolution of galaxies. The Cloud seems
to show us a very early stage in the development of a large
barred spiral galaxy.

Studies made in 1964 confirm an earlier suspicion that
the Cloud is actually a much-flattened system whose plane
lies about 40° or 50° from the edge-on position. The total
mass of the Large Cloud is estimated to be about 10% the
mass of our own Galaxy, or about 25 billion suns. With an
absolute magnitude of about -18 the Cloud shines with the
total light of some 2 billion suns.

Radial velocity measurements reveal a motion of about
170 miles per second in recession, but this is almost
entirely the result of the motion of the Sun around the
center of our own Galaxy. Applying this correction, we
find that the Large Cloud is not measurably changing its
distance from the Milky Way system, but is very likely in
orbit about it in some multi-million year period. It also
appears that the Small Cloud forms a gravitationally revolving pair with its larger neighbor. The two Clouds seem to
be loosely linked by a number of scattered stars and clusters, and radio observations indicate that the gaseous halo
surrounding each cloud is extensive enough to merge gradually into that of the other. Within this common envelope, the
more striking condensations of dust and gas appear to be
distributed rather unevenly. The Small Cloud may contain as
much gas as the Large one, but appears to be relatively
free of obscuring dust clouds, and contains only a few nebulous regions. In the Large Cloud, the most striking areas
are the great glowing "H-II regions" where vast complexes
of nebulosity are illuminated by groups of supergiant stars.

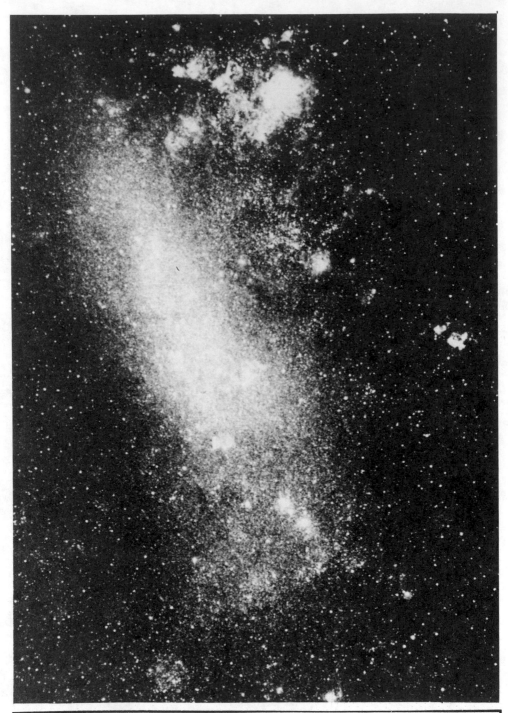

THE LARGE MAGELLANIC CLOUD. One of the nearest members of
the Local Group of Galaxies. The bright object near the
top is the vast Tarantula Nebula NGC 2070. Photograph in
the light of hydrogen with the Mt.Wilson 10-inch refractor.

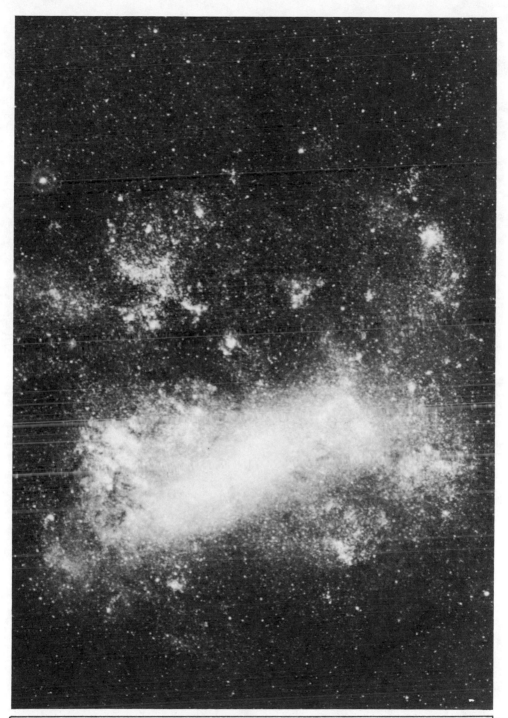

OUTER REGIONS OF THE MAGELLANIC CLOUD. Many outlying star groups, clusters, and nebulous regions are shown in this photograph, made at the southern station of Harvard Observatory.

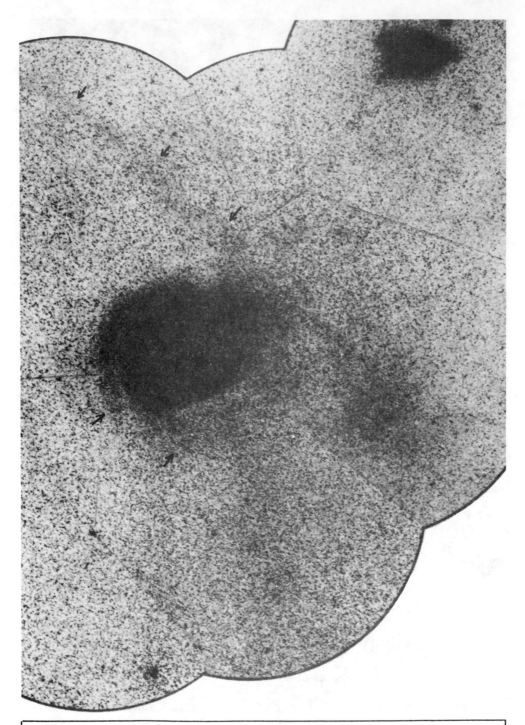

SPIRAL STRUCTURE IN THE LARGE MAGELLANIC CLOUD. A mosaic photograph obtained with the 2-inch Aero-Ektar lens at Mt. Stromlo in Australia.

DESCRIPTIVE NOTES (Cont'd)

A survey of the giant stars in the Large Cloud strikingly illustrates the great richness of this neighboring galaxy. There are at least 200,000 stars brighter than absolute magnitude -1.5 (600 times the light of the Sun) and at least some 15,000 which exceed absolute magnitude -4.0 (3600 times the luminosity of the Sun). For the true supergiants the survey is essentially complete. There are about 750 stars in the Cloud which exceed absolute magnitude -6.5 (36,000 times the solar luminosity). The most brilliant of these stars - usually claimed to be the most luminous star known at present anywhere in the Universe - is the famous star S Doradus, the chief star of the cluster NGC 1910. The cluster is located on the northern rim of the central bar of the cloud (See photograph, page 844) and is a brilliant aggregation nearly 250 light years in diameter, containing over 100 giant and supergiant stars. S Doradus is a variable of a peculiar sort, with an odd spectrum rather resembling that of P Cygni; the spectral type is near A0, but with bright hydrogen lines. The apparent brightness changes irregularly from magnitude 8.4 to about 9.5; the mean luminosity thus averages about 500,000 times the light of the Sun, and has occasionally exceeded one million times. However it should be remembered that there is no positive proof that S Doradus is actually a single star; it is very possible that we are dealing with a double or multiple system.

In a study of over 1100 plates in the Harvard collection, S.Gaposchkin (1943) obtained a light curve which could be interpreted as that of an eclipsing binary with a period of 40.2 years, an eccentricity of 0.4, and showing primary minima in 1900 and 1940, and secondary minima in 1890 and 1930. The two minima differ only slightly in depth, with an amplitude of about 0.65 magnitude. Each minimum lasts for over three years. The chief difficulty in the interpretation lies in the abnormally large diameters required to explain the length of the eclipses; Gaposchkin derived diameters of 1400 and 1260 times the Sun for the two stars, and masses of 60 and 55 suns. At the present time, the validity of the eclipsing binary interpretation remains uncertain, although the expected minimum occurred reasonably on schedule in the late 1960's; the star returned to normal in 1973. S Doradus, in any event, shows other variations which cannot be attributed to eclipses, changes

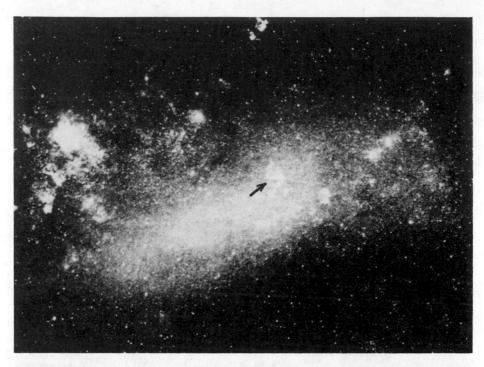

S DORADUS. Top: The position of cluster NGC 1910 in the
large Magellanic Cloud. Below: The cluster, with the super-
luminous star encircled.

                    Boyden Station, Harvard Observatory

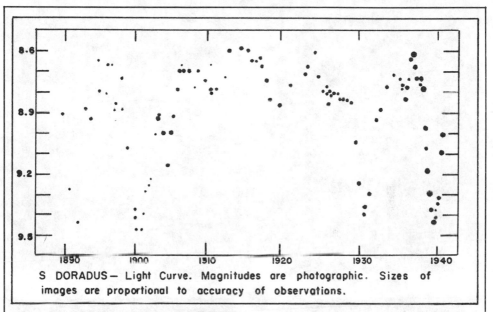

S DORADUS— Light Curve. Magnitudes are photographic. Sizes of images are proportional to accuracy of observations.

of 0.3 magnitude were clearly evident between 1910 and 1930.
In recent years the star has shown a gradual fading, following its return from the deep minimum of 1940, and has
occasionally surrendered its title of "most luminous star"
to one of the other super-luminaries in NGC 1910.

 There are eight other stars in the Large Cloud which
are brighter than the 10th apparent magnitude, and which
must therefore exceed the Sun in luminosity by a factor of
250,000 times. Four of these objects are blue supergiants,
and four are red supergiants, rapidly evolving stars which
must possess masses near the theoretical limit. As a standard of comparison, it may be remembered that our Sun, at
a distance of 190,000 light years, would appear as a star
of magnitude 23.6. It is almost overwhelming to realize
that these enormous supergiants must be greatly outnumbered
by stars of the solar type and fainter which cannot be seen
at all at such a distance; that for every bright giant there
must be thousands of ordinary stars which go undetected; and
that the total population of the Cloud must therefore be at
least 25 or 30 billion stars.

 The study of variable stars in the Clouds is a fascinating chapter in astronomical history. With the 24-inch
Bruce refractor at Harvard's Arequipa station in Peru, early
photographic studies of the Magellanic Clouds were begun in

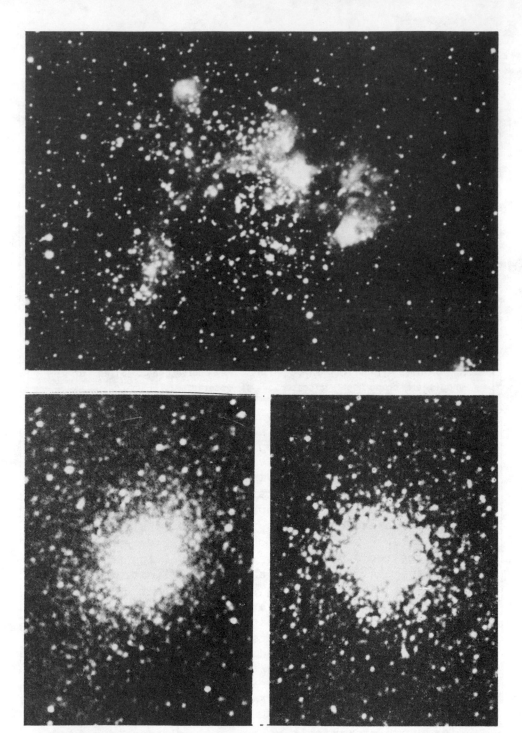

STAR CLUSTERS IN THE LARGE MAGELLANIC CLOUD. Above: The
stellar association NGC 1936 photographed at Mt.Stromlo.
Below: The globular clusters NGC 1978 and NGC 1866, photographed at the southern station of Harvard Observatory.

DESCRIPTIVE NOTES   (Cont'd)

1899. Many variables were detected on the Bruce plates by Henrietta Leavitt of the Harvard Staff. In 1906 a published catalog listed 808 variables in the Large Cloud and 969 in the Small Cloud. Light curves and periods were eventually derived, and these stars are known today to be typical cepheids of Population I, comparable in type and luminosity to the cepheids in our own Galaxy. It was the finding of such pulsating stars in the Small Cloud which led to the discovery of the period-luminosity relation for cepheid variables, and gave astronomers the key to the measurement of vast stellar distances. (Refer to page 591)

At the distance of the Clouds, the search for the fainter RR Lyrae type stars (cluster variables) is more difficult, and has been unsuccessful until relatively re- cently. Such stars, at apparent magnitude 19.5, are at the limit of detection with either of the two 74-inch reflect- ors of the southern hemisphere. In the years between 1951 and 1958 a study at Radcliffe finally identified a number of RR Lyrae type stars associated with the cloud's globular star clusters. Thus all speculation concerning their sup- posed absence in the Magellanic Clouds is now ended. This also demonstrates that the Clouds are not absolutely pure Pop.I systems, as has sometimes been claimed. Several novae have been seen in the Cloud as well, and the luminosities are comparable to those in our own Galaxy.

The cluster NGC 1910, mentioned in connection with the super-luminous star S Doradus, is only one of a vast multitude of clusters which ornament the Large Cloud. In a survey made in 1960, Shapley and Lindsay listed some 700 open clusters, a number of them connected with nebulous regions. Some of these aggregations are so enormous that the term "cluster" is probably inapplicable ; they are more nearly comparable to the large "associations" such as the whole Orion complex in our own Galaxy. One of these objects is concentrated in the nebulosity NGC 1936; this is the group which Shapley has called "Constellation I". This great association has a diameter of about 500 light years and a total mass of about 24,000 solar masses, plus another 60,000 solar masses in the ionized gases of the surrounding nebulosity. In addition, radio observations indicate that a mass of about 5 million suns is present in the form of neutral hydrogen, in a region 1000 light years in diameter centered on the association. From these gas clouds, new

stars are being formed, and it is an exciting fact that
the several hundred members of Constellation I are all blue
giants. Six stars of the group have luminosities higher
than 100,000 suns (absolute magnitude -7.7). Here undoubtedly, is one of the regions of space where stars are born.
    In additon to the open clusters of the Large Cloud,
about 60 clusters of the globular type have been identified
and the majority seem to have stellar populations similar
to the globulars of our own Galaxy. A few, however, are
distinctly abnormal, resembling globulars in appearance,
but containing bright blue stars and showing a color-magnitude diagram suggesting a galactic cluster. It is tempting
to identify these objects as "newborn" globulars, but it
is not certain that the full explanation is quite that
simple. The brightest globular in the Large Cloud is NGC
1866, shown on page 846. This cluster is one of the anomalous "blue" types, and also differs from an orthodox globular by containing at least a dozen classical cepheid variables.
    Turning now to the diffuse nebulae, there are at least
50 examples bright enough to be visible in the Cloud with
moderate telescopic equipment. Some 400 planetary nebulae
have been recognized, and one peculiar object which from
its appearance and strong radio emission is a probable
supernova remnant. It is a ring-shaped nebulosity about
7' in diameter with a central knot of faint stars. It is
#70 in a list of nebulae in the Magellanic Clouds published
by K.G.Henize in 1956, and is located well away from the
main mass of the Cloud, to the north, at 05434s6755. Though
the appearance is suggestive of a planetary nebula, the
computed diameter of over 400 light years makes such an
identification extremely unlikely. The most probable explanation for such an object is the explosion of a supernova,
which must have occurred in the very distant past.

THE TARANTULA NEBULA. Among the 50 or so bright diffuse
nebulae which appear scattered across the face of the Large
Magellanic cloud, one object stands forth without an equal;
not only in the Cloud, but, as far as we know, in the entire
Universe. NGC 2070, also called "30 Doradus", the Great
Looped Nebula, or the Tarantula Nebula, is visible to the
naked eye at a distance of 190,000 light years. If such an

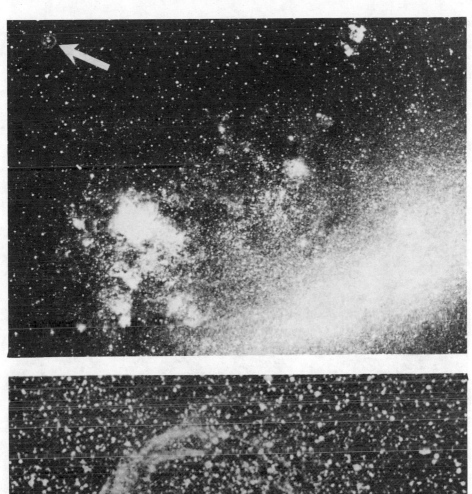

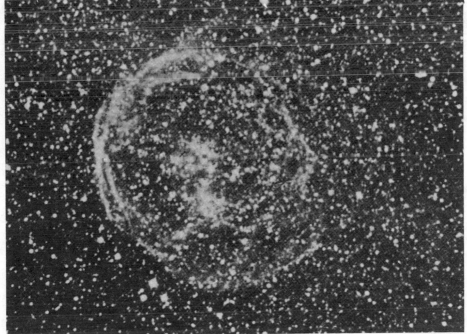

THE RING NEBULOSITY IN THE LARGE MAGELLANIC CLOUD. The
position is indicated in the upper photograph; the close-
up of the Great Ring was obtained with the 74-inch telescope at Radcliffe Observatory.

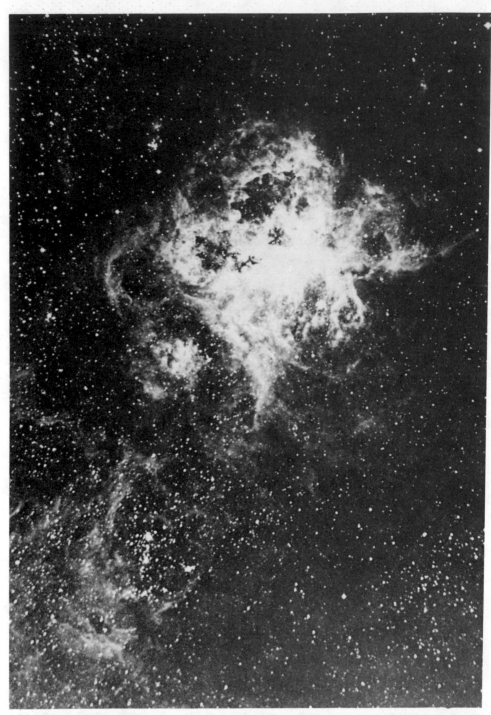

THE TARANTULA NEBULA. NGC 2070, the greatest known diffuse nebula, and the most striking single feature of the Large Magellanic Cloud.

Radcliffe Observatory, 74-inch reflector

DESCRIPTIVE NOTES (Cont'd)

object were as close to us as the Orion Nebula, it would
cover some 30° of the sky, and shine with a total brightness three times greater than that of Venus. The diameter
of the Tarantula is some 800 light years, while the outer
streamers increase the total size to about 1800 X 1700
light years. From radio observations the total mass is
known to approach 500,000 solar masses, and to a radio
telescope the Tarantula stands out more clearly than any
other feature of the Cloud. This nebula is extremely complex in form with much structural detail in the shape of
extending filaments and streamers. In the center lies a
cluster of over 100 supergiant stars, covering an area some
100 light years in diameter. In this vast nebula, as in
Shapley's "Constellation I", the drama of star birth and
evolution is taking place before our eyes.

The Tarantula is the largest diffuse nebula known anywhere in the Universe. Our own Galaxy contains nothing
comparable, but similar objects on a slightly smaller scale
are known in a few other external galaxies. The nebulosity
NGC 604 in the "Pinwheel" Galaxy M33 is one of the better
known examples. It is, of course, much more distant than
the Tarantula, but the true size is still only about half
as great. (Refer also to the Small Magellanic Cloud in
the constellation of Tucana)

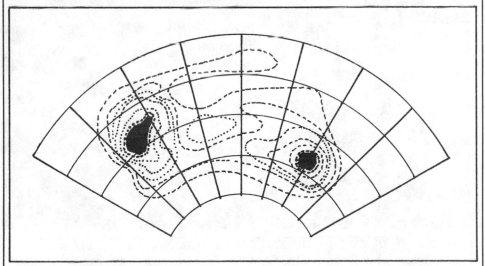

RADIO OBSERVATIONS reveal the common hydrogen envelope of the two
Magellanic Clouds. Wavelength = 21 centimeters

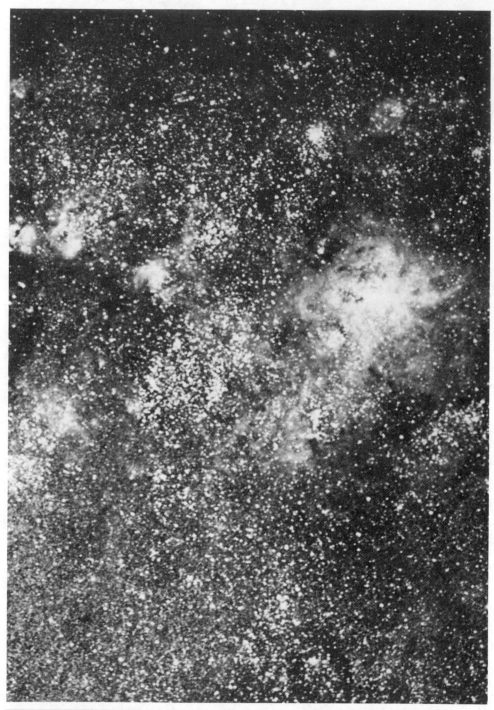

STAR FIELD IN THE LARGE MAGELLANIC CLOUD. A brilliant
field near the Tarantula Nebula, showing great numbers of
supergiant stars.

Southern Station, Harvard Observatory

DRACO

LIST OF DOUBLE AND MULTIPLE STARS

NAME	DIST	PA	YR	MAGS	NOTES	RA & DEC
Σ1362	4.8	129	63	7 - 7	relfix, spect F0	09332n7318
Σ1408	3.5	12	38	8½ -9½	cpm pair, relfix, spect F2	10071n7317
Σ1437	23.5	290	55	7 - 9½	relfix, spect A3	10302n7406
Σ1516	36.2	102	40	7 - 7½	(OΣ539) AB pair	11121n7345
	7.1	306	35	- 11	optical, dist inc; AC binary, PA inc; spect K5	
Σ1573	11.2	178	53	6½- 7½	cpm pair, relfix; spect F8	11465n6736
β794	0.4	118	62	7 - 8½	binary, 79 yrs;	11510n7402
	5.7	72	00	- 13½	PA inc, spect F7	
A75	0.3	37	24	7½- 8	binary, 79 yrs; PA dec, spect F5	11514n7213
A1088	0.3	252	66	7 - 8	PA inc, spect F0	11581n6928
Σ1590	5.1	235	44	7 - 10	relfix, spect K0	11590n7107
Σ1588	12.9	45	58	8½- 8½	PA & dist dec, K2	11597n7238
Σ1599	10.2	346	66	7 - 10	relfix, spect K5	12030n6904
Σ3123	0.3	100	62	7 - 7	binary, about 120	12035n6859
	3.0	309	24	- 15	yrs; PA dec, spect	
	26.0	181	53	- 8	F5; D= 0.2" pair.	
Σ1602	17.1	180	54	7½- 9	dist inc, spect G5	12046n6921
Σ1654	3.7	24	38	7½- 8½	relfix, spect K0; cpm pair	12300n7505
OΣΣ123	68.9	147	24	6½- 7	field glass pair; cpm; Spect F2, F3	13250n6500
Σ1767	4.2	347	63	8 - 8½	PA slow dec; cpm; spect G0	13327n6801
Σ1860	1.0	108	60	7½- 8½	PA slight inc; spect A5	14323n5527
Σ1872	7.6	47	63	7 - 8	PA inc, spect K0	14396n5811
	76.0	58	25	- 11		
Σ1878	4.1	322	58	7 - 8½	PA dec, dist inc; spect F2	14408n6129
Σ1882	12.2	2	31	7 - 8	slight dist inc;	14428n6119
	9.1	83	16	- 11	spect F2	
Σ1918	17.9	21	19	6 - 10	relfix, spect F2	15069n6318
Σ1927	16.1	354	14	7 - 8	relfix, spect G0	15109n6203
0 Σ294	3.2	248	58	6½- 11	relfix, spect A2	15116n5614
Hu 149	0.5	275	66	7 - 7	PA dec, dist inc; spect K0	15232n5423

DRACO

LIST OF DOUBLE AND MULTIPLE STARS (Cont'd)

NAME	DIST	PA	YR	MAGS	NOTES	RA & DEC
Σ1948	12.5	50	15	8- 8½	relfix, spect A5	15253n5504
β945	16.2	49	13	6½- 12½	optical, PA inc; spect F8	15277n5737
TW	3.3	25	62	8 - 9½	(0Σ299) slight PA inc; primary is ecl. variable	15331n6404
Σ1969	0.1	250	62	8 - 8½	PA inc, dist dec; spect K0	15404n6008
β946	1.8	136	58	6 - 11	PA dec, spect A2	15464n5532
Σ1984	6.5	273	44	6 - 8	relfix, spect A0	15497n5303
Σ2006	1.6	186	63	7½- 9	PA dec, spect A3;	15594n5904
	46.2	214	55	- 8½	AC dist inc.	
Es---	12.3	140	58	7- 11½	spect A0	16081n5804
Σ2045	2.5	182	55	8 - 8½	relfix, spect F8	16196n6137
Σ2054	1.0	353	68	6 - 7	PA slow dec; both spectra gG7	16231n6149
η	5.3	143	62	3 - 8	(14 Draconis) (*) (0Σ312) relfix	16233n6138
Hu 748	5.9	82	58	6 - 12½	relfix, spect gK1	16275n5131
Σ2060	3.7	246	50	9 - 9	relfix, spect G5	16275n5651
16 + 17	90.3	194	56	5 - 5½	cpm pair, relfix; spect both A0	16350n5301
17	3.2	109	62	5½- 6½	(Σ2078) slow PA dec; spect A0, A0	16350n5302
Σ2092	8.2	5	62	7½- 8½	relfix, spect G5	16384n6048
Stn 805	6.8	169	03	7 - 9	spect B8	16406n6224
Hu 664	0.4	304	59	8 - 8	relfix, spect F5	16425n5138
20	1.0	72	66	7 - 7½	(Σ2118) binary, about 730 yrs; PA dec, spect F0	16562n6507
A1144	5.8	303	46	7 - 14	cpm; spect F5	16576n7422
Σ2128	12.1	47	53	8 - 9	PA dec, spect K0	17027n5939
μ	1.8	66	66	5½- 5½	(21 Draconis)	17043n5432
	13.2	175	58	- 13	binary, PA dec; spect dF6, dF6, ABC all cpm (*)	
Es 77	18.3	274	59	6½- 11	optical, dist inc, spect B9	17071n5054
Σ2138	22.3	135	49	8 - 8½	relfix, spect K0, K8	17091n5433

LIST OF DOUBLE AND MULTIPLE STARS   (Cont'd)

NAME	DIST	PA	YR	MAGS	NOTES	RA & DEC
OΣ327	0.1	65	62	8½- 9	binary, 91 yrs; spect dF4	17132n5611
Σ2155	9.9	114	55	6 - 9½	relfix, spect F0; VW Drac in field; 4' to SE.	17155n6046
Stn 821	6.6	298	02	7½-11½	spect A0	17186n6346
Σ2180	3.2	262	55	7 - 7	relfix, spect F0	17278n5055
β	4.2	13	34	3 - 11	(β1090) relfix; spect G2 (*)	17293n5220
ν	61.9	312	55	5 - 5	easy cpm pair; (ΣI 35) (*)	17312n5513
26	1.1	141	61	5½- 8	(β962) binary; 74 yrs; PA dec (*)	17345n6155
Σ2207	0.6	120	55	8 - 8½	PA dec, spect A2	17371n6709
Σ2199	1.8	66	68	7 - 7½	PA dec, spect F5	17377n5547
Σ2218	1.7	326	66	6½- 7½	PA & dist dec; spect F5	17400n6342
ψ	30.3	15	58	5 - 6	(Σ2241) easy cpm pair, relfix, spect F5, dF8	17428n7211
T	15.6	222	36	var-10	slight dist inc; primary N-type variable	17557n5814
Σ2261	9.5	262	31	7½- 9½	relfix, spect A2	17570n5213
Σ2273	21.1	284	30	7 - 7½	relfix, cpm pair; spect F2, F2	17590n6409
	23.7	266	21	- 12½		
Σ2271	3.0	267	59	7½- 8½	dist inc, spect G0	17592n5251
OΣ349	0.3	44	62	8 - 8½	binary, PA dec, about 400 yrs; spect G5	18009n8354
Σ2284	3.5	190	53	7½- 9	relfix, spect F5	18013n6557
Σ2278	36.9	26	49	6½- 7	slight dist dec; spect F0, A3	18020n5625
Σ2278b	6.1	146	49	8 - 8½	relfix, spect A3	18020n5625
Σ2279	13.2	181	25	8½- 9	relfix	18034n5052
40 + 41	19.3	232	55	5½- 6	(Σ2308) fine cpm pair, relfix, spect dF6, dF5	18039n8000
Σ2302	5.8	247	37	7 - 10	relfix, spect A0	18047n7547
	23.1	280	25	- 9½		

LIST OF DOUBLE AND MULTIPLE STARS (Cont'd)

NAME	DIST	PA	YR	MAGS	NOTES	RA & DEC
Σ2326	16.4	196	40	$7\frac{1}{2}$- $8\frac{1}{2}$	PA slow dec, spect A5	18102n8128
Σ2307	4.3	205	52	$8\frac{1}{2}$- $8\frac{1}{2}$	relfix, spect G0	18116n6914
β1274	95.6	239	25	$6\frac{1}{2}$- 10	spect dF1	18138n5634
β1274b	0.9	146	15	10- $10\frac{1}{2}$		
	5.4	7	24	- 10		
ϕ	0.2	299	62	$4\frac{1}{2}$- 6	(43 Draconis) PA dec, spect A0	18215n7119
39	3.7	352	62	5 - 8	(Σ2323) slow PA dec, spect A1, F5, F8; all cpm	18232n5846
	89.0	21	56	- $7\frac{1}{2}$		
Hu 66	0.4	271	60	8 - 8	(0Σ351) PA dec; AC PA dec, spect G5	18240n4844
	0.7	18	60	- $8\frac{1}{2}$		
Hu 932	2.7	89	21	7 -$12\frac{1}{2}$	spect F2	18309n6230
Σ2348	0.3	78	62	6 - 6	(A1377) all cpm; AB binary, about 185 yrs; PA inc, spect K0	18328n5219
	25.7	272	58	- 8		
Σ2368	1.9	323	51	7 - $7\frac{1}{2}$	PA dec, spect A3	18377n5218
	36.6	125	20	- 11		
Σ2377	16.5	340	20	$7\frac{1}{2}$- 9	spect K0	18380n6329
Σ2384	0.9	310	62	$8\frac{1}{2}$- 9	binary, about 150 yrs; spect G1	18385n6704
0Σ363	0.2	128	60	$7\frac{1}{2}$- $7\frac{1}{2}$	PA inc, spect F0	18398n7738
Σ2398	15.3	163	62	8 - $8\frac{1}{2}$	AB= rapid proper motion pair (*)	18425n5930
	67.4	194	26	- 12		
Σ2403	1.2	270	62	$6\frac{1}{2}$- 9	PA inc, slight dist dec, spect sgG7	18437n6100
Σ2410	1.6	87	58	8 - $8\frac{1}{2}$	PA dec, spect F8	18462n5916
β971	0.2	355	54	$6\frac{1}{2}$- $8\frac{1}{2}$	spect F5	18462n4923
O	34.2	326	49	$4\frac{1}{2}$- $7\frac{1}{2}$	(47 Drac)(Σ2420) optical, dist inc; orange & blue pair Spect K0	18505n5920
β1255	1.5	95	37	6 - 12	PA inc, spect dF4	18535n4848
Σ2452	5.7	218	51	$6\frac{1}{2}$- $7\frac{1}{2}$	relfix, spect A0, A2	18552n7543
Σ2433	7.6	124	37	7 - 10	relfix, cpm pair; spect F2	18560n5641
	36.1	127	19	- 12		

LIST OF DOUBLE AND MULTIPLE STARS (Cont'd)

NAME	DIST	PA	YR	MAGS	NOTES	RA & DEC
Σ 2438	0.7	8	62	7 - 7	binary, about 260 yrs; spect A2; PA dec.	18566n5809
Σ 2440	17.0	123	36	6½- 9	relfix, cpm pair; spect G5, G8	18568n6220
Hu 757	5.2	300	62	7 - 9½	(Σ 2450) relfix; spect K2	19009n5211
Hu 757b	0.2	353	60	9½- 10	PA dec	
BH	10.5	58	19	8 - 8	(Arg 33) Spect A0; Ecl.bin; 1.8172 d.	19028n5723
Es---	12.3	130	11	6½- 12	spect gK5	19039n4951
Stn 873	6.6	41	03	8½- 11½	spect A2	19051n5913
0Σ 369	0.7	20	58	7 - 7½	PA dec, spect F8	19078n7200
Σ 2509	1.7	330	59	7 - 8	Dist inc; PA dec, spect F5	19164n6307
Σ 2550	2.0	249	56	8 - 8	relfix, spect F2	19279n7316
Σ 2549	2.0	331	24	7½- 12	(β655) relfix;	19307n6312
	25.8	286	49	- 9	spect K0; AC & AD	
	50.7	276	24	- 8	dist both inc.	
Σ 2571	11.3	20	49	7½- 8	relfix, spect F0	19318n7809
Σ 2553	1.0	113	57	8½- 9	PA inc, spect G0	19327n6157
Σ 2554	18.5	196	49	8 - 8½	relfix, spect A2	19332n6010
Ku 2	0.7	252	60	7 - 9	PA & dist dec; spect F2	19350n7130
Σ 2573	18.2	27	55	6 - 8½	relfix, spect A2	19394n6023
Σ 2574	0.3	230	59	8 - 8	PA inc, dist dec; spect F5	19400n6233
ε	3.1	15	62	4 - 7½	(63 Drac) (Σ 2603) PA inc, spect G8 & dF6	19484n7008
Σ 2604	27.8	184	25	6½- 8½	relfix, spect G5	19522n6403
Σ 2640	5.4	18	61	6 - 10	PA dec, spect A2	20041n6345
β470	2.6	214	54	9½- 11	relfix, spect G	20046n6337
Σ 2642	2.0	181	62	8½- 8½	PA inc, spect G0; β470 in field	20049n6333
Σ 2650	21.8	228	25	7½- 10½	spect A0	20066n6610
Σ 2652	0.3	238	58	7½- 7½	PA dec, spect A0	20082n6156
Σ 2694	4.0	346	36	6½- 10	relfix, spect A0	20178n8023
β1134	4.2	81	25	6 - 12	cpm; spect gK5	20205n6349

DRACO

LIST OF VARIABLE STARS

NAME	MagVar.	PER	NOTES	RA & DEC
R	7.1--13.0	246	LPV. Spect M5e--M7e	16325n6651
S	8.2---9.5	136	Semi-reg; spect M6	16419n5500
T	7.2--13..	422	LPV. Spect N0e; visual double star	17557n5814
U	9.1--14.5	317	LPV. Spect M6e--M8	19099n6712
V	9.5--14.5	278	LPV. Spect M4e	17573n5453
W	9.0--15.0	262	LPV. Spect M3e--M4e	18055n6557
X	9.9--15..	257	LPV. Spect M5e	18068n6609
Y	7.8--14.5	326	LPV. Spect M5e	09369n7805
RR	9.9--13.2	2.831	Ec.Bin.; spect A2	18414n6237
RS	9.0--12.0	280	Semi-reg; spect M5e	18389n7417
RU	9.4--13..	297	LPV. Spect M5e	18187n5931
RV	8.4--14.2	208	LPV. Spect M1e - M3e	12354n6551
RX	9.9--10.4	3.786	Ecl.Bin; spect F	19019n5839
RY	6.0---8.2	170:	Semi-reg; spect N4p	12545n6616
RZ	9.8--10.7	.5509	Ecl.Bin; lyrid, spect A5	18223n5852
SS	8.4--10.5	52	Semi-reg; spect M5	12241n6858
SU	9.4--10.3	.6604	Cl.Var.; spect A2--A5	11351n6737
SV	9.1--15.0	257	LPV. Spect M7e	18324n4920
SW	9.3--10.4	.5697	Cl.Var; Spect F4	12155n6947
SX	9.8--11.6	5.169	Ecl.Bin; spect A7	18038n5823
SZ	8.5---9.5	Irr	Spect M5	19099n6601
TT	9.4--11..	110	Semi-reg; spect M6	17121n5755
TU	9.9--13..	345	LPV. Spect M4e	18502n4851
TV	8.6--10..	Irr	Spect M8p	17081n6423
TW	7.7--10.0	2.807	Ecl.Bin; spect A5+K2	15331n6404
TX	6.9--8.2	79	Semi-reg; spect M4e--M5	16343n6034
TY	8.8---9.9	Irr	Spect M8	17362n5746
TZ	9.6--10.5	.8660	Ecl.bin; Spect A7	18208n4732
UU	9.0--11.5	120	Semi-reg; spect M8e	20249n7506
UV	8.6---9.8	77	Semi-reg; spect M5	14425n5619
UW	7.0--8..	Irr	Spect K5p	17565n5440
UX	6.2--7.0	170:	Spect N0	19234n7628
UZ	9.5--10.5	3.261	Ecl.Bin;	19261n6850
VV	9.5--10..	Irr		12151n6932
VW	6.0---6.5	Irr	Spect K0	17159n6043
WW	8.0---9.2	4.630	Ecl.Bin; spect sgG2 + sgK0	16384n6048
WZ	8.5--14.0	402	LPV. Spect M6e	16587n5225

LIST OF VARIABLE STARS (Cont'd)

NAME	MagVar.	PER	NOTES	RA & DEC
XZ	9.2--10.3	.4765	Cl.Var; Spect A6--F6	19094n6446
ZZ	9.0--14..	275	LPV. Spect M7e	19406n6739
AA	9.0--14..	338	LPV. Spect M6	20054n6609
AC	6.0---6.3	Irr	Spect M5	20199n6844
AF	5.2--5.24	20.27	(73 Drac) Alpha Canum Venaticorum type; Spect A3p	20322n7447
AG	8.2--10..	Irr	Erratic, spect Gep	16014n6656
AH	7.5---8.5	110:	Semi-reg; spect M7	16474n5754
AI	7.2--8.2	1.1988	Ecl.Bin; spect A0	16552n5246
AZ	7.0---7.8	Irr	Spect M2	16415n7245
BH	8.0---8.6	1.8172	Ecl.Bin; spect A0. Double star Arg 33	19028n5723
BS	8.8---9.7	1.6820	Ecl.bin; spect F5	19572n7329
BY	8.1-- 8.5	3.8134	Ellipsoidal var? Also flare star; spect K7e	18327n5141

LIST OF STAR CLUSTERS, NEBULAE, AND GALAXIES

NGC	OTH	TYPE	SUMMARY DESCRIPTION	RA & DEC
3147	79[1]	⊖	Sb; 11.3; 3.0' x 2.3' vB,L,lE,vgvsmbM (*)	10128n7339
3329		⊖	S0; 12.9; 1.6' x 1.0' pB,S,lE,psmbM	10406n7705
3403	335[2]	⊖	Sb; 12.9; 1.9' x 0.6' pF,L,E,psmbM	10501n7357
3735	287[1]	⊖	Sb; 12.6; 3.8' x 0.6' pB,L,mE,mbM, nearly edge-on	11331n7048
4125		⊖	E5/S0; 11.1; 2.1' x 1.1' pB,pL,cE,mbM	12057n6527
4128	263[1]	⊖	Sa/S0; 12.9; 1.8' x 0.4' cB,cL,E,bM; spindle-shaped	12061n6903
4133	278[1]	⊖	SBb; 13.1; 1.0' x 0.6' pB,cL,lE, gmbM	12065n7511

LIST OF STAR CLUSTERS, NEBULAE, AND GALAXIES (Cont'd)

NGC	OTH	TYPE	SUMMARY DESCRIPTION	RA & DEC
4236	51[5]	⊖	SB; 10.7; 22' x 5' vF,eL,mE; large dim loose-structured spiral	12143n6945
4256	846[2]	⊖	Sb; 13.0; 4.0' x 0.5' pB,L,cE,bM,BN; edge-on; spindle shaped streak	12164n6611
4291	275[1]	⊖	E1; 12.4; 0.7' x 0.5' pB,vS,R,lbM. NGC 4319 is 6' to SE	12181n7540
4319	276[1]	⊖	SBc; 13.0; 1.1' x 0.7' pB,pS,vlE,sbM	12197n7536
4386	277[1]	⊖	E6/S0; 12.8; 0.7' x 0.3' pB,cL,E,psmbM	12224n7548
4589	273[1]	⊖	E2/S0; 12.1; 0.9' x 0.7' cB,L,1E,pgmbM	12356n7428
4750	78[4]	⊖	Sb; 12.2; 1.8' x 1.3' pB,L,R,vgvsmbM; dark ring surrounds nucleus	12484n7309
5678	237[1]	⊖	Sc; 12.1; 2.4' x 1.1' B,L,1E,vgmbM	14307n5808
5866	215[1]	⊖	E6p; 11.1; 2.9' x 1.0' vB,cL,pmE,gbM; edge-on with thin equatorial dust band	15051n5557
5879	757[2]	⊖	Sb; 12.2; 3.9' x 1.2' cB,S,E,mbMN; 7^m star 7' N	15084n5712
5905	758[2]	⊖	SBb; 13.1; 4.4' x 3.2' pF,pS,E; NGC 5908 is 12.5' to SE	15141n5542
5907	759[2]	⊖	Sb; 11.0; 11.0' x 0.6' cB,vL,vmE, long thin streak; edge-on spiral (*)	15146n5631
5908	760[2]	⊖	Sb; 13.0; 2.4' x 0.4' pF,pS,mE; nearly edge-on; thin equatorial dust lane; pair with NGC 5905	15154n5536
5949	906[2]	⊖	SD; 12.9; 2.0' x 0.9' F,S,1E,vglbM	15272n6455
5982	764[2]	⊖	E3; 12.4; 1.1' x 0.7' cB,S,R,psbM; NGC 5985 is 7' to E.	15376n5932

LIST OF STAR CLUSTERS, NEBULAE, AND GALAXIES   (Cont'd)

NGC	OTH	TYPE	SUMMARY DESCRIPTION	RA & DEC
5985	766^2	⊘	Sb; 12.0; 4.3' x 2.1' pB,cL,E (*)	15386n5930
6015	739^3	⊘	Sc; 11.8; 5.5' x 1.9' vF,pL,mE,vgbM; multiple-arm spiral	15507n6228
6340	767^2	⊘	Sa; 12.4; 1.6' x 1.0' cF,pL,R,vgmbM	17111n7222
6412	41^6	⊘	Sc; 12.2; 1.9' x 1.5' cL,R,vgbM	17308n7545
6503		⊘	Sb; 11.0; 4.8' x 1.0' pF,L,mE; 9^m star foll 4'	17499n7010
6543	37^4	◎	vB,S, mag 8.6; diam 22" x 16"; bluish disc with 10^m central O-type star (*)	17588n6638
6643		⊘	Sc; 12.0; 3.0' x 1.3' pB,pL,E,vSN. multiple arm spiral (*)	18212n7433

DESCRIPTIVE   NOTES

ALPHA   Name- THUBAN, from the Arabic name for the
             entire constellation. Mag 3.64; spectrum A0
III.   Position 14030n6437. Thuban is easily located about
midway between the bowl of the Little Dipper and the famous
double star Mizar at the bend of the Great Dipper's handle.
The measured parallax of Thuban indicates a distance of
about 215 light years, giving an actual luminosity of about
135 times that of the Sun. (Absolute magnitude -0.5)
    Thuban is a spectroscopic binary with a period of
51.38 days; the components move in an orbit of 0.38 eccentricity, with a separation of about 20 million miles. The
brighter star has a somewhat unusual spectrum showing
strong lines of silicon. Thuban shows an annual proper
motion of 0.055"; the mean radial velocity of the system is

about 10 miles per second in approach. The star has been suspected of variability by various observers in the past, but there appears to be no present evidence for any real light changes. Admiral Smyth, however, writing in 1844, stated that the star had been "suspected of variability, for Ptolemy, Ulugh Beigh, and Lacaille mark it as of 3rd magnitude; and Pigott as a bright 4th; Tycho Brahe, Hevelius, and Bradley rank it of the 2nd; and although marked of the latter size in the British Catalogue, Mr. Baily found that in the original entries it is designated once of the 3rd, and once of the 4th. I have had it in view many times, and always looking like a small third; though Baron de Zach but shortly before classed it as 2.3." On Johann Bayer's charts of 1603 the symbol for Thuban is larger than that of Gamma Draconis, though Gamma now outshines Thuban by about 1.4 magnitudes.

The most famous and often-mentioned fact about Thuban is that it was the Pole Star some 4800 years ago, during the "Old Kingdom" period of Ancient Egypt. According to a recent calculation, Thuban was nearest to the true Pole about 2830 B.C.; the minimum distance being less than 10'. This date coincides closely with the beginning of the great age of the pyramid builders of the 4th Dynasty, and much has been written about the probable connection between this star and the Great Pyramid of Khufu at Gizeh, the most gigantic, puzzling, and mathematically perfect structure of the ancient world. It is undoubtedly true that the wonderfully accurate orientation of the pyramid was achieved by observations of the stars, and it appears that the descending passage of the pyramid was constructed to point directly at Thuban as the star passed below the true Pole at its lower culmination, but the purpose of this alignment still remains obscure. The modern traveller, particularly the astronomically minded one, is still touched by the awe which the pyramid has inspired in all ages. Charles Barns, author of "1001 Celestial Wonders" (1929) described his sensations during a visit to Khufu's gigantic tomb:

"I myself, some years ago, crept down into the sepulchral chambers deep in the solid masonry of this most ancient of tomb observatories, and gazing obliquely up through the murky rift, beheld a rectangular patch of blue Egyptian sky where Thuban once reigned in solemn grandeur- a thrill-

DESCRIPTIVE NOTES (Cont'd)

ing moment!" Certainly erroneous, however, is the popular
supposition, often repeated, that the star could have been
seen by day from the interior of the pyramid. Actual tests,
made from the bottoms of deep wells and chimneys have shown
that even 1st magnitude stars cannot be detected by the unaided eye under such conditions.

The probable connection between Thuban and the Great
Pyramid is a topic of great interest to astronomers and
historians alike, since it offers the dramatic possibility
of fixing the precise date of the pyramid's construction
through astronomical calculation. In a fascinating discussion of this problem, written in 1912, Percival Lowell
pointed out that the descending passage of the pyramid is
oriented toward a point 3° 34' below the Pole, the location
which Thuban would have occupied, allowing for refraction,
some 645 years before or after the time of its closest
approach to the true Pole. Lowell in this way obtained two
possible dates: 3400 B.C. or 2140 B.C. Newer calculations
of the time of the nearest approach of Thuban to the Pole
now permit a slight revision of these figures, giving us a
choice of 3475 B.C. or 2185 B.C. Lowell himself pointed
out that the more recent date "is negated by what we know
of Egyptian history, and we are thus left with the other.."

The puzzling feature of the problem now becomes evident; both dates are completely incompatible with the presently accepted chronology of the 4th Dynasty. The reign of
Khufu is placed by the majority of modern archeologists
somewhere in the period 2700- 2600 B.C. Professor John A.
Wilson of the Oriental Institute of the University of
Chicago dates the beginning of the 4th Dynasty at about
2650 B.C. and the reign of Khufu about 50 years later.
According to Wilson, these dates are probably accurate to
within 75 years. In any case, it does not appear even re-
motely possible that the chronology can be in error by a
margin so enormous as 800 years, or that the reign of Khufu
can be pushed back to the supposedly pre-Dynastic Egypt of
3475 B.C. Wherein, then, lies the error? The possibility
that the pyramid has shifted, or the alignments changed,
appears to be ruled out by modern measurements which show
that the orientation is still phenomenally accurate. The
maximum error, on the east side, is only 5.5 minutes of
arc. Evidently, no measureable shifting of the land or of

DESCRIPTIVE NOTES (Cont'd)

the pyramid has occurred since the days of construction.
Three possibilities then remain:

1. The astronomical evidence is accepted; the date of
construction is then fixed at about 3475 B.C. or some 800
years before the time of Khufu. Egyptologists find no his-
torical or archeological evidence to support this theory.

2. The archeological evidence is accepted; the pyramid
was constructed during the reign of Khufu and therefore
dates to about 2600 B.C. The supposed alignment of the
descending passage toward Thuban must then be dismissed as
a meaningless coincidence; in Khufu's time Thuban was very
near the true Pole, but the passage itself pointed toward
an apparently blank piece of sky some 3° below.

3. The only remaining possibility, which would recon-
cile both views, is the supposition that the Egyptian archi-
tects were aware of the fact of precession, and that the
pyramid was aligned on a spot where Thuban had been at some
date in the past considered highly significant, or where it
would be at some expected time in the future.

In this maze of conflicting facts and contradictory
opinions, the author will, I hope, be excused for giving
the last word to an astronomer. Lowell wrote: "We are not
here dealing with conjectures as to when a certain king or
dynasty can be made to fit into a general chronological
scheme by the relics it has left us of itself. Calculations
from known astronomic data can tell us to an exactness
guaged only by the size of the opening of the passage as
seen from below precisely when the pyramid was built, with
only the choice above described. For that such a pointing
can be but the sport of chance, the whole structure of the
pyramid emphatically denies.."

BETA   Name- RASTABAN. Mag 2.78; spectrum G2 II.
          Position 17293n5220. The computed distance of
this star is some 300 light years; the actual luminosity
about 600 times that of the Sun. (Absolute magnitude about
-2.1) Beta Draconis shows an annual proper motion of 0.02"
and the radial velocity is 12 miles per second in approach.

A faint companion, estimated to be of the 14th magnitude, but now measured photoelectrically at about 11.5, was
discovered by S.W.Burnham in 1889. The two stars probably
form a physical system, separated by about 370 A.U.

DESCRIPTIVE NOTES (Cont'd)

GAMMA
Name- ELTANIN. Mag 2.22; spectrum K5 III.
Position 17554n5130. The distance of the
star is about 110 light years; the actual luminosity about
145 times that of the Sun. (Absolute magnitude -0.4) The
star shows an annual proper motion of 0.025" and the radial
velocity is 16 miles per second in approach.

A rich mythology surrounds this star due to its position in the head of the Dragon; R.H.Allen states that a number of temples in Ancient Egypt were oriented toward it.
It was also through observations of Gamma Draconis that the
discovery of the aberration of light was made by Bradley in
1729. Recent photoelectric measurements reveal a slight
variability of the star with a total range of 0.08 magnitude.

Several faint optical companions to the star are listed by R.G.Aitken, including a 13th magnitude star at 21",
two stars of magnitude 12.5 at 48" and 56", and a star of
11.4 at 97".

DELTA
Mag 3.06; spectrum G9 III. Position is
19125n6734. The distance of this star is
about 120 light years, giving an actual luminosity of 75
times that of the Sun and an absolute magnitude of +0.2.
Delta Draconis shows an annual proper motion of 0.13"; the
radial velocity is 15 miles per second in recession.

ZETA
Mag 3.20; Spectrum B6 III. The position is
17086n6547. Zeta Draconis is computed to be
about 600 light years distant. The actual luminosity is
1500 times that of our Sun, the absolute magnitude about
-3.2. The annual proper motion is 0.025"; the radial
velocity is 8 miles per second in approach.

ETA
Mag 2.74; spectrum G8 III. The position is
16233n6138. Eta Draconis is approximately
75 light years distant; the actual luminosity is about 40
times that of the Sun, and the absolute magnitude is about
+0.9. The measured annual proper motion is 0.06" and the
radial velocity is about 8 miles per second in approach.

The star is a close double and not an easy object
owing to the large difference in brightness between the

components. There has been no definite change in separation
or P.A. since discovery in 1843. The projected separation
of the pair is about 125 A.U. and the faint companion is a
dwarf of spectral type dK1 with about one-fourth the solar
luminosity.

The closer double star Σ2054 lies in the same field,
about 11' to the north.

IOTA   Name- ED ASICH. Mag 3.30; spectrum K2 III.
Position 15238n5908. The computed distance of
the star is about 100 light years, the actual luminosity
about 45 times that of the Sun, and the absolute magnitude
about +0.8. The annual proper motion is only 0.01"; the
radial velocity is 6.5 miles per seoond in approach.

MU   Mag 5.06; spectrum dF6; position 17043n5432.
The distance of the star is about 100 light
years; the total luminosity is about 8 times that of our
Sun. Mu Draconis shows an annual proper motion of 0.11"
and a radial velocity of 10 miles per second in approach.

This star is a long period binary, first seen by W.
Herschel in 1779. Although the period is still uncertain,
and published values range from 650 to 4000 years, recent
computations by A.Fantoli (1957) indicate a probable period
of about 1090 years with an eccentricity of 0.57 and a
semi-major axis of 5.2". Periastron occurred in 1934 when
the apparent separation was just over 2", corresponding to
a projected separation of about 70 A.U. During the last
180 years the P.A. of the pair has decreased through an arc
of about 180°, or close to 1° per year. The observed arc,
however, is only the periastron end of an elongated ellipse
and the orbital motion in the next few centuries will become noticeably slower as the separation of the two stars
steadily increases. The predicted positions up to 1990 are
shown in the orbital diagram opposite.

In the Mu Draconis system, the two visible stars are
of nearly equal magnitude and both spectra are dF6. There
is evidence for the existence of a third star in the system
with a period of about 3 years, a mass of 0.6 Sun, and an
orbit 3 AU in radius with an eccentricity of about 0.4. In
addition, a 13th magnitude companion at 14" was discovered

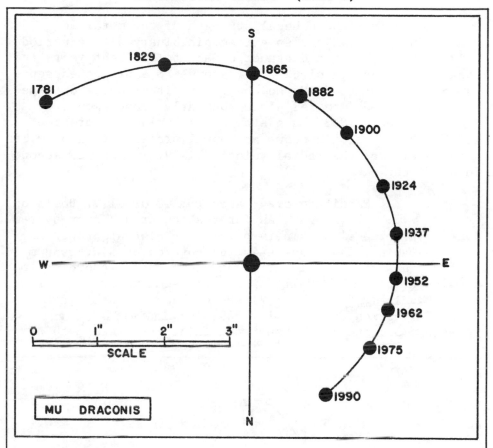

MU DRACONIS

by S.W.Burnham in 1889. The AB-C system, sometimes listed
under Burnham's catalog number β1088, is a physical triple
as the faint star is now known to share the proper motion
of the bright pair; the A-C projected separation is about
400 AU, with a very slowly decreasing PA.   A dwarf star
of uncertain type, the faint companion has a computed luminosity of about 1/190 that of the Sun.

NU   (24 + 25 Draconis) Position 17312n5513. This
               is the faintest of the four stars which form
the little quadrangle marking the head of Draco.   A fine
wide double star with common proper motion, Nu Draconis is
one of the easiest pairs in the sky for very small telescopes, and is generally divisible in good binoculars. The
magnitudes are 4.95 and 4.98; both spectra are close to A5

but peculiar for the unusual strength of the metallic
lines. The primary is a spectroscopic binary with a period
of 38.5958 days. At a distance of about 120 light years the
apparent separation of 62" corresponds to a projected separation of about 2300 AU. No change in either angle or separation has been detected since the early measurements of
F.G.W.Struve in 1833. Each star has a luminosity of about
11 suns. Nu Draconis shows an annual proper motion of 0.15
seconds of arc; the radial velocity is 9.5 miles per second
in approach.

26   Magnitude 5.24; spectrum G0 or G1 V. Position
             17345n6155. An interesting triple star system
consisting of a close binary pair and a distant proper motion companion. The close pair, discovered by S.W.Burnham
in 1879, has an orbital period of 74.16 years according to
a computation by R.G.Hall (1949). The semi-major axis of
the orbit is 1.5" or about 21 AU, with an eccentricity of
0.19. Periastron occurred in 1949. The individual magnitudes are 5.3 and 8.1, though the faint star seems 9th or

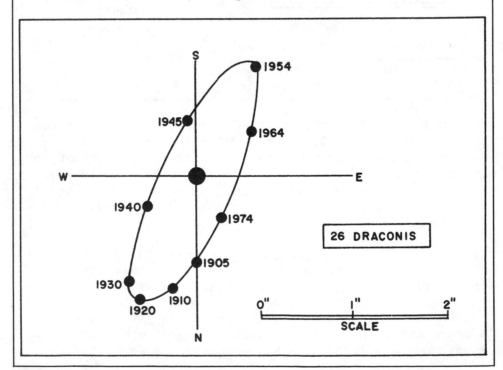

26 DRACONIS

DESCRIPTIVE NOTES (Cont'd)

10th to most visual observers. This star is a fairly close neighbor in space at a distance of about 45 light years, and shows an annual proper motion of 0.57" in PA 154°. From the computed orbit the masses are found to be 0.84 and 0.72 the solar mass, and the absolute magnitudes are +4.5 and +7.3. The brighter star is a main sequence object very much like our Sun; the companion is a red dwarf.

At a distance of 12.3' in PA 162° is the third member of the system, detected through a comparison of proper motion plates in 1921. It is a star of magnitude 9.9, spectrum dM1, computed absolute magnitude about +9.2. Although this star shares the motion of the binary pair, no sign of orbital revolution is to be expected; the separation from the AB pair is estimated to be about 0.17 light year! The radial velocity of the whole system is 7.8 miles per second in approach.

Σ 2398   (ADS 11632) Position 18425n5930. This is one of the closest double stars to our Solar System, a red dwarf pair which might be called a miniature copy of 61 Cygni. It is located about 1° west and slightly north from the bright optical color contrast pair Omicron Draconis. It was first measured by F.G.W.Struve in 1832 when the separation was about 12" in PA 134°. This had increased to 15.3" in PA 163° in 1962, according to measurements by Van den Bos. Orbital elements of the system are

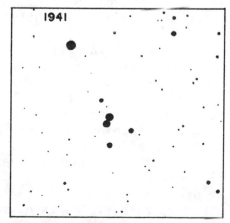

1941

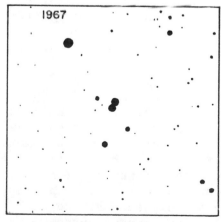

1967

THE PROPER MOTION OF Σ2398.   Total Displacement in 26
Years = 59"

still uncertain. W.Rabe in 1958 obtained a period of 346 years with periastron in the year 2113 AD; the orbit has a semi-major axis of 11.8" and an eccentricity of 0.55. These elements are subject to considerable revision since the observed arc constitutes only a small portion of the entire orbit. To illustrate the uncertainty in the elements, an orbit computation by J.Hopmann in 1954 gave a period of about 2500 years, with a semi-major axis of 44.4". Evidently, it will not be possible to compute a definitive orbit until the star has been observed for a great many years.

The distance of this system is 11.3 light years, virtually the same as 61 Cygni; the annual proper motion is 2.28" in PA 324°. There is a large difference in the radial velocities of the two stars, owing to the orbital motion; the measured values are 0.6 and 8.4 miles per second, both in recession. Facts of interest concerning the two stars are given in the following table.

	MAG.	SPECT.	ABS.MAG.	LUM.	MASS
A	8.9	dM4	+11.2	0.0027	0.29
B	9.7	dM5	+12.0	0.0013	0.25

NGC 6543   Position 17588n6638. A small bright planetary nebula located almost exactly at the north pole of the Ecliptic, about midway between Delta and Zeta Draconis. This is one of the dozen or so most conspicuous planetaries in the sky, appearing nearly stellar with low magnifications but revealing a blurred disc with higher powers. T.W.Webb, using a 3-inch telescope gives its diameter as 15"; H.D.Curtis at Lick Observatory (1918) measured it as 22" x 16" with the longer dimension oriented toward PA 35°. Lick photographs show it as a helix with two turns, the luminous shells of gas overlapping in a curious manner so as to give a vague impression of intertwining spiral arms. A somewhat similar effect is seen in the much larger Helical Nebula (NGC 7293) in Aquarius. The surface brightness of NGC 6543 is very high, and Curtis found that all the main features were recorded in a ten-second exposure at the Lick 36-inch refractor. "An exposure of a few minutes burns out all the detail in the central portion."

DEEP SKY OBJECTS IN DRACO. Top: The planetary nebula NGC 6543. Below: The edge-on galaxy NGC 5866.
Lick Observatory photographs

DESCRIPTIVE NOTES   (Cont'd)

The appearance in the small telescope is very much as Webb described it- "a luminous disc, much like a considerable star out of focus". The total magnitude is about 8.5 and the central star is about 9.5 visually and 10.2 photographically; the star is, however, not easy to distinguish in the bright mass of nebulosity.   The description in Norton's Atlas is "very bright, oval disc, like a star out of focus, bluish, with a central 9.6 magnitude star".   To Admiral Smyth also the color appeared distinctly bluish - a "very fine pale blue". Others see the tinge as more of a bluish-green, becoming strikingly vivid in large telescopes The distinctive color is the result of the radiation of doubly ionized oxygen, at 5007 and 4959 angstroms, the so-called "forbidden lines" characteristic of many of the planetary nebulae.

Photographs obtained at Palomar Observatory with the 48-inch Schmidt telescope reveal that the nebula is centered in an extremely faint shell or ring of gas about 4' in diameter, the outer edges showing a serrated or scalloped appearance with several small bright condensations of material on the west side. This large diffuse wreath probably represents a very ancient outburst or period of gas ejection in the central star, ages before the onset of the activity which created the present nebula.

NGC 6543 was the first object of its kind to be analyzed with the spectroscope, by William Huggins, on August 29, 1864. This observation solved the riddle of the nebulae "at a single glance" since Huggins saw immediately that the spectrum was that of a tenuous luminous gas, rather than a mass of unresolved stars as some observers had expected.

The distance, as in the case of all the planetaries, is not precisely known, various catalogs offering values ranging from 500 to over 1000 parsecs. In his listing of the brighter planetaries, C.R.O'Dell (1963) gives a distance of about 3200 light years for NGC 6543, making the true diameter about 20,000 AU or nearly one-third of a light year. The fainter outer shell is some 3.5 light years in diameter, and the central star is a super-hot O-type dwarf with a computed temperature of about 35,000°K and a luminosity of nearly 100 times the Sun. (Refer also to M57 in Lyra, NGC 7293 and NGC 7009 in Aquarius, and NGC 7662 in Andromeda)

SPIRAL GALAXY NGC 5985 in DRACO. A fine multiple-arm spiral, photographed with the 61-inch astrometric reflector at the U.S. Naval Observatory in Flagstaff, Arizona.

GALAXIES IN DRACO. Top left: The many-armed spiral NGC 6643. Top right: The more compact spiral NGC 3147. Below: A nearly edge-on system, NGC 5907.   Palomar Observatory

LIST OF DOUBLE AND MULTIPLE STARS

NAME	DIST	PA	YR	MAGS	NOTES	RA & DEC
ε	0.9	288	61	$5\frac{1}{2}$- 6	(1 Equl)(Σ2737)	20566n0406
	10.6	70	59	- 7	All cpm; AB binary,	
	74.8	280	24	-$12\frac{1}{2}$	101 yrs (*)	
Kui 102	0.3	20	39	$6\frac{1}{2}$- $7\frac{1}{2}$	Binary, 54 yrs,	20576n0719
					Spect A5, PA dec.	
2	2.8	218	55	7 - 7	(λ)(Σ2742) neat	20598n0659
					pair, slight PA dec	
					Spect F8	
Σ2749	3.5	168	61	8 - 9	PA inc, spect G0	21022n0320
Σ2749b	0.5	17	61	10- 10		
OΣ527	0.2	200	61	$6\frac{1}{2}$- 8	PA dec, spect A2	21055n0457
5	1.9	268	58	4 - 11	(γ) (ΣT 54) (β71)	21079n0957
	47.7	5	25	- 12	slow PA dec, spect	
	352	153	22	- 6	Fp, AB cpm, C opti-	
					cal	
Ho 151	1.6	196	62	9 - 9	relfix	21084n0340
Σ2765	2.7	81	62	8 - 8	relfix, spect A3	21086n0921
J576	2.2	232	54	9 - $9\frac{1}{2}$	relfix, spect F8	21100n0733
β270	0.3	1	61	$7\frac{1}{2}$- $9\frac{1}{2}$	PA inc, spect F0	21110n0701
	32.5	32	16	- $12\frac{1}{2}$		
δ	0.3	25	61	5 - 5	(Σ2777) rapid	21120n0948
	47.7	14	25	- 10	binary (*)	
β682	5.6	103	43	$7\frac{1}{2}$- 12	relfix, spect K5	21120n0429
	91.5	176	13	- 10		
β163	0.6	251	67	7 - 9	binary, 72 yrs;	21162n1122
					PA dec, eccentric	
					orbit, spect F8	
Σ2786	2.6	188	61	7 - 8	slight PA inc,	21172n0919
					spect A2	
β838	1.5	132	65	$7\frac{1}{2}$- $9\frac{1}{2}$	PA inc, spect F8	21184n0254
OΣ435	0.5	227	67	8 - $8\frac{1}{2}$	PA inc, spect K0;	21189n0240
					One component may	
					be var.	
β	34.4	257	12	5 - 13	(h3023) optical	21204n0636
	69.3	307	12	- 11	group, spect A2	
Σ2791	2.5	104	55	$8\frac{1}{2}$ - 9	relfix, spect F0	21212n0409
Σ2793	0.4	229	57	8 - $8\frac{1}{2}$	(β164) PA dec,	21226n0910
	26.6	242	54	- $8\frac{1}{2}$	spect A2	

LIST OF VARIABLE STARS

NAME	MagVar	PER	NOTES	RA & DEC
R	8.7--15.0	261	LPV. Spect M3e	21108n1236
S	8.0--10.0	3.436	Ecl.Bin; spect B8	20547n0453
T	9.2--10.9	147	Semi-reg; spect M3	21133n0904
RU	8.5--9..	65?	Semi-reg; spect M6	21158n0733

DESCRIPTIVE NOTES

DELTA   Position 21120n0948. Magnitude 4.49; spectrum F7 V. Delta Equulei is a noted double star whose period was for many years the shortest known for any visual binary. It was discovered by O.Struve in 1852 and the system has made 19 revolutions in the last 112 years. The orbital period is 5.70 years. Always a close and difficult pair, with a separation never exceeding 0.35", this star is a fine test object for larger instruments. Widest separation occurs near PA 25° as in 1921 and 1927. According to computations by W.J.Luyten and E.Ebbighausen (1934) the semi-major axis of the orbit is 0.26" and the eccentricity is 0.42; periastron occurs in September 1969.

The components of this system are nearly equal in brightness and type, with a total luminosity of about eight times the Sun; the absolute magnitude of each star is close to +4. The actual separation averages about 4.6 AU, or somewhat less than the separation of Jupiter and the Sun. In addition to the orbiting pair, there is a third visible star at 60" distance (1967) discovered in 1833 when the separation was 27". This star, of the 10th magnitude, has no real connection with Delta, and the distance is steadily widening from the proper motion of Delta itself.

The computed distance of Delta Equulei is about 55 light years and the pair shows an annual proper motion of 0.31" in PA 172°. The radial velocity is 9 miles per second

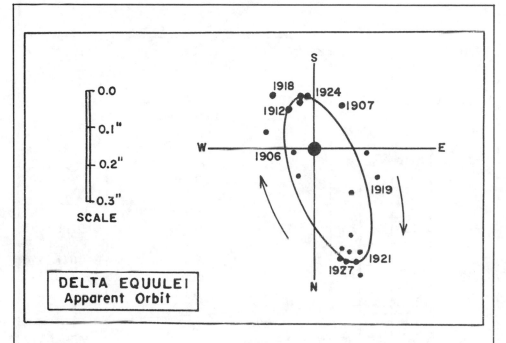

DELTA EQUULEI
Apparent Orbit

in approach, the figure varying somewhat as the two stars move in their orbits. (At the present time, the shortest period known for any visual binary is 1.725 years for the star Wolf 630 or BD−8°4352 in Ophiuchus. Another star, B427 in Sagittarius has a period of 2.68 years)

EPSILON   Magnitude 5.29, spectrum F5 IV; position 20566n0406. Epsilon Equulei is a close binary star with a third more distant component, discovered by F.G.W.Struve in 1835. The apparent orbit of the closer pair is a very narrow and elongated ellipse, oriented only 3° from the edge-on position; the apparent separation varies from 0.1" to a maximum of 1.1" in an orbital period of 101 years. Periastron occurs in 1920. According to an orbit computation by Van den Bos (1933) the semi-major axis is 0.66" or about 40 AU; the eccentricity is 0.70. The 7th magnitude star at 10.9" is a physical member of the system (mag 7.2, spectrum dF4) at a projected separation of 670 AU from the bright pair. The estimated distance of the whole system is about 200 light years; the annual proper motion is 0.19" in PA 218° and the radial velocity is 11 miles per second in recession.

ERIDANUS

LIST OF DOUBLE AND MULTIPLE STARS

NAME	DIST	PA	YR	MAGS	NOTES	RA & DEC
I 264	0.8	63	59	8 - 8½	(h3444) PA dec,	01296s5338
	38.8	7	30	- 10½	spect F5	
Hu 1345	5.6	201	22	7 - 13	spect B9	01322s5716
h3449	26.1	179	33	7½- 12½	Optical, PA & dist inc, spect K0	01337s5327
△4	10.4	104	34	7½- 8	cpm pair, relfix, spect F0	01369s5342
p	9.6	206	40	6 - 6	(△5) binary, about 480 yrs; PA slow dec, spect K2, K5	01379s5627
✗	4.6	202	44	3½- 11	cpm, slow PA inc, spect gG5	01540s5151
I 455	5.7	202	45	7½- 11	Spect K0	02065s5546
h3527	2.0	41	54	7 - 7	no certain change, spect both A0	02414s4044
Σ315	2.3	158	59	7½- 8½	relfix, cpm, spect G0	02469s1045
β10	2.7	99	39	7 - 11	relfix, spect A0	02479s0512
Hu 811	1.8	226	46	7½- 11	cpm, PA inc, spect G0	02557s1603
Kui 10	2.3	237	55	5 - 12	dist dec, spect A2	02562s0259
θ	8.2	88	52	3½- 4½	fine pair, PA slow inc, spect A3, A2; cpm; (*)	02564s4030
A2610	0.4	360	34	8½- 8½	PA inc, spect F5	02570s1028
A209	1.7	76	42	8½- 9½		02572s0244
ρ²	1.9	77	61	5½- 10	(β11)(9 Eridani) PA & dist dec, spect G5	03002s0753
Σ341	8.6	226	36	7½- 9½	relfix, spect F5	03005s0217
β1174	0.8	293	48	7½- 11½	cpm, PA dec, spect F8	03011s1110
h3548	12.2	123	36	7 - 11	cpm, relfix, spect G0	03015s2133
β527	1.1	90	59	8 - 8½	PA inc, spect F5	03037s1338
Σ356	15.7	12	19	7½- 11	relfix, spect F5	03043s1331
β528	0.5	13	61	8½- 8½	PA dec, spect F5	03059s0347
	48.8	99	17	- 11		

ERIDANUS

LIST OF DOUBLE AND MULTIPLE STARS   (Cont'd)

NAME	DIST	PA	YR	MAGS	NOTES	RA & DEC
β400	24.4	55	56	6½- 11	Optical, dist inc, spect gM1	03088s0400
h3556	0.5	1	44	6½- 7	(Jc 8) binary, 44	03107s4436
	3.5	202	44	- 9½	yrs; Spect F6, all cpm. both PA dec	
β84	0.9	8	66	7 - 7½	PA dec, spect B9	03136s0606
15	0.2	331	63	4½- 7½	(λ23) PA dec, spect G6	03162s2242
h3565	7.2	118	54	5 - 8½	PA & dist inc, cpm; spect F2	03164s1844
τ⁴	5.7	288	37	4 - 10	(16 Erid) (Jc 1)	03173s2155
	39.2	112	55	- 10½	AB cpm, relfix; spect gM3; C is optical, PA inc.	
h3570	34.2	256	20	6½- 12½	spect A0	03194s2030
β531	3.9	48	45	6½- 12	PA dec, dist inc, spect G0	03208s0758
I 468	2.1	222	27	7½- 10½	spect G0	03217s4015
A2909	0.1	199	46	7½- 7½	binary, 25 yrs;	03221s1550
	16.8	252	21	-13	spect G0, PA inc.	
β12	2.3	274	53	7½- 10½	relfix, spect A0	03221s1410
A2911	1.9	114	41	6½- 13½	spect A0	03254s1432
Σ408	1.3	329	63	8 - 8	PA dec, spect A3	03282s0427
Σ411	19.1	88	36	7 - 8	spect F8, relfix	03298s0715
	38.2	28	00	- 11		
β534	2.5	194	43	7½- 11	relfix, spect G5	03364s0840
Δ15	7.8	328	52	7½ - 8	relfix, spect A2	03380s4031
Σ436	40.1	236	35	7 - 8	optical, dist inc, spect F2	03385s1246
Hu 436	1.4	298	59	7½- 9	PA inc, spect F5	03398s1718
h3589	5.2	349	51	6½- 10	relfix, cpm pair; spect K0	03423s4049
Rst4760	0.7	288	52	7 - 9½	spect G5	03467s0136
f	7.9	212	57	5 - 5½	(Δ16) fine pair; PA inc, spect A0; A0; cpm.	03467s3746
β401	4.5	254	53	7 - 11	relfix, cpm pair;	03477s0140
	40.6	289	13	- 11½	spect F2	
30	8.2	135	52	5½- 10	(h338) relfix, spect B8	03502s0531

LIST OF DOUBLE AND MULTIPLE STARS (Cont'd)

NAME	DIST	PA	YR	MAGS	NOTES	RA & DEC
32	6.9	347	63	5 - 6	(w Erid) (Σ470) fine cpm pair with color contrast; spect G8 & A1	03518s0306
β543	11.1	22	58	8½- 10	cpm; PA slow dec; spect K2	03550s0118
γ	52.8	242	09	3½- 11	(34 Erid) (h3608) optical (*)	03557s1339
β1042	55.6	93	13	7½- 8½	spect G5	03561s0248
	39.2	250	10	- 11		
β1042b	1.2	34	43	8½- 9½	relfix	
Σ487	11.9	9	58	8½- 9	relfix, spect F0	03585s1035
	21.7	236	04	- 10½		
Σ489	3.2	197	51	8½- 8½	relfix, spect G0	03599s0709
β1004	1.5	106	59	7 - 7½	binary, PA dec, AC	04001s3438
	47.9	149	56	- 11½	dist dec, spect G0	
δ165	0.4	88	59	8½- 8½	(Daw 79) PA dec, spect F8	04003s2840
I 152	0.7	67	54	8 - 8½	relfix, spect G0	04031s3535
Hu 1363	0.3	121	59	7½- 7½	PA dec, spect A3	04048s2208
Σ501	29.2	296	16	8½- 9½	relfix, spect K0	04061s0249
I 153	0.9	343	55	8 - 8	PA slow inc, spect A3	04063s3259
A469	0.2	37	58	8 - 8	PA inc, spect A2	04069s0804
A2801	0.2	27	64	8 - 8	binary, 20 yrs; PA inc, spect gG0	04082s0500
Σ514	7.9	75	12	8½- 10	relfix, spect K2	04102s0658
h3628	50.3	50	33	7 - 8½	relfix, spect both F5	04107s3617
Hu 30	5.1	175	28	7 - 13	spect F0	04109s2315
Hwe 10	2.0	44	55	8 - 8	PA inc, spect F8	04110s2841
39	6.4	146	51	6 - 9	(Σ516) PA dec, cpm pair; spect K3, G2	04120s1023
O²	82.8	105	40	4½- 9½	(40 Erid) cpm;	04130s0744
O²b	7.6	347	61	9½- 11	remarkable triple system (*)	
h3632	11.0	163	37	7½- 10	relfix, spect A2	04131s3012
β548	6.2	345	39	7 - 11½	relfix, spect A2	04142sI013
41	0.4	148	14	4 - 5	(υ⁴) (I 270) cpm; spect B8	04160s3355

ERIDANUS

LIST OF DOUBLE AND MULTIPLE STARS   (Cont'd)

NAME	DIST	PA	YR	MAGS	NOTES	RA & DEC
Σ527	6.0	193	30	8 - 10½	relfix, spect F0	04166s0733
h3642	6.0	159	48	6½- 8½	relfix, spect A2	04171s3402
Hu 438	4.3	162	23	6½- 14	spect B9	04179s1633
Rst4769	0.4	18	52	7½- 7½	spect A3	04179s0126
λ36	8.0	346	33	7 - 13½	spect G0	04186s1927
β744	0.7	290	62	7½- 7½	binary, 77 yrs;PA	04194s2551
	38.6	3	51	- 12	inc, spect F2	
	44.5	41	51	- 8½	(h3644) ABD cpm	
Σ536	1.6	178	62	8 - 8½	PA inc, spect A5	04197s0448
h342	17.4	235	16	8 - 9	relfix, spect K2	04208s0507
	28.6	83	04	- 12½		
I 384	18.7	187	00	6½-13½	spect G5	04213s3539
Σ544	2.8	351	42	8½- 9	slow PA dec, spect G0	04223s0852
I 59	42.2	198	28	6½- 10	spect F5	04231s3452
I 59b	3.7	281	28	10- 11½		
β403	1.5	92	59	7½- 9	PA dec, spect F8	04232s0220
β311	0.4	182	26	6½- 7	binary, 175 yrs; spect A2, PA inc.	04248s2411
β184	1.5	252	57	6 - 7	slow PA dec, spect F5	04258s2137
Stn 8	2.4	354	32	8 - 13½	(B69) spect G5	04268s2518
	7.0	351	32	- 9		
h3659	36.6	36	19	6 - 14	relfix, spect K0	04288s3546
Σ560	29.8	44	19	6½- 9½	Spect A2	04291s1345
46	1.5	56	36	6 - 11	(β881) cpm; spect B9	04315s0651
h3664	21.6	192	00	7½- 11	spect F5	04328s2509
Σ570	12.8	259	55	7 - 8	relfix, cpm pair; spect both A0	04329s0950
Σ571	17.7	258	04	6½- 11	cpm; spect B9	04335s0343
B70	1.3	66	38	7 - 11½		04346s2708
51	29.9	79	63	5½- 12	(β88) spect A4	04351s0234
Σ576	12.4	172	52	6½- 7	relfix, spect A0	04357s1308
53	0.7	52	61	4 - 7	PA inc, spect K2	04359s1424
β1236	1.4	99	51	7½- 10½	AB PA dec, spect G5	04375s2121
	40.4	314	38	- 8½		
Hwe 11	3.4	99	36	8½- 9	relfix, spect G0	04376s1959
54	0.3	161	23	5½- 6	(Stn 9) spect gM4, presently single.	04382s1946

ERIDANUS

LIST OF DOUBLE AND MULTIPLE STARS   (Cont'd)

NAME	DIST	PA	YR	MAGS	NOTES	RA & DEC
Don 75	0.2	59	54	8½- 8½	PA inc, spect F5	04405s2105
55	9.2	317	52	7 - 7	(Σ590) relfix,	04412s0853
					cpm, spect G8, F2	
Σ596	1.1	263	53	7½- 11½	(Hu 104) relfix,	04435s1202
	10.4	292	23	- 10	spect F0; AC PA inc	
					dist dec.	
β186	1.1	192	45	8 - 11	PA inc, dist dec,	04436s0705
					spect G0	
β316	1.1	181	56	8 - 8	relfix, spect F8	04503s0522
62	67.3	75	13	6 - 8	(Hh 138) optical,	04539s0515
					spect B9	
Σ631	5.5	106	37	7½- 8	relfix, spect A0;	04584s1324
					on Eridanus-Lepus	
					border	
Σ624	28.6	89	15	9 - 9½	relfix	05002s0550
Σ636	3.7	102	49	7½- 8½	relfix, spect A0	05006s0844
A481	0.3	324	63	7 - 8	PA dec, spect B9	05018s0606
66	52.8	9	22	6 - 9	(Σ642) spect B9	05043s0443
Σ649	21.6	72	62	7 - 8½	relfix, spect B8	05059s0844

ERIDANUS

LIST OF VARIABLE STARS

NAME	MagVar	PER	NOTES	RA & DEC
ν	3.4---3.5	.1735	Beta Canis Majoris type spect B2	04338s0327
O^1	4.02± .02	.0815	δ Scuti type, spect F2	04094s0658
R	5.4---6.0	?	Variability uncertain, Spect G4 (*)	04531s1630
S	4.8---5.7	?	Class uncertain (*)	04576s1237
T	7.4--13..	252	LPV. Spect M3e--M5e	03531s2411
U	8.5--14..	274	LPV. Spect M4e	03484s2506
V	8.0---9.3	97:	Semi-reg; spect M6	04020s1552
W	7.6--14.4	376	LPV. Spect M7e	04094s2516
Y	9.5--12..	302	LPV. Spect M7e	02039s5723
Z	6.5---7.4	80	Semi-reg; spect M4	02455s1240
RR	6.8--8.0	97	Semi-reg; spect M5	02498s0828
RS	8.2--12..	296	LPV. Spect Me	04157s1838
RT	7.5--12..	371	LPV. Spect M7e	03319s1620
RU	9.9--10.7	.6322	Ecl.Bin; lyrid, spect F0	03524s1505
RW	9.2--10.5	91	Semi-reg; spect M6	04201s0537
RX	8.5---9.5	.5872	Cl.Var; spect A3	04475s1550
RZ	8.0---9.5	39.28	Ecl.Bin; spect A5+sgG8	04413n1046
SS	9.4--13..	314	LPV.	03094s1204
SU	8.5--10..	112	Semi-reg; spect M4	03489s0131
SV	9.1---9.7	.7137	Cl.Var; spect A4--A8	03095s1133
SW	9.8--14..	401	LPV.	03506s0919
SX	9.5--13..	282	LPV.	04529s0700
SY	9.0--10..	96	Semi-reg; spect N0	05073s0534
TZ	9.8--12.5	2.606	Ecl.Bin; spect F	04192s0608
UU	9.6--11..	340	Semi-reg; spect M7	04345s2741
UV	9.5--16..	467	LPV.	04561s0609
UW	9.0--10..	96	Semi-reg; spect M	03014s1425
VW	8.8--10.5		Irr or semi-reg; spect K5	03193s2138
VX	9.0---9.5		Semi-reg; spect M3	03228s1232
VY	8.5--10..	103	Semi-reg; spect M	03389s1054
WX	9.3--10.0	.8233	Ecl.Bin; spect A7+F6	03218s0053
YY	7.6--8.3	.3215	Ecl.Bin; W Ursae Majoris type, spectra G5+G5	04098s1036
AS	8.3---9.0	2.664	Ecl.Bin; spect A0	03299s0329
AU	9.5--10.5	60	Semi-reg;	04150s2508
BC	9.5--10.6	.2639	Cl.Var; spect A6--A7	04447s1443

LIST OF VARIABLE STARS (Cont'd)

NAME	MagVar	PER	NOTES	RA & DEC
BH	9.9--12..	90	Semi-reg	03213s2004
BM	8.0---9.0		Period uncertain, very long, spectrum M6; probably Ecl.Bin.	04111s1031
BR	8.0---9.0	175	Semi-reg; spect M5	03463s0710
CO	8.3---8.6	5.7875	Ecl.Bin; spect G0	02338s4517
CS	8.8---9.3	.3113	Cl.Var; spect A2	02344s4311
CW	8.1---8.6	2.7284	Ecl.bin; spect F0	03017s1756
DL	6.21± .03	.1562	δ Scuti type, spect F0	03542s0954
DM	4.32± .04	30:	Semi-reg; spect gM4 (54 Erid) Double star?	04382s1946

LIST OF STAR CLUSTERS, NEBULAE, AND GALAXIES

NGC	OTH	TYPE	SUMMARY DESCRIPTION	RA & DEC
685		(galaxy)	SBc; 12.7; 2.3' x 2.3' F,vL,R, vglbM	01459s5302
782		(galaxy)	SBb; 12.7; 2.0' x 2.0' pB,pL,lE, 12^m star 0.5' nf	01561s5801
1084	64[1]	(galaxy)	Sc; 11.1; 2.1' x 1.0' vB,pL,lE, pmbM	02435s0747
1140	470[2]	(galaxy)	I/S0 pec; 12.8; 1.1' x 0.5' pB,S,lE	02522s1014
1172	502[2]	(galaxy)	E1; 13.0; 0.8' x 0.7' pF,pL,R, psbM	02593s1502
1179		(galaxy)	Sc/pec; 13.0; 5.4' x 3.1' cF,pS,gbM	02597s1906
1187	245[3]	(galaxy)	SBc; 11.3; 5.5' x 4.0' pF,cL,pmE,gbM; multiple arm spiral	03004s2304
1199	503[2]	(galaxy)	E2; 12.7; 0.9' x 0.7' cB,pS,R, smbM; brightest member of small group	03013s1548
1209	504[2]	(galaxy)	E5; 12.6; 0.8' x 0.4' B,S,cE,psbM	03038s1548

LIST OF STAR CLUSTERS, NEBULAE, AND GALAXIES (Cont'd)

NGC	OTH	TYPE	SUMMARY DESCRIPTION	RA & DEC
1232	258[2]	⊖	Sc; 10.7; 7.0' x 6.0' pB,cL,R,gbM; fine many-arm spiral	03075s2046
1241	286[2]	⊖	SBb; 13.0; 1.8' x 1.0' F,pL,R,vglbM	03088s0907
1291	△487	⊖	SB; 10.2; 5.0' x 2.0' vB,pL,mbM	03155s4117
1297		⊖	E2; 13.0; 1.0' x 0.8' F,pS	03170s1916
1300		⊖	SBb; 11.3; 6.0' x 3.2' cB,vL,vmE,svmbM (*)	03175s1935
1309	106[1]	⊖	Sc; 11.8; 1.6' x 1.5' cB,cL,R,gbM; 8[m] star 4' sp	03198s1535
1325	77[4]	⊖	Sb; 12.5; 3.0' x 1.0' F,mE, 9[m] star on NE edge	03223s2143
1332	60[1]	⊖	E7/S0; 11.2; 3.0' x 0.8' vB,S,E,smbMN	03241s2131
1337		⊖	Sc; 12.4; 5.4' x 0.9' cF,vL,mE; nearly edge-on	03256s0834
1353	246[3]	⊖	Sb; 12.4; 2.5' x 1.0' pB,cL,E,mbM	03298s2100
1357	290[2]	⊖	Sa; 12.5; 1.2' x 1.0' pF,pL,R,lbM; 9[m] star 2.2' nf	03309s1350
1358	446[3]	⊖	Sb; 13.1; 2.0' x 1.2' vF,S, faint outer ring	03312s0516
----	I.1953	⊖	SBc; 12.5; 2.4' x 2.4' vF,cL,R,bM	03314s2139
1359		⊖	SB/pec; 12.5; 1.6' x 1.3' F,L,R,lbM; ragged structure	03315s1941
1376	288[2]	⊖	Sc; 12.9; 1.4' x 1.3' eF,pL,R,bM	03347s0512
1386		⊖	S0; 12.4; 2.5' x 1.0' F,pL,E, SBN	03350s3610
1389		⊖	E2; 12.8; 1.0' x 0.8' F,pS,1E	03353s3555
1395		⊖	E3; 11.3; 1.5' x 1.0' B,pS,E,psmbM	03363s2311
1400	593[2]	⊖	E1/S0; 12.4; 0.7' x 0.7' cB,pS,R,sbM; NGC 1407 nf	03372s1851

LIST OF STAR CLUSTERS, NEBULAE, AND GALAXIES (Cont'd)

NGC	OTH	TYPE	SUMMARY DESCRIPTION	RA & DEC
1407	107[1]	⊘	E0; 11.4; 0.8' x 0.8' vB,L,R,svmbMN; NGC 1400 sp 11.6'	03379s1844
1415	267[2]	⊘	Sa/Sb; 12.8; 1.8' x 0.9' pB,S,1E,pglbM	03387s2243
1417	455[2]	⊘	Sb; 12.9; 1.6' x 1.1' pF,pL,1E,lbM	03395s0452
1421	291[2]	⊘	Sb; 12.0; 3.0' x 0.6' F,cL,mE; nearly edge-on	03402s1340
1426	248[3]	⊘	E3; 12.7; 0.8' x 0.6' pF,S,1E,bM	03406s2216
1437		⊘	SBa; 12.9; 2.0' x 1.5' F,vL,R,glbM, vSBN	03417s3601
1439	249[3]	⊘	E1; 12.9; 0.8' x 0.7' F,pS,gpmbM	03426s2205
1440	458[2]	⊘	Sa/S0; 13.0; 1.0' x 0.7' pB,pS,R,smbM	03428s1827
1452	459[2]	⊘	SBa; 13.0; 1.7' x 0.9' F,L,R,lbM	03431s1847
1453	155[1]	⊘	E2; 12.8; 0.8' x 0.6' pB,S,R,	03440s0408
1461	460[2]	⊘	Sa/S0; 12.8; 1.9' x 0.6' pB,S,1E,mbMN; lens-shaped	03461s1632
----	I.2006	⊘	E2; 12.8; 0.4' x 0.3' pB,S,R,BM	03522s3608
1487	△480	⊘	I/pec; 12.6; 1.2' x 1.0' pB,pL,R,gbM; double galaxy or collisional system ?	03541s4231
1507	279[2]	⊘	SB; 12.9; 2.9' x 0.5' vF,pL,mE, nearly edge-on	04018s0220
1518		⊘	SBc/pec; 12.2; 2.6' x 0.9' B,L,pmE,gbM; 9^m star 2' sp	04047s2118
1521		⊘	E3; 13.0; 0.7' x 0.5' pB,R,bM	04062s2111
1531		⊘	E3; 13.0; 0.5' x 0.3' pB,pL,E,bM; pair with 1532	04101s3259
1532	△600	⊘	Sb; 11.8; 5.0' x 1.0' B,vL,vmE,psmbM, nearly edge-on; NGC 1531 np 1.6'	04102s3300

ERIDANUS

LIST OF STAR CLUSTERS, NEBULAE, AND GALAXIES (Cont'd)

NGC	OTH	TYPE	SUMMARY DESCRIPTION	RA & DEC
1535	26[4]	◎	vB,S,R; Mag 9; diam 20" x 17"; pale bluish disc with 11.5 mag central star	04121s1252
1537		⊘	S0; 12.0; 1.2' x 0.6' vB,pS,1E, vmbM	04118s3141
1600	158[1]	⊘	E2/S0; 12.4; 1.0' x 0.7' pB,pL,R,gbM	04292s0510
1625		⊘	SBb; 13.0; 1.8' x 0.3' vF,E,sbM; two other dim nearly edge-on spirals in field with Nu Eridani	04346s0324
1637	122[1]	⊘	Sc/pec; 11.4; 2.7' x 2.0' cB,L,R,vgbM	04389s0256
1638	525[2]	⊘	E2/S0; 13.1; 0.6' x 0.5' F,pL,1E	04391s0153
1640		⊘	SBb; 12.5; 1.8' x 0.9' vF,pS,E,gbM; faint outer arms	04401s2032
1659	589[3]	⊘	Sc; 13.0; 1.4' x 0.8' pF,pS,E,hM	04440s0453
1667		⊘	Sb/Sc; 12.9; 1.0' x 0.9' pF,pS,1E	04462s0624
1700	32[4]	⊘	E1; 12.2; 0.9' x 0.6' cB,S,mbM	04545s0456
1726		⊘	E2; 13.0; 0.6' x 0.5' F,R, 13^m star 1' south	04573s0749
----	I.2118	□	eL,eF; 140' x 40'; large reflection nebula, evidently illuminated by Rigel in Orion, 2.5° distant.	05045s0717

DESCRIPTIVE NOTES

ALPHA   Name- ACHERNAR, "The Star at the End of the River"; Magnitude 0.53; Spectrum B5 IV or V; Position 01359s5730. This is the 9th brightest star in the sky, lying at the very southernmost extremity of the constellation Eridanus; so far south in fact that it is never seen in most of the United States. From the lower parts of Florida and Texas it may be observed on autumn evenings, a short distance above the southern horizon. Opposition date (midnight culmination) is October 20.

     Achernar is a star of the "Orion" type, a very hot and luminous bluish giant. It is located at a distance of about 120 light years and has an actual luminosity of about 650 times that of our Sun; the surface temperature is close to 14,000°K. From the known spectral type and the total luminosity, the actual diameter is estimated to be about 7 times the size of the Sun. The annual proper motion of the star is about 0.01"; the radial velocity is 11.5 miles per second in recession.

BETA   Name- CURSA. Magnitude 2.79; spectrum A3 III. Position 05054s0509. The star is located some 3° northwest of Rigel in Orion. Cursa is about 80 light years distant and has an actual luminosity of about 40 times that of our Sun. (Absolute magnitude = +0.8.) The annual proper motion is 0.12"; the radial velocity is 5 miles per second in approach.

     The wide double star 66 Eridani is in the same low-power field. Refer to page 882.

GAMMA   Name- ZAURAK. Magnitude 2.98; Spectrum M0 III Position 03557s1339. This star is also known as 34 Eridani. The distance is about 260 light years; the annual proper motion about 0.13", and the radial velocity about 37 miles per second in recession. Gamma Eridani has an actual luminosity of some 330 times that of the Sun; the computed absolute magnitude is -1.5.

     The small companion at 53" was first observed by J. Herschel in 1834; it evidently has no real connection with the bright star.

EPSILON   Magnitude 3.73; spectrum K2 V. Position 03306s0938. One of the nearest of the naked-eye stars, probably ranking third on the list; only Alpha Centauri and Sirius are known to be closer. At a distance of 10.8 light years the star is just a shade nearer than 61 Cygni. Epsilon Eridani has about a third the luminosity of our Sun (absolute magnitude +6.1) and the diameter and mass are about .9 and 0.75 the solar value, respectively. The star shows a large proper motion of 0.97" annually in PA 271°; the radial velocity is 9 miles per second in recession. This is one of the nearest stars which is reasonably comparable to our Sun in type and luminosity.

In an analysis of 860 plates made at Sproul Observatory, P.van de Kamp (1973) finds evidence for an unseen companion to the star, with an orbital period of 25 years, and a semi-major axis of 7.7 AU. The orbital eccentricity is 0.5 and the computed mass of the companion is less than 0.05 Sun, definitely among the smallest stellar masses known. (Refer also to Tau Ceti and Epsilon Indi)

THETA   Name- ACAMAR. Double star, combined magnitude 2.90; spectrum A3 V and A2. Position 02564s 4030. An attractive pair, the components differing by about one magnitude in brightness, and forming a common proper motion pair with a constant separation of 8.2". E.J.Hartung (1968) refers to this as "a brilliant white pair, one of the gems of the southern sky". According to R.H.Allen, the star was considered by Baily to have possibly decreased in luminosity since ancient times, since Ptolemy (about 130 AD) ranked it as first magnitude.

Very slow orbital motion may be indicated by the slight change in PA: from 82° in 1835 to 88° in 1952; the period evidently must amount to many thousands of years, and the projected separation of the pair is about 285 AU. Direct parallax measurements indicate a distance of about 115 light years; the individual luminosities are then about 50 and 20 suns; the absolute magnitudes are +0.6 and +1.6. The annual proper motion is only 0.08"; the radial velocity of the primary is about 7 miles per second in recession (slightly variable) and that of the companion about 12 miles per second, also in recession. Evidently the brighter star is a spectroscopic binary, but the period is unknown.

DESCRIPTIVE NOTES   (Cont'd)

OMICRON 2   (40 Eridani) Magnitude 4.48; Spectrum
K1 V. Position 04130s0744. A remarkable
triple star system, containing the classic example of a
white dwarf star for the small telescope. The apparent
separation of the AB pair, discovered by William Herschel
in 1783, is 82.8", corresponding to an actual separation
of about 400 AU or nearly 40 billion miles. Since the PA
has changed by only 2° since the first accurate measures
in 1836, the period of this very wide binary is uncertain
but is estimated to be some 7000 to 9000 years. At a distance of only 16 light years this is the 8th nearest of the
naked eye stars (See list on page 640) and shows the very
large annual proper motion of 4.08" in PA 213°. The true
space velocity is about 62 miles per second, and the radial
velocity is 25 miles per second in approach.

The companion star is itself a visual double for
higher powers, detected by O.Struve in 1851. It consists
of a remarkable combination; a white dwarf (B) and a red
dwarf (C) forming a binary pair with a separation of 7.6"
(1961) and an orbital period of about 248 years. According
to computations by Van den Bos the semi-major axis of the
system is 6.89" or about 34 AU with periastron in 1848;
the eccentricity is 0.40. At present the BC separation is
increasing toward a maximum of 9" about 1990; the orbital
motion is retrograde.

The "C" component, a faint red dwarf, is noted for
its unusually small mass of only 0.2 the solar mass. Only

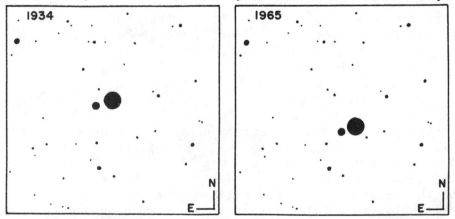

OMICRON 2 ERIDANI— The Proper Motion over a 31 year Interval. The
Total Displacement is 126". From Lowell Observatory photographs.

DESCRIPTIVE NOTES   (Cont'd)

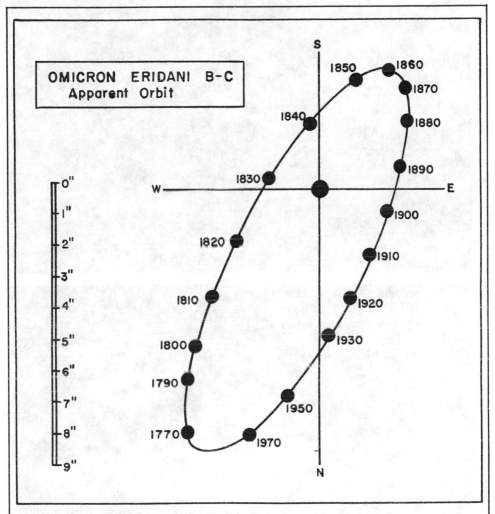

OMICRON ERIDANI B-C
Apparent Orbit

three visible stars are definitely known to have smaller masses: UV Ceti, Ross 614B in Monoceros, and Krueger 60B in Cepheus. The "B" component is, however, the most interesting member of the system, being the only white dwarf star which can honestly be called an easy object for the small telescope. It was the first star of the type to be recognized, in 1910. (Although Sirius B was discovered in 1862, and is a much more famous star, a satisfactory spectrum was not obtained until 1915.) The diameter of Omicron Eridani B is computed to be about 17,000 miles, or just a little more than twice the size of the Earth. From

DEEP SKY OBJECTS IN ERIDANUS. Top: The wide double Omicron 2 Eridani AB, photographed at Lowell Observatory. Arrow indicates the white dwarf. Below: The fine spiral galaxy NGC 1232 as portrayed with the 200-inch telescope, Palomar.

DESCRIPTIVE NOTES   (Cont'd)

the orbital elements, the mass is known to be nearly half
that of the Sun; as a result the average density of this
star is about 65,000 times that of the Sun. Its substance
weighs nearly 2 tons to the cubic inch. This is about 90
thousand times the density of water; the surface gravity on
such a body would be about 37,000 times that of the Earth.

	MAG.	SPECT.	LUM.	DIAM.	MASS.	ABS.MAG.	DENS.
A	4.48	K1 V	0.30	0.90	0.75:	+6.0	1.04:
B	9.7	DA	.0027	0.019	0.44	+11.2	65,000
C	10.8	dM4e	.0008	0.43	0.20	+12.3	1.8

The fantastic density of a white dwarf star, which
once seemed quite unbelievable, is now known to be perfectly compatible with present knowledge of the structure of
matter. White dwarfs are "dying stars" which have consumed
their nuclear fuel supplies and are no longer producing
energy. The great density is the result of gravitational
contraction, which in these stars is unopposed by any
internal nuclear reactions. (For a summary of facts and
statistics concerning white dwarf stars, refer to the
article on Sirius B in the constellation Canis Major)

e   (82 Eridani) (LFT 277) Magnitude 4.28;
             Spectrum G5 V. Position 03179s4316. This
small star is mentioned here for its unusually large proper
motion of 3.14" annually in PA 76°. Only 20 light years
distant, the star closely resembles our own Sun but is
somewhat smaller and about 70% as bright. The computed
absolute magnitude is +5.3. 82 Eridani shows a radial
velocity of about 52 miles per second in recession; the
true space velocity is about 80 miles per second.

R and S   Spectra G4 and F0. Positions 04531s1630 and
                04576s1237. The variability of these two
stars was originally reported by B.Gould in 1879, but
apparently never confirmed; it is uncertain that either
star is actually variable.

SPIRAL GALAXY NGC 1300 in ERIDANUS. One of the most perfect examples of a barred spiral. This photograph was made with the 200-inch reflector at Palomar Observatory.

FORNAX

LIST OF DOUBLE AND MULTIPLE STARS

NAME	DIST	PA	YR	MAGS	NOTES	RA & DEC
B24	6.1	122	27	6½- 13	spect G5	01562s3318
Stn 5	2.7	201	42	8 - 8½	relfix, spect F5	02153s3057
β738	0.1	297	60	7½- 8	PA & dist dec; spect F8	02211s3006
B38	1.1	252	48	7½- 10	spect A2	02294s2743
ω	10.8	244	52	5 - 8	(h3506) cpm pair; relfix, spect B9	02316s2827
h3509	23.5	59	18	7½- 11½	spect F0	02320s3145
λ19	0.5	308	56	8 - 8½	PA dec, spect F5	02382s2421
β261	3.0	100	43	8 - 9½	relfix, spect G5	02416s2806
43G	12.1	191	54	6 - 8	slight PA inc; spect G0	02420s2543
B1436	1.1	30	47	8½- 11½	spect F5	02444s3708
h3532	5.4	145	52	7 - 8	relfix, spect F2	02466s3737
γ¹	12.0	145	54	6 - 12	(β877) (h2161)	02476s2445
	40.9	143	54	- 11	relfix, spect G5; AC dist & PA dec.	
η²	5.0	14	33	6 - 11	(h3536) slight PA inc, spect K0	02482s3603
δ163	0.3	73	42	8 - 8½	(Daw 77) spect F5	02543s2909
Cor 17	3.4	188	54	8½- 9	relfix, spect G0	02546s3939
β741	0.9	180	25	7½- 8	binary, 137 yrs;	02550s2510
	28.6	224	54	- 9	PA inc, spect G5	
α	1.9	294	63	4 - 7	(h3555) binary, spect F8 (*)	03099s2912
I 464	0.6	217	24	8 - 8½	spect K5	03139s3914
λ25	10.2	18	36	6½- 12	spect K0	03247s2845
χ³	6.3	248	39	6½- 11	(I 58) relfix; spect A0	03263s3602
B52	0.2	311	65	7 - 7½	binary, 19.4 yrs; nearly edge-on; spect F5	03319s3115
B53	1.4	225	42	6½- 10	spect K0	03325s3202
σ	5.0	180	02	6 - 11	spect A2	03444s2930
h3596	9.2	137	53	8½- 8½	relfix, spect A3	03465s3156

FORNAX

LIST OF VARIABLE STARS

NAME	MagVar	PER	NOTES	RA & DEC
R	7.5--13.0	387	LPV. Spect Ne	02270s2619
S	5.6---8.5		Uncertain (*)	03441s2433
T	9.5--11..	92	Semi-reg; spect M2	03275s2833
U	9.1--11..	318	LPV.	03423s2523
X	8.5--10..	75:	Semi-reg; spect M3	02407s2620
RZ	8.0--9..	65	Semi-reg; spect M5	03303s2549
SS	8.5--9.5	.4954	Cl.var; spect A3--G0	02056s2706
ST	7.7--9..	277	Semi-reg; spect M5	02422s2925
SU	9.5--11..	2.435	Ecl.bin; spect A2	02195s3726

LIST OF STAR CLUSTERS, NEBULAE, AND GALAXIES

NGC	OTH	TYPE	SUMMARY DESCRIPTION	RA & DEC
922	239[3]	⊘	SBc; 12.3; 1.5' x 1.0' cF,pL,1E,pgbM	02229s2501
986		⊘	SB; 11.8; 1.5' x 0.8' pB,L,pmE,sbM	02316s3915
---	F.S.	⊘	vvL,eF,R, diam 65'; unusual dwarf galaxy; Fornax System (*)	02370s3440
1049		⊕	Mag 13, diam 0.4'; vS,F; extra-galactic cluster; member of Fornax System (*)	02377s3429
1079		⊘	Sa; 12.6; 1.1' x 0.8' B,pL,pmE, sbM	02416s2913
1097	48[5]	⊘	SBb; 10.6; 9.0' x 5.5' vB,L,vmE,vBN (*)	02443s3029
1201	109[1]	⊘	S0; 11.8; 1.5' x 1.0' cB,pS,1E,	03020s2615
1255		⊘	Sc; 12.1; 3.5' x 2.2' F,pL	03114s2558
1288		⊘	Sc; 13.0; 1.1' x 1.0' vF,L,R,vglbM	03153s3246

LIST OF STAR CLUSTERS, NEBULAE, AND GALAXIES (Cont'd)

NGC	OTH	TYPE	SUMMARY DESCRIPTION	RA & DEC
1292		⊘	Sb; 12.8; 1.5' x 0.7' F,pS,lE,vgbM	03160s2748
1302		⊘	SBa; 11.4; 2.4' x 2.1' S,R,psvmbM	03177s2614
1316	△548	⊘	S0p; 10.1; 3.5' x 2.5' vB,cL,lE, vsvmbMN; radio source (*)	03207s3725
1317	△547	⊘	SB; 12.2; 0.7' x 0.6' pB,pS,psbM	03208s3717
1326		⊘	SB; 11.8; 3.0' x 2.5' pS,psvmbMN	03220s3639
1339		⊘	E0; 12.8; 1.0' x 1.0' cB,pS,R,psbM	03261s3227
1341		⊘	SBa; 13.1; 0.8' x 0.7' F,S,R	03261s3719
1344	257[1]	⊘	E5; 11.6; 2.0' x 1.0' cB,pL,R,vgbM	03267s3114
1350	△591	⊘	Sb; 11.8; 3.0' x 1.5' B,L,mE,vmbMN	03291s3347
1351		⊘	E4; 12.8; 0.8' x 0.6' pB,pS,R,psbM	03286s3502
1360		◎	vL,F; 6.0' x 4.5'; with 9^m star in center; class uncertain (*)	03310s2600
1365		⊘	SB; 11.2; 8.0' x 3.5' vB,vL,mE, BN (*)	03318s3618
1366	857[3]	⊘	E6; 13.0; 0.9' x 0.4' vF,S,lbM	03320s3123
1371	262[2]	⊘	SB; 12.2; 2.0' x 1.4' pB,pL,vlE,psbM	03328s2506
1374		⊘	E0; 12.4; 0.8' x 0.8' vB,pL,lE,gmbM	03334s3524
1379		⊘	E0; 12.3; 0.6' x 0.6' B,pL,R,gpmbM	03342s3537
1380	△574	⊘	E7/S0; 11.4; 3.0' x 1.0' vB,L,R,psbM; lenticular	03346s3509
1381		⊘	E7/S0; 12.6; 2.0' x 0.5' F,pL,E	03347s3528

LIST OF STAR CLUSTERS, NEBULAE, AND GALAXIES   (Cont'd)

NGC	OTH	TYPE	SUMMARY DESCRIPTION	RA & DEC
1385	263[2]	⊖	Sc; 11.8; 2.6' x 1.6' pB,pS,1E,gpmbM	03352s2440
1387		⊖	S0; 12.1; 1.0' x 0.9' vB,pL,R,bM	03351s3541
1398		⊖	SBb; 10.7; 4.5' x 3.8' cB,cL,R,vmbM (*)	03368s2630
1399		⊖	E0; 10.9; 1.4' x 1.4' vB,pL,R,psbM	03366s3537
1404		⊖	E1; 11.5; 1.0' x 1.0' vB,pL,R,psmbM	03370s3545
1406		⊖	S ; 12.7; 3.0' x 0.8' F,cL,vmE,vglbM	03375s3128
1425	852[2]	⊖	Sb; 12.1; 3.5' x 1.7' F,pL,E,gbM	03401s3004
1427		⊖	E3; 12.4; 1.4' x 1.0' pF,S,1E, psbM	03404s3534

DESCRIPTIVE NOTES

ALPHA   Magnitude 3.86; Spectrum F8 IV; position
03099s2912. Alpha Fornacis is a long period
binary, first measured by J.Herschel in 1835 when the separation was 5.3" in PA 310°. During the next century the
pair closed in to about 0.9" and then began to widen out
again; a measurement by Van den Bos (1963) gave the separation as 1.9" in PA 295°. Period and orbital elements are
still uncertain as shown by the differences in the results
obtained by Woolley and Van den Bos:

	Period	Semi-major axis	Eccentr.	Periastr.
Woolley	408 yr	3.5"	0.68	1942
Van den Bos	155 yr	2.7"	0.68	1949

The faint star has an apparent magnitude of about 7, but
has been suspected of variability by Innes, Van den Bos and
others. In 1895 and in late 1925 it was a faint and very

difficult object, while in 1913 and again in 1917 it could
not be detected at all by the experienced observer R.T.
Innes. The range appears to be at least two magnitudes and
the star in recent years seems to have remained near maximum light, with a photometrically determined magnitude of
6.47.

Parallax measurements give the distance as about 40
light years; the resulting absolute magnitudes are +3.5
and about +6.5, the primary being a subgiant with about 3
times the solar luminosity. The annual proper motion is
0.72" in PA 27°; the radial velocity is 12.5 miles per
second in approach.

S   Position 03441s2433. The suspected variability
of this star appears to rest upon the observations of Hartwig, Holetschek, and Abetti in March 1899. The
star was then claimed to be about magnitude 5.5, but no
variations have been detected since. Spectrum F8; present
magnitude 8.5.

**FORNAX
GALAXY
CLUSTER**   This is a compact group of 18 bright galaxies
and a number of fainter ones, located on the
border of Fornax and Eridanus at about 3h 35m
-35°40'. If an ocular with a one-degree field
is centered at this position it is possible to include nine
galaxies in the view at one time. These nine objects are
marked by the symbol (*) in the following list of members:

NGC 1316	NGC 1351	NGC 1381*	NGC 1386*
NGC 1317	NGC 1365	NGC 1387*	NGC 1389*
NGC 1326	NGC 1374*	NGC 1399*	NGC 1437
NGC 1341	NGC 1379*	NGC 1404*	
NGC 1350	NGC 1380*	NGC 1427	

The last three objects listed actually lie across the Fornax border, in Eridanus. For descriptive data, refer to
pages 885 and 886.

The brightest galaxy of the Fornax group is the 10th
magnitude object NGC 1316 which appears to be an elliptical
or S0 galaxy, although it is listed in the Shapley-Ames
Catalogue as a spiral of uncertain classification. This

GALAXY NGC 1365 in FORNAX. One of the most prominent members of the Fornax Galaxy Cluster and a fine example of a barred spiral.   Radcliffe Observatory photograph.

peculiar system is identified with the strong radio source
called "Fornax A". Resembling the unusual galaxy NGC 5128
in Centaurus, the system has sometimes been regarded as a
pair of colliding galaxies. Modern photographs show that
this interpretation is almost certainly incorrect; the object is evidently a single elliptical or S0 system spotted
with a number of dust clouds and dark lanes. NGC 1275 in
Perseus has a very similar appearance, and it now seems
that violent outbursts of some sort have occurred in the
nuclei of such galaxies, releasing great clouds of gaseous
material. Masses of several hundred million suns may be
involved in outbursts on this scale.

In the "Atlas of Southern Galaxies" (Cordoba, 1968)
NGC 1316 is referred to as "a spheroidal object of smooth
distribution of brightness and ellipticity E2.5. A chain of
dark globules symmetrically aligned with respect to the
nucleus produces an absorption estimated to be less than
0.05 magnitude in the total photographic magnitude, whose
revised value is 9.56. The maximum diameter of these clouds
of obscuring material is 10" of arc. Using special photographic techniques, H.Arp has detected faint extensions of
this object whose structure is related to the radio distribution...."   The observations give a distance modulus
of about 31 magnitudes for the galaxy, corresponding to a
distance of 17 megaparsecs or some 55 million light years.
A companion galaxy, NGC 1317, lying 6' to the north, is a
rather tight spiral with a very luminous center; it is
very probably somewhat more distant than NGC 1316.

Second in brightness in the Fornax Cluster is the
nearly spherical system NGC 1399, an E0 galaxy of magnitude
10.9 with a small bright nucleus. The fainter elliptical
NGC 1404 lies in the same field, 8' to the south.

Third in brightness in the group is the wonderful
barred spiral NGC 1365, probably the finest object of its
type in the southern sky. The bright central bar, measuring
about 3' in length, has an actual extent of some 45,000
light years, while the long faint curving spiral arms sweep
out for even greater distances at both ends of the bar.
With a computed absolute magnitude of about -20, this is
one of the most luminous of all known barred spirals. A
supernova was recorded in the system in 1957. (Mean red-
shift of the Fornax Galaxy Cluster = 1080 miles per second)

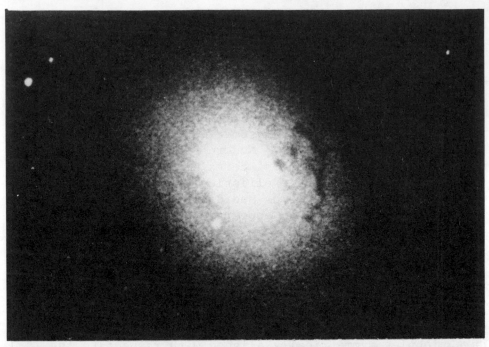

PECULIAR GALAXIES IN FORNAX. Top: The strong radio source NGC 1316. Below: The "Fornax System", one of the nearby dwarf elliptical members of the Local Group of Galaxies.

DESCRIPTIVE NOTES   (Cont'd)

FORNAX
SYSTEM
Position 02370s3440. In addition to the members of the Fornax Galaxy Cluster (Page 899) the constellation also contains a very peculiar galaxy called the "Fornax System", an example of a freak type of galaxy which may actually be fairly common in the Universe, but almost impossible to detect at distances beyond the Local Group. Another similar object has been discovered in Sculptor, and is called the "Sculptor System"; both were discovered at Harvard in 1938.

These objects are spherical swarms of exceedingly faint stars, somewhat resembling our own globular clusters except for the great difference in size, brightness, and density. The Fornax System for example, is roughly 50 times the size of the largest globular cluster, but is still far beyond the reach of any amateur telescope because of its extremely low density and open structure. Even with powerful instruments it was first detected as a mere hazy spot on the negative, resembling an imperfection on the plate rather than the true image of a real celestial object. The brightest individual stars were only 19th magnitude! The outline of the system is elliptical, oriented NE-SW, with a major diameter of about one degree. The distribution of the stars is very uniform, with no nuclear condensation and no nebulosity. Five faint globular star clusters, however, appear to be members of the system. The brightest of these is NGC 1049, with an apparent visual magnitude of 12.9 and an integrated spectral type of F0; it is the brightest object right of center in the photograph opposite. Partial resolution of this cluster has been achieved with the 200-inch telescope. The other four globulars of the Fornax System have apparent magnitudes of 13.7, 13.9, 14.1, and 16.6. A study of these clusters leads to a distance of about 630,000 light years. It is an interesting thought that if the Fornax System was at the distance of the great Andromeda spiral M31 its group of five globulars could be detected, but the galaxy itself would not be seen at all!

The Fornax System is now recognized as a member of the Local Group of Galaxies, at about three times the distance of the Magellanic Clouds. Some facts about this phantom galaxy are given here: Total apparent magnitude about 9; diameter = 65' or 15,000 light years; total luminosity about 20 million suns; absolute magnitude= -13.5.

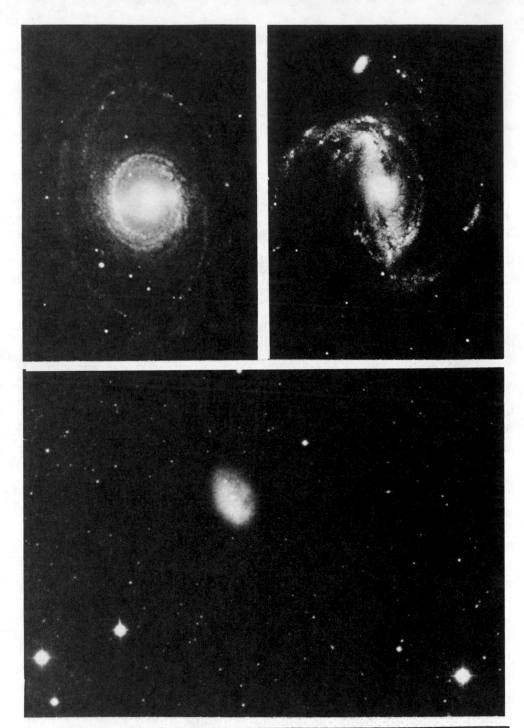

DEEP SKY OBJECTS IN FORNAX. Top left: The barred spiral galaxy NGC 1398. Top right: Another barred system, NGC 1097. Below: The peculiar nebula NGC 1360, usually classed as a planetary.   Palomar Observatory photographs

GEMINI

LIST OF DOUBLE AND MULTIPLE STARS

NAME	DIST	PA	YR	MAGS	NOTES	RA & DEC
Ho 21	9.3	240	40	6½ – 12½	relfix, spect B9	05579n2734
Σ830	12.0	254	21	8 – 8½	spect F2	06003n2739
	25.6	188	10	– 11½		
1	0.2	184	62	5 – 5½	binary, 13.2 yrs;	06011n2316
					PA inc, spect G5	
OΣ134	31.0	188	24	7 – 8	relfix, spect G0;	06062n2426
					in field with M35	
3	0.5	339	58	6 – 10	(β1241) relfix;	06067n2307
	18.4	63	58	– 14	spect B2	
4	0.3	252	60	7 – 7½	PA dec, spect B9	06075n2301
					(β1058)	
η	1.5	266	62	3 – 6½	(β1008) primary	06119n2231
					variable, spect M3,	
					G8 (*)	
Σ889	21.3	237	57	7 – 9½	PA inc, spect K2	06168n2502
S513	59.4	258	20	8 – 9	relfix, spect B8	06182n2109
	42.0	247	08	– 11		
Σ897	18.0	348	16	8 – 8½	relfix, spect A0	06193n2642
Σ899	2.0	19	61	7 – 8	relfix, spect A0	06199n1736
μ	122	141	00	3 – 9½	(β1059) spect M3	06199n2233
μb	0.8	260	39	9½ – 10½	slow PA dec (*)	
OΣ139	0.3	238	59	7 – 9½	PA dec, dist dec;	06226n2229
					spect A3	
β1191	2.0	304	66	7 – 13½	spect K0	06232n1848
OΣ140	2.8	118	43	7 – 9½	relfix, spect B9	06238n1533
15	27.1	204	57	7 – 9	(Hh 223) dist dec;	06248n2049
	77.7	34	11	– 12½	spect K0	
ν	113	329	24	4 – 8 ½	(OΣΣ77) (18 Gem)	06260n2015
ν a	0.2	129	44	4 – 4½	PA inc, spect B7	
ν b	0.2	282	44	8½ – 9	(β1192) PA dec.	
OΣ141	2.2	142	53	7½ – 9½	relfix, spect A0	06271n1756
Ho 340	6.5	22	27	7 – 13	spect G5	06275n1759
OΣ143	7.9	103	58	7 – 10	relfix, cpm; spect	06283n1659
	44.8	344	56	– 11	K2, K0; C optical,	
					dist inc.	
OΣ519	8.2	78	34	8 – 10½	relfix, spect G5	06284n1546
20	20.0	210	56	6 – 7	(Σ924) relfix,	06294n1749
					cpm; spect F8	
OΣ145	1.5	338	57	7 – 9½	relfix, spect F5	06295n1545

LIST OF DOUBLE AND MULTIPLE STARS   (Cont'd)

NAME	DIST	PA	YR	MAGS	NOTES	RA & DEC
Σ932	1.8	317	55	8 - 8	PA dec, spect F2	06315n1448
Ho 341	1.4	134	00	7 - 12	spect A0	06332n1344
	65.0	155	00	- 9		
0Σ149	0.5	1	61	6½- 9	binary, 116 yrs;	06333n2720
					PA dec, spect G2	
Σ942	3.2	246	62	9 - 9	relfix, spect F0	06346n2342
Ho 625	13.4	194	06	7 - 12	spect A5	06352n2430
	47.4	251	06	- 9½		
β571	2.9	342	62	6 - 12	PA inc, spect A3	06370n1302
A2728	5.0	232	53	7½- 13	spect A5	06385n1529
ε	110	94	25	3 - 9	spect G8, K2 (*)	06408n2511
Σ957	3.5	93	42	7½- 9	relfix, spect A0	06420n3053
	18.9	101	07	- 13		
0Σ155	15.5	261	25	7 - 10	relfix, spect K2	06423n2444
0Σ156	0.5	263	61	6½- 7	binary, about 450	06445n1815
					yrs; PA dec, spect	
					A2	
0Σ160	1.5	181	56	7 - 10	PA inc, spect G5	06514n2114
38	7.0	151	62	5½- 7½	(Σ982) PA dec;	06518n1315
					attractive pair;	
					binary, about 3000	
					yrs; spect F0, dG4	
0Σ161	20.1	171	12	6½- 11	relfix, spect B9	06520n2138
Σ981	2.1	140	59	8 - 8	cpm; slow PA & dist	06522n3014
					dec, spect G0	
A2833	0.6	169	42	8½- 9	spect G5	06522n1202
Σ991	3.8	167	37	8 - 9	relfix, spect A0	06540n2501
	134	202	07	- 10½		
Σ991c	13.2	359	07	10½- 11		
Σ1000	0.3	341	55	8½- 8½	(A1061) PA inc,	06563n2518
	22.3	67	55	- 8½	spect G; AC relfix	
0Σ162	0.3	320	47	7½- 9	(A2461) PA inc	06570n1601
	21.4	155	24	- 10	spect G0	
β100	3.3	258	44	7 - 10½	relfix, spect K0	06582n1228
0Σ163	0.2	4	58	7 - 8½	PA inc, dist dec;	06584n1151
	14.2	158	00	- 12	spect A2	
Ho 342	1.0	85	67	8 - 9	PA inc, dist slight	07000n1310
					inc, spect F5	
ζ	87.0	84	24	4 - 10½	primary cepheid;	07012n2039
	96.5	350	25	- 8	AC dist inc (*)	

LIST OF DOUBLE AND MULTIPLE STARS (Cont'd)

NAME	DIST	PA	YR	MAGS	NOTES	RA & DEC
Σ1014	2.0	36	53	8½- 8½	relfix, spect F8	07027n2613
OΣ164	13.7	50	05	6 - 10	spect K2	07031n2456
Σ1017	12.3	254	34	8½- 9	relfix, spect G	07043n1655
45	10.4	14	63	5 - 10	(OΣ165) optical;	07055n1601
	57.1	331	63	- 13	dist inc, PA dec;	
					spect gG8	
Σ1027	6.9	355	38	8 - 8	relfix, spect K2	07059n1659
τ	1.9	178	25	5 - 11½	(46 Gem) (β1009)	07080n3020
	60.9	345	63	- 12½	relfix, AB cpm;	
					spect K2	
Σ1035	8.7	41	49	7½- 7½	relfix, spect F5	07090n2222
Σ1037	1.2	329	62	7 - 7	binary, 117 yrs;	07097n2719
	14.1	88	62	13	very eccentric	
					orbit, PA dec,	
					spect dF6; AC is	
					optical, PA dec.	
OΣ168	24.0	67	00	6½- 10½	spect G5	07098n2126
Wei 14	2.2	157	53	8 - 8½	relfix, spect A0	07100n1516
OΣ167	5.3	158	33	7 - 10½	relfix, spect A5	07103n3214
OΣ520	0.6	12	59	8 - 9	PA inc, spect A0	07107n2835
Σ1047	22.4	25	23	7½- 10	slow dist inc;	07115n1551
Σ1047b	0.5	336	22	11- 11	spect F2 (A2526)	
52	23.9	265	10	6 - 12½	(Ho 343) optical;	07116n2459
					PA & dist inc;	
					spect gM1	
A2527	1.9	289	27	8½- 13½	spect A5	07124n1909
Σ1053	13.8	310	00	7½- 10	spect A0	07138n2438
Σ1054	18.6	291	30	7½- 8½	relfix, spect F2	07148n3503
	79.4	268	17	- 9		
λ	9.6	33	53	3 - 10	(Σ1061) relfix;	07152n1638
					cpm; spect A3	
Σ1068	4.0	352	40	8½- 9	relfix, spect A2	07170n1328
δ	6.3	218	62	3½- 8	(Σ1066) PA inc,	07171n2205
					binary, spect F0,	
					dK6 (*)	
Σ1070	1.7	321	39	8 - 9	relfix, spect A0	07181n3408
	87.5	122	08	- 10½		
56	16.5	202	35	6 - 13	optical. dist dec.	07190n2033
					spect gM0	
Σ1081	1.6	232	62	8 - 8½	PA inc, spect B9	07212n2133

LIST OF DOUBLE AND MULTIPLE STARS (Cont'd)

NAME	DIST	PA	YR	MAGS	NOTES	RA & DEC
Σ1083	6.4	44	50	6½- 7½	relfix, spect A5	07226n2036
Ho 346	12.8	58	30	7 - 12	relfix, spect G5	07230n1815
Σ1087	21.3	40	13	8 - 11	slight dist inc; spect A2	07232n1412
Σ1088	11.1	195	14	7 - 9	relfix, spect B9 Σ1087 at 113"	07232n1412
Σ1089	7.2	8	38	8½- 8½	relfix, spect A2	07234n1457
0Σ171	1.1	132	53	7 - 10	relfix, spect G5	07235n3143
Σ1090	60.8	97	21	7 - 8	relfix, spect F0	07236n1837
Σ1090b	19.9	320	15	8 - 10		
Σ1094	2.5	96	53	7½- 8½	relfix, spect B8	07246n1525
S548	11.3	22	01	7½- 12	spect K5	07247n2215
	35.6	276	21	- 9½		
63	42.9	324	17	5½- 9½	cpm; spect F5; neb NGC 2392 is 40' SE	07248n2133
ρ	3.4	8	35	4½- 12	cpm; spect F0	07259n3153
	214	291	09	- 10½		
Σ1102	7.6	47	32	7½- 9	relfix, spect F8	07276n1358
β579	1.0	216	45	7 - 11½	(0Σ173) spect A2	07280n3314
	18.7	232	16	- 12		
Σ1106	10.6	32	19	8½- 8½	cpm; spect G0, G0	07285n1625
Σ1108	11.5	178	34	6½- 8½	relfix, cpm pair; spect K0, F8	07298n2300
α	1.8	140	67	2½- 3½	CASTOR (Σ1110)	07314n3200
	72.5	164	55	- 9	multiple star (*)	
Σ1116	1.8	99	57	7 - 7½	PA dec, spect B8	07318n1225
0Σ175	0.3	330	62	6 - 6½	dist dec, spect K0	07320n3104
	81.3	195	08	- 10		
Σ1124	19.4	326	34	8 - 8	relfix, spect A0	07380n2155
Σ1129	0.8	196	45	8½- 10	(A2876) spect F5	07388n1810
	21.7	64	45	- 9		
κ	7.0	239	62	4 - 10	(0Σ179) slight PA inc, cpm; spect G8	07414n2431
A674	1.0	128	34	7½- 10	spect A5	07415n3114
β	29.6	280	00	1 - 14	POLLUX. Optical	07423n2809
	201	73	22	- 9	spect K0 (*)	
	234	90	24	- 10½	C= β580	
β580	1.4	135	22	9 - 12	PA slight inc.	
Ho 247	0.3	206	63	8 - 8½	binary, about 170 yrs; PA inc; spec F	07432n2115

GEMINI

LIST OF DOUBLE AND MULTIPLE STARS (Cont'd)

NAME	DIST	PA	YR	MAGS	NOTES	RA & DEC
π	21.0	214	23	5 - 11	(Σ1135) optical,	07443n3332
	91.9	341	22	- 10	dist dec, spect gM0	
AGC 2	0.8	116	37	8 -11	relfix, spect G5	07448n2848
Σ1140	6.3	273	42	7 - 8½	relfix, spect G5	07455n1828
82	0.2	71	58	7 - 7	PA dec, spect F2, A	07456n2316
	4.0	34	38	- 13	(β1062)	
Σ1147	2.3	169	54	9 - 9	slight PA inc;	07473n2439
					spect G0	
A2881	4.2	324	22	8 - 13	spect K0	07505n1551
OΣ183	15.9	20	01	7 - 11	spect A2	07511n1610
Wei 17	7.0	96	38	8 - 8½	relfix, spect F0	07512n1505
Ho 250	0.7	158	44	7 - 9	relfix, spect G5	07542n2106
	9.4	154	00	- 13		
OΣ187	0.3	18	62	7 - 7½	binary, 160 yrs;	08010n3310
					PA dec, spect A0	

LIST OF VARIABLE STARS

NAME	MagVar	PER	NOTES	RA & DEC
ζ	3.7---4.1	10.152	Cepheid; spect F7--G3 (*)	07011n2039
η	3.1---3.9	233	Semi-reg; spect M3 also visual double (*)	06119n2231
R	5.9--13.9	370	LPV. Spect S3e--S6e	07043n2247
S	8.3--14.6	294	LPV. Spect M4e--M5e	07400n2334
T	8.1--14..	288	LPV. Spect S4e--S9e	07463n2352
U	8.9--14..	Irr	Peculiar, Nova-like; SS Cygni type (*)	07521n2208
V	7.9--14.3	276	LPV. Spect M4e--M5e	07203n1312
W	6.8---7.5	7.9147	Cepheid, spect F6--G5	06321n1522
X	7.9--13.4	263	LPV. Spect M5e	06439n3020

LIST OF VARIABLE STARS (Cont'd)

NAME	MagVar	PER	NOTES	RA & DEC
Y	9.0--11..	160:	Semi-reg; spect M6e--M7	07382n2033
RS	9.8--11.5	140:	Semi-reg; spect M8	06584n3036
RT	9.9--15..	350	LPV. Spect N	06436n1840
RW	9.9--11.7	2.865	Ecl.Bin; spect B5+F5	05584n2309
RX	9.2--11.1	12.209	Ecl.Bin; spect A4	06469n3318
RY	8.5--11.0	9.301	Ecl.Bin; spect A2e+K2	07245n1546
RZ	9.6--11.0	5.530	Cepheid, spect F5--G5	05596n2214
SS	9.0--10..	89.2	RV Tauri type, spect F8--G5	06055n2238
ST	9.2--13..	246	LPV. Spect M6e--M8e	07359n3436
SU	9.9--12.2	50.1	RV Tauri type, spect F5--G6	06109n2743
SW	8.9--10.3	680	Semi-reg; spect M5	06564n2607
TU	7.5--8.4	230:	Semi-reg; spect N3	06078n2602
TV	7.3--8.2	182	Semi-reg; spect M1	06088n2153
UZ	9.3--13..	350	LPV. Spect M9	07100n1745
VV	9.5--13..	253	LPV.	06228n2534
VW	8.7--9.1	Irr	Spect N	06389n3130
WW	9.9--10.5	1.2378	Ecl.Bin; spect B6	06090n2331
WY	7.5---8.0	±50 yr?	Ecl.Bin. Spect M2ep+B	06089n2313
WZ	9.5--14..	332	LPV. Spect M3e	07135n2605
XX	8.5--13.2	382	LPV. Spect M10e	07229n3327
YY	9.1--9.6	.8143	CASTOR C; Ecl.Bin; (*)	07314n3159
AA	9.3--10.2	11.302	Cepheid; Spect F6--G4	06035n2620
AD	9.5--10.5	3.788	Cepheid; spect F5--G2	06402n2059
AG	9.4--10..	Irr	Spect M6	06499n3525
AM	9.5--13..	355	LPV. Spect M10	07042n2823
BK	9.2--9.5	Irr	Spect K5	06271n1339
BN	6.0---6.6	Irr	Spect 07e	07342n1701
BQ	6.9---7.3	Irr	Spect M4	07105n1615
BU	6.1--7.5	Irr	(6 Geminorum) Spect M1	06093n2255
CR	8.5---9.5	Irr	Spect N	06315n1607
DE	8.6--10..	Irr	Spect M4	06407n2259
DM	4.8--16.5	---	Nova 1903	06410n3000
DN	3.3--15.0	---	Nova 1912	06517n3213
FK	9.4--10..	Irr	Spect M6	06466n1214
IS	5.4---6.2	47:	Semi-reg; spect gK4	06464n3240
LR	9.0 ±0.06	.23887	β Canis Maj type, spect B0	06122n2219
LT	9.0---9.2	1.0748	Ecl.bin; spect B1	06154n2336

LIST OF STAR CLUSTERS, NEBULAE, AND GALAXIES

NGC	OTH	TYPE	SUMMARY DESCRIPTION	RA & DEC
2129	26^8	⊙	pL, diam 5'; about 50 stars mags 8...15; class D	05580n2318
----	I.2157	⊙	S, diam 4'; about 20 faint stars; class D	06016n2400
2158	17^6	⊙	pS,mC,vRi; mag 11, diam 4'; about 150 faint stars, class G; on edge of M35 (*)	06043n2406
2168	M35	⊙	vL,cRi,pC, mag 5.5; diam 30'; about 120 stars mags 8..... class E (*)	06057n2420
----	I.443	□	vF,vL,curved arc 25' x 5'; with filamentary structure; probably supernova remnant	06139n2248
----	I.444	□	7th mag star (12 Gem) with faint neby, diam 4'	06175n2319
----	J900	◎	vS,B, mag 12, diam 10"; nearly stellar	06230n1749
2266	21^6	⊙	pS,eC,Ri; diam 5'; stars mags 11....15; class F (*)	06405n2702
2304	2^6	⊙	pL,Ri,mC; mag 10; diam 4'; 20 faint stars; class D	06523n1805
2339	769^2	⊘	Sc; 12.5; 2.0' x 1.4' pB,pL,1E,glbM	07054n1852
2355	6^6	⊙	Mag 12, diam 6'; about 30 faint stars, class G	07142n1352
2371-2372	316^2	◎	B,pL,E,bMN; mag 12.5 with F central star; diam 50" x 30"; brighter ends produce appearance of double nebula	07224n2935
2392	45^4	◎	B,S,R, Mag 8, bluish disc 40" diam, surrounding 10th mag star spect 08e (*)	07262n2101
2420	1^6	⊙	cL,Ri,C; diam 7'; mag 9; 30 stars mags 10...18; class E (*)	07354n2141

DESCRIPTIVE NOTES

ALPHA   Name- CASTOR, "The Horseman". The 23rd brightest star in the sky; magnitude 1.59; spectrum A1 V and A5. Position 07314n3200. Opposition date (midnight culmination) is about January 12.

Castor and Pollux form a prominent pair just 4½° apart; Castor is the northern star and the slightly fainter of the two, shining with a diamond whiteness in contrast to the bright golden tint of Pollux. The sparkling pair have since remote times suggested the concept of Heavenly Twins; in Greek legend they were the sons of Leda and Zeus, and are referred to in various writings as the *"Twin Laconian Stars"*, the *Spartan Twins*, the *Ledaean Lights* or *Ledaean Stars*, the *Gemini Lacones* or *Geminum Astrum*. The Roman poet Manilius in his *Astronomica*, composed probably during the reign of Augustus, calls them *Phoebi Sidus* in reference to the legend that the Twins were under the protection of the god Apollo; in other ancient Latin writings they are called the *Ledaeum Sidus* or *Ledaei Juvenes*. In Rome they were titled the *Dioscuri*- the Sons of Zeus, and honored as the guardians of the Eternal City.

The Heavenly Twins are depicted frequently on coins of the ancient Greek and Roman world. They appear on fine silver tetradrachms of the Greek kings of Bactria, on coins minted in Bruttium in Southern Italy in the 3rd century BC, on the tetradrachms of Rhegium, and on silver denarii of the early Roman Republic, dating from about 265 BC. The Twins are traditionally shown as two youths on horseback, charging at full gallop; often the design includes two stars placed over the heads of the figures to indicate their heavenly nature. The coin shown in Figure 2 above is a silver tetradrachm minted by King Eucratides of Bactria about 170 BC. In the Phoenician city of Tripolis (modern

CASTOR AND POLLUX. Castor, visually the fainter of the two stars, appears brighter on the photographic plate, owing to its bluer color.

Tripoli in Lebanon, about 85 miles north of Tyre on the
Mediterranean coast) an extremely attractive series of
silver tetradrachms were issued showing the heads of the
Twins in classic style; the specimens shown in figures 1
and 3 were minted about 100 BC.

In the Greek world, Castor and Pollux were venerated
by mariners, and were invoked for protection against storms
and the perils of the seas. In the legend of the Argonauts
we find them guiding and protecting the adventurers in
their quest for the Golden Fleece. Shelley's version of the
Homeric *Hymn to Castor and Pollux* refers to this ancient
tradition:

> *"When wintry tempests o'er the savage sea*
> *Are raging, and the sailors tremblingly call*
> *On the Twins of Jove with prayer and vow......"*

The electrical glow sometimes seen in a ship's
rigging in stormy weather, and often called "St.Elmo's Fire"
was in classical times associated with the guiding spirit
of the Twins, and called the *Ledaean Lights*. In the early
Christian scripture *Acts of the Apostles* (XXVIII, 11) Paul
states that on the final portion of his journey to Rome in
62 AD, *"we departed in a ship of Alexandria which had win-*
tered in the isle, whose sign was Castor and Pollux......"
though modern translators of the Scriptures differ as to
whether the ship's name was the *Castor and Pollux*, or if
the account means that the ship's figurehead was a statue
of the Twins.

In Roman legend the Twins were honored by a place in
the heavens as a "reward for their brotherly love, so
strongly manifested while on Earth". In some versions of
the story, however, only Pollux was immortal, and he gener-
ously traded his heavenly place periodically with his mortal
brother. Another tradition makes Castor the son of Leda and
Tyndareos of Sparta; according to this version Castor was a
brother of both Clytemnestra and Helen of Troy. On some
antique star maps the Twins are identified with Apollo and
Hercules; occasionally with Romulus and Remus, the legendary founders of Rome. O.Seyffert in his *Dictionary of Clas-*
sical Antiquities (1882) states that a "considerable temple
was built to them near the Forum (414 BC) in gratitude for
their appearance and assistance at the Battle of   Lake

GEMINI

DESCRIPTIVE NOTES (Cont'd)

Regillus twelve years before. In this building, generally
called simply the Temple of Castor, the Senate often held
its sittings. It was in their honor too, that the solemn
review of the Roman *equites* was held on the 15th July.....
At Athens too they were honored as gods under the name of
*Anakes (Lords Protectors)....They are the ideal types of
bravery and dexterity in fight.......the tutelary gods of
warlike youth...."*

Arabian astronomers called them *Al Tau'aman,* or the
Twins, and *Al Burj al Jauza,* the Constellation of the Twins
while Castor was *Al Ras al Taum al Mukaddim,* the Head of
the Foremost Twin. Some Medieval Arabian maps, however,
show them as a pair of Peacocks. The Babylonian astrologers
called the pair *Mastab-ba-galgal,* the "Great Twins", while
Castor was an object of special veneration in Assyria under
the title *Son of the Supreme Temple.* In India they were
Acvini, "The Horsemen", though in western tradition it is
Castor who is identified as a horseman; Pollux excelled
at boxing. Both brothers were believed to guide Roman arm-
ies to victory, a tradition made famous in western literature by Lord Macaulay's account of the famous Battle of
Lake Regillus in his *Lays of Ancient Rome.* Another well
known reference to Castor and Pollux occurs in Tennyson's
Maud:

> *"It fell at the time of year,*
> *When the face of night is fair on the dewy downs,*
> *And the shining daffodil dies, and the Charioteer*
> *And starry Gemini hang like glorious crowns*
> *Over Orion's grave low down in the west,*
> *That like a silent lightning under the stars*
> *She seemed to divide in a dream from a band*
> *of the blest..."*

Castor is located at a distance of 45 light years
and has a total luminosity of about 36 times that of our
Sun. The annual proper motion is 0.20" in PA 238°; the
radial velocity is about 3 miles per second in recession.
This star is probably the finest double in the north
sky for a fair-sized glass. It was probably resolved first
by G.D.Cassini as early as 1678, but not re-discovered
until J.Bradley began his series of measurements in March
1718. In the years between 1719 and 1759, a change of about

915

DESCRIPTIVE NOTES   (Cont'd)

30° was noted in the PA. In 1803, Sir William Herschel announced that the components form a system in which the two stars are gravitationally connected and revolve about each other in space. According to R.H.Allen, however, this possibility had been suggested by the Reverend John Michell as early as 1767. Castor, in any case, was the first true physical binary to be recognized, and the first object beyond our own Solar System in which the force of gravitation was shown to be operating, as it does in the planetary system.

The two stars visible in the amateur telescope, Castor A and B, are magnitudes 2.0 and 2.8; their period of revolution is about 4 centuries and their mean separation approximately 90 AU or about 8.4 billion miles, a little more than the distance across the entire Solar System. The pair was widest (6.5") in 1880; they have since been slowly closing their separation toward a minimum of about 1.8" (about 55 AU) at periastron about 1965. The exact period and orbital elements are not precisely determined. In most current texts, the period of 380 years, derived by K.A. Strand some 30 years ago, is still quoted; more recent calculations by P.Muller (1956) and W.Rabe (1958) are compared in the following table.

	Period	Semi-major axis	Eccentr.	Periastron
Strand	380 yr	5.941"	0.37	1968.8
Muller	511 yr	7.370"	0.36	1950.6
Rabe	420 yr	6.295"	0.33	1965.3

The orbital motion is retrograde, with an inclination of about 115°, or about 25° from the edge-on position. A third faint star (Castor C) accompanies the main pair at a separation of 72.5" in PA 164°. It is magnitude 9.1 and is in slow orbital revolution around the rest of the system in a period probably exceeding 10,000 years.

The interesting fact about Castor is that each of the three visible stars is itself a spectroscopic binary; the entire system of six components forming one of the most remarkable examples of a multiple star in the heavens. From spectroscopic analysis, the following picture of the system has been developed.

DESCRIPTIVE NOTES (Cont'd)

CASTOR A, Magnitude 1.99, consists of two stars revolving in a rather eccentric elliptical orbit with a period of 9.2128 days, eccentricity 0.499, separated by about 4 million miles. The components are both A-type main sequence stars, nearly identical in size and luminosity, each about twice the diameter of the Sun and 12 times its brightness. The total mass of the system is about 3.2 solar masses.

CASTOR B, magnitude 2.85, consists of two stars revolving in a nearly circular orbit in a period of 2.9283

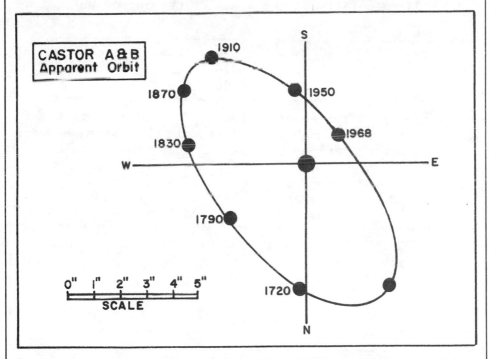

days, with a separation computed to be less than 3 million miles. The spectral classes are both A5 or possibly slightly later. Each star is about 1.5 times the solar diameter and approximately 6 times the solar luminosity; the total mass of the system is 2.3 solar masses.

CASTOR C, also designated YY Geminorum, is a faint red dwarf, located at the immense distance of about 100 billion miles from the main pair A&B. At this distance of more than 1000 AU the star is still gravitationally associated with the rest of the system, but the orbital motion is

so slow that no period can yet be derived; as a guess it
may be something like 10,000 years or more.

Castor C is of special interest to the astronomer
since it is a spectroscopic double in which the orbit lies
very nearly in the line of sight, the components therefore
eclipsing each other as they revolve. Both of the stars are
red dwarfs; a revised spectral class of dK6 now replaces
the earlier estimate of dM1. The stars revolve with an orbital velocity of 70 miles per second, completing one orbit
in 0.8143 days, or about 19.5 hours. Their separation is
1.67 million miles, and the visual range during an eclipse
is 9.1 to 9.6.

CASTOR C	Mass	Diam.	Lum.	Dens.
1	0.63	0.76	0.025	1.4
2	0.57	0.68	0.025	1.8

Below: The relative sizes of the six components of Castor
with the Sun for comparison. To complete this scale model
the separations between the various components should be as
follows:

A1---A2 = 2.3 inches   C1---C2 = 0.9 inch
B1---B2 = 1.7 inches   AB---C = 4500 feet
A----B = 340 feet

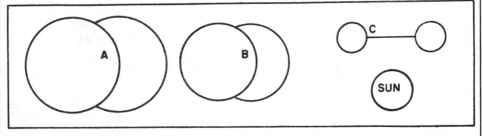

Gemini lies on the east edge of the winter Milky Way
and the fields of Castor and Pollux are rich in numerous
faint and distant stars. At least 5 very faint galaxies
lie within 1° of Castor, but are beyond the range of the
small telescope. The difficult double star Rho Geminorum is
1° to the west, and another close pair, OΣ175 is just 1°
south. The radiant point of the Geminid meteor shower is
quite close to Castor, at about 7^h28^m +33°; the maximum is
December 10- 12 each year.

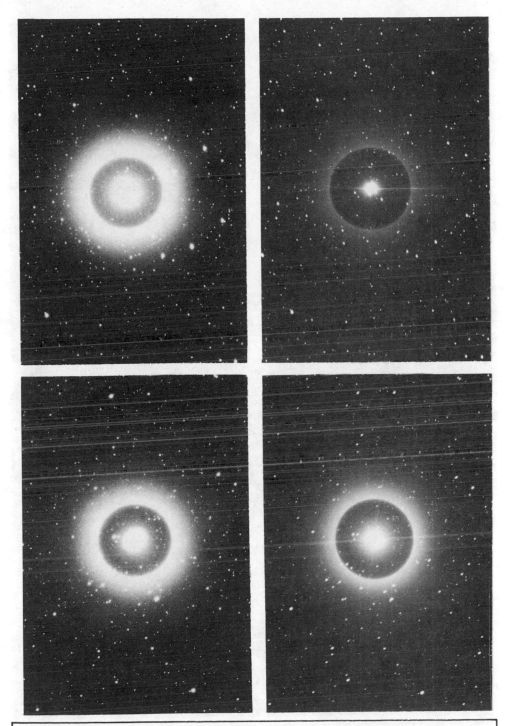

CASTOR AND POLLUX, photographed with the 13-inch camera at Lowell Observatory. Castor is at top, Pollux below. Each star is shown in blue light (left) and in red light (right)

DELTA GEMINORUM and PLUTO. The new planet (arrow) is shown here shortly after discovery at Lowell Observatory, in February 1930. Date of Photograph is March 4, 1930.

DESCRIPTIVE NOTES (Cont'd)

BETA
Name- POLLUX, from the Greek *Polluces* or *Polledeuces*, the "Boxer" or "Pugilist". The 17th brightest star in the sky; magnitude 1.16, spectrum K0 III. Position 07423n2809, about $4\frac{1}{2}°$ SSE from Castor. Opposition date (midnight culmination) is January 15. The star, although designated Beta, is actually brighter than Alpha Geminorum (Castor) and it has been suggested that one of these stars has changed in luminosity in the last few centuries.

The distance of Pollux is about 35 light years; very close to the standard 10-parsec distance for calculating absolute magnitudes; thus the apparent and absolute magnitudes of the star are nearly the same. Pollux is a star of the Arcturian type, yellowish in color, with a surface temperature of about 4500°K; luminosity about 35 suns, and actual diameter about eleven times the sun. The star shows an annual proper motion of 0.62" in PA 265°, and a radial velocity of 1.9 mile per second in recession; the true space velocity is about 19 miles per second.

Aitken's "ADS" Catalog lists a number of faint companions to Pollux, but it is now known that none of them are physically related to the bright star. There is a 10th magnitude star at 201" and another at 234" (1923 measurements) but both separations are increasing from the proper motion of the primary. The closer star, however, is itself a double, and probably a physical pair. Designated β580, it has a separation of 1.4", magnitudes 9 and 12, a rather difficult object.

GAMMA
Name- ALHENA or ALMEISAM. Magnitude 1.93; Spectrum A0 IV. Position 06348n1627. The computed distance of this star is 105 light years; the true luminosity about 160 times that of the Sun, and the absolute magnitude about -0.7. The annual proper motion is 0.07"; the radial velocity is 7.7 miles per second in approach.

DELTA
Name- WASAT. Magnitude 3.51; spectrum F0 IV; position 07171n2205. Delta Geminorum is an interesting binary of very slow motion, the faint companion having been first measured by the elder Struve in 1829. The small star is a K6 dwarf of about one-ninth the luminosity

of the Sun. In the last 130 years the PA between the stars
has increased by only 20°; the present separation (1964)
is 6.3" in PA 218°. According to preliminary orbit calculations by Hopmann (1959) the period of this pair is on
the order of 1200 years, with periastron around the year
1437 AD. Hopmann's orbit has a semi-major axis of 6.898"
and an eccentricity of 0.11. The primary star is probably
a close binary itself, since there are small variations in
the measured radial velocity. The present projected separation of the components is about 95 AU.

It was near Delta Geminorum that the ninth planet
of the Solar System, Pluto, was discovered at the Lowell
Observatory in February 1930. Curiously, the planet Uranus
was also in Gemini, near the star Eta Geminorum, at the
time of its discovery by Herschel in 1781.

The distance of Delta Geminorum is computed to
be about 53 light years; the actual luminosity is about 8
times that of the Sun, and the absolute magnitude is +2.5.
The star shows an annual proper motion of 0.02"; the radial
velocity is 1.5 mile per second in recession.

EPSILON   Name- MEBSUTA. Magnitude 2.98; spectrum G8
               Ib. Position 06408n2511. This is a super-
giant G star at a distance of about 1100 light years. It
has an absolute magnitude of about -4.6 and an actual luminosity of 5700 times that of the Sun. The annual proper
motion is only 0.015"; the radial velocity is about 6 miles
per second in recession. The 9th magnitude companion at
110" is probably not physically related to the primary; no
relative motion has been detected since discovery.

Epsilon Geminorum was occulted by Mars on April 7,1976,
an extremely rare occurrence.

ZETA   Position 07011n2039. One of the brightest of
            the cepheid variable stars, a pulsating giant
with a period of 10.15172 days, discovered by J.Schmidt
from observations made between 1844 and 1847. The photographic range is from 4.4 to 5.2 with a spectral change of
F7 Ib at maximum to G3 Ib at minimum. There are several
peculiarities about this star. The light curve is distinctly abnormal, with a symmetrical shape, so that the times of
rise and fall are nearly equal; and there is a noticeable

"hump" on the ascending branch. Recent studies show that the period is decreasing at the rate of about 3.6 seconds per year. For visual observers, the stars Kappa Geminorum (mag 3.57) and Upsilon (mag 4.07) closely match the maxima and minima of Zeta.

The distance of the star is computed to be about 1500 light years; the visual luminosity at maximum is about 5700 times that of the Sun. The annual proper motion is less than 0.01"; the radial velocity averages 4 miles per second in recession.

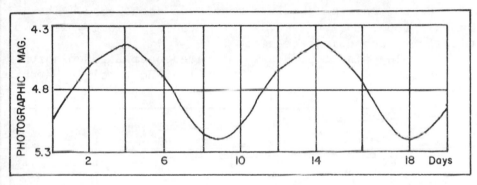

The 8th magnitude star at 96.5" is not a true companion to Zeta, and the distance is increasing from its proper motion of 0.12" per year, much greater than the motion of Zeta itself. There is also another optical companion of the 11th magnitude at 87.7" in PA 84°.

ETA   Name- PROPUS. Magnitude 3.33 (slightly vari-
               able); Spectrum M3 III. Position 06119n2231.
Eta Geminorum is a well known double star, though difficult for telescopes smaller than 12 inches in aperture. The companion, of magnitude 6.5, was first seen by S.W.Burnham in 1881; the separation was then 1.0" in PA 301°. In the last 80 years the pair has widened somewhat, and a measurement by Van den Bos in 1962 gave 1.5" in PA 266°. The small star is a G8 subgiant with a luminosity of about 7 times the Sun.

The red giant primary is a semi-regular variable with an average period of 233 days and a range of about 0.9 magnitude; the variability was discovered by J.Schmidt in

1844. The star is also a spectroscopic binary with a long period of 2983 days or 8.2 years; the unseen companion is believed to be another M-giant of probably somewhat later spectral type. An analysis of the light curve indicates that faint minima may recur near the same phase of the 8-year cycle of revolution; thus it seems possible that the two stars form an eclipsing system. All three stars lie within a cloud of cool gas at least 300 AU in diameter.

The computed distance of Eta Geminorum is about 200 light years, the actual luminosity about 160 times that of the Sun, and the absolute magnitude about -0.7. The annual proper motion is 0.07"; the radial velocity is 11.5 miles per second in recession.

The planet Uranus was located near Eta Geminorum at the time of its discovery by William Herschel in March 1781.

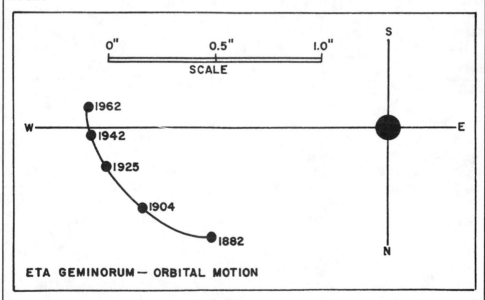

ETA GEMINORUM — ORBITAL MOTION

MU   Magnitude 2.86; Spectrum M3 III. Position 06199n2233. Mu Geminorum is an irregular variable star of small range; the maximum amplitude appears to be less than 0.2 magnitude. The star is about 160 light years distant; the actual luminosity is about 145 times that of the Sun and the absolute magnitude about -0.6. The

DESCRIPTIVE NOTES   (Cont'd)

annual proper motion is 0.13"; the radial velocity is 34
miles per second in recession.

Aitken's "ADS" Catalog lists a distant companion of
the 10th magnitude, 122" distant in PA 141°. This star,
which probably has no real connection with Mu itself, is a
close double with a separation of 0.8" and a slowly diminishing PA.

XI
Magnitude 3.38; spectrum F5 IV. Position
06425n1257. The computed distance is about 65
light years, the actual luminosity about 13 times that of
the Sun, and the absolute magnitude +2.0. Xi Geminorum
shows an annual proper motion of 0.22" in PA 210°; the
radial velocity is 15 miles per second in recession.

U
Variable. Position 07521n2208. A remarkable
dwarfish variable star discovered by J.R.Hind
in 1855. It is the typical example of a rare type of nova-
like eruptive star sometimes called "cataclysmic variables"
or "miniature novae". The star is normally a 14th magnitude
object, but at intervals of several months it undergoes
sudden outbursts in which the light increases approximately
100 times. The complete rise from the 14th to the 9th magnitude usually takes not more than two days, and is often
accomplished in a mere 24 hours. U Geminorum is thus an
exciting object to study, and the field is kept under nearly constant survey by many observers in the hopes of de-
tecting an outburst. The hourly increase in light during a
rise to maximum is a fascinating sight to watch. U Geminorum and SS Cygni (See page 774) are the two outstanding
examples of this rare type of variable.

For owners of small telescopes without setting
circles, U Geminorum is most easily located by first turning the instrument toward the star 85 Geminorum (mag 5.3)
located at 07527n2001. From this star, the telescope is
slowly moved north about 2.1° using the photographic chart
on page 930 as a guide.

U Geminorum exhibits two types of maxima which
often occur alternately: the "long maximum" of about 17
days duration and the "short maximum" of about 9 days duration. Both types are shown on the accompanying light curves
on pg 929. Outbursts occur at an average interval of about

100 days, but the figure has varied between 62 and 257 days in a completely unpredictable manner. Three outbursts of the star were recorded in 1966, and two in 1967; the explosions are thus not as common as those of SS Cygni, but the magnitude range is correspondingly higher.

U Geminorum has a peculiar spectrum showing double emission lines and a background continuum with the energy distribution corresponding to a late type G dwarf. As the light increases to maximum the emission lines weaken and the spectrum appears nearly continuous; the color then is found to resemble class B or A. From observations made at Mt.Wilson it is now known that U Geminorum is actually an extremely close binary system of exceptional interest. The hotter component is a bluish, underluminous star classed as sdBe (subdwarf B-type with emission lines) and the other star is apparently a late type G dwarf or subdwarf. The two stars have one of the shortest orbital periods known, now determined to be 0.177808 day, or about 4.5 hours.

In a comprehensive study of the star in 1965, W. Krzeminski found that the components form a rapid eclipsing binary system in which the primary minimum lasts approximately 15 minutes; the light diminishing about 0.9 magnitude to a minimum of 14.7. The form of the light curve seems to indicate a total or near-total eclipse, and the extreme shortness of the period implies unusually small diameters for the stars and a separation of only a few hundred thousand miles.

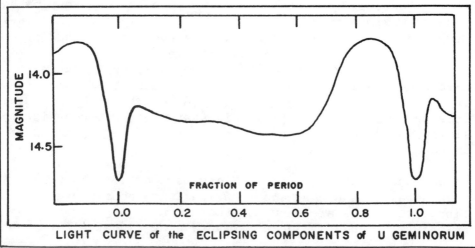

LIGHT CURVE of the ECLIPSING COMPONENTS of U GEMINORUM

OUTBURST OF U GEMINORUM. The famous eruptive variable star is shown here in one of its sudden outbursts, in January 1930. These photographs were made with the 13-inch telescope at Lowell Observatory.

U Geminorum shows both a smaller proper motion (about 0.04" per year) and a fainter maximum magnitude than SS Cygni, implying that it is at a two or three times greater distance. Parallax measurements indicate a distance of some 300 light years; the resulting absolute magnitudes for the system are:

    At minimum =   +9.1; Luminosity = 0.02 X Sun
    At maximum =   +4.1; Luminosity = 1.9 X Sun

The investigations of the very similar star SS Cygni lead to an almost identical result. The individual stars appear to have absolute magnitudes of about +10, which seemingly places the blue component very near the category of white dwarfs. In all the well studied stars of this type, good evidence for binary motion is found, either in the form of composite spectra, periodic variations in radial velocity, or in an eclipse type light curve. The U Geminorum stars thus appear to form a very well defined class of objects. As a working hypothesis we may accept the idea that all these stars are extremely close and rapid binaries in which at least one component is near the white dwarf state. The sudden outbursts are undoubtedly connected in some way with the duplicity of the system, but the exact details are un-certain. It may be that the partly degenerate star is ab-sorbing material from the close companion, as has been sug-gested; but it is more likely that the actual mechanism of the outbursts is rather more complex than this simple pic-ture suggests. Spectroscopic studies reveal the existence of a rotating cloud or ring of gas around the blue star, and it appears that the explosive increase in light is due not only to the brightening of the star, but to a large in-crease in radiation from the cloud.

    The idea that the outbursts of the U Geminorum type stars originate in the bluer component is an assumption which has gone virtually unchallenged until a thorough study of the system was made by W.Krzeminski in 1965. By making extremely accurate measurements of the light varia-tions, Krzeminski found that the eclipses of the system can no longer be detected when the star is undergoing an out-burst. Obviously, if the 100-fold rise in light is due to an eruption of the blue star, a drop of some 5 magnitudes should be expected when this flaring object is eclipsed by

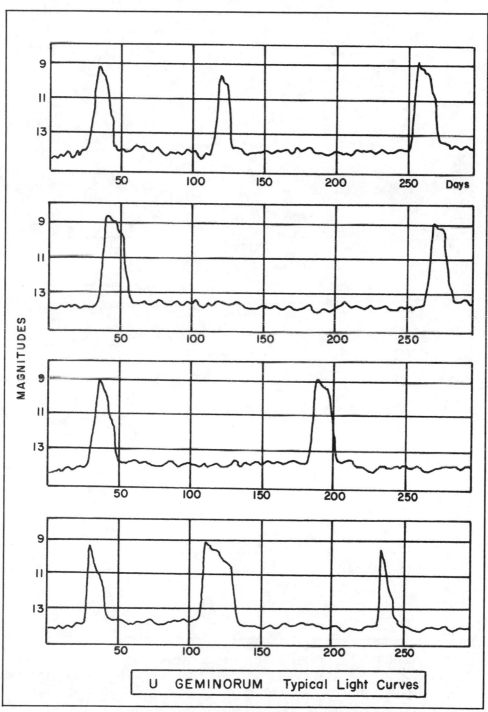

U GEMINORUM   Typical Light Curves

929

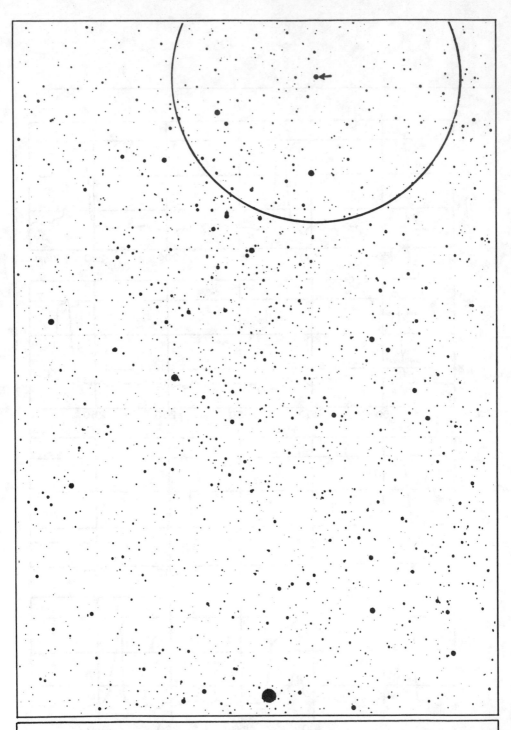

U GEMINORUM Field, from a Lowell Observatory 13-inch telescope plate. The circle is 1° in diameter with north at the
top. Bright star 85 Geminorum is near bottom of the chart.

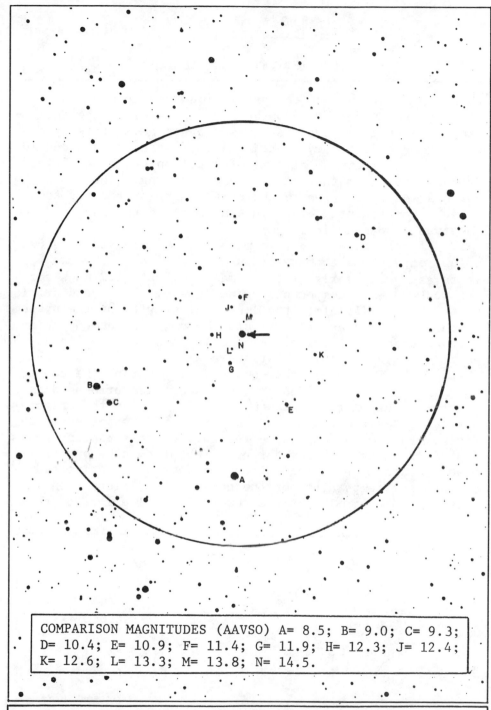

COMPARISON MAGNITUDES (AAVSO) A= 8.5; B= 9.0; C= 9.3;
D= 10.4; E= 10.9; F= 11.4; G= 11.9; H= 12.3; J= 12.4;
K= 12.6; L= 13.3; M= 13.8; N= 14.5.

U GEMINORUM FIELD, from a Lowell Observatory 13-inch telescope plate. Circle diameter = 1°; north is at the top.
Limiting magnitude about 14.

DESCRIPTIVE NOTES (Cont'd)

the companion. Even on the assumption that much of the
light increase during an outburst originates in the encircling gaseous ring, and that this ring is never totally hidden during an eclipse, Krzeminski finds that only about 10%
of the light increase at time of eruption can be explained
by radiation from the hot star and its ring. "The remaining
90% must be attributed to the brightening of the red component. Consequently, the red star must be the seat of the
observed eruptions in U Geminorum."

A tentative schematic model for the system is shown
below, based on a summary of current evidence. The two
stars are very nearly of equal mass; 1.3 and 1.2 suns for
the red and blue components respectively. The blue dwarf is
some 700,000 miles from the center of gravity of the system
which is located actually within the body of the companion,
just beneath its surface. Comparable in size to a typical
white dwarf, the bluish star has a diameter of about 0.025
that of the Sun, while the computed diameter of the cooler
star is 0.7 times the Sun. The hot dwarf is surrounded by
a rotating asymmetrical gaseous ring, of which the brighter
portion faces the companion star. Matter is supplied to the
ring by the companion, the efficiency of this capture process increasing markedly during an eruption. The outburst
itself consists of a rapid increase in the effective radius
and surface temperature of the cooler star, brought on in
some way, presumably, by the powerful tidal effect of the

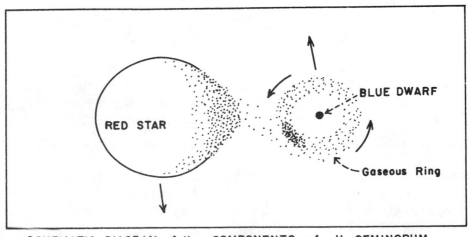

RED STAR

BLUE DWARF

Gaseous Ring

SCHEMATIC DIAGRAM of the COMPONENTS of U GEMINORUM

dense companion. During this process the hotter and deeper layers of the star are exposed. Also, some of this matter goes into orbit around the hot star, temporarily increasing the brightness of the encircling ring, and adding to the total light increase.

U Geminorum and SS Cygni are the brightest and best known stars of the class. Among other examples are SS Aurigae, UU Aquilae, X Leonis, SU Ursae Majoris, RU Pegasi, and SW Ursae Majoris. Although about 100 stars of the type have been identified, the majority are very faint objects and only some twenty have been observed with any degree of thoroughness. There is also a small sub-type of which RX Andromedae and Z Camelopardi are the chief examples; these stars show more erratic activity than the classic SS Cygni stars, and also maintain occasional periods of constant brightness at some level between maximum and minimum. In addition, the peculiar star AE Aquarii should be mentioned; it is also more erratic and irregular than any of the true U Geminorum stars, but resembles them physically. It is a close dwarf binary with a period of about 10 hours.

Many of the U Geminorum stars display individual peculiarities. SS Aurigae on occasion goes through periods of accelerated activity when numerous short maxima rapidly follow each other at only a fraction of the normal interval. SU Ursae Majoris may act in a similar manner, but at other times has shown "supermaxima" during which the star remains at peak brightness for a time equal to a whole average period. Although the average interval for this star is about 16 days, the period has occasionally been as short as 8 days; this is the shortest interval recorded for any star of the type.

The average interval between maxima is known only roughly for many of the U Geminorum stars, but seems to range from 12 days (AB Draconis) to about 450 days (SW Ursae Majoris). The greatest observed magnitude range approaches 6 magnitudes, and in general the stars of greater range show longer average periods. This "period-brightness relation" is of great theoretical interest, since it seems to link these stars with the recurrent novae, stars which explode with much greater violence but also display proportionately longer periods. It is encouraging to note that the standard examples (T Coronae Borealis, RS Ophiuchi,

DESCRIPTIVE NOTES (Cont'd)

and WZ Sagittae) are all close binaries, as is the classic full-scale nova DQ Herculis. We thus have three classes of objects to consider: the U Geminorum stars, the recurrent novae, and the full-scale or "classical" novae. What is the relationship of the three types? Do they form an evolutionary sequence? If so, is duplicity an essential feature in all the classical novae, as it appears to be in the other two classes? Will U Geminorum eventually develop into a real nova? Or does the evolution run in the opposite direction, in which case the star was once a nova in the remote past? Finally, is it possible that the evolutionary hypothesis is in error, in which case the three classes may be related only in the sense that the same general mechanism is operating? None of these questions can be definitely answered at the present time.

R.P.Kraft (1961) has suggested that the U Geminorum stars are descendants of the close dwarf binary stars of the W Ursae Majoris type. Studies by Kraft and Luyten in 1966 showed that the two classes of objects have similar masses and periods, are very similarly distributed in space and have comparable motions in the Galaxy. It is suggested that the primary of a W Ursae Majoris system, as it expands in the course of its evolution, loses mass to the companion; the evolution of the primary toward the hot subdwarf state is thus accelerated. In its own subsequent expansion the cooler companion loses mass, some of which forms the ring around the hot primary. At least one W Ursae Majoris type star has shown minor eruptions- perhaps the first symptoms of its coming transformation into a U Geminorum system. (Refer also to SS Cygni, SS Aurigae, SU Ursae Majoris, AE Aquarii, T Coronae Borealis, RS Ophiuchi, and WZ Sagittae. For an account of novae in general, see Nova Aquilae 1918)

M 35   Position 06057n2420. A fine galactic star cluster, visible in an opera glass as a nebula, and possibly detectable by a keen eye without optical aid on the best of nights. It is located about 2½° NW from Eta Geminorum. The exact circumstances of the discovery of M35 are obscure, though it is certain that the cluster was known before the time of Messier, who recorded it in 1764. In a list prepared in 1745, the Swiss astronomer

STAR CLUSTER M35 in GEMINI. A fine group for the small telescope. The smaller cluster in the field is NGC 2158. Lowell Observatory photograph with the 13-inch telescope.

de Cheseaux seems to include the object, referring to it as
a cluster "above the northern feet of Gemini". Although no
coordinates are given, the identification with M35 appears
definite. Messier himself remarked that M35 was shown on
the star atlas of John Bevis in 1750, along with a few
other bright nebulae and star clusters known at the time
including M1, M11, M13, M31, and M42.

        M35 is an excellent object for any small tele-
scope; while particularly effective with a 6 or 8-inch
glass with low power, something of its beauty can be appre-
ciated in even a 2-inch instrument. Curving rows of bright
stars give an impression of rows of glittering lamps on a
chain; fainter stars form a sparkling background with an
orange star near the center. S.Raab in his paper "Research
on Open Clusters" describes it as "a splendid specimen,
a very large thin and circular cluster, without sharply
conspicuous condensation. On the contrary, the centre seems
to be rather void of stars. In the vicinity of the centre
is a gathering of 5 or 6 stars....the cluster passes imperceptibly into the environs.." Lord Rosse counted 300
stars in the group, out to a radius of 13'. Lassel, observing with a 24-inch reflector, called this cluster "a
marvelously striking object- the field of view is perfectly
full of brilliant stars, unusually equal in magnitude and
distribution over the whole area. Nothing but a sight of
the object itself can convey an adequate idea of its exquisite beauty." W.T.Olcott also classed M35 as one of the
finest clusters in the heavens, and several observers have
commented on the tendency of the stars to occur in curving
rows, reminding one of the bursting of a sky rocket.

        An apparent diameter of 30' is given by Bailey
(1908) while H.Shapley (1930) obtained 46' and a value of
45' was adopted by W.M.Smart in his study of the cluster.
From his observations in 1930 R.J.Trumpler derived a distance of 840 parsecs or about 2700 light years; a more recent study by K.A.Barkhatova (1950) gives 2200 light years.
The true diameter of the cluster is about 30 light years;
the total brightness is equal to about 2500 suns. Spectral
types of the stars range from B3 to G0 for main sequence
objects; M35 also contains several yellow and orange giants
of late G and early K type. The brightest star of the group
is a B3 main sequence star with an apparent magnitude of

STAR CLUSTER M35 in GEMINI, photographed with an 8-inch
Celestron telescope by David Healy of Manhasset, New York.
The faint cluster at top is NGC 2158.

NGC 2158. The unusually rich galactic cluster appears at upper right; some of the stars of the large group M35 are seen at lower left. Photograph made with a 12½-inch reflector by Evered Kreimer of Prescott, Arizona.

7.5; the computed absolute magnitude is about -1.7 and the
luminosity about 400 times the Sun. The brightest yellow
giant is magnitude 8.6 apparent, absolute magnitude about
-0.6, and luminosity about 150 times the Sun; the color is
equivalent to spectral class K0. Studies of these stars
suggest a distance modulus of about 9.2 magnitudes.

About half a degree SW of M35 is the rich little
galactic cluster NGC 2158, described below.

NGC 2158   Position 06043n2406. A very rich and distant
galactic star cluster, located about half a
degree southwest of M35, looking like a faint nebulosity in
a 6-inch glass, and mentioned in the new edition of Webb's
"Celestial Objects" as a "faint dim cloud of minute stars".
Only some 4' in diameter, and with a total magnitude of
about 11, the brightest individual stars are ±16th magnitude. Inconspicuous in the small telescope, resolvable only
in large instruments, and relatively unknown to the amateur
until quite recently, this cluster would nevertheless rank
among the finest of all galactic clusters if it was as near
to us as M35 itself. The unusual richness suggests that NGC
2158 may be a transition case between the galactic clusters
and the globulars; the density and symmetrical distribution
of stars is almost comparable to some of the less-condensed
globulars, but the cluster does not show the degree of
crowding toward a thick central mass which usually characterizes the globulars. In a study by H.Arp and J.Cuffey
(1962) it was found that the stellar population is intermediate between the normal galactic clusters and the unusually ancient ones such as NGC 188 in Cepheus which have H-R
diagrams resembling the globulars. NGC 2158 is thus another
addition to the short list of "intermediate-age star clusters" which include NGC 7789 in Cassiopeia and NGC 752 in
Andromeda. The three clusters all have very similar color-
magnitude diagrams; the diagram for NGC 2158 shows that all
stars brighter than absolute magnitude +2.5 have begun
evolving toward the giant stage, and the resulting probable
age of the group is about 800 million years. The brightest
members of the cluster are red giants with absolute magnitudes of about -2.5, comparable to the brightest stars in
a typical globular star cluster.

If the classification as a galactic star cluster is retained, in spite of the definite peculiarities, NGC 2158 is one of the most remote of all known clusters. Arp and Cuffey find the distance to be 4900 parsecs, or about 16 thousand light years, more than six times the distance of M35. This places it far out near the outer rim of the Galaxy, and offers some verification of Arp's suggestion that the "intermediate-age" clusters were formed in the outer regions of the Galaxy where the interstellar material is less dense than in the solar neighborhood, and less rich in the atoms of the metals. (Refer also to NGC 752 in Andromeda and NGC 7789 in Cassiopeia)

NGC 2392   Position 07262n2101. A small bright planetary nebula discovered by William Herschel in 1787 and located about midway between the stars Kappa and Lambda Geminorum, about 2/3° southeast of the wide double star 63 Geminorum. It was described by John Herschel as a 9th magnitude star exactly in the center of an exactly round and bright atmosphere equally dispersed all around. A small telescope shows it as an 8th magnitude star-like object, distinguishable by its soft fuzzy glow and perceptible disc which measures about 40" across. A ghostly bluish-green color becomes evident with larger telescopes. T.W.Webb in 1852 described this nebula as "quite like a telescopic comet" while Lord Rosse (1850) spoke of it as "a wonderful object as seen with the 6-foot telescope; it has been several times examined and as yet we have not seen the slightest indication of resolvability. The outer ring is seen on a pretty good night completely separated from the nucleus surrounding the brilliant point or star. The light is very bright and always appears to be flickering, owing no doubt to the unsteadiness of the atmosphere. There is a small dark space to the right of the star, which indicates a perforation similar perhaps to that discovered in Nos. 838, 2050, and others. The annular form of this object was detected by Mr. Johnstone Stoney, my assistant, when observing alone..." (Nos. 838 and 2050 are Herschel's numbers for the planetaries now known as M97 in Ursa Major and NGC 6826 in Cygnus).

NGC 2392 is an impressive sight in powerful telescopes. The central star, of about 10th magnitude, is first

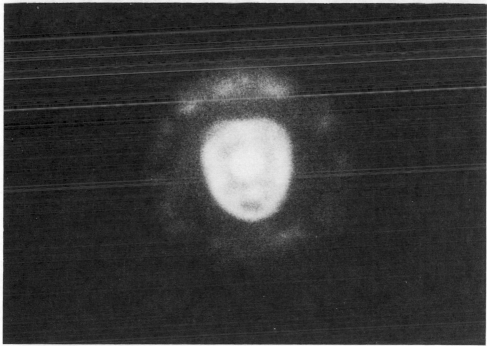

PLANETARY NEBULA NGC 2392 in GEMINI. The "Eskimo Nebula" or "Clown-Face Nebula" photographed in red light (top) and ultraviolet (below). Palomar Observatory photographs with the 200-inch reflector.

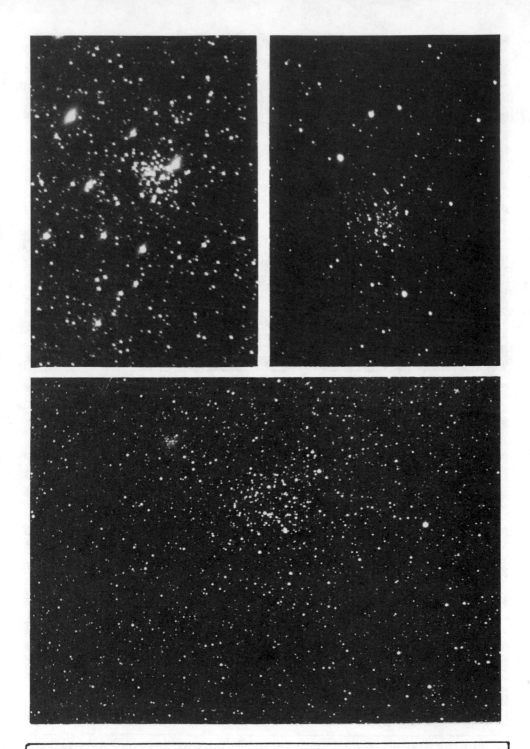

STAR CLUSTERS IN GEMINI. Top left: NGC 2266. Top right:
NGC 2420. Lowell Observatory photographs. Below: M35 and
NGC 2158; photograph by Wendell Shields of Prescott, Ariz.

encircled by a bright inner ring 19" X 15" in size; a dark zone divides the inner nebulosity from an outer ring about twice the size, but noticeably fainter. There are several bright condensations in the inner shell which make a pattern resembling a human face; the outer shell evidently suggests a parka-like hood to the imaginative observer since the nebula is popularly called the "Eskimo Nebula". To the author of this book, the whole nebula irresistibly suggests the classic and unforgettable features of W.C.Fields.

Widely different distances are given for the nebula in standard catalogs, a situation which seems to be true of planetary nebulae in general. The Skalnate Pleso Catalog (1951) has 1370 light years, L.Kohoutek (1962) has 1760 light years, I.S.Shklovsky (1956) has 3000, L.Berman (1937) obtained 2800, and C.R.O'Dell (1963) has 3600 light years. Tentatively accepting a compromise figure of about 3000 light years, the actual diameter is found to be about 36 thousand AU for the outer ring, or about 0.6 light year. The central star, one of the brightest known in any of the planetaries, is an O8 type dwarf with about 40 times the visual luminosity of the Sun; the absolute magnitude may be about +0.7 and the surface temperature about 40,000°K. Very strong radiation from the central star excites the bright fluorescent glow of the nebulosity; the bluish-green tint is produced by two strong spectral lines at 5007 and 4959 angstroms, the so called "forbidden lines" of doubly ionized oxygen.

Careful radial velocity measurements of NGC 2392 have revealed that the diameter of the cloud is growing at the rate of about 68 miles per second. At the estimated distance, the expected increase in size amounts to about 1" in 30 years. Curiously, however, although photographs made 60 years apart have been critically compared, no definite increase in size has been detected. The answer to this mystery probably lies in the fact that the apparent edge of the nebula is not the true boundary, but merely marks the edge of the zone of excitation; material continues to expand beyond this radius but can no longer be seen because the distance from the central star has become too great to allow the illuminating process to operate efficiently. F.L. Whipple (1938) has obtained a probable age of only 1700 years for this nebula, making it one of the youngest known. (See also M57 in Lyra, NGC 7662 in Andromeda, etc.)

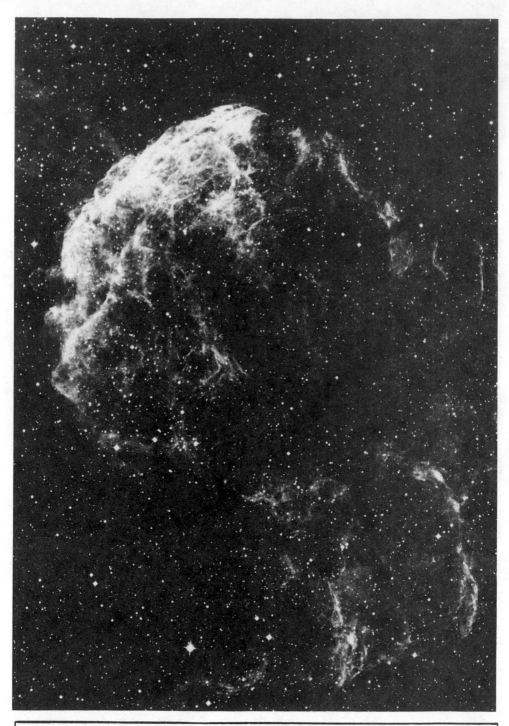

NEBULA IC 443 in GEMINI. This expanding gas cloud is a moderately strong radio source, and is probably a supernova remnant. Palomar Observatory, 48-inch Schmidt telescope.

GRUS

LIST OF DOUBLE AND MULTIPLE STARS

NAME	DIST	PA	YR	MAGS	NOTES	RA & DEC
B530	1.0	36	42	7½- 10½	Spect F5	21273s3852
h5275	0.1	331	52	8½- 8½	(B1008) AB binary;	21279s3646
	40.8	201	19	-11½	spect F2	
h5288	19.7	60	52	7½- 8½	relfix, spect F0	21395s3810
L8912	55.0	356	31	5½- 8½	(Brs 15) optical; dist inc, PA dec; spect G2	21450s4732
a	28.3	149	47	2 - 12½	(Rst 5483) spect B5 (*)	22051s4712
h5319	2.1	127	55	7½- 7½	PA slow inc, spect F2	22090s3833
Cor250	5.5	353	33	7½- 10	spect G5	22099s4918
Hd 298	3.4	32	59	5½- 10½	PA inc; cpm pair; spect G1	22150s5352
I 134	0.4	296	47	8 - 8½	PA dec, spect A3	22193s5624
π^1	2.7	201	56	6½- 11	(I 135) relfix; A is variable; spect S5, G0. π^2 is 4' E.	22197s4612
π^2	4.6	214	53	6 - 12	PA slow inc; cpm; (I 382) Spect F0	22201s4611
Jc 19	24.8	74	52	6½- 7½	optical, dist dec; spect both F5	22217s4142
I 136	1.7	276	58	7½- 9	PA slow inc, spect K0	22229s4522
h5338	30.3	183	16	7 - 11	spect K5	22253s5203
B562	4.1	360	27	7½- 14	spect G5	22268s4620
h5344	5.0	168	55	8 - 11	relfix, spect F2	22327s3859
β771	2.7	263	34	6 - 10½	relfix, cpm; spect A2 (σ^2)	22341s4051
I 138	3.4	278	34	7 - 11	spect F8	22349s4007
B569	1.7	178	59	8½- 11	PA dec, spect F0	22357s5400
Cor252	7.8	129	43	6½- 10	(Marked Cor 63 in Norton) cpm; spect G1	22396s4728
h5362	10.5	141	37	7 - 10	relfix, spect A5	22438s4712
h5366	14.8	252	50	8 - 9	relfix, spect A2	22497s4303
τ^2	0.3	166	59	7½- 8	(I 22) PA inc, spect G0	22524s4844
Hu1335	0.5	100	09	8 - 8½	binary, 87 yrs; PA inc, spect G5	22557s4547

LIST OF DOUBLE AND MULTIPLE STARS   (Cont'd)

NAME	DIST	PA	YR	MAGS	NOTES	RA & DEC
β1011	2.1	298	36	6½- 10	no certain change; spect G9	22598s3641
υ	1.1	211	48	5½- 9	(β773) spect A0	23041s3910
θ	1.1	75	59	4½- 7	(Jc 20) PA inc; cpm; spect F6	23041s4347
△246	8.6	256	52	6 - 6½	slight PA dec, nice pair; cpm, spect F7	23044s5057
I 1467	0.5	9	36	7 - 8½	spect G5, PA dec.	23104s4953
h5390	22.5	44	33	6 - 10½	spect K0	23139s4446
△248	0.5	243	47	6½- 8	(Rst 5560) spect A	23180s5035
	16.8	211	52	- 9		
△249	26.5	212	51	6½- 7½	relfix, spect A5, A5, cpm.	23211s5405

LIST OF VARIABLE STARS

NAME	MagVar	PER	NOTES	RA & DEC
π'	5.8---6.4	Irr	Spect S5; also visual double star I 135	22197s4612
R	7.5--14.8	332	LPV. Spect M5e	21453s4709
S	6.0--15..	401	LPV. Spect M5e--M7e	22230s4841
T	6.8--11..	137	LPV. Spect M0e	22228s3750
V	9.0--9.3	4.4943	Ecl.Bin; spect F5	21488s4237
W	8.9---9.5	1.4843	Ecl.Bin; spect F5	22384s4406
RS	7.6---8.2	.1470	Cl.Var; spect A5	21398s4825

LIST OF STAR CLUSTERS, NEBULAE, AND GALAXIES

NGC	OTH	TYPE	SUMMARY DESCRIPTION	RA & DEC
7070		⊖	SD; 12.6; 1.9' x 1.5' F,cL,lE,gvlbM; foll at 4.7' by NGC 7072, F,S,R	21273s4319
7079		⊖	S0; 12.3; 0.5' x 0.5' B,R,cS, psbM	21293s4418
7097		⊖	E5; 12.6; 0.6' x 0.4' B,S,vlE, mbM	21371s4246
7107		⊖	SB; 13.1; 1.7' x 1.5' cL,vF,R,vglbM	21392s4502
7119		⊖	Sc; 13.1; 1.0' x 1.0' F,S,R, gbM	21431s4645
7144		⊖	E0; 12.2; 1.4' x 1.4' vB,pS,R,mbM	21495s4829
7145		⊖	E0; 12.7; 1.2' x 1.2' B,S,R	21501s4807
7162		⊖	Sc; 13.1; 1.7' x 0.6' cF,cL,cE, glbM	21567s4333
7166		⊖	S0; 12.6; 1.5' x 0.5' cB,S,vlE,smhM, BN	21576s4339
7213		⊖	Sa; 11.8; 1.0' x 1.0' vB,pS,R, gbM; 16' SE from Alpha Gruis	22062s4725
----	I.5181	⊖	S0; 12.6; 1.6' x 0.5' cB,pL,E, foll at 2' by NGC 7232	22103s4616
7232		⊖	SBa; 13.0; 1.4' x 0.3' pB,S,pmE,psbM; foll at 2' by NGC 7233	22126s4605
7233		⊖	S0/Sa; 13.5; 1.4' x 0.3' F,S,R	22127s4606
----	I.5186	⊖	Sb; 12.5; 2.0' x 0.8' eF,S,R	22134s3705
----	I.5201	⊖	SBc; 12.8; 8.0' x 4.0' L,vF	22183s4619
7307		⊖	Sc; 13.1; 3.0' x 0.8' F,pL,pmE	22309s4112
----	I.5240	⊖	SBa; 12.6; 2.0' x 1.5' pF,pL,R	22390s4504

LIST OF STAR CLUSTERS, NEBULAE, AND GALAXIES (Cont'd)

NGC	OTH	TYPE	SUMMARY DESCRIPTION	RA & DEC
7410		⊖	SB; 11.8; 4.0' x 1.1' cB,L,vmE,mbM; nearly edge-on	22521s3956
7412		⊖	SBb; 12.2; 3.0' x 2.0' eF,vL	22530s4255
7418		⊖	SBc; 11.8; 2.8' x 2.5' cB,vL,v1E,vg1bM	22538s3717
7421		⊖	SBa; 12.8; 1.8' x 1.5' cB,L,v1E,gpmbM	22541s3737
----	I.5267	⊖	Sa; 11.8; 2.0' x 1.5' pB,S,R,mbM	22544s4343
----	I.1459	⊖	E3; 11.3; 1.0' x 0.7' F,pS,bM	22545s3641
7424		⊖	Sc; 12.0; 6.0' x 6.0' F,cL,v1E,vgmbM	22545s4120
----	I.5273	⊖	Sc; 12.0; 1.8' x 1.4' vF,cL,1E	22567s3758
7456		⊖	Sc; 12.5; 4.5' x 1.5' vF,L,mE,vgbM	22593s3951
7462		⊖	SBc; 12.7; 3.0' x 1.0' cF,pS,vmE	23000s4106
7496		⊖	SBb; 12.2; 2.0' x 1.0' pB,cL,1E,vgmbM	23070s4342
7531		⊖	Sb; 12.5; 1.5' x 0.5' pB,S,1E,pgbM	23121s4353
7552	△475	⊖	SBa; 11.6; 3.0' x 3.0' B,S,mE, vsbM	23135s4253
7582	△476	⊖	SBb; 11.8; 3.0' x 2.0' pB,L,mE,gbM; group with NGC 7590, 7599	23158s4238
7590	△477	⊖	Sb; 11.9; 2.2' x 0.8' pB,pL,pmE,gbM; foll by 7599	23163s4231
7599		⊖	SBc; 12.0; 3.8' x 1.2' F,pL,pmE,gbM	23167s4232

DESCRIPTIVE NOTES

ALPHA   Name- AL NA'IR. Magnitude 1.76; spectrum
B5 V. Position 22051s4712. The computed
distance is about 65 light years, the actual luminosity
about 70 times that of the Sun, and the absolute magnitude
+0.2.   The annual proper motion is 0.19" in PA 140°; the
radial velocity is 7 miles per second in recession. The
faint companion star at 28" is probably not a true physical
companion to Alpha; projected separation =565 AU.

BETA   Magnitude 2.17; spectrum M3 II; position
22397s4709. The distance of Beta Gruis is
approximately 280 light years, giving an actual luminosity
of about 800 times that of the Sun, and an absolute magnitude of -2.5.   The annual proper motion is 0.13"; the radial velocity is about 1 mile per second in recession.   Small
magnitude variations have been recorded, with a total range
of about 0.1 magnitude.

GAMMA   Magnitude 3.03; spectrum B8 III; position
21509s3736. The computed distance is
about 540 light years; the actual luminosity about 1450
times that of the Sun, and the absolute magnitude about
-3.1. The star shows an annual proper motion of 0.10"; the
radial velocity is 1.2 miles per second in approach.

HERCULES

LIST OF DOUBLE AND MULTIPLE STARS

NAME	DIST	PA	YR	MAGS	NOTES	RA & DEC
β621	0.6	38	60	8 - 9½	PA dec, spect A0	15482n4440
κ	28.2	12	62	5 - 6	(Σ2010)(7 Herc) optical, dist slow dec, PA inc, spect gG8, gK2	16058n1711
Σ2015	2.9	160	62	7½- 8½	relfix, spect F5	16074n4529
	12.7	98	14	- 13		
0Σ307	17.5	202	46	7 - 10	spect K0	16091n4756
Σ2025	2.6	164	36	7½- 11	relfix, spect F0	16097n4741
Σ2024	23.6	44	58	6½- 10½	relfix, spect gK4, F6, cpm.	16101n4230
Σ2030	5.6	238	24	7½- 11	relfix, spect A0	16110n4054
Σ2021	4.0	347	62	7½- 8	PA inc, spect K0	16110n1340
Σ2037	1.2	249	62	9 - 9	PA inc, spect G5	16166n1731
0Σ309	0.4	272	60	7½- 7½	PA inc, spect A5	16176n4147
Kui 72	2.0	137	60	5½- 10½	PA dec, spect A9	16182n3950
τ	6.7	146	58	4 - 14½	(22 Herc) (β1198) relfix, cpm pair; spect B5	16182n4626
γ	41.6	233	38	3½- 9½	optical, PA dec; spect A9	16197n1916
Σ2040	6.7	313	43	8 - 10	relfix, spect F2	16208n1357
ω	1.1	216	62	4½- 12	(24 Herc) (β625)	16231n1409
	28.4	96	57	- 11	cpm; dist dec; AC optical, PA dec; spect A0	
Σ2049	1.0	198	66	6½- 7½	relfix, spect A3	16258n2606
Σ2052	1.0	147	67	7½- 7½	binary, about 220 yrs; PA dec, spect dK2	16267n1831
Σ2051	13.7	19	34	7 - 8½	relfix, spect K0	16270n1042
Σ2056	6.3	315	45	8 - 9	relfix, spect A3	16292n0532
Σ2059	0.9	204	62	8 - 8	relfix, spect F5	16292n3810
31	5.2	233	24	6½- 12	(β816) spect A0	16296n3337
Σ2063	16.4	195	58	5½- 8	relfix, spect dF9	16303n4542
0Σ313	0.7	137	67	7 - 7½	PA dec, spect G5	16309n4013
Σ2061	2.5	25	44	7 - 10	relfix, spect F2	16313n3101
32	3.8	34	26	6½- 13½	(β818) spect F0	16315n3036
Σ2062	2.5	113	35	8 - 10	cpm, relfix, spect K0	16321n0847

HERCULES

LIST OF DOUBLE AND MULTIPLE STARS (Cont'd)

NAME	DIST	PA	YR	MAGS	NOTES	RA & DEC
Σ2068	5.0	253	62	8½- 8½	relfix, spect F5	16325n4722
Σ2069	27.7	73	22	7 - 10½	spect K0	16345n3355
42	25.6	92	58	4 - 10½	optical, spect gM2	16374n4901
					(Σ2082)	
Σ2079	16.8	91	30	7 - 8	relfix, spect F0	16375n2306
36 + 37	69.8	230	32	6 - 6½	cpm; spect A0, dA5	16382n0419
ζ	1.4	56	61	3½- 5	(Σ2084) binary,	16394n3142
					34 yrs; PA dec (*)	
Σ2085	6.1	309	41	7½- 9	relfix, spect A0	16403n2141
Σ2083	12.7	335	15	8½- 9	cpm, spect both K0	16404n1342
Σ2087	5.7	290	44	8 - 8	relfix, spect G5	16405n2346
Σ2091	0.6	314	68	7½- 8	slow PA inc,	16405n4117
					spect F0	
Σ2094	1.1	76	66	7½- 8	slight PA dec,	16421n2336
	24.8	312	62	- 11	spect F2	
D 15	1.3	152	68	8 - 8	binary, about 115	16424n4334
					yrs; PA dec, spect	
					dK6	
Σ2097	2.0	81	52	8½- 8½	PA dec, spect G5	16430n3550
46	5.1	162	37	7 - 9	(Σ2095) relfix	16431n2827
					spect F5	
Σ2098	14.3	146	15	8 - 9	relfix, spect G5	16438n3006
	64.6	135	15	- 10		
Σ2101	4.2	52	60	6½- 9	slight PA dec,	16440n3543
					spect G0	
Σ2104	5.8	19	55	6 - 8	relfix, spect F2	16469n3600
Σ2103	5.3	43	58	6 - 10	slight PA inc,	16473n1320
					spect A1	
52	1.5	10	62	5- 10	(β627) (A1866)	16478n4604
	66.2	230	11	- 12	AB cpm, spect A2	
52b	0.2	300	62	10-10½	binary, 56 yrs	
Σ2107	1.3	77	61	6½- 8	binary, dist inc,	16498n2845
					about 260 yrs; PA	
					inc, spect dF5	
Σ2109	5.9	314	34	7 - 10	relfix, spect K0	16516n2115
56	18.1	93	58	5½- 10	(Σ2110) cpm,	16530n2549
					relfix, spect gG5	
54	2.5	183	58	6 - 12	(β954) slow PA	16532n1831
					inc, cpm, spect K4	
OΣ 318	2.7	249	21	6½- 9½	relfix, spect K0	16544n1413

LIST OF DOUBLE AND MULTIPLE STARS   (Cont'd)

NAME	DIST	PA	YR	MAGS	NOTES	RA & DEC
Σ2112	2.0	261	63	8½- 9½	relfix	16564n3152
A1874	4.4	57	58	7½- 11½	dist & PA inc, all	16565n4726
	114	263	20	-7½	cpm; spect K0	
					(Σ32)	
A2085	5.4	355	27	7 - 13½	spect A0	16588n1640
A2085b	1.5	327	27	13½- 14		
Σ2115	19.0	239	58	6 - 10	relfix, spect Ap	16593n1501
Es 633	6.3	259	08	7 - 11½	spect G5	17000n4248
β822	1.6	227	43	7 - 11	cpm; spect K4	17017n1946
Pry---	1.8	231	58	7 - 10	relfix, spect A0	17025n1940
Σ2120	16.8	234	62	6½- 9	optical, dist inc.	17028n2809
					spect K0	
OΣ324	3.8	220	45	6½- 11	relfix, spect K2	17061n3116
Hu 1176	0.1	275	63	6 - 6	binary, 8 yrs; PA	17063n3600
	20.0	136	58	- 12	dec, spect A5	
					(Ho 412) (c Herc)	
Σ2135	7.9	187	62	7 - 8½	PA inc, spect K0	17100n2117
Σ2142	5.2	114	58	6 - 10	cpm, relfix, spect	17104n4948
					A2	
Σ2137	4.1	144	43	8 - 9	relfix, spect A2	17116n1600
α	4.6	110	62	3 - 5½	(Σ2140) fine ob-	17124n1427
	81.2	39	60	- 11	ject; PA slow dec,	
					red & greenish	
					color contrast pair	
					Spect M5, G5 (*)	
δ	8.8	241	61	3 - 8½	(Σ3127) optical,	17130n2454
					dist inc, spect A3	
					(*)	
68	4.5	60	43	5 - 10	(u Herc) (0Σ328)	17155n3309
					relfix, probably	
					cpm; A is ecl.bin;	
					spect B3 (*)	
Σ2147	6.2	94	24	7 - 11	relfix, spect M	17156n2858
Kui 80	0.7	157	58	6½- 8½	spect F8	17197n2848
A2183	1.0	128	44	7½- 10½	spect A0	17213n1657
ρ	4.1	317	62	4½- 5½	(Σ2161) fine pair	17220n3711
					slow PA inc, spect	
					Ap, B9	
Σ2162	1.4	279	56	8½- 9	relfix, spect A2	17222n3630

LIST OF DOUBLE AND MULTIPLE STARS (Cont'd)

NAME	DIST	PA	YR	MAGS	NOTES	RA & DEC
Σ2160	3.9	66	58	6 - 10	relfix, spect B9	17223n1539
Hu 1179	0.1	229	63	7 - 7	PA dec, both spect F7	17224n3838
Σ2165	9.4	59	62	7½- 8	PA & dist slow inc, spect F0	17243n2930
Σ2168	2.2	202	63	7½- 8	relfix, spect F5	17249n3548
A2184	1.3	20	60	7 - 10½	PA inc, spect F0	17253n1630
0Σ330	14.3	57	12	7 - 10½	relfix, spect F5	17276n1600
Σ2178	10.7	130	34	7 - 8½	relfix, spect K0	17277n3459
Σ2182	5.2	359	37	8 - 9	relfix, spect G0	17304n2354
Σ2189	21.0	100	19	8 - 10	spect A0	17315n4755
	65.1	360	19	- 8½		
Σ2190	10.3	23	58	6 - 9½	relfix, cpm pair; spect A7	17338n2102
Σ2192	12.9	45	57	7½- 10	PA dec, spect K0	17381n2916
Σ2194	16.3	8	58	6 - 8½	relfix, cpm pair; spect gK1, F4	17390n2432
Σ2203	0.6	303	66	7½- 7½	PA dec, spect A2	17397n4141
β1251	0.6	4	48	6 - 11	PA & dist dec, spect F1	17397n1558
h1303	39.7	151	59	7 - 10	optical, dist inc spect A3	17420n1426
Σ2213	4.5	329	63	7½- 8	relfix, spect F8	17430n3109
Σ2224	7.6	350	31	7 - 10	relfix, spect K0	17443n3920
μ	34.0	247	55	3½- 9½	(Σ2220) slight dist inc, cpm; spect G5, M3 (*)	17445n2745
μb	1.1	236	64	10 - 10	(AC7) binary, 43 yrs; PA inc, spect dM3 (*)	
Σ2215	0.6	272	62	6 - 7½	PA dec, spect A0	17449n1743
0Σ336	5.6	343	26	6½- 12	(β632) relfix; spect B3	17461n3418
	32.0	349	21	- 13		
Σ2232	6.4	141	41	7 - 8½	relfix, spect A2	17482n2518
0Σ338	0.7	354	67	7 - 7½	PA dec, spect K0	17497n1520
	28.0	217	00	- 13		
	95.2	251	06	- 10		
Σ2242	3.5	327	54	8 - 8	relfix, spect F0	17497n4455
Σ2239	2.4	319	48	8½- 9	relfix, spect F8	17498n2815
Σ2243	1.5	45	56	8½- 9	relfix, spect G5	17515n3606

LIST OF DOUBLE AND MULTIPLE STARS   (Cont'd)

NAME	DIST	PA	YR	MAGS	NOTES	RA & DEC
90	1.6	116	58	5 - 8	(β130) cpm, slow PA dec, spect K3	17517n4001
Σ2246	5.5	102	37	8½- 9	relfix	17538n3930
Σ2245	2.4	294	63	7 - 7	relfix, cpm pair;	17542n1820
β417	1.4	292	60	8 - 10	PA inc, spect G0	17545n3927
Σ2258	2.5	222	37	8 - 8½	relfix, spect G5	17554n4838
Ho 73	2.3	216	46	9 - 9	(Σ2256) both PA inc, spect G5	17554n3548
	8.4	318	24	- 13		
Σ2257	21.6	150	25	8 - 11	spect K0 (Ho 73 in field)	17556n3541
Hu 235	1.6	276	62	7 - 9½	PA inc, Spect F5	17557n4551
Σ2259	19.6	278	49	7 - 8	relfix, spect A0	17572n3003
Σ2247	11.5	190	25	8½- 9	relfix, spect F5	17575n2930
Σ2263	7.5	161	35	8 - 9	relfix, cpm pair	17589n2633
95	6.3	258	58	5 - 5	(Σ2264) neat pair; cpm, relfix, spect A7, G5 (*)	17594n2136
Σ2267	0.8	255	62	8 - 8	PA inc, spect A3	18001n4011
β1127	0.9	94	60	8 - 9½	PA dec, spect F2	18010n4414
Σ2277	27.0	125	58	6 - 8	spect A0; both optical, both PA inc, AC dist inc.	18018n4828
	94.0	292	59	- 10		
Ho 426	12.0	204	22	7 - 12	PA inc, spect F2	18019n2639
Es 471	18.2	271	07	7 - 14	spect A2	18026n2707
	30.4	44	07	- 10		
Ho 78	7.7	200	25	7 - 13	spect G0	18030n3316
OΣ341	0.3	78	61	7 - 8½	binary, 20 yrs; highly eccentric orbit, spect dG1	18037n2126
	28.2	172	33	- 9		
	38.4	100	33	- 9		
	62.8	38	33	- 9		
Σ2282	2.5	86	60	7 - 8	slight PA dec; spect A0	18049n4021
99	1.4	332	61	5½- 8½	Binary, 54 yrs; PA dec, spect F7	18051n3033
OΣ524	0.2	266	62	7½- 9	binary, about 250 yrs; PA dec, spect A2	18053n1939
100	14.2	183	55	6 - 6	(Σ2280) easy cpm pair, relfix, spect both A3	18058n2606

HERCULES

LIST OF DOUBLE AND MULTIPLE STARS   (Cont'd)

NAME	DIST	PA	YR	MAGS	NOTES	RA & DEC
0Σ344	2.3	145	58	6½– 11	slow PA dec, spect AO	18059n4942
Σ2289	1.1	223	67	6½– 7½	PA dec, spect F2, AO	18079n1628
Hu 674	0.5	243	58	7½– 8	PA dec, spect A2	18084n5023
Ho 82	0.1	238	62	6½– 6½	PA inc, spect A2;	18099n3326
	0.8	222	62	– 10	AC PA inc	
Σ2292	0.9	270	68	8 – 8	PA inc, spect A2	18101n2738
0Σ346	5.4	329	49	7½– 8½	relfix, spect F2	18132n1945
Σ2309	3.6	352	63	8½– 9	relfix, spect AO	18181n2530
Σ2310	5.1	236	35	6½– 10	relfix, spect B9	18186n2746
β640	1.9	320	57	7½– 12	PA dec, spect G5	18188n2730
β641	0.9	342	56	7 – 9	slow PA dec, spect B8	18197n2129
Σ2315	0.5	141	62	6½– 7½	binary, about 775 yrs; PA dec, spect A2	18230n2722
Σ2317	24.8	224	25	8 – 10	spect KO	18233n2603
	44.6	190	25	– 9½		
Σ2317b	1.1	323	54	10– 10		
β1326	5.5	105	58	7 – 13½	relfix, spect B3	18247n2625
Σ2319	5.4	191	56	7 – 7½	relfix, spect F5	18256n1916
	41.2	278	56	– 10		
Σ2320	1.3	3	56	7 – 9	PA dec, spect B9	18257n2440
Σ2330	17.6	169	52	8 – 9	PA & dist dec, spect KO	18289n1309
Σ2339	0.1	59	59	8 – 8	PA dec, spect F5;	18316n1742
	2.3	272	60	– 8	AC relfix	
0Σ359	0.5	15	61	6½– 6½	binary, 190 yrs; PA dec, spect gG8	18334n2334
0Σ358	1.6	168	67	6½– 7	binary, about 290 yrs; PA dec, spect dG0, dF8	18337n1656
h2834	20.2	250	00	7 – 12	spect B9	18345n2204
Ho 87	0.2	186	61	8 – 8	binary, about 130 yrs; spect G5	18364n1630
Σ2360	2.4	357	50	7½– 8½	PA dec, spect B8	18372n2053
A2988	0.2	216	61	8½– 8½	binary, 52 yrs; PA dec, spect A5	18389n2447
β645	9.3	305	24	7 – 12	spect B9	18411n1925

LIST OF DOUBLE AND MULTIPLE STARS (Cont'd)

NAME	DIST	PA	YR	MAGS	NOTES	RA & DEC
110	48.2	73	35	4½- 13	Both dist inc, both	18435n2030
	63.8	75	35	- 11	PA dec, spect F6	
					(h2839)	
Σ 2400	9.7	164	62	8 - 11	optical, dist inc,	18467n1612
Σ 2400b	1.0	201	62	11- 11	spect G0; BC relfix	
Σ 2401	4.3	38	51	7 - 8½	relfix, spect B5	18468n2107
Σ 2399	15.8	119	19	8 - 8½	relfix, spect A0	18468n1309
	31.5	48	19	- 10		
Σ 2411	13.5	95	37	7 - 10	relfix, spect K0	18500n1428
Σ 2415	1.9	290	68	6½- 8½	slight PA dec,	18524n2033
					spect A0	
113	35.4	32	24	5- 12	(β646) composite	18526n2235
					spect= G0+A3	
113b	7.0	145	15	12- 12	PA dec	
Hu 676	1.4	80	59	7 - 10	relfix, spect G5	18549n1446

LIST OF VARIABLE STARS

NAME	MagVar	PER	NOTES	RA & DEC
α	3.1--3.9	90:	Semi-reg; spect gM5 (*)	17124n1427
o	4.1--4.2	Irr	Spect B9	18056n2845
30	4.7---6.0	70	(g Herculis) Semi-reg;	16270n4159
			spect gM6	
68	4.7--5.4	2.0510	Ecl.Bin; lyrid, spect	17155n3309
			B3+B5 (*)	
89	5.4---5.5	70:	(V441 Herc) Semi-reg;	17534n2603
			spect F2p	
R	8.4--15.0	318	LPV. Spect M6e	16040n1830
S	6.5--13.7	307	LPV. Spect M5e--M7e	16496n1501
T	7.1--14.0	165	LPV. Spect M2e--M4e	18072n3101
U	6.5--13..	406	LPV. Spect M7e--M8e	16236n1900

HERCULES

LIST OF VARIABLE STARS (Cont'd)

NAME	MagVar	PER	NOTES	RA & DEC
W	7.7--14..	280	LPV. Spect M3e	16334n3727
X	6.3---7.5	95:	Semi-reg; spect M6e	16011n4723
Z	7.0---7.9	3.9928	Ecl.Bin; spect F4	17559n1509
RR	7.8--10..	240	Semi-reg; spect K5--N0e	16028n5038
RS	7.5--12.5	219	LPV. Spect M5e--M6	17196n2258
RT	8.6--15.1	298	LPV. Spect M4e	17088n2707
RU	7.0--14.0	484	LPV. Spect M7e	16082n2512
RV	9.0--15.5	205	LPV. Spect M2e	16587n3118
RX	7.3---8.0	1.779	Ecl.Bin; spect A0+A0	18283n1235
RY	8.4--14..	221	LPV. Spect M4e--M6e	17576n1929
RZ	9.0--15..	329	LPV. Spect M5e	18347n2600
SS	8.6--12.5	107	LPV. Spect M3e	16305n0658
ST	7.2--8.5	148	Semi-reg; spect M7	15493n4838
SU	9.0--12.2	334	LPV. Spect M6e	17469n2233
SV	9.2--14..	239	LPV. Spect M5e	18243n2500
SX	7.8---9.5	103	Semi-reg; spect G3e--K0	16053n2503
SY	8.0--13..	117	LPV. Spect M1e	16594n2233
SZ	9.9--11.0	.81810	Ecl.Bin; spect A0	17378n3258
TT	9.5--10.3	.9121	EclBin ; lyrid, spect A2	16521n1655
TV	9.0--14.6	303	LPV. Spect M4e	18128n3147
TX	8.2---8.9	2.0598	Ecl.Bin; Spect A4+A4	17170n4156
UU	8.0--10..	Irr	Spect F2---F8	16342n3804
UV	8.0--13..	342	LPV. Spect M6e	16432n1213
UW	7.5---8.6	100:	Semi-reg; Spect M5e	17127n3626
UX	8.8---9.7	1.5489	Ecl.Bin; spect A0	17519n1657
UZ	8.5--11..	264	LPV. Spect M5e	17282n1757
VX	9.7--10.6	.45537	Cl.Var; spect A3--F0	16285n1828
VY	9.0--13..	300	LPV.	17051n1714
WW	9.5--12..	311	LPV. Spect M2	17231n4610
WY	9.5--14..	376	LPV. Spect M5e	17580n2336
XZ	9.0--13..	172	LPV. Spect M0	18079n1805
YZ	9.5--10..	102	Semi-reg; spect M4	18160n2123
AC	7.1--9.4	75.46	RV Tauri type, Spect F2--K4e	18281n2150
AD	9.5--11.0	9.7666	Ecl.Bin; spect A4--K2	18479n2040
AE	9.0--13..	249	LPV. Spect M4e	18411n2256
AI	9.0--13..	405	LPV. Spect M6e	16524n4902
AK	8.1---9.6	.42152	Ecl.Bin; W Ursae Majoris type, spect F2 + F6	17117n1625

LIST OF VARIABLE STARS (Cont'd)

NAME	MagVar	PER	NOTES	RA & DEC
AN	9.0--10..	65:	Semi-reg; spect M5	17334n2045
AS	8.6--14..	268	LPV. Spect M3e	16366n1410
AW	9.5--10.9	8.8009	Ecl.Bin; spect K2+G4	18234n1816
AZ	9.5--15..	269	LPV. Spect M4	18206n2843
BB	9.5--10.3	7.5072	Cepheid, W Virginis type, spect G5	18436n1217
BE	9.0--10..	120:	Semi-reg; spect M4	16231n2922
BG	8.2--11..	348	LPV. Spect M3e	17072n1844
BL	9.4--10.2	1.3075	Cepheid; W Virginis type, spect A9--G0	17590n1915
CF	9.0--13..	306	LPV. Spect M0	17429n2131
CX	9.5--11..	125:	Semi-reg; spect M7	17086n2739
DE	9.8--11..	171	Semi-reg; spect K0	18044n2053
DH	9.4--12.0	4.7792	Ecl.Bin; spect A5	18455n2247
DI	8.4---9.2	10.550	Ecl.Bin; Spect B6+B6	18514n2413
DN	9.5--11..	227	LPV.	16098n1100
DO	9.7--13..	217	LPV.	16285n2333
DQ	1.3--15	---	Nova Herculis 1934 (*)	18061n4551
DZ	9.9--11..	120	Semi-reg; spect M0	18052n1733
GN	8.0--10..	Irr	Spect M4	16302n3858
GO	9.8--11..	50:	Semi-reg; spect M5	17313n1635
HS	8.3---8.7	1.6374	Ecl.Bin; spect B6	18488n2440
IQ	7.3---8.2	75	Semi-reg; spect M4	18157n1758
LQ	6.2---6.4	Irr	Spect gM4 (10 Herc)	16095n2337
MM	9.2--10.1	7.9604	Ecl.Bin; spect G0	17565n2209
MZ	9.1--10..	100:	Semi-reg; spect M6	18461n1903
NQ	8.0---8.6	.87022	Ecl.Bin; spect A0	18094n1819
NT	9.0---9.5	Irr	Spect M7	17305n2753
OP	6.2---6.9	Irr	Spect gM6	17554n4521
V336	9.0---9.5	53	Semi-reg	16059n1857
V337	8.5---9.2	280	Semi-reg; spect M8	17474n4543
V338	9.7--10.6	1.3057	Ecl.Bin; spect A9	17517n4347
V342	9.9--10.6	.85173	Ecl.Bin; lyrid, spect F2	18220n2503
V350	9.0--10..	Irr	Spect M5	17197n2448
V352	9.4--10..	350:	Semi-reg; spect M4	17409n3033
V360	6.3--15.5	---	Nova 1892	17145n2430
V446	3.0--	---	Nova 1960	18551n1307
V451	3.60 ±.04	6.0075	Alpha Canum type, spect A2p	16593n1501

LIST OF VARIABLE STARS   (Cont'd)

NAME	MagVar	PER	NOTES	RA & DEC
V522	9.4--10..	Irr	Spect M4	17457n4002
V529	9.2--10..	400	Semi-reg; spect M9	18064n4213
V533	3.0--14.5	---	Nova 1963	18128n4150
V535	6.49 ±0.08	10.1	Alpha Canum type, spect A0p	18443n2156
V600	7.0 ±.02	.2058	β Canis type, Spect B0	16347n1434
V620	6.2 ±.02	.0797	δ Scuti type, spect A8	17090n2417
V642	6.4--6.55	12:	Semi-reg; spect gM4	17315n1452
V644	6.34±.02	.1151	δ Scuti type, spect F2	16530n1342
V645	7.3 ±.07		δ Scuti type? spect A7	16531n2907

LIST OF STAR CLUSTERS, NEBULAE, AND GALAXIES

NGC	OTH	TYPE	SUMMARY DESCRIPTION	RA & DEC
6052		⊘	Sp/Irr; 13.0; 0.8' x 0.6' F,pL	16031n2041
6058	637[3]	◎	pF,vS,R, Mag 12, diam 25" x 20"; with faint central star	16028n4049
----	I.4593	◎	S,F, Mag 11, diam 13" x 10" with faint central star	16094n1212
6106	151[2]	⊘	Sb; 12.9; 1.9' x 0.9' F,pL,E, vgbM	16163n0731
6181	753[2]	⊘	Sc; 12.5; 2.0' x 0.8' pB,pL,vlE, pgmbM	16301n1956
6205	M13	⊕	!! Magnitude 5.7; diam 23'; class V; eB,eRi,eCM; stars mags 11..... superb object; NGC 6207 in field (*)	16399n3633
6207	701[2]	⊘	Sc; 12.3; 2.0' x 1.0' pB,pL,E,vgmbM. ½° NE from M13	16413n3656
6210	Σ5	◎	vB,vS,R, Mag 9.7; diam 20" x 16"; bluish disc with 06-type central star mag 12½	16425n2353

LIST OF STAR CLUSTERS, NEBULAE, AND GALAXIES (Cont'd)

NGC	OTH	TYPE	SUMMARY DESCRIPTION	RA & DEC
6229	50^4	⊕	Mag 8.7; diam 3.5'; class VII; vB,L,R, stars faint	16456n4737
6239	727^3	⊘	SBb; 12.9; 2.1' x 0.8' vF,E, bM	16484n4250
6341	M92	⊕	Mag 6.5; diam 8'; class IV; vB,vL,eCM, stars mags 12.. fine object (*)	17156n4312
6482		⊘	E3; 12.2; 0.7' x 0.5' vF,S,1E, vsvmbMN, vSBN	17498n2305
6574		⊘	Sc/Sd; 12.8; 1.0' x 0.7' pB,S,R	18095n1458

DESCRIPTIVE NOTES

ALPHA   Name- RAS ALGETHI. Position 17124n1427. One
of the brightest of the irregular variable
stars, reddish in color, spectrum M5 II, magnitude 3.1 to
about 3.9. The name is derived from the Arabic phrase *Ras
al Jathiyy*, the "Head of the Kneeler", though the star was
also known as *Al Kalb al Ra'i*, the Shepherd's Dog. In China
it was the *Throne of the Emperor*. The "Kneeler" is of
course the great figure of the boisterous ancient Greek
hero and unruly demigod *Heracles*, the Hercules of the Romans. The star pattern is one of the earliest to be defined
and named, and from the most ancient times was identified
with great national heroes or gods. In Babylonian lands it
was associated with the obscure sun-god *Izhdubar*, with the
mighty hunter *Nimrod*, and with *Gi-il-ga-mes* or *Gilgamesh*
of the Flood legend. To the ancient Phoenicians it represented the sea god *Melkarth*. Eudoxus, in the days of Plato,
and Hipparchus (about 140 BC) both refer to it under titles
which may be translated as "The Kneeling One" or the "Man
on his Knees", while the Greek poet Aratus (about 260 BC)
calls it οκλαζων or the "Kneeling One" and ειδωλον which

seems to mean "the Phantom". Aratus describes the figure
as:

> *"Laboring on his knees....*
> *.......and his right foot*
> *is planted on the writhing serpent's head..."*

Aratus seems to admit, however, that the origin and
meaning of the "Kneeling One" has been lost even to the
Greeks of his time, for *"of it no one can clearly speak"*.
In art, Heracles is often depicted kneeling in the act of
stringing his great bow, and so he appears on a silver coin
of wonderfully archaic style (Fig.1) minted at Thebes in
Boeotia, north-west of Athens, about 430 BC. His portrait,
bearded and wearing a lion-skin headdress, is seen on a
fine silver tetradrachm of Camarina in Sicily (Fig.2); the
coin dates from about 450 BC.

A different style portrait appears on virtually all
the coins of Alexander the Great, who claimed descent from
Hercules and from Zeus as well. The classic features seen
in Fig.4 are thought to be a thinly-disguised portrait of
Alexander himself. The reverse of this coin, a silver tetradrachm minted about 330 BC, shows an impressive portrayal
of the Olympian Zeus, one of the ancient Seven Wonders. On
the famous *Shekel of Tyre* (Fig.5) Hercules appears as the
hybrid god of the Phoenicians, *Heracles-Melkarth*.   This

variety, first issued about 126 BC, was the most commonly
used silver coin throughout the Judaean world in the 1st
Century, and is probably the coin referred to in the New
Testament story of the *Thirty Pieces of Silver*.

On ancient Roman coins, and on some issues of Parthia
and Bactria, Hercules is usually shown standing, carrying
his great club and often a lion-skin shield. The coin shown
in Fig.3 is a silver tetradrachm struck by Demetrius of
Bactria about 180 BC.

Hercules appears to us as a combination divinity, who
gradually assumed the attributes and gained credit for the
exploits of other ancient heroes. In Greek tradition he was
born in Thebes in Boeotia shortly before the time of the
Trojan War; he was the son of Zeus and the mortal woman
Alcmene, which aroused the jealousy and enmity of the goddess Hera. Curiously, however, his name seems to mean *"The
Glory of Hera"*; Greek tradition gives his original name as
Alceides. He was a great-grandson of the Greek hero Perseus
and displayed tremendous strength from his earliest youth.
Eventually he came to play a part in many of the epic tales
of Greece, including the expedition of the Argonauts, the
deliverance of Prometheus from his bondage high atop a crag
in the Caucasus, and the legend of the *Twelve Labors* which
suggest some possible connection with the symbolism of the
Twelve Signs of the Zodiac.

6   7   8   9

Many of the Labors appear on classic coins. On a fine
silver didrachm from Selinus in Sicily, dating from 460 BC
(Fig.6) we see the hero subduing the Cretan Bull; on a
classic coin of Heraclea, minted about 360 BC (Fig.7) he is
shown battling the Nemaean Lion; while on a silver stater
of Stymphalus in Arcadia, dating from about the same year,

DESCRIPTIVE NOTES (Cont'd)

he is brandishing his club against the attack of the dread
Stymphalian Birds. (Fig.8) After these turbulent exploits
it is a pleasure to find him in a rare pose on a silver
stater from Croton in Bruttium, seated in quiet rest on his
lion-skin. The coin (Fig.9) date to about 410 BC.

It was in the course of the Tenth Labor, the quest of
the Cattle of Geryones, that Hercules passed through the
Straits of Gibraltar, and set up there the two *Pillars of
Hercules* on the western edge of the known world. On antique
charts of the Medieval Age, the spot was traditionally
labeled with the legend *"Ne Plus Ultra"*, meaning "nothing
more beyond". After the discovery of the Americas, and the
consequent rise of Spanish power and influence, Spain chose
the design of the *Pillars of Hercules* for its coinage,
changing the motto to *"Plus Ultra"*.

This device and legend is seen here on the large
silver 8-real coins of the Spanish colonies, the almost
legendary *Pieces-of-Eight* of pirate lore. Specimen #10 was
minted at Potosi, Bolivia, in 1682, while the very interesting coin shown in Fig.11 was a portion of the treasure
recovered from the wreck of the Dutch ship *Hollandia*, which
sank off the Scilly Isles in 1743.

Hercules, although potentially immortal, eventually
accepted death to end the torment caused by the poisoned
robe of Deianira which had been stained with the deadly
blood of the centaur Nessus. A funeral pyre was built for
him atop Mt. Oeta, from whence he was carried to the Olympian Heavens and welcomed as one of the immortals, causing

Atlas, as Bulfinch slyly says, to feel the added weight.
Here he was finally reconciled with Hera, and won her
daughter Hebe, goddess of eternal youth, as his wife. In
the sky he appears in the traditional kneeling pose, with
his left foot (Iota Herculis) on the Head of Draco, his
head marked by Alpha, and his upraised club, we may imagine
indicated by Beta and Gamma. Most modern sky-watchers find
it easier to remember the "Butterfly" figure formed by the
six stars Beta, Delta, Epsilon, Zeta, Eta, and Pi; the last
four outline an uneven quadrilateral called the "Keystone"
of Hercules, the guide to locating the fabulous Hercules
Star Cluster M13. It may have been a mistake, one might
think, to place Hercules in the heavens; Edith Hamilton in
her book *Mythology* speculates that *"it is not easy to imagine him contentedly enjoying rest and peace, or allowing
the blessed gods to do so either."*

        The variability of Alpha Herculis was discovered by
Sir William Herschel in 1795; the average period of the
star appears to be about 90 days, with individual periods
varying between rather wide limits. It is also one of the
largest known red giant stars, and one of the few whose
apparent angular size has actually been measured with the
interferometer used on the 100-inch reflector at Mt.Wilson;
a value of 0.03" was obtained. To convert this to an actual
diameter in miles it is necessary to know the distance to
the star. Alpha Herculis, however, is unfortunately too
remote to show any measurable trigonometric parallax; in
fact all attempts have given a <u>negative</u> <u>parallax</u> which must
indicate that the star is more distant than the supposedly
"background" stars which are being used as reference objects! This alone indicates a distance well in excess of
300 light years. The spectroscopic parallax method gives a
distance modulus of about 5.6 magnitudes, equivalent to
some 430 light years; the resulting actual diameter is then
just over 400 times the Sun, or about 350 million miles.
Although these figures are somewhat uncertain, this seems
to be the largest size known for any star visible to the
naked eye, with the probable exception of Betelgeuse in
Orion, and possibly the "Garnet Star" Mu Cephei. A recent
revision of the distance of Betelgeuse increases its computed diameter to over 800 times the size of the Sun, when

DESCRIPTIVE NOTES   (Cont'd)

at maximum. Betelgeuse definitely appears to be the larger
of the two, though it is probable that the size of both
giants varies during the light cycle.

If the computed distance of about 430 light years is
accepted, the actual luminosity of the star is found to be
about 830 times the Sun; the absolute magnitude is about
-2.5. The annual proper motion is 0.03"; the radial velocity is some 20 miles per second in approach. In its physical characteristics , Alpha Herculis is a typical red
giant. With a mass of only a few suns, but a diameter a
few hundred times greater, the density works out to less
than 0.0000001 the solar density, a virtual vacuum by any
ordinary earthly standards. The temperature varies from
2400°K to about 2650°, making this one of the coolest of
all known stars. Much of the radiation is emitted in the
infrared; if our eyes were sensitive to radiation at all
wavelengths, Alpha Herculis would appear as one of the most
brilliant stars in the sky.

For the amateur telescope, this is one of the finest
colored double stars in the heavens. Mary Proctor, in her
delightful book *Evenings With the Stars* (1924) describes
it as "a splendid double when seen with a powerful telescope, the components being orange and emerald green. The
writer can never forget the feeling of amazement with which
she gazed at this celestial gem through the 12-inch lens at
the Yerkes Observatory, nor the exclamation of delight when
it was glimpsed by an observer who told her to come quickly and look at Alpha Herculis.....this was one of the red-
letter day experiences, indelibly recorded in her memory.."

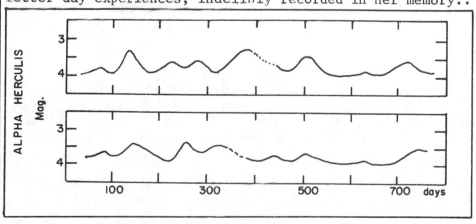

The companion is magnitude 5.39 and of a bluish-green hue, a striking contrast with the bright orange of the primary. Since the large star is variable, the pair will be found easier to divide at some times than at others. It is now certain that the two stars form a physical system with a period of something like 4000 years. The separation is at least 700 AU and may be much more if one star is closer to us than the other. The apparent separation of 4.6" has not changed significantly since 1830, but the PA appears to be decreasing slowly at a rate of 7° per century.

The discovery of the companion is usually credited to Sir William Herschel in August 1779; Herschel himself however, stated that the duplicity of the star had been noted by Nevil Maskelyne about two years earlier.

The small star, which appears single in even the largest telescopes, is actually a close binary with a period of 51.59 days and a nearly circular orbit (eccentricity of 0.028). The spectrum is composite, G5 III and about F2.

An interesting feature of the system is the presence of a vast expanding gaseous shell which clearly originates in the red giant primary, but extends out far enough to actually envelope the companion star. Since the rate of expansion of this shell (about 6 miles per second) exceeds the escape velocity of the system, it is evident that the substance of the star is gradually being dispersed into space, at the rate of about one-millionth of its mass in a thousand years. This is evidently one of the factors in stellar evolution, and it is now known that many other red giants are surrounded by similar expanding clouds. Eta Geminorum, VV Cephei, and Zeta Aurigae are other typical examples.

BETA   Name- KORNEPHOROS. Magnitude 2.78; spectrum G8 III. Position 16281n2136. The distance of Beta Herculis is estimated to be about 105 light years; the actual luminosity is then about 65 times that of the Sun and the absolute magnitude about +0.3. The annual proper motion is 0.10"; the radial velocity is 15 miles per second in approach.

The spectroscope reveals the star to be a close binary with a period of 410.575 days and an eccentricity of 0.550.

DELTA   Magnitude 3.14; spectrum A3 IV. Position
17130n2454. The distance is about 95 light
years, the actual luminosity about 40 times that of our Sun
and the absolute magnitude about +0.8. The star shows an
annual proper motion of 0.16"; the radial velocity is about
24 miles per second in approach.

Delta Herculis is an interesting example of an
optical double. The 8th magnitude companion has no real
connection with Delta, but merely happens to lie nearly in
the same line of sight. The two stars are moving in different directions: Delta almost due south, and the faint star
nearly due west. As early as 1824, John Herschel wrote
"There can be no doubt of a material change both in position and distance having taken place in this star.....
quantities too large to leave any room for doubt. As the
change is contrary to what the presumed proper motion of
the large star would alone produce, this star merits particular attention".

When measured by F.G.W.Struve in 1830 the separation was 25.8". By 1960 it had decreased to the minimum
value of about 9" in PA 241°; the pair will henceforth continue to widen. The two stars have a peculiar contrast in
colors, often described as "greenish and pale violet"; F.G.
W.Struve, however, called them "green and ashy white", while

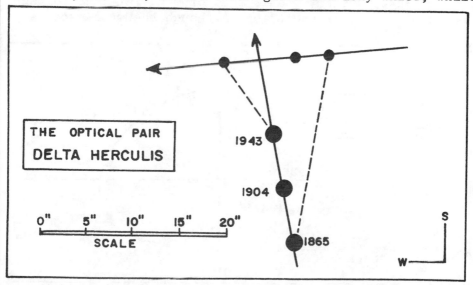

THE OPTICAL PAIR
DELTA HERCULIS

1943

1904

1865

0"   5"   10"   15"   20"
SCALE

S

W

T.W.Webb quoted a variety of color impressions of various other observers, ranging from "pale yellow and bluish-green" to "usually white and azure" to "pale yellow and ruddy purple". The spectral class of the small star is dG4; the true color, then, is yellowish or pale orange. Delta itself shows a variable radial velocity and is probably a spectroscopic binary, but no orbit has yet been determined.

ZETA   Magnitude 2.81; spectrum G0 IV; position 16394n3142. Zeta Herculis is an interesting double star system, discovered by William Herschel in 1782. One of the closer binaries, it is 30 light years distant and has an orbital period of 34.38 years. According to computations by P.Baize (1949) the semi-major axis of the system is 1.369" or about 12 AU; the eccentricity is 0.47, and periastron occurred in late 1967. The apparent separation varies from 0.4" to about 1.6" with the widest separation in 1956 and 1990.

	MAG.	SPECT.	MASS.	ABS.MAG.	LUM.
A	2.8	G0 IV	1.14	+2.8	6
B	5.5	dK0	0.82	+5.5	0.5

The bright star is the classical case of a subgiant in a visual binary system, and is thus of much interest to the students of stellar evolution. Although the mass exceeds

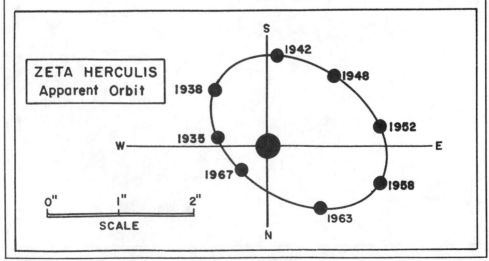

ZETA HERCULIS
Apparent Orbit

DESCRIPTIVE NOTES   (Cont'd)

that of our Sun only slightly, the star has already evolved to the right of the Main Sequence, and now has about six times the luminosity of the Sun. A few other such cases are now known.

The annual proper motion of Zeta Herculis is 0.61" in PA 309°; the radial velocity is 42 miles per second in approach.

ETA   Magnitude 3.46; spectrum G7 III or IV. The position is 16412n3901. The distance of this star is about 60 light years; the actual luminosity is about 11 times that of the Sun, and the absolute magnitude about +2.1. The annual proper motion is 0.10"; the radial velocity is 4.5 miles per second in recession.

This is the nearest naked-eye star to the great globular star cluster M13, located about 2½° toward the south. (Refer to page 978)

MU   Magnitude 3.42; spectrum G5 IV; the position is 17445n2745. Mu Herculis is a well-observed triple star system. The star at 34" is magnitude 9.8 and was first detected by William Herschel in 1781. In 1856 it was found to be a close double itself with individual magnitudes of 10.3 and 10.8 and a combined spectral class of dM3. These two faint red dwarfs are in fairly rapid orbital motion in a period of 43.20 years. According to computations by P.Couteau (1960) the semi-major axis of the system is 1.36" with an eccentricity of 0.178; periastron occurs in 1965 and widest separation in 1989. The two stars have actual luminosities 175 and 250 times less than that of the Sun; the 3rd magnitude primary is a subgiant with about three times the solar luminosity. No estimate of period can yet be made for the revolution of the dwarf pair around the bright star; the PA has changed by only 7° since the time of Herschel.

Mu Herculis is 30 light years distant and shows a fairly large annual proper motion of 0.81" in PA 203°; the radial velocity is about 9 miles per second in approach.

It is an interesting comment on the difficulty of making accurate color estimates of double stars, that both Smyth and Webb saw the red dwarf companion of Mu Herculis as "cerulean blue".

PI　　　　Magnitude 3.13; spectrum K3 II; position 17133n
　　　　3652. The computed distance is about 400 light
years; the actual luminosity about 750 times that of the
Sun, and the absolute magnitude -2.4. Pi Herculis shows an
annual proper motion of 0.03"; the radial velocity is 16
miles per second in approach.

68　　　　(u Herculis) Position 17155n3309. This is a
　　　　rapidly revolving eclipsing binary star, discovered by J.Schmidt in 1869. It is a typical "lyrid" type
system in which two bright giant stars revolve nearly in
contact, mutually eclipsing each other at every revolution
and resulting in a continually varying light curve. Maxima
of the system are equal (mag 4.7) and the minima alternate
between 5.0 and 5.4.   The exact period of the system is
2.051027 days and the orbit is nearly circular with the

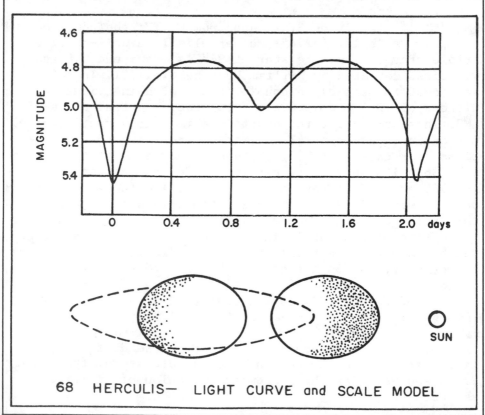

68   HERCULIS— LIGHT CURVE and SCALE MODEL

slight eccentricity of 0.053. A distance of about 600 light years has been obtained for this star through the spectroscopic parallax method; this gives a total luminosity of about 360 times that of the Sun. The annual proper motion is only 0.015"; the radial velocity about 12.5 miles per second in approach.

	ABS.MAG.	SPECT.	DIAM.	LUM.	MASS
A	-1.2	B3 III	6	260	7.5
B	-0.2	B5	6	100	2.9

68 Herculis is a system which resembles the giant pair Beta Lyrae, though on a somewhat smaller scale. Both stars are B-type giants, and are so close to each other as to be distorted into oval shapes by tidal attraction and rapid rotation. Each star is some 5 million miles in diameter and their separation, center-to-center, is just over 6 million miles. The center of gravity of the system is located within the body of the brighter star, about 1.7 million miles from its center.

The orbit of 68 Herculis is not seen precisely edge-on, but is inclined about 75° from the plane of the sky; the eclipses are thus partial obscurations and each star is about 60% hidden by its companion during each of the eclipses. Slight irregularities in the light curve are caused by reflection and tidal effects.

68 Herculis is also the visual double star O∑328, discovered by O.Struve in 1847. The companion is magnitude 10.2 and the separation is 4.5" with no appreciable change in the last century. The two stars possibly form a true physical system with a projected separation of 840 AU.

95   (∑2264) Magnitude 4.42; spectra A7 III and G5 III
Position 17594n2136. This is a famous little double star, first measured by F.G.W.Struve in 1829 and notorious among observers for discordant estimates of color. The magnitudes are 5.13 and 5.21; the components have remained relatively fixed for over a century at 6.3" in PA 258°. A brief survey of the supposed color fluctuations of this pair is given by Miss Agnes Clerke (1905) in her popular work "The System of the Stars", as follows:

"Familiar with them as vividly tinted objects, Professor
Piazzi Smyth was astonished, on pointing his telescope
toward them from the Peak of Teneriffe, July 29, 1856, to
perceive them both white. In the following year, nevertheless, they shone as before in "apple green and cherry red",
and were so observed by Admiral Smyth, Dawes, and others.
Captain Higgins actually watched these colors fade and then
revive in 1862-63, in the course of about a year; but no
trace of them has been seen of late; the stars of 95 Herculis are now of an identical pale yellow. Their spectra are
not identical. Dr.Vogel, in 1899, classed one as solar, the
other as Sirian in type. The history of these stars goes
back to 1780 when Herschel observed tham as bluish-white
and white; J.Herschel and South called them "bluish-white
and reddish" in 1824; Struve, 1828-32, greenish yellow and
reddish yellow, in precise agreement with Pickering's
appraisement in 1878. Thus the magnificent tints of orange
and green which Secchi admired in 1855, and Piazzi Smyth
missed in 1856, were of a transitory character."

　　　To which the modern observer can add only the
question: "transitory...or illusionary?" Despite the mass
of reports and the reliability of the observers, it seems
quite unlikely that color changes of such rapidity could be
physically real. Star colors, even in the case of strong
contrast pairs, are delicate and elusive; disagreements
among the most experienced observers seem the rule rather
than the exception. In any case, similar changes have been
reported for many other well-known pairs including Gamma
Leonis, 70 Ophiuchi, Delta Herculis, Gamma Delphini and
Zeta Cancri. None of these stars have shown any noticeable
spectral changes which would, necessarily, accompany any
real change in color.

　　　The estimated distance of 95 Herculis is about
400 light years, giving a projected separation of 775 AU
and true luminosities of 120 and 110 times the Sun. Both
stars show an annual proper motion of 0.03" and a radial
velocity of 18 miles per second in approach. The components are both giant stars of luminosity class III, with
computed absolute magnitudes of about −0.4 and −0.3. (See
also Gamma Leonis, 70 Ophiuchi, Delta Herculis, and the
remarks concerning the color estimates of the companion to
Regulus)

DQ   Nova Herculis 1934, one of the brightest novae of the 20th Century. Position 18061n4551, near the Hercules-Lyra border, about midway between Vega and the head of Draco.   The nova was first noticed by the British amateur J.P.Prentice, on the night of December 13, 1934. At that time it was of the 3rd magnitude, but slowly brightened to 1st magnitude in time for Christmas, reaching maximum brilliancy of magnitude 1.3 on December 22.   Previous to the explosion the star had been a 15th magnitude object; the light increase thus amounted to about 400,000 times. At its estimated distance of about 1200 light years the actual peak luminosity was about 65,000 times that of the Sun; the absolute magnitude was about -7.2.

The light curve of Nova Herculis was striking for its similarity to that of Nova T Aurigae 1891. Both curves show a long protracted maximum, a sudden drop at about 100 days with a slow recovery to a lower secondary maximum, followed by a very slow fading to the pre-outburst level. Exploding stars of this type have been called "slow novae". The light curves of the two stars are compared on page 976.

Modern observations reveal an oval nebulosity of about 3" in extent, surrounding the star. When photographed in red light, this nebulosity shows several peculiar con-

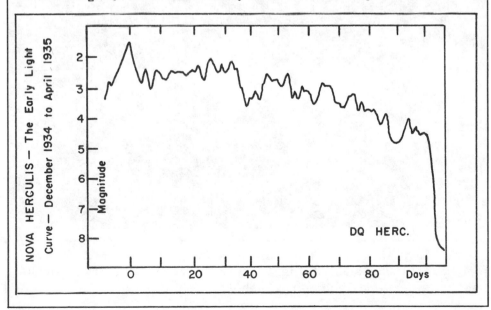

NOVA HERCULIS — The Early Light Curve — December 1934 to April 1935

DQ HERC.

NOVA HERCULIS 1934. The star is shown near maximum (below) in March 1935, and fading to normal (top) about two months later. Lick Observatory photographs.

densations which have led some to believe that the nova
actually split into several parts when it exploded. In the
case of Nova Pictoris, which exploded in 1925, a similar
appearance has been detected. However, it now seems certain
that these objects are merely bright knots of nebulous
matter, and not actual fragments of the exploded star. The
expansion velocity of the nova-shell of DQ Herculis was
measured at 480 miles per second at peak light intensity;
a month later it had risen to about 600 miles per second.
The highest velocity measured in this nova occurred about
100 days after maximum, just before the great drop in magnitude; some of the hydrogen absorption lines then showed
a displacement equivalent to about 1000 miles per second.

A most interesting and significant discovery has
been made concerning this nova. In 1954, Dr.M.F.Walker
found it to be an eclipsing binary system with a period of
only 4 hours and 39 minutes, and a range of 1.3 magnitude
in yellow light. This is one of the shortest periods known
for any eclipsing double, comparable to the periods of the
erratic SS Cygni and U Geminorum stars. Primary eclipse
lasts about an hour and is probably total in the sense that
the nova star itself may be entirely hidden. A surrounding
nebulous disc is eclipsed only partially.

The light curve of this remarkable system does not
repeat exactly time after time, but shows small changes and
irregularities. There are also rapid "flickerings" in the
light outside of eclipse. In addition, a periodic waver of
about 1/20 magnitude has been detected; the period is an
incredibly short 71 seconds. Walker has suggested that
this oscillation may represent an actual pulsation of the
nova-component; this theory is supported by the fact that
the oscillation vanishes during eclipse.

The observed star, then, appears to be the nova-
component itself, and is a bluish dwarf with an absolute
magnitude of about +8. The eclipsing component has not yet
been identified spectroscopically, but is considerably
fainter than the nova-star and is probably a red dwarf with
an absolute magnitude of about +9. The distance between the
two stars is something like 200,000 miles, comparable to
the separation of the Earth and Moon; the computed masses
are each about 0.25 the solar mass. This is a surprising
feature since a degenerate star of such small mass should

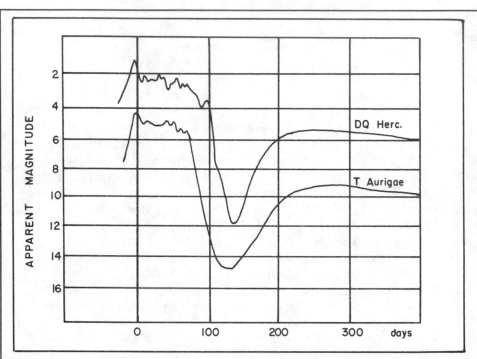

ABOVE— The Light Curves of NOVA HERCULIS and NOVA AURIGAE Compared.
BELOW— The LIGHT CURVE of the ECLIPSING COMPONENTS of DQ HERC.

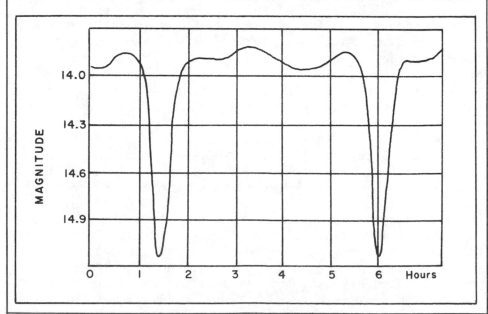

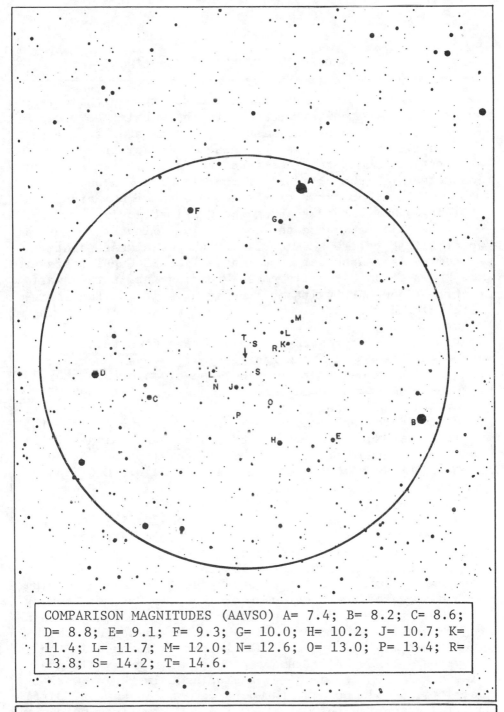

COMPARISON MAGNITUDES (AAVSO) A= 7.4; B= 8.2; C= 8.6; D= 8.8; E= 9.1; F= 9.3; G= 10.0; H= 10.2; J= 10.7; K= 11.4; L= 11.7; M= 12.0; N= 12.6; O= 13.0; P= 13.4; R= 13.8; S= 14.2; T= 14.6.

DQ HERCULIS. Identification chart for Nova Herculis 1934. Circle diameter = 1° with north at the top; limiting magnitude about 15. Chart made from a Lowell Observatory 13-inch telescope plate.

be quite stable. It seems impossible to explain the outburst of Nova Herculis according to the usual theory of
instability and resulting mass loss as the star approaches
the white dwarf state. In this case at least, it appears
necessary to admit that the nova explosion was in some way
connected with the presence of the close companion.

The light curve of the components of DQ Herculis is
strikingly similar to that of the dwarf eclipsing system
UX Ursae Majoris, whose period is only 4 minutes longer; it
appears that we are here dealing with two almost identical
systems. Also, a number of the nova-like SS Cygni stars are
now known to be extremely close binaries, as is the dwarf
recurrent nova WZ Sagittae. The latter star has the shortest period now recognized for any eclipsing binary: 81.6
minutes.

Many interesting questions have been raised about
the significance of these facts. Are all nova outbursts
triggered by some interaction between close double stars?
The question seems reasonably well answered for the SS
Cygni stars and possibly for the recurrent novae; nor is
DQ Herculis the only known case of duplicity among the full
scale novae. Nova T Aurigae (1891) and Nova Aquilae (1918)
seem to be rather similar systems. (See also Nova Persei
1901, UX Ursae Majoris, WZ Sagittae, SS Cygni, U Geminorum
and AE Aquarii)

M13   (NGC 6205) Position 16399n3633. This is the
Great Globular Star Cluster in Hercules, the
finest cluster of its type in the north half of the sky and
one of the most spectacular telescopic objects in the
heavens. It was first mentioned by Halley in 1715, having
been discovered the previous year. Charles Messier, coming
across it in 1764, described it as a round "nebula containing no stars", a comment which adequately describes the
quality of his telescope. The cluster is located in the
"Keystone" of Hercules, about a third of the way along a
line drawn from Eta to Zeta. Exploring this region with a
pair of good binoculars, the observer should have no difficulty in locating the object, which may sometimes be seen
even with the naked eye as a hazy-looking "star" of the 6th
magnitude. In the small telescope it appears as a bright
roundish nebula about 10' in apparent size; the diameter,

THE FIELD OF THE HERCULES STAR CLUSTER. M13 as it appears
on a plate made with the 13-inch telescope at Lowell Observatory. The faint galaxy at upper left is NGC 6207.

however, is more than doubled on the best photographic
plates. A 4 to 6-inch telescope begins to resolve the misty
glow into hundreds of tiny star-points, but for a really
fine view a 12-inch or larger glass is desirable. In the
greatest telescopes the cluster is an incredibly wonderful
sight; the vast swarm of thousands of glittering stars,
when seen for the first time or the hundredth, is an absolutely amazing spectacle.

Halley himself, as the discoverer, was also the first
to note that M13 "shows itself to the naked eye when the
sky is serene and the Moon absent." Messier, in June 1764,
found it "round and brilliant" with a brighter center, but
described it as "a nebula which I am sure contains no star"
and gave the apparent size as about 6'. J.E.Bode, in 1774,
also seems to have found M13 irresolvable, and referred to
it as "a very distinguishable nebula which appears as a
pretty lively and round nebulous patch. In the centre is a
bright nucleus". Sir William Herschel, with his great reflectors, disclosed the true nature of M13 by resolving it
into "a most beautiful cluster of stars exceedingly compressed in the middle and very rich". Admiral Smyth found
it to be "an extensive and magnificent mass of stars with
the most compressed part densely compacted and wedged together under unknown laws of aggregation". To T.W.Webb it
was "spangled with glittering points in a 5½-ft. achromatic
and becomes a superb object in large telescopes." Both
John Herschel and Lord Rosse noted the curvilinear streams
of stars which seem to radiate out into space from the
cluster's outer edges; a similar appearance has been noted
in other bright globulars, as M3 and M5.

"It is the finest of all the clusters in the northern
skies," states Mary Proctor in her book *Evenings With The
Stars* (1924) "and is just visible to the unaided eye on a
dark night. With an opera-glass, the cluster presents the
appearance of a faint hazy speck, with a little star on
each side, but when Herschel examined it with his great reflector, he computed that the 'speck' must contain at least
fourteen thousand stars. Nowadays, by means of photography,
it has been possible to obtain a close-up view, as it were,
of what may be termed literally a ball composed of thousands of suns, with outlying streamers curving outward as
though wafted by a celestial breeze......the cluster is a

"....As if the light of a thousand suns should suddenly appear in the heavens.... such would be the glory of the Eternal One!"

Bhagavad-Gita XI, 12

mass of glittering starlight.....A glance at the magnificent photograph of this cluster....will convey an idea of
the wonders of this display far more than any descriptive
words. The glamour of the light from these intermingled
stars remains for the fortunate observer who has the privi-
ledge of actually *seeing* them through the giant reflector."
Florence A. Grondal in *The Romance of Astronomy* (1937)
called this cluster *"A Treasure-Trove of 50,000 Stars"*.

Sir William Herschel had estimated that M13 contained
some 14,000 stars, a figure which was more than doubled by
actual survey at Mt.Wilson. More than 30,000 star images
were counted, down to the 21st magnitude. No very reliable
count can be made in the rich central core where the innumerable images merge into a great glowing mass, but it
seems certain that the total population cannot be less than
a million stars. The total luminosity of M13 is over 300
thousand times the Sun, and the mass is perhaps equal to
half a million solar masses. Red giants of the 11th magnitude are the brightest members of the cluster; each of
these stars has an actual luminosity of about 2000 suns.
As a standard of comparison it may be remembered that the
Sun would appear as a 19th magnitude object at the distance
of M13. The integrated visual magnitude of the group is
+5.7, the absolute magnitude about -8.7.

The integrated spectral type of the Hercules Cluster
is about F5; there are no blue giants known in the group
with the possible exception of a single star of type B2,
whose membership is uncertain. Clusters such as this are
classed as pure Population II systems. A few small patches
of obscuring material, however, seem to be present, though
it is not known whether these, even if real, actually be-
long to the cluster. The most prominent of these dark blots
may be seen in the photograph on page 985, near the right
edge of the central mass. There also appear to be several
ill-defined dark lanes at various positions in the cluster;
three of these meet on the south-east side of the central
core to form a Y-shaped pattern which was noted by Lord
Rosse more than a century ago, and which is very evident
in medium-size telescopes today.

Only a few variable stars have been detected in M13;
in contrast some globulars contain more than a hundred of
them. The reason for this difference is still not quite

M13- THE GREAT GLOBULAR STAR CLUSTER IN HERCULES, one of
the most splendid objects in the heavens. This photograph
was made with a 12½-inch reflector by Evered Kreimer of
Prescott, Arizona.

DESCRIPTIVE NOTES (Cont'd)

clear. Four cluster-type variables of the RR Lyrae class
have been identified with apparent magnitudes of 14.6; at
present three cepheids of longer period are known, and a
few long period pulsating red stars of the Mira type.

The distance of the Hercules Cluster has long been
given as about 32,000 light years; more recent studies with
the 200-inch telescope have revised this figure down to
about 25,000 light years or possibly a little less. From
the color-magnitude diagram the apparent distance modulus
of M13 is about 14.4 magnitudes, equivalent to 24,600 light
years; if a light loss of about 0.3 magnitude through space
absorption is allowed, the true modulus is reduced to 14.1
magnitudes, or about 21,000 light years. The true distance
undoubtedly lies somewhere in this general range; perhaps
no greater degree of accuracy can be expected at present.
The spectroscope indicates a radial velocity of nearly 150
miles a second in approach. This motion is a combination of
at least three different velocities: the rotation of the
Galaxy, the Sun's motion in space, and the revolution of
the cluster itself around the galactic center. At present,
M13 is about as far from the galactic center as we are
(some 30,000 light years) and the period for one galactic
revolution would be on the order of 200 million years.

The diameter of the Hercules Cluster is about 160
light years; the figure is necessarily somewhat uncertain
owing to the indefinite boundaries of the object and the
gradual thinning out of its members in the outer portions.
Most of the stars are concentrated in the central core,
less than 100 light years in diameter, but some of the
outer wanderers extend through a region over 200 light
years in extent. The true density of such a cluster is a
most interesting question. Photographs give the impression
of incredibly thick crowding and suggest that the stars are
packed virtually in contact. This is an illusion, due to
the vast distance of the group and the merging of the num-
berless star images. The central region covers an area
some 100 light years across, or roughly a million cubic
light years in volume. Assuming that a million stars populate this region it is evident that the density is no
greater than about one star per cubic light year. In the
actual center of the cluster the star density may be several times greater, but in no case would it approach actual

GLOBULAR STAR CLUSTER M13. The Hercules Cluster as it appears on a plate made with the 200-inch telescope at Palomar.

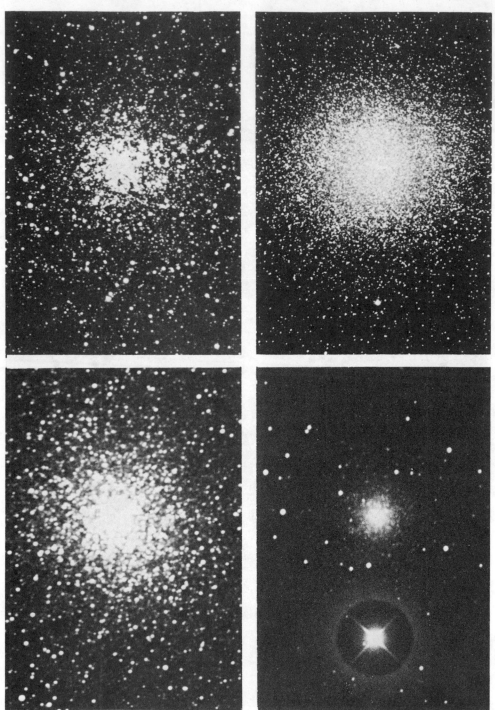

NOTABLE GLOBULAR STAR CLUSTERS. Upper Left: NGC 6397 in Ara, probably the nearest to Earth. Upper Right: Omega Centauri, the largest example known. Lower left: M22 in Sagittarius. Lower right: The very remote NGC 2419 in Lynx.

DESCRIPTIVE NOTES (Cont'd)

crowding. This fact is better understood by constructing an
imaginary scale model of the cluster. On such a model the
stars would be represented by a million grains of sand,
distributed throughout a spherical volume of space some 300
miles in diameter. Each grain would be 0.03 inch in diameter, and separated from the next nearest grain by a distance
of three miles! Even in the most closely packed central
mass the grains would still be separated from each other by
the greater part of a mile. Thus even the globular cluster, which appears to us as the most densely packed mass of
stars to be found anywhere in the Universe, is shown to be,
by earthly standards, almost empty space.

      The appearance of the heavens from a point within
the Hercules Cluster would be a spectacle of incomparable
splendor; the heavens would be filled with uncountable numbers of blazing stars which would dwarf our own Sirius and
Canopus to insignificance. Many thousands of stars ranging
in brilliance between Venus and the full moon would be continually visible, so that there would be no real night at
all on a planet in a globular cluster. Inhabitants of such
a planet would probably know nothing of other clusters, of
the Galaxy, and of the other galaxies, as their view would
be completely blocked by the brilliance of their own skies.
To them, the Hercules Cluster would be "the Universe".

      THE FAMILY OF GLOBULAR CLUSTERS. Approximately
100 similar clusters are now known, seemingly surrounding

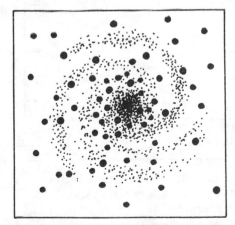 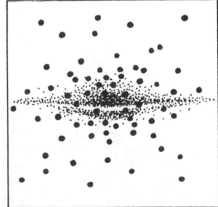

**DISTRIBUTION OF THE GLOBULAR CLUSTERS around the edges
of the Galaxy — face-on and edge-on views.**

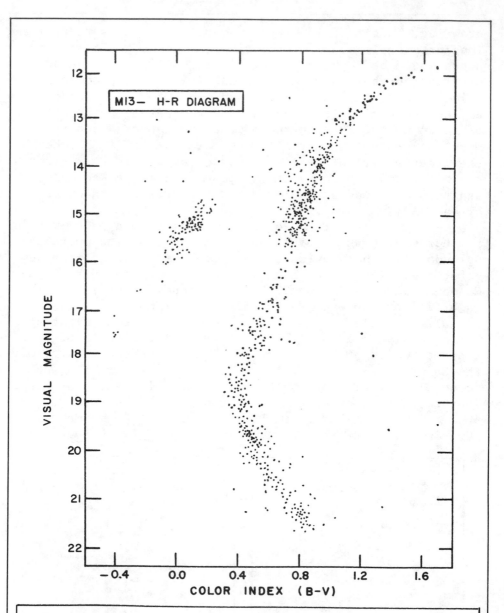

THE COLOR—MAGNITUDE DIAGRAM for the Hercules Cluster, from observations made at Palomar, McDonald, and Lowell Observatories. Stars fainter than about 18.5 are still Main Sequence objects; giant stars populate the rest of the Diagram. For explanation, refer to text, page 990.

our Galaxy on all sides. When the distribution of these
objects is plotted, it is found that they form a nearly
spherical system and that the center of this system is
identical with the center of our Galaxy. This discovery was
made by Harlow Shapley. It has since been found that other
galaxies are accompanied by their own globular cluster
families; the Andromeda spiral for example has about 140,
and the great elliptical galaxy M87 in Virgo possesses well
over a thousand.

Since our Solar System is some 30,000 light years
from the Galactic Center, it is evident that the globular
clusters should not appear distributed uniformly around the
sky, but should appear mainly in the half of the sky which
is centered on the galactic nucleus. Such is found to be
the case. The region of Sagittarius, Scorpius, Ophiuchus,
and Ara is a rich field for the globular cluster hunter;
the opposite part of the sky is nearly devoid of them.

The greatest of the globular star clusters, and one
of the nearest to the Earth, is the magnificent Omega Centauri, some 17,000 light years distant and visible to the
naked eye. NGC 104 in Tucana, in the far southern heavens,
ranks second in size and brightness. M22 in Sagittarius is
another fine example. In the northern sky M5 in Serpens and
M3 in Canes Venatici are the two finest globulars after M13
itself. The nearest of the globulars is probably NGC 6397
in Ara, believed to be some 8200 light years distant; the
most remote is NGC 2419 in Lynx, some 180,000 light years
away, but still visible in amateur telescopes as an 11th
magnitude object.

Compared with each other, most globulars are nearly
identical in appearance except for differences in brightness and apparent size, due chiefly to the difference in
distance. The average globular has a star population of at
least 100,000, and an absolute magnitude of -7 or -8. A
typical diameter is about 150 light years and a typical
luminosity is about 100,000 suns. The chief actual difference among globulars is in the richness and degree of condensation which is expressed in twelve classes ranging from
the most highly compressed clusters (Class I) to the most
loosely condensed ones (Class XII). The Hercules Cluster
itself has a moderate degree of condensation, designated
Class V.

DESCRIPTIVE NOTES   (Cont'd)

THEORETICAL CONSIDERATIONS. As previously mentioned, the globular clusters are "Population II" systems, a fact now known to be indicative of great age. To understand why the globulars are regarded as being extremely ancient we must know something about the way in which star types change with time. The simplest way to demonstrate this is by use of the famous "H-R Diagram", a graph in which stars are plotted by luminosity and spectral type. A typical diagram of this sort is shown on page 84. The vertical coordinate represents actual luminosity in terms of the Sun, with the most luminous stars at the top; the horizontal coordinate represents color, spectral class, or temperature, all of which are obviously interrelated. The hottest stars are plotted at the left and the coolest at the right.

　　　　Let us now begin to plot various groups of stars beginning with a "young" star cluster which has existed just long enough so that all its members have reached a stable state. The resulting plot is shown below at left. All these stars are operating on the same nuclear reaction, the hydrogen-helium cycle; therefore the position of each star in this case is determined entirely by its mass, the more massive stars being hotter and more luminous. Thus we see why the plotted stars lie in a narrow band running diagonally across the graph. Bright hot blue stars are in the upper left corner, and faint red cool ones are in the lower right corner. An average star such as our Sun would

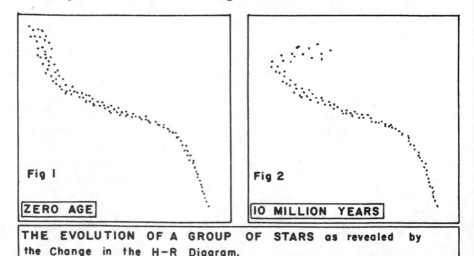

Fig 1

ZERO AGE

Fig 2

10 MILLION YEARS

THE EVOLUTION OF A GROUP OF STARS as revealed by the Change in the H-R Diagram.

DESCRIPTIVE NOTES   (Cont'd)

lie about midway on the band - which from now on we will
designate by its proper name - the Main Sequence.

Now as stars become older, their evolution reveals
itself by a change of position on the H-R Diagram. Herein
lies our clue to determining stellar ages. For our present
purpose, we need only remember two main facts: (1) Evolution of a main sequence star causes an expansion which lowers the surface temperature while the total radiation increases; thus the star moves to the right and upward on the
H-R Diagram. (2) The most massive stars evolve at the most
rapid rate, and are also the first to begin their evolution
away from the main sequence. Thus the stars near the top of
the diagram will be the first to show any change.

Keeping these facts in mind, we refer to Figure 2,
which shows the same star group of Figure 1, but after a
lapse of some 10 million years. The evolution of the more
luminous stars is evident, as we expected, but the rest of
the main sequence has shown no change. In Figure 3 the same
star group is shown again, after an interval of about 300
million years, and in Figure 4 after about one billion
years. In this last diagram, it is apparent that all stars
brighter than absolute magnitude +2 have evolved to the
right of the main sequence. Finally, in Figure 5, our star
cluster is approaching an age of 10 billion years, and the
"turn-off" point from the main sequence is near absolute
magnitude +4.

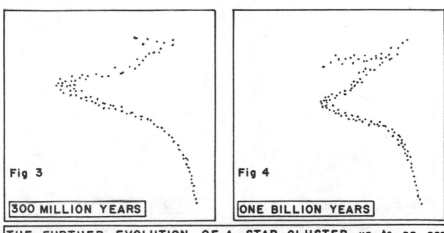

Fig 3

300 MILLION YEARS

Fig 4

ONE BILLION YEARS

THE FURTHER EVOLUTION OF A STAR CLUSTER up to an age
of one billion years.

DESCRIPTIVE NOTES   (Cont'd)

        As we see, the "turn-off" point moves steadily
down the main sequence as the star group grows older. This
provides a yardstick for the calibration of stellar ages,
and some of the calculated results are presented in the
brief list which ends this article. But before applying
this method to M13, certain difficulties must be acknow-
ledged. To begin with, the fainter stars of the cluster are
much too dim for actual spectra to be obtained, and we must
instead plot the stars by apparent color and magnitude. The
resulting diagram must then somehow be "fitted" to the
standard main sequence so that the true position of the
turn-off point may be determined. Obviously it is important
to know the distance of the cluster as accurately as possible; and in studying some clusters it may be necessary to
correct the observations for dimming and reddening of the
stars by interstellar dust. Finally, of course, there remains the fundamental problem of converting the known turn-
off point to a definite age in years; for this information
we must rely on the theoretical astrophysicists who are
still adding details to the general picture of stellar
evolution as new facts emerge.
        The most thorough observations of M13 have been
made with the 200-inch telescope at Palomar Observatory
and with the 82-inch and 42-inch reflectors at McDonald
and Lowell Observatories, respectively. The combined results are shown on page 988 and on a smaller scale below

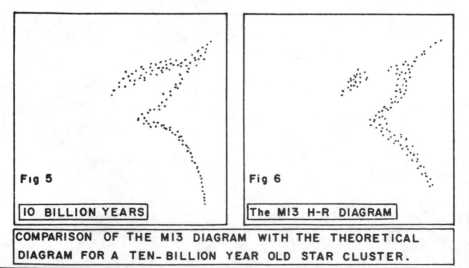

Fig 5

IO BILLION YEARS

Fig 6

The MI3 H-R DIAGRAM

COMPARISON OF THE MI3 DIAGRAM WITH THE THEORETICAL
DIAGRAM FOR A TEN-BILLION YEAR OLD STAR CLUSTER.

DESCRIPTIVE NOTES   (Cont'd)

(Figure 6) to permit a more direct comparison with Figures
1 through 5. The turn-off point from the main sequence
appears to be at about absolute magnitude +4.1; which is
believed to indicate an age of about 10 billion years. This
is comparable to the ages deduced for many other globulars,
and is equalled among the galactic clusters only by M67 in
Cancer and NGC 188 in Cepheus. As of 1968 this latter
cluster still remains the most ancient star group known.
Results for some other clusters are given in the following
table.

CLUSTER	CONSTELLATION	"TURN-OFF"	AGE
NGC 2362	Canis Major	−8	1 million
Double Cluster	Perseus	−7	2 "
Pleiades M45	Taurus	−2.5	20 "
M11 Scutum	Scutum	−1.3	60 "
Hyades	Taurus	+0.8	400 "
Praesepe M44	Cancer	+0.8	400 "
M67	Cancer	+3.5	10 billion
NGC 188	Cepheus	+4.5	13 "

M92   (NGC 6341) Position 17156n4312. Globular
               star cluster, discovered by J.E.Bode in
December 1777; Messier's independent discovery occurred
in March 1781. This is a beautiful rich globular cluster
which in almost any other constellation would be considered
a major show object; in Hercules it has been somewhat over-
shadowed by the splendor of the fabulous M13. The two star
clusters are only some 9° apart, M92 lying to the northeast
of the greater cluster, and some 6° north of Pi Herculis
which marks the northeast corner of the large trapezoid
called the "Keystone" of Hercules. M92 is rather easily
located in binoculars as a fuzzy starlike object of magnitude 6½, and rather small telescopes permit some resolution
of the outer edges. The view in large instruments is stun-
ning beyond words; the countless star images run together
into a dazzling central blaze which is equalled by only a
few of the globulars. This rich nuclear mass invariably
appears burned-out on photographs due to strong over-exposure; so we have here one case, at least, where the eye at

GLOBULAR CLUSTER M92. A fine bright globular cluster in Hercules, photographed with the 120-inch reflector at Lick Observatory.

DESCRIPTIVE NOTES (Cont'd)

the telescope has a distinct advantage over the camera.
Lord Rosse, with his great reflector at Parsonstown, Ire-
land, believed that the nuclear region presented some indication of spiral structure and mentioned the appearance of
occasional dark spaces between bright groupings of minute
stars. Isaac Roberts, using his 20-inch reflector in 1891,
stated that his photographs "showed the cluster to be involved in dense nebulosity, which on the negative almost
prevents the stars from being seen through it"... This was,
of course, an erroneous impression undoubtedly resulting
from the unresolved luminous background of innumerable
faint stars.

According to Sawyer's "Bibliography of Individual
Globular Clusters" (First Supplement, 1963) M92 has an
integrated spectral type of F2, a total photographic magnitude of 7.3, and an apparent distance modulus of 15.2
magnitudes, equivalent to a distance of about 35,000 light
years. The cluster contains 16 known variable stars, 14 of
which are definitely short period pulsating variables of
the RR Lyrae type; one object, however, appears to be a
short period eclipsing binary resembling W Ursae Majoris.
For some reason not fully understood, eclipsing binaries
appear to be very rare in globular star clusters; even the
great Omega Centauri, with its hundreds of thousands of
stars, seems to contain only a single known eclipsing
binary system. In all the globulars, short period pulsating variables of the RR Lyrae class outnumber all other
types; these stars are thus often called "cluster variable"
stars. Periods are generally less than one day, and the
spectral types seem restricted to A and F; these stars are
usually regarded as a sub-class of the cepheids.

The color-magnitude diagram obtained for M92 by
H.Arp, W.Baum, and A.Sandage (1953) shows that the most
luminous members are red giant stars with apparent magnitudes of about 12 and absolute magnitudes of about -3; the
total luminosity of the cluster is close to 250,000 times
the light of the Sun. M92 shows a radial velocity of 73
miles per second in approach.

This cluster may be slightly less ancient than the
great M13; the position of the turn-off point seems to be
a few tenths of a magnitude higher on the H-R Diagram.
(Refer to discussion on Page 990)

DISTANT FIELD OF GALAXIES in HERCULES. A very remote group of galaxies, showing a variety of types in a single photograph.   Palomar Observatory 200-inch telescope.

LIST OF DOUBLE AND MULTIPLE STARS

NAME	DIST	PA	YR	MAGS	NOTES	RA & DEC
h3497	34.2	82	13	$5\frac{1}{2}$- $9\frac{1}{2}$	spect K5	02183s5610
h3503	0.8	219	48	8 - $8\frac{1}{2}$	PA dec, spect F8	02265s5822
	17.7	300	32	- $10\frac{1}{2}$		
h3520	21,8	204	31	8 - $8\frac{1}{2}$	perhaps slight dist inc.	02374s5503
△7	36.7	96	53	7 - 7	spect K0, A5 B= I386	02383s5946
I386	0.3	1	59	8 - 8	PA inc, spect A5	
△10	38.3	70	16	$7\frac{1}{2}$- $8\frac{1}{2}$	relfix, spect G0, K0	03030s5131
h3559	43.1	40	17	$6\frac{1}{2}$- 11	spect A2	03092s6406
h3562	34.4	331	17	9 - $9\frac{1}{2}$	relfix, spect F8	03112s6431
Slr 4	1.1	252	14	$9\frac{1}{2}$- $9\frac{1}{2}$	perhaps slow PA inc, spect F5	03119s4724
Hu 1353	2.6	191	22	$8\frac{1}{2}$- 11		03124s5615
h3576	3.0	342	53	$7\frac{1}{2}$- 9	spect A2	03228s4551
h3575	34.9	45	16	$6\frac{1}{2}$- 10	relfix, spect A0	03230s5115
λ29	14.6	259	55	$7\frac{1}{2}$-$12\frac{1}{2}$	dist inc, spect G0	03354s5007
Slr 5	1.7	190	43	$7\frac{1}{2}$- $9\frac{1}{2}$	relfix, spect F0	03403s4823
Hu 1361	4.2	77	45	$7\frac{1}{2}$- 11	PA dec, spect F8	03546s4755
h3611	4.2	140	45	8 - $8\frac{1}{2}$	relfix, spect A3	03549s4003

LIST OF VARIABLE STARS

NAME	MagVar	PER	NOTES	RA & DEC
R	5.0--14..	403	LPV. Spect M7e	02524s5006
S	9.0--12..	336	LPV. Spect M6e	02238s5948
T	7.3--13.6	217	LPV. Spect M4e	02593s5050
U	7.0--14..	348	LPV. Spect M6e	03512s4559
V	7.8---8.9		Semi-reg; spect M	03022s5908
W	9.5--10.6	137	Semi-reg; spect M	02427s5431
X	9.2--11.3	280	Semi-reg; spect M	02464s5916
RS	8.5--14..	203	LPV. Spect M3e	02346s6248
RT	9.0--13..	335	LPV. Spect M5e	03275s5609
TV	6.6---6.8	30:	Semi-reg; spect M5	02288s5802
TW	5.2---5.8		Semi-reg; spect N0	03113s5731

HOROLOGIUM

LIST OF STAR CLUSTERS, NEBULAE, AND GALAXIES

NGC	OTH	TYPE	SUMMARY DESCRIPTION	RA & DEC
1249		⊘	SBc; 12.3; 4.0' x 2.0' B,L,vmE, vgbM	03086s5332
1261		⊕	Mag 8, diam 3'; class II; B,L,R, stars faint	03109s5525
----	I.1933	⊘	Sc/Sd; 13.2; 1.8' x 1.1' F,S, 1E	03243s5257
----	I.1954	⊘	SBb; 12.2; 2.5' x 1.2' F,pL,R	03302s5205
1411		⊘	S0; 12.0; 1.4' x 1.1' B,pS,R,smbM; lenticular	03371s4415
1433	△426	⊘	SBa; 11.4; 7.0' x 6.0' vB,L,E,vsvmbM	03404s4724
1448		⊘	Sc/Sd; 11.8; 6.0' x 1.1' pB,L,vmE, nearly edge-on	03429s4448
1493	△438	⊘	SBc; 11.8; 1.7' x 1.5' F,cL,R,vglbM	03559s4621
1494		⊘	Sd; 12.2; 2.5' x 2.0' F,L,R, vgvlbM	03562s4903
1512	△466	⊘	S0/SB; 11.8; 3.0' x 2.5' B,cL,R,bM; lenticular or early barred spiral	04023s4329
1527		⊘	S0; 12.1; 1.4' x 0.6' pB,pS,E, vsmbMN	04069s4801
----	I.2035	⊘	E0/S0; 12.6; 0.6' x 0.6' F,vS,R	04076s4538

HYDRA

LIST OF DOUBLE AND MULTIPLE STARS

NAME	DIST	PA	YR	MAGS	NOTES	RA & DEC
β1244	0.9	26	57	8 - 8	PA dec, spect A0	08112n0208
Σ1210	15.5	113	29	7 - 9½	relfix, spect A0	08132n0257
Hwe 21	1.4	241	45	7½- 11	spect F8	08140s0304
β102	3.3	120	44	7 - 10½	relfix, spect A0	08144s0852
A337	0.4	130	62	8 - 8½	binary, 89 yrs; PA dec, spect F2	08148s0513
Σ1216	0.6	262	62	7½- 8	binary, about 435 yrs; PA inc, spect A0	08188s0126
Rst5297	39.0	84	46	7½- 10½	(Scj 10) PA & dist dec, spect K0; B= 0.3" pair, PA= 75°	08237s0015
A550	0.1	189	62	7½- 7½	binary, 40 yrs; PA dec, spect F0	08252s0415
Σ1233	0.4	65	62	7 - 7	All cpm; AB binary, 53 yrs; orbit high-ly inclined, PA inc, spect dF0 (A551)	08260s0221
	18.1	330	30	- 11		
Σ1243	2.0	228	43	8 -10½	relfix, spect A0	08313n0145
h99	61.0	202	18	6½- 7½	optical, PA dec, AB	08354s0638
h99b	9.8	211	03	7½-12	spect G0, A0	
Σ1255	26.6	31	25	7 - 8	relfix, cpm pair;	08370n0557
Σ1255b	44.5	14	28	-13½	AB spect both G5	
	56.9	46	28	-13	BC PA dec, dist inc	
Σ1260	5.1	304	47	7½- 8	relfix, spect A2	08383s1200
A3065	0.2	273	60	8½- 8½	PA inc, spect A0	08386s1710
9	31.0	111	20	5- 11½	(h4124) slight dist dec, PA dec, spect K1, optical	08394s1546
Σ1270	4.7	262	55	6½- 7½	cpm pair, relfix, spect F5	08428s0225
β1069	2.4	68	58	7½- 11	PA inc, spect K5	08437s1050
ε	0.3	183	68	3½- 5	(Σ1273) all cpm;	08442n0636
	3.1	271	59	- 8	AB binary, 15 yrs;	
	19.3	195	38	- 12	AC PA inc (*)	
β586	0.3	81	60	6½- 9	PA inc, composite spect= G0+A2	08451s1652
β335	2.7	269	36	7 - 10½	relfix, spect F5	08456n0246
OΣ194	12.6	56	27	7 - 10½	relfix, spect K2	08457n0044

LIST OF DOUBLE AND MULTIPLE STARS (Cont'd)

NAME	DIST	PA	YR	MAGS	NOTES	RA & DEC
ρ	12.4	145	35	$4\frac{1}{2}$- 12	(AGC 3) relfix, cpm pair, spect A0	08458n0601
A2552	0.2	70	66	$8\frac{1}{2}$- $8\frac{1}{2}$	PA dec, spect F5	08461n0108
h107	27.3	52	11	$7\frac{1}{2}$- 11	PA dec, spect K0	08479s0400
15	0.9	125	58	6 - 9	(β587) AB cpm, PA	08491s0659
	45.9	360	24	- 10	dec, dist inc,	
	51.9	54	24	- 11	spect A2; C&D are optical.	
Σ1290	3.0	320	30	8 - 10	relfix, spect A2	08494n0439
A2900	0.6	301	49	$7\frac{1}{2}$- 10	slight PA inc, spect A3	08498n0532
Σ1292	6.5	188	22	9 - 9	cpm pair, relfix; spect A5	08512s0023
A2554	0.3	255	41	8 - 10	PA inc, spect F0	08513n0201
β24	1.1	175	60	8 - 9	relfix, spect F0	08518s0834
17	4.3	359	53	7 - 7	(Σ1295) relfix, cpm, spect A3, A5	08530s0747
β210	2.9	183	53	8 - 8	(Arg 72) relfix, spect F2	08545s1714
Ho 358	1.9	298	33	7 - 12	spect A0	08556s1841
Kui 36	2.5	197	58	$6\frac{1}{2}$- 12	spect A0	08556n0144
β409	9.7	185	01	8 -$10\frac{1}{2}$	spect A0	08583s0900
Σ1302	2.4	232	43	$8\frac{1}{2}$- 9	relfix, cpm,	08586n0256
	32.0	269	04	-$12\frac{1}{2}$	spect F8	
Σ1308	10.5	85	36	8 - 9	relfix, cpm, spect F5	09025s0347
Σ1309	11.5	273	37	8 - $8\frac{1}{2}$	relfix, spect F5	09040n0301
h804	12.8	329	12	8 - 11	spect A0	09053s1017
Σ1316	7.1	138	59	8 -$11\frac{1}{2}$	AB cpm, slight PA	09054s0656
	4.3	232	22	-$10\frac{1}{2}$	dec, AC optical, dist dec, spect G0	
19	1.5	296	58	$5\frac{1}{2}$- 10	(Kui 38) spect B8	09063s0823
OΣ197	1.4	62	55	$7\frac{1}{2}$- 9	relfix, spect F0	09069n0309
h4182	25.2	107	15	7 -$10\frac{1}{2}$	spect F0	09078s1639
β104	2.5	103	32	7 - 12	dist dec, slow PA dec, spect K0	09089n0030
β336	2.0	238	54	$8\frac{1}{2}$- $9\frac{1}{2}$	relfix, spect A0	09094s1636
A2973	2.6	7	29	$7\frac{1}{2}$-$12\frac{1}{2}$	spect K0	09100s0643
θ	29.4	197	53	4 - 10	(h2489) optical, PA inc, dist dec; B9	09118n0232

HYDRA

LIST OF DOUBLE AND MULTIPLE STARS   (Cont'd)

NAME	DIST	PA	YR	MAGS	NOTES	RA & DEC
Wei 21	25.8	14	18	$7\frac{1}{2}$- 9	spect A0	09124s0833
Ho 363	2.4	181	57	7 - 9	perhaps slight PA inc, spect A0	09127s1955
Σ1329	14.5	253	54	$8\frac{1}{2}$- $8\frac{1}{2}$	optical, dist dec, spect both G5	09132s0102
β212	1.3	209	60	$7\frac{1}{2}$- 8	PA dec, spect A5	09136s0808
β588	2.2	142	53	$6\frac{1}{2}$-11	cpm, PA inc, spect F5	09141n0056
h4193	3.1	118	60	8 - 12	relfix, spect A5	09141s2255
23	1.6	282	59	5 - 11	(Kui 40) spect K0	09142s0609
A125	2.9	29	59	$7\frac{1}{2}$- 11	slight PA inc,	09163s1012
	50.9	82	13	$-10\frac{1}{2}$	spect A2	
A3077	4.1	190	43	6 - 11	spect K0	09172s1537
Σ1343	9.8	272	27	$8\frac{1}{2}$- 9	relfix, probably cpm, spect G5	09174n0513
27	229	211	23	5 - 7	(Sh 105) probably cpm, spect G8, F5	09180s0920
27b	9.6	196	37	7 - 9	relfix, cpm pair; spect F5, K2	
HW 6	1.5	45	54	8 - $9\frac{1}{2}$	PA inc, spect F0	09180s2316
β337	8.3	334	32	7 - 11	PA inc, spect F2	09202s1741
A1342	0.1	5	67	7 - 7	binary, 15 yrs; PA	09204s0937
	1.8	204	58	- $11\frac{1}{2}$	dec, spect A2; ABC all cpm.	
Σ1347	21.1	311	37	$6\frac{1}{2}$- 8	cpm pair, relfix, spect F0	09207n0343
Hd 123	4.4	5	57	$7\frac{1}{2}$- 10	relfix, spect F2	09216s2327
	54.5	191	57	$-10\frac{1}{2}$		
Σ1348	1.7	317	67	$7\frac{1}{2}$- $7\frac{1}{2}$	cpm, PA dec, spect F5	09218n0634
Σ1355	2.6	345	57	7 - 7	PA inc, cpm, spect F5	09247n0627
29	0.2	184	56	7 - 7	(A1588) relfix, all	09248s0900
	10.8	174	34	$-11\frac{1}{2}$	cpm, spect A0 (β590)	
Σ1357	7.4	54	44	7 -$10\frac{1}{2}$	relfix, spect K0	09259s0946
A1763	1.3	115	49	$6\frac{1}{2}$-$11\frac{1}{2}$	PA inc, cpm, spect A5	09265s0102
β591	0.8	26	66	$7\frac{1}{2}$- $8\frac{1}{2}$	relfix, spect F5+A3	09271s0254
β339	1.3	236	59	9 - $9\frac{1}{2}$	PA inc, spect G0	09286s1531

LIST OF DOUBLE AND MULTIPLE STARS (Cont'd)

NAME	DIST	PA	YR	MAGS	NOTES	RA & DEC
Σ1365	3.4	156	57	7 - 8	slight PA dec, spect F8	09290n0141
Σ1367	5.4	183	37	8 - 9½	relfix, spect G0	09298s1037
β910	6.8	305	42	7½-10	relfix, cpm pair;	09305s1347
	166	278	13	- 9	spect G5	
S604	51.5	91	15	6 -9½	relfix, spect A1	09334s1921
A2557	1.3	60	57	7 - 12	cpm, spect G5	09351n0155
Stn 19	2.7	266	43	7½- 9½	relfix, spect F2	09377s1656
	24.1	290	25			
I	54.4	292	04	5 - 8	spect B2p	09390s2322
β214	3.2	250	59	7 - 11	cpm, PA dec, spect F8	09392s1815
h4261	8.2	84	41	8 - 10	relfix, spect G5	09512s1915
β592	9.9	192	47	6½- 12	relfix, spect G0	09526s1557
β216	3.4	156	57	6 - 11	slow PA dec, spect A2	09545s2619
Hd 124	13.2	8	27	8 - 10	spect A0	09589s2231
	42.0	50	28	- 12		
Hu 1253	0.5	84	45	7½- 9	PA dec, spect A0	09599s1448
β1072	11.9	48	59	6 - 12	AC spect both A0;	10017s1751
	21.0	274	35	- 7	AC cpm, relfix; AB optical, slight PA inc.	
h4285	8.7	358	59	8 -10½	relfix, spect F5	10043s2254
S607	9.5	146	59	8½- 9½	probably cpm; slight dist dec, spect G5; galaxy NGC 3124 is 4' N.	10043s1904
β217	1.9	121	59	8 - 8	PA inc, spect F8	10045s2428
β218	0.6	132	59	8 - 8½	PA inc, spect A0	10050s1928
β911	4.6	312	59	7½-11	relfix, AB cpm; AC	10060s1930
	66.4	62	59	-9½	optical, dist inc; spect G0	
Σ1416	11.5	277	40	6½- 8½	relfix, spect A2	10099s1550
Hn 101	1.3	110	59	6 - 10	cpm, slight PA dec, spect F5	10144s2025
h4305	17.5	216	21	7½- 9½	relfix, spect K0	10182s2323
β219	1.9	187	60	7½- 9	relfix, spect A0	10192s2217
h4311	4.2	122	44	7½-10½	relfix, cpm pair; spect F8	10209s1307

LIST OF DOUBLE AND MULTIPLE STARS   (Cont'd)

NAME	DIST	PA	YR	MAGS	NOTES	RA & DEC
B201	1.7	68	38	7½- 12	spect F5	10287s2355
44	18.3	64	46	5 - 13	(β1269) no certain change, spect K4	10316s2329
β411	0.7	338	63	6½- 8	binary, about 210 yrs; dist inc, PA dec, spect F6	10337s2625
β1075	3.3	280	59	6 -12	(φ²) cpm, relfix, spect gM1	10338s1605
I 857	0.3	284	54	7½-7½	PA dec, spect A0	10343s2831
I 502	0.3	21	59	7½- 9	PA dec, spect K0	10432s2446
	1.9	137	59	-13½		
Σ1473	30.7	10	16	8 -8½	relfix, spect F8	10452s1520
	98.7	326	08	- 10		
Σ1474	69.4	25	33	7 - 8	relfix, spect A0	10452s1500
Σ1474b	6.7	196	49	7½- 7½	relfix	
I 503	1.2	115	59	7½- 9	PA inc, spect F2	10469s2633
I 211	1.9	207	60	5½- 9½	cpm pair, PA inc, spect dA8	10569s3328
B208	0.1	10	37	6½- 7	binary, spect F0	11000s2634
Hwe 25	2.5	339	59	8 - 9	slow PA inc, spect A0	11009s2715
χ¹	0.2			6 - 6	(φ47) rapid binary 7.4 yrs; spect dF4 (*)	11029s2701
h4412	12.7	266	54	8 -8½	relfix, spect A0	11076s2920
I 1539	0.1		37	8 - 8½	PA uncertain, too close, spect A3	11187s3515
N	9.1	210	52	6 - 6	(Hh376) cpm, spect dF6, dF7	11298s2859
h4455	3.3	243	54	6 - 8½	AB cpm, relfix, spect K0	11341s3318
	47.6	84	00	-13		
I 232	2.0	162	60	7 - 9	slight PA inc, spect K0	11375s3310
h4465	27.3	345	34	5½- 13	(λ133) AC cpm,AC spect K5, F7	11392s3213
	67.0	44	19	- 7½		
β	0.9	8	59	4½- 5	(h4478) dist dec, PA inc, cpm, spect A0	11504s3338
Δ116	19.6	263	59	7 - 7	cpm pair, relfix, spect both G0	11542s3159
	24.3	333	59	- 11½		

LIST OF DOUBLE AND MULTIPLE STARS   (Cont'd)

NAME	DIST	PA	YR	MAGS	NOTES	RA & DEC
I 510	0.1	119	26	7 - 7½	PA dec, spect A0	11564s2538
I 215	0.6	121	60	7½- 8	PA dec, spect G0	11593s3422
h4495	6.6	317	40	6½- 9	relfix, cpm,	12035s3241
	26.2	64	00	- 13	spect G0	
Jc 17	3.4	20	51	6½- 8½	relfix, cpm, spect A0 (marked Jc 8 in Norton's Atlas)	12075s3426
h4505	10.2	271	47	7½-10½	relfix, spect K0	12091s3019
Hwe 72	1.3	164	59	6½- 8½	PA dec, spect B9 (marked He 19 in Norton's Atlas)	12110s3331
h4513	47.0	99	19	7½- 10	spect K0	12164s3259
B228	0.1	280	59	8 - 8	PA inc, spect F0	12248s2826
I 514	0.1	135	49	8 - 8	spect A3	12282s3024
Rst1675	0.5	282	45	8½- 10	spect K0	12291s3139
h4528	26.8	145	60	7 -11½	PA dec, dist inc, spect A5	12315s3149
B230	1.3	170	59	5½- 12	cpm, PA inc, spect F2, globular star cluster M68 0.6° NE	12350s2652
B231	4.3	82	32	7½- 13½	spect B9	12382s2726
Es 439	2.1	64	37	9 - 9½	spect F5	12490s2800
Stn 26	3.0	33	48	7½- 9½	relfix, spect G5	12512s2902
h3556	6.0	82	54	8 -9	relfix, spect G0	12516s2741
I 1225	0.1		59	8 - 8	PA uncertain, too close, spect A2	12575s2340
B239	4.0	256	32	7 -13½	spect A0	12597s2900
342G	0.2		60	7 - 7	(φ305) spect A3	13089s2617
Hwe 27	2.7	293	34	7½- 9	relfix, spect A0	13102s2850
	38.0	281	16	- 13		
Stn 28	0.1	23	59	7 - 7½	(φ297) PA inc, spect A2	13117s2401
	12.4	333	31	-11½		
Hu 1503	1.2	192	59	7 -10½	spect F5	13187s2232
R	21.2	324	31	var-12½	(Ho 381) primary variable, spect M7 (*)	13270s2301
SS	13.3	196	03	7½-13	(Ho 540) primary variable, spect B9	13277s2323
S 651	10.9	192	38	5 - 7	relfix, cpm pair; spect A3, A5	13340s2614

LIST OF DOUBLE AND MULTIPLE STARS   (Cont'd)

NAME	DIST	PA	YR	MAGS	NOTES	RA & DEC
h4606	0.1	327	60	7 - 7	(ϕ352) (335G)	13388s2312
	31.1	352	31	- 9	spect A0	
h4617	4.9	261	54	8 - 9½	relfix, spect G5	13479s2938
λ200	9.3	98	00	7 - 12	spect K0	14084s2933
Rst5382	0.4	245	50	7½- 9	spect G5	14101s2516
β1246	2.4	188	00	5½- 12	AB cpm, AC optical,	14162s2536
	64.0	112	59	- 11	AC dist inc, spect dF4	
Stn 31	0.6	271	56	8 - 9	PA dec, spect K2	14223s2754
52	4.0	279	35	5 - 11	(β940) relfix,	14252s2916
	40.8	282	14	- 12	spect B8; primary is 0.1" pair	
Cor 172	12.6	58	32	8 - 10½	spect K0	14270s2519
β805	23.6	134	00	7 - 13	spect F5; in field	14369s2629
	124	42	00	- 9½	of globular NGC 5694	
β805c	1.1	285	59	9½-11½	PA inc	
β806	0.7	97	38	7½- 9½	spect A2; in field	14375s2603
	71.6	67	31	- 9	with globular NGC	
	17.8	329	00	-13½	5694; C= 1.2" pair	
β345	0.9	292	60	7 - 7½	PA dec, spect F0	14388s2929
54	8.8	126	54	5 - 7	cpm, slight PA & dist dec, spect dF1 & dF9	14431s2514
59	0.8	335	53	6 - 6	(β239) cpm; PA inc spect A5	14557s2727

HYDRA

LIST OF VARIABLE STARS

NAME	MagVar	PER	NOTES	RA & DEC
R	4.0--10.0	386	LPV. Spect M7e (*)	13270s2301
S	7.5--13.2	257	LPV. Spect M4e	08510n0316
T	7.2--13.1	288	LPV. Spect M3e--M4e	08532s0857
U	4.7---6.2	Irr	Spect N2; very red, Carbon star	10351s1307
V	6.5--12..	533	Semi-reg; spect N6e (*)	10492s2059
W	6.7--9..	382	Semi-reg; spect M8e	13462s2807
X	8.2--13..	302	LPV. Spect M7e	09331s1428
Y	6.9---9..	303:	Semi-reg; spect N3	09488s2247
RR	8.5--13..	342	LPV. Spect M4e	09427s2347
RS	9.0--13..	334	LPV. Spect M6e	10489s2822
RT	7.1--10.2	253	Semi-reg; spect M6e--M7	08272s0609
RU	7.3--14..	334	LPV. Spect M6e	14087s2839
RV	7.6---8.5	116:	Semi-reg; spect M5	08373s0924
RW	9.5--10.5	376	Z Andromedae type, Spect gMep	13316s2507
RX	9.0--10.5	2.282	Ecl.Bin; spect A8	09032s0804
RZ	9.2--12..	334	LPV. Spect M4e	09224s0635
SS	7.7---8.0		Type uncertain, see note on R Hydrae	13277s2323
ST	8.8--14.4	304	LPV.	09356s2026
SU	9.5--10.5	95:	Semi-reg; spect M4	09514s2137
SV	9.9--11.0	.47856	Cl.Var.	12279s2546
SW	9.0--12..	219	LPV. Spect M2e	13006s2850
SX	8.5--10.5	2.896	Ecl.Bin; spect A3+K5	13418s2632
TT	7.5--9.0	6.5934	Ecl.Bin; spect A3e+dG6p	11108s2612
TU	9.5--14..	278	LPV. Spect M4e	08557s0038
TV	7.9---8.1		Type uncertain, spect A3	13344s2322
UZ	9.0--14.5	260	LPV.	09143s0424
VV	9.5--14..	155	LPV.	09088s0910
VY	8.9--11.2	2.0012	Ecl.Bin; spect A3	10179s2254
VZ	9.0---9.7	2.9043	Ecl.Bin; spect F5+F5	08292s0609
WW	9.5--12..	311	LPV.	08553s0305
WX	9.5--12..	234	LPV. Spect M3e	09111s1411
AI	9.0---9.5	8.29	Ecl.Bin; spect F0+F5	08162n0026
AK	6.8---7.3	112:	Semi-reg; spect M4	08376s1707
BD	9.5--11.3	117	Semi-reg	11137s2955
BK	9.5--10.3	Irr		11176s3459
CZ	8.5--14..	442	LPV. Spect Ne	10250s2518
EP	9.5--14..	165	LPV.	12490s2631

LIST OF VARIABLE STARS   (Cont'd)

NAME	MagVar	PER	NOTES	RA & DEC
EY	8.5--11..	Irr	Spect M8	08438n0149
FF	7.0--9..	20?	Semi-reg; spect M	10354s1114
FH	9.0---9.5	Irr		09315s0646
FI	8.5--15..	324	LPV. Spect M4e	12372s2624
FK	8.0--9..	Irr	Spect M	08221s0821
FO	9.5--10.0	1.159	Ecl.Bin.	09574s1854
FP	9.0--13..		LPV.	11107s2821
FS	8.5--10.5	167	Semi-reg	10052s1626
FT	9.0--11..	300:	LPV	13539s2518
FU	9.5--12.0		LPV.	14355s2929
FV	9.5--10..	Irr		08405s0619
FZ	8.0--9..	Irr	Spect M6	08189n0507
GK	8.3---9.1	3.5870	Ecl.Bin; lyrid, Spect G4	08282n0227
IN	6.3--6.8	45:	Semi-reg; spect gM4	09180n0025
IO	6.9--7.0	80:	Semi-reg; spect gM4	10052s2215

LIST OF STAR CLUSTERS, NEBULAE, AND GALAXIES

NGC	OTH	TYPE	SUMMARY DESCRIPTION	RA & DEC
2548	22[6] M48		vL,Irr,pRi; Mag 5.5; diam 40'; about 50 stars mags 9...13; Class F (*)	08112s0538
2610	35[4]		F,S, Mag 13, diam 35" x 30"; 16^m central star; 13^m star on NE rim; 7^m star 3.3' NE	08312s1558
2642			SBb/SBc; 12.7; 1.6' x 1.6' vF,pL,gbM	08383s0357
2713			SBa; 12.7; 3.2' x 1.0' pB,cE,mbM	08548n0308
2763	275[3]		Sc; 12.7; 1.5' x 1.5' vF,pS,bM; asymmetrical form	09045s1517
2781	66[1]		SO/Sa; 12.7; 2.0' x 0.7' B,S,cE,smbM; dim outer ring	09091s1436

LIST OF STAR CLUSTERS, NEBULAE, AND GALAXIES   (Cont'd)

NGC	OTH	TYPE	SUMMARY DESCRIPTION	RA & DEC
2784	59[1]	⊘	E8/S0; 11.8; 3.0' x 1.0' B,L,mE, gmbM	09101s2358
2811	502[2]	⊘	SBa; 12.4; 1.7' x 0.4' pB,S,mE,psmbM	09139s1606
2815	242[3]	⊘	SBb; 12.9; 3.0' x 0.8' F,S,mE,gbM	09141s2324
2835		⊘	SBc/pec; 12.0; 5.9' x 3.0' L,F,lE; fine many-armed spiral	09157s2208
2848	488[3]	⊘	Sc; 12.8; 2.1' x 1.4' vF,cL,E,glbM; 11^m star 3' nf 14^m lens-shaped galaxy NGC 2851 is 5.2' ENE	09178s1618
2855	132[1]	⊘	Sa/E1; 12.5; 1.2' x 1.1' pB,pL,lE,gmbM	09191s1141
2865		⊘	E3; 12.5; 0.8' x 0.6' B,S,lE, gbM	09212s2258
2889	555[2]	⊘	Sc; 12.4; 1.4' x 1.3' pF,S,vlE,vglbM	09248s1125
2902	276[3]	⊘	E0/S0; 13.1; 0.5' x 0.5' vF,vS	09285s1430
2907	506[2]	⊘	Sa/Sb; 12.9; 1.0' x 0.8' pF,S,cE; nearly edge-on; equatorial dust band; som- brero structure	09293s1632
2924		⊘	E0; 13.2; 0.5' x 0.5' pB,S,R	09328s1611
2935	556[2]	⊘	SBb; 12.4; 3.2' x 2.4' pB,pS,vlE,gmbM	09345s2054
2962		⊘	S0/Sa; 12.9; 1.8' x 1.0' F,vS,cE,psbM; dim outer ring	09383n0524
2983	289[3]	⊘	SBa; 12.6; 1.4' x 1.0' F,pS,lE,bM	09413s2015
2986	311[2]	⊘	E1; 12.2; 1.0' x 1.0' pB,pS,R,mbM	09420s2103
2989		⊘	Sb; 13.1; 1.0' x 0.7' F,R,gbM	09431s1809
2992	277[3]	⊘	Sa/pec; 13.0; 1.0' x 0.5' cF,S,E,bM; long streamer extends 2' to N	09433s1406

LIST OF STAR CLUSTERS, NEBULAE, AND GALAXIES   (Cont'd)

NGC	OTH	TYPE	SUMMARY DESCRIPTION	RA & DEC
2993	278[3]	⊖	S/pec; 13.0; 0.4' x 0.4' cF,S,R,bM; pair with 2992, 2.9' to NW	09434s1408
3052	272[3]	⊖	Sc; 13.0; 1.8' x 1.2' F,pL,lE, glbM	09520s1824
3054		⊖	Sb; 12.6; 3.1' x 1.4' pB,pL,cE	09521s2528
3078	268[2]	⊖	E3; 12.2; 0.6' x 0.4' pB,S,lE,mbM	09562s2641
3081	596[3]	⊖	S0/Sa; 12.8; 1.3' x 0.9' vF,cS,E,lbM; encircling outer ring	09568s2233
3091	293[2]	⊖	E2; 12.7; 1.0' x 0.9' pB,pS,lE,bM; brightest member of small group	09578s1923
3109		⊖	I; 11.2; 11.0' x 2.0' cF,vL,vmE,lbM (*)	10008s2555
3124		⊖	Sb/Sc; 12.8; 2.5' x 2.0' F,pL,R,lbM; fine face-on spiral; double star S607 in field, 4' to south	10042s1900
3145	518[3]	⊖	SBb; 12.5; 2.4' x 1.0' F,pL,E,vgslbM; 7.9' SW from Lambda Hydrae (*)	10077s1210
3200		⊖	Sa/Sb; 12.8; 3.6' x 0.9' pB,E,bMN; nearly edge-on; many-armed spiral	10162s1744
3203		⊖	E8/S0; 13.2; 2.0' x 0.3' pB,S,mE,gbM; lens-shaped edge-on	10163s2627
3242	27[4]	◎	vB,L,lE; Mag 8.9; diam 40"; bluish disc with central 11^m star (*)	10224s1823
3285		⊖	SBc; 13.2; 1.3' x 0.8' pB,S,E,gbM	10313s2712
3309		⊖	E0; 12.7; 0.7' x 0.7' B,L,R; 1.8' pair with 3311	10343s2716
3311		⊖	E2; 13.5; 0.6' x 0.5' F,S,lE; small group with 3309 & 3312	10344s2717

LIST OF STAR CLUSTERS, NEBULAE, AND GALAXIES (Cont'd)

NGC	OTH	TYPE	SUMMARY DESCRIPTION	RA & DEC
3312		⊖	SB/pec; 13.1; 2.4' x 0.9' cF,E,gbM; spiral structure dim; group with 3309, 3311	10348s2720
3390		⊖	Sb; 13.2; 3.0' x 0.4' F,S,E north to south; edge-on spiral	10458s3117
3464		⊖	SBb; 13.2; 2.6' x 1.4' eF,pL,E; Red variable V Hydrae is 43' WSW	10522s2049
3585	269[2]	⊖	E5/E6; 11.3; 1.5' x 0.8' B,pL,E, vsmbMN; Variable TT Hydrae 17' to north.	11109s2629
3621	241[1]	⊖	Sc/Sd; 10.6; 5.0' x 2.0' cB,vL,E, coarse structure	11159s3232
3673		⊖	SBb; 12.9; 2.5' x 1.8' F,vL,gvlbM; 7^m star 6' S	11228s2628
3717		⊖	Sb; 12.6; 2.0' x 1.0' pB,S,mE; edge-on spiral	11290s2959
3885	828[3]	⊖	S0/Sa; 12.9; 1.0' x 0.7' cF,vS,E,bM	11443s2739
3904	864[2]	⊖	E3; 11.9; 1.5' x 1.0' pB,S,cE,mbM	11467s2902
3923	259[1]	⊖	E4; 11.1; 2.0' x 1.2' B,pL,E,gmbM	11485s2833
3936		⊖	Sb; 13.0; 3.3' x 0.3' vF,cL,vmE; edge-on spiral	11499s2637
----	I.2995	⊖	SBc; 12.7; 3.2' x 1.0' vF,L,cE	12030s2739
4105	865[2]	⊖	E3; 12.0; 1.5' x 1.5' pF,pS,1E,psbM; interacting pair with 4106; sep= 1.1'	12041s2930
4106	866[2]	⊖	E2/S0; 12.5; 1.0' x 0.8' pF,pS,1E,pgbM; faint extension on E side	12042s2931
----	I.764	⊖	Sc; 12.9; 5.0' x 2.0' eF,pL,E north to south	12076s2928
4304		⊖	SBb; 12.4; 1.0' x 1.0' vF,L,R,vgvlbM; reversed-S spiral	12196s3312

LIST OF STAR CLUSTERS, NEBULAE, AND GALAXIES   (Cont'd)

NGC	OTH	TYPE	SUMMARY DESCRIPTION	RA & DEC
4590	M68	⊕	L,eRi,R,rrr; Mag 8, diam 9'; Class X; stars mags 13.... (*)	12368s2629
5061	138[1]	⊖	E2; 11.7; 1.3' x 1.0' vB,S,vlE,vsmbM; 10^m star 2.5' east	13153s2636
5078	566[2]	⊖	Sa; 13.2; 1.4' x 0.5' pB,pS,cE,psbM; nearly edge-on; dust lane, "sombrero" structure	13171s2709
5085	780[2]	⊖	Sb; 12.3; 2.9' x 2.5' F,L,R, vglbM; very regular face-on spiral	13176s2409
5101	567[2]	⊖	SBa; 12.5; 3.0' x 1.0' cB,pS,lE,psbM	13190s2711
5135		⊖	SBa; 12.8; 1.8' x 0.5' pB,S,E	13229s2934
5150		⊖	E2; 13.1; 0.7' x 0.6' cF,S,R,pslbM; vF E0 galaxy 5153 is 5' to SE	13249s2918
5236	M83	⊖	Sc; 8.0; 10.0' x 8.0' ! vB,vL,lE,esbMN (*)	13343s2937
5328	923[3]	⊖	E2; 12.9; 1.0' x 0.9' pB,pS,lE,slbM; Variable W Hydrae is 50' to west	13500s2814
----	I.4351	⊖	Sb; 12.8; 5.0' x 0.8' L,F,SN,vmE; edge-on spiral	13549s2905
5464		⊖	Sd/S pec; 13.1; 0.5' x 0.5' pF,S,lE,pslbM	14042s2946
5556		⊖	SBc/Sd; 12.5; 2.5' x 2.2' eF,L,R, spiral form faint	14176s2901
5592	924[3]	⊖	Sa/Sb; 13.1; 0.7' x 0.5' F,S,lE, gvlbM	14210s2827
5694	196[2]	⊕	cB,cS, Mag 11, diam 2'; vRi, stars eF; Class VII; one of most distant globular clusters (*)	14367s2619

HYDRA

DESCRIPTIVE NOTES

ALPHA   Name- ALPHARD, the "Solitary One", also known as COR HYDRAE, the "Dragon's Heart". Magnitude 1.97; spectrum K3 or K4 III. Position 09251s0826. The star is possibly slightly variable, since there is a spread of about 0.2 magnitude in the measurements. Alphard is at a distance of about 95 light years and has an actual luminosity of about 110 suns; the absolute magnitude is about -0.3. The annual proper motion is 0.04"; the radial velocity is 2.5 miles per second in approach.

There is a 10th magnitude bluish star at 281" in PA 153°; the star has no real connection with Alphard. The triple star 29 Hydrae lies half a degree to the south.

GAMMA   Magnitude 3.02; spectrum G8 III. Position 13162s2254. The computed distance is about 115 light years, the actual luminosity about 65 times that of the Sun, and the absolute magnitude about +0.3. The star shows an annual proper motion of 0.09"; the radial velocity is about 3 miles per second in approach.

This is the nearest bright star to the interesting long period variable R Hydrae which lies 2.6° to the east. (See page 1015)

EPSILON   Magnitude 3.36; spectrum G0 III. Position 08442n0636; the northernmost star in the little circlet which marks the head of Hydra. Epsilon Hydrae is one of the outstanding examples of a multiple star system, with four components visible and a fifth known to exist. The close pair, A and B, were discovered in 1888 by Schiaparelli; they form a rapid binary in direct motion with a period of only 15.3 years and an average separation of about 0.2" or 8½ AU. This pair is not an easy object in any telescope, and is almost impossible to divide even in large instruments when the stars are nearest each other at about PA 10°, or in fact within 60° of this PA. According to orbital computations by B.Adams (1939) the semi-major axis of the system is about 0.21", the eccentricity is 0.61, and periastron occurs in early 1977. The individual magnitudes are 3.7 and 4.8, the masses are 1.75 and 1.60, and the total luminosity is about 70 times the Sun. The spectral class of the companion is still uncertain.

1012

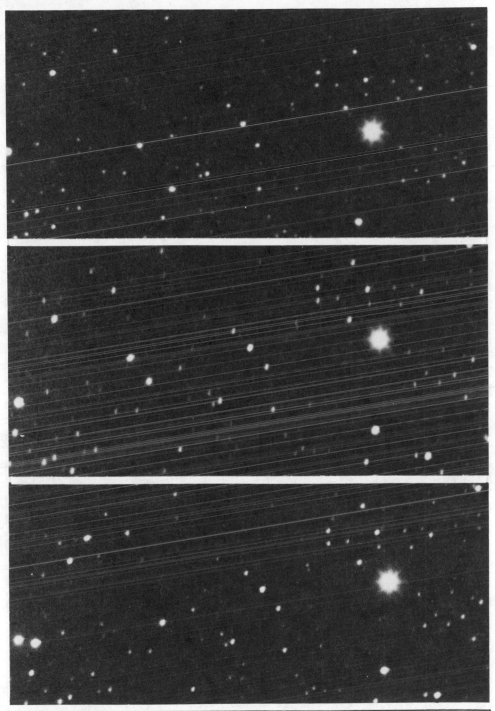

FIELD OF ZETA HYDRAE, photographed by the author on April 15, 16, and 17, 1977, with a 2.4-inch aerostigmat lens. The moving object at left center is the asteroid Pallas, one of the largest of the minor planets.

DESCRIPTIVE NOTES   (Cont'd)

The third star, C, at 3.1" is magnitude 7.8 and was discovered by F.G.W.Struve in 1830. It has a spectral type of dF7 and a luminosity which exceeds the Sun only slightly; this star is in slow revolution around the bright pair with a period estimated at 650 years or more. The projected separation of the A-C system is about 130 AU and the stars have shown a steady increase in PA, from 195° in 1830 to 271° in 1960. Finally, the fourth component, D, shares the proper motion of the other stars but no orbital motion has yet been detected; the apparent separation is 19.3" which corresponds to a projected separation of some 825 AU. The star is a K0 dwarf with about 2.5% the luminosity of the Sun.

The computed distance of the Epsilon Hydrae system is about 140 light years. All the stars show the same annual proper motion of 0.20"; the radial velocity is about 21.5 miles per second in recession.

ZETA   Magnitude 3.12; spectrum K0 II or III. The position is 08528n0608. Zeta Hydrae has a computed distance of about 220 light years; this gives an actual luminosity of 230 times the Sun and an absolute magnitude of -1.1. The annual proper motion is 0.10"; the radial velocity is 13.5 miles per second in recession.

This is the brightest star of the six that compose the ring-shaped asterism marking the head of Hydra. The other members are Epsilon, Delta, Rho, Eta, and Sigma.

NU   Magnitude 3.10; spectrum K3 III. Position 10472s1556. The star is approximately 150 light years distant and has an actual luminosity of about 100 times the Sun (absolute magnitude = -0.2.) The star shows an annual proper motion of 0.22" in PA 33°; the radial velocity is very nearly zero.

PI   Magnitude 3.25; spectrum K2 III. Position 14035s2627. The computed distance is about 85 light years, the actual luminosity about 28 times that of the Sun, and the absolute magnitude about +1.2. The annual proper motion is 0.16"; the radial velocity is 16 miles per second in recession.

CHI-1   Magnitude 5.06; spectrum dF4; position
                   11029s2701. This is a binary system with one
of the shortest periods known for any visual pair, but too
close for small telescopes. The two stars are magnitudes
5.7 and 5.9 and the average separation of the pair is less
than 0.2". Orbital elements are given by Van den Bos (1957)
as follows: Period = 7.4 years; semi-major axis = 0.14" or
about 4 AU; eccentricity = 0.28; periastron = January 1968;
inclination = 94.7° or less than 5° from the edge-on orientation. The two stars appear to be nearly equal in type and
magnitude, with a total luminosity of about 6 suns. Chi-1
Hydrae is approximately 90 light years distant; the radial
velocity is about 10 miles per second in recession.

R   Variable. Spectrum gM/e. Position 13270s2301
                   or about 2.6° east from Gamma Hydrae, which,
in turn, is the bright star some 12° south of Spica. This
was the third of the long period variable stars to be discovered; the first two were Omicron Ceti (Mira) and Chi
Cygni. The discovery of R Hydrae is credited to Maraldi in
1704, though it seems that the star was actually seen by
Hevelius as early as 1662, but not recognized as a variable
star. This is one of the easiest of the long period pulsating stars for observation by amateurs, often reaching 4th
magnitude at maximum. When at minimum it is usually some
250 times fainter, and can be identified only when its
position is accurately known. Like Mira itself, R Hydrae
is an M-type giant and distinctly reddish in color. In the
Arizona-Tonantzintla Catalog (1965) a color index (B-V) of
+1.52 is given; the visual magnitude was 4.98 at the time

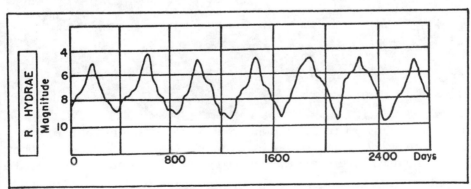

R HYDRAE. The star is shown here during the bright maximum of March 1977. The comparison plate (top) was made in April 1934. Lowell Observatory photographs with the 13-inch telescope.

DESCRIPTIVE NOTES (Cont'd)

the measurement was made. The star, unfortunately, appears
at its ruddiest when near minimum, the color fading somewhat during the rise to peak luminosity.

A peculiar feature of R Hydrae is an apparent slow
shortening of the period which was near 500 days in the
early 18th Century. In Aitken's "ADS" Catalog (1932) it was
reported as about 425 days, and has now decreased still
further to less than 400 days. R Aquilae and R Centauri are
two other cases in which a slow change in period has been
detected over an interval of many years. The explanation is
not yet known, but is believed to indicate some sort of
change in the star's internal structure; R Hydrae may be
passing through a phase where its evolution proceeds with
abnormal rapidity.

The estimated distance of the star is about 325
light years; the maximum luminosity is thought to be about
250 times that of the Sun. The star shows an annual proper
motion of 0.06"; the radial velocity is 6 miles per second
in approach.

R Hydrae is also the visual double star Ho 381; the
12th magnitude companion having first been measured by G.W.
Hough in 1891. No definite change in either separation or
PA is reported in the standard catalogs; this suggests the
possibility of a true physical system since the PA should
have increased by more than 10° in the last 70 years if the
small star does not share the proper motion of R itself.
The apparent separation is 21", corresponding to a projected separation of about 2100 AU; if it is at the same distance as R Hydrae the companion must be a dwarf star with
about 1/12 the solar luminosity, and an absolute magnitude
of +7.5. The spectral type seems undetermined.

Observers who wish to make magnitude estimates of
R Hydrae should note that the brightest field star, marked
"SS" on the chart (page 1018) is itself a variable with a
range of about 0.3 magnitude and spectral type B9. R.G.
Aitken, in the ADS Catalogue, calls it an eclipsing binary
of the Algol type, period 8.20 days, with two spectra visible. In the newer Moscow General Catalogue, however, it is
merely identified as "Ecl.Bin?" and no period is given. SS
Hydrae is also a visual double with a 13th magnitude companion at 13".

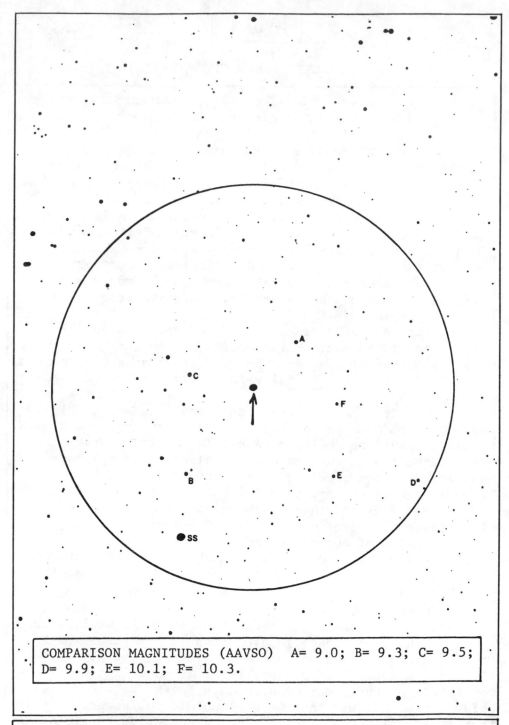

COMPARISON MAGNITUDES (AAVSO)   A= 9.0; B= 9.3; C= 9.5;
D= 9.9; E= 10.1; F= 10.3.

R HYDRAE.   Identification field from a Lowell Observatory
13-inch telescope plate. The circle diameter is 1° with
north at the top; limiting magnitude about 14.

V   Variable. Position 10492s2059, about 5°
south of Nu Hydrae. This is a semi-regular
red variable star discovered by S.C.Chandler at Harvard in
1888. It is one of the rare "carbon stars" resembling the
noted R Leporis and S Cephei; these are all low temperature
giants whose spectra show lines of carbon compounds.   The
spectral type is usually given as N6, but on the newer
"carbon-star" classification it would be called $C6_3$. (The
subscript indicates the relative carbon abundance)

   V Hydrae appeared in several lists of unusually
red stars long before the discovery of its variations in
light. Miss Agnes Clerke wrote in 1905 "....now known as V
Hydrae, is No.16 of Lalande's, No.136 of Schjellerup's Red
Stars, and was recorded by Dr. Copeland at Dunsink, March

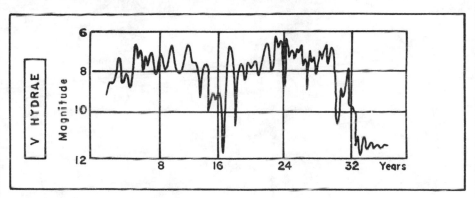

22, 1876, as "brown red" and of 7.2 magnitude. But three
years later, Dr. Dreyer found it risen to the sixth magnitude, and of a "most magnificent copper red," while Bir-
mingham observed it in 1874 as of the eighth, Duner in 1884
as faded to 9.5 magnitude. Its fluctuations are comprised
in a nominal period of 575 days."

   Modern catalogues give the period as about 530
days, but this cycle is superimposed on a much longer
fluctuation with a period of about 18 years. Thus the star
at one time, as in 1916, may show an unusually high maximum, while some 9 years later, as in 1925, it will show an
unusually low minimum. The greatest amplitude of these
variations is about 6 magnitudes, or a difference of about
250 times in light intensity. The above light curve shows
the recorded variations from 1894 to 1926.

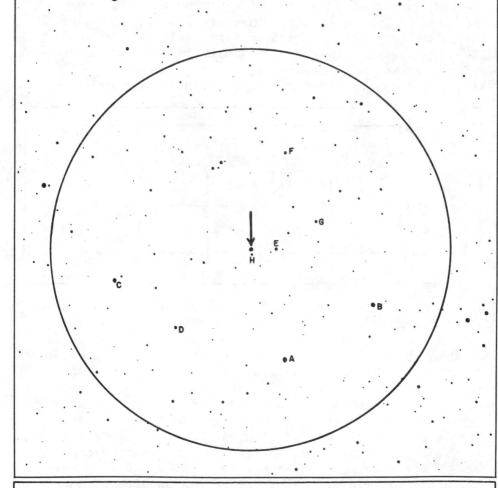

V HYDRAE– AAVSO COMPARISON MAGNITUDES: A= 8.2;
B= 9.4; C= 10.0; D= 10.6; E= 11.1; F= 11.5; G=
12.8; H= 12.9.

V HYDRAE. Identification field from a Lowell Observatory
13-inch telescope plate. Circle diameter = 1° with north
at the top; limiting magnitude about 14.

V HYDRAE. The unusual color of the star is here illustrated by a comparison of blue and red plates. Photographs made with the 48-inch Schmidt at Palomar Observatory.

NGC 2548. This galactic cluster in Hydra is now believed
to be identical with the missing Messier object M48. This
photograph was made with a 5-inch lens. Lowell Observatory

DESCRIPTIVE NOTES   (Cont'd)

The absolute magnitude of an N-type variable is believed to be about -1.5 at maximum; this leads to a distance of about 1300 light years for V Hydrae. The star shows an annual proper motion of 0.02"; the radial velocity is 9 miles per second in approach. (For a general account of the long period variables, refer to Omicron Ceti)

M48   (NGC 2548) Position 08112s0538. Galactic star cluster, located in a rather isolated position near the western border of Hydra, some 14° southeast of Procyon, and well off the main stream of the Milky Way. M48 has long been regarded as one of the mysterious missing Messier numbers, since no cluster exists at the position he gave, which is 4° north of NGC 2548. Norton's Atlas, in fact, charts M48 at the spurious Messier position although no cluster of any sort exists at that spot. From Messier's description there is little doubt that his object is identical with NGC 2548, allowing for a 4° error in the declination. It was discovered by Messier in 1771.

The cluster does not appear to be a particularly well known object though it is easily located in binoculars and the total magnitude of about 5½ suggests naked-eye visibility under good conditions. It is a large group which more than fills a 30' field; the over-all angular size is about 42' according to P.Doig (1926) or 38' according to K.A.Barkhatova (1950). Generally triangular in outline, the cluster is dominated by a central chain-like grouping of ten or so 10th and 11th magnitude stars; several dozen fainter members increase the total population to about 50 stars down to the 13th magnitude. M48 contains three yellow giants of type G or K, but virtually all the other members are A-type main sequence stars; the brightest of these has an apparent magnitude of 8.8 and a spectral class of A2. The true luminosity of this star is about 70 times that of the Sun.

E.G.Ebbighausen has derived a distance of about 1700 light years for M48, in reasonably good agreement with the value of 1530 light years obtained previously by R.J. Trumpler (1930). The resulting true diameter of the group is about 20 light years, and the true luminosity about 1400 times the Sun.

DESCRIPTIVE NOTES   (Cont'd)

M68   (NGC 4590) Position 12368s2629. Globular
star cluster discovered by Messier in 1780;
a rich group for large telescopes, not to be compared with
such clusters as M13 and M5, but a pleasing object in
adequate instruments. To locate, draw an imaginary line
from Delta through Beta in Corvus, extend it out 3.8° to
the 5th magnitude double star B230, the only naked-eye star
in the area. The cluster lies about 0.6° northeast of this
star.
       Although Messier found M68 unresolvable, any good
modern 6-inch reflector should reveal multitudes of its
stellar components, while a 10-inch glass begins to give a
hint of the true splendor of the group which undoubtedly
contains over 100,000 stars. This is a very rich cluster,
the innumerable faint stars gradually concentrating to a
thick central mass about 2' in diameter; the total diameter
of the group is about 9' or slightly over 100 light years.
Recent studies of the H-R Diagram of M68 indicate a distance of 46,000 light years. The total luminosity of the
cluster is then about 100,000 times the Sun, and the total
absolute magnitude about -7.7.
       According to Sawyer's "Bibliography of Individual
Globular Clusters" (First Supplement, 1963) M68 contains a
total of 38 known variable stars. The integrated spectral
type of the cluster is A6; the radial velocity is about 72
miles per second in approach.

M83   (NGC 5236) Position 13343s2937. Large spiral
galaxy discovered by Lacaille in 1752; it is
located on the Hydra-Centaurus border about 18° south of
Spica. This is one of the brightest galaxies of the southern sky, a magnificent system whose dynamic appearance
conveys a strong impression of whirling motion. The two
principal arms of the spiral pattern form a reversed letter
S, and there is a third fainter arm segment starting from
the south side of the nucleus and sweeping out toward the
southwest. M83 is thus sometimes described as a "three-
branch" spiral, while other astronomers, from the structure
of the central mass, suggest a classification among the
barred types. With a total magnitude of about 8, this is
one of the 25 brightest galaxies in the heavens.

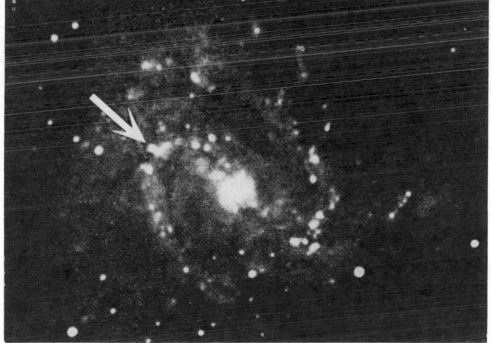

DEEP SKY OBJECTS IN HYDRA. Top: The Globular Cluster M68, photographed with the 13-inch telescope at Lowell Observatory. Below: Spiral Galaxy M83, showing the supernova of 1923; Lowell Observatory 42-inch reflector photograph.

DESCRIPTIVE NOTES (Cont'd)

The spiral arms of M83 are very well defined and are bordered on their inside edges by slender dark dust lanes, an arrangement which appears to be characteristic of many of the nearby spirals. The arms are marked by a rich series of star clouds, a profusion of hot giant stars, and bright clumps of nebulosity. On short exposures the dust lanes can be traced deep into the nuclear regions, and it appears that one dust lane originates in front of the nuclear mass while the other passes behind it. Small but very bright, the nucleus itself measures about 20" across and shows a strong emission spectrum.

The distance of M83 appears to be fairly well determined, at about 10 million light years. Observations of the giant stars in the spiral arms suggest a modulus of about 27.4 magnitudes, equivalent to 9.8 million light years; S.van den Bergh (1976) has obtained about 8.5 million light years; both results are well below the value implied by the observed red shift of 198 miles per second. The total luminosity of M83 appears to be about 5 billion times the Sun, and the visible diameter about 30,000 light years. M83 is a member of a small group which includes the unusual system NGC 5253 in Centaurus, and NGC 5128, the peculiar radio-galaxy characterized by a strong central dust band. (See page 566)

M83 has shown a remarkable number of supernovae in the last 50 years. The first of these, an object of the 13th magnitude, was discovered by C.O.Lampland at Lowell Observatory in 1923 (photograph, page 1025); it was located on the northeast side of the galaxy on the outer edge of one of the two major spiral arms. A second fainter supernova appeared in 1950 and a third one in 1957. The fourth known supernova in M83, and the brightest one to date, was detected in July 1968 by J.C.Bennett, an amateur observer of Praetoria, South Africa. Nearly central in the nucleus, the object had an apparent photographic magnitude of about 12; the true absolute magnitude was about -15.3, or close to 100 million times the light of the Sun. The appearance of four supernovae in 50 years is exceptional, since the expected frequency of these outbursts is usually thought to be about one per three hundred years per galaxy. At least one other nearby spiral, however, NGC 6946 in Cepheus, has shown an abnormal frequency of supernovae.

SPIRAL GALAXY M83 in HYDRA. One of the brightest and most impressive galaxies of the southern sky, photographed with the 74-inch reflector at Radcliffe Observatory in South Africa.

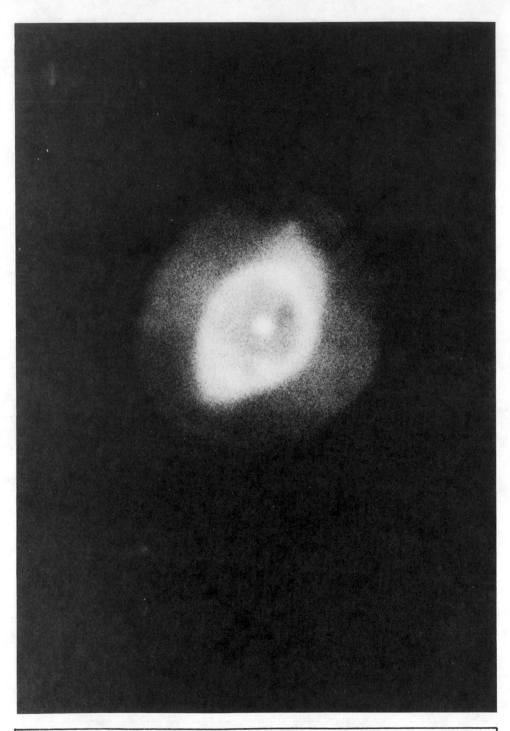

PLANETARY NEBULA NGC 3242 in HYDRA, showing the inner and outer rings and the central star. This photograph, in red light was made with the 200-inch reflector at Palomar.

NGC 3242   Position 10224s1823. This is a fine planetary nebula, easily located about 1.8° south of the star Mu Hydrae. In the small telescope it shows as a pale bluish softly glowing disc measuring about 40" x 35", appearing like a "ghost of Jupiter". The total magnitude is about 9; the central star is 11.4 visually.

A peculiar and interesting structure is shown in larger instruments. There is a bright, strongly elliptical inner ring which strikingly resembles the outline of a human eye; this feature measures 26" x 16" and is oriented southeast to northwest. The "eye" is enclosed by a fainter outer shell of nearly spherical form, some 40" in diameter. In the exact center lies the illuminating star, a hot blue dwarf with a nearly continuous spectrum and a surface temperature of about 60,000°K. Much of the light of the cloud may be termed fluorescence, caused by the strong ultra-violet radiation of the central star. The bluish-green tint is due to the strong emission of doubly ionized oxygen (5007 and 4959 angstroms) but color photographs show that other fainter tints are also present. The surface brightness of the nebula is high, and bears magnifying well. All the main features may be discerned with a good 10-inch glass but the central star always seems fainter than the measured magnitude would indicate. This is generally true of all the planetaries, and illustrates the difficulty of observing a faint star against a background of luminous nebulosity.

The distance of NGC 3242 is not known accurately, though it is undoubtedly among the dozen nearest of the planetaries. The Skalnate Pleso Catalog (1951) has 1900 light years, L.Kohoutek (1961) obtained 2050 light years, and C.R.O'Dell (1963) obtained 3300 light years. The most recent of these determinations would suggest an actual diameter of about 0.6 light year or some 36,000 AU for the nebula, consistent with the results obtained for other bright planetary nebulae such as NGC 7662 in Andromeda and NGC 7009 in Aquarius. The computed luminosity of the central star is about 25 times that of the Sun; the absolute visual magnitude may be about +1.3.(For a summary of facts about the planetary nebulae, refer to M57 in Lyra)

GLOBULAR STAR CLUSTER NGC 5694 in HYDRA. One of the most distant globular clusters. Lowell Observatory photograph with the 42-inch reflecting telescope.

DESCRIPTIVE NOTES   (Cont'd)

NGC 5694   Globular star cluster. Position 14367s2619, at the eastern tip of the tail of Hydra, about 1½° SW of the small group consisting of 54, 55, 56, and 57 Hydrae. This pattern is the site of the obsolete asterism *Noctua*, the "Night-Owl", which appears on the star charts of E.H.Burritt and on other maps of the early 19th century but is no longer recognized today. The loss of this little asterism might seem unfortunate to astronomers, since no other celestial creature so appropriately honored their profession. There is, however, an *Owl Nebula* (M97) in the constellation Ursa Major.

NGC 5694 is one of the more remote globular star clusters, discovered by Sir William Herschel in May 1784, and first recognized as a globular by C.O.Lampland and C.W. Tombaugh at Lowell Observatory in 1932. W.Baade at Mt. Wilson in 1934 made the first detailed study of the group, obtaining a distance of 39,300 parsecs and a linear diameter of about 35 parsecs. Recent studies at Cerro Tololo indicate a somewhat lesser distance of slightly over 100000 light years; the cluster lies on the far side of the Galactic Center, about 85,000 light years from the Center.

From photometric measurements, the total integrated magnitude (pg) is about 10.9 and the integrated spectral type is A9. With an apparent diameter of less than 2' this is one of the most difficult clusters to resolve; the ten brightest stars are all about magnitude 16½. No variable stars are known in this cluster. The radial velocity is about 116 miles per second, in approach.

From current studies at Cerro Tololo (1976) the lowest possible true space motion for NGC 5694, with respect to the Galactic Center, is 273 km/sec, which is well in excess of the escape velocity for the Galaxy (190 km/sec) at that position. This seems to be one definite case in which an object has a velocity so high as to permit its eventual escape from the Galaxy. Is NGC 5694 perhaps a true intergalactic wanderer, a temporary visitor from another galaxy? Or is it simply a Milky Way cluster which has been accelerated to a high velocity, possibly by passing at some time near a massive body such as one of the Magellanic Clouds? Accurate studies of the cluster's space motion may ultimately solve this problem. (Refer also to NGC 2419 in Lynx)

GALAXIES IN HYDRA. Top: The irregular system NGC 3109.
Below: The spiral galaxy NGC 3145. Mt.Wilson Observatory
photographs made with the 60 and 100-inch reflectors.

LIST OF DOUBLE AND MULTIPLE STARS

NAME	DIST	PA	YR	MAGS	NOTES	RA & DEC
h3420	22.6	32	18	$7\frac{1}{2}$- $10\frac{1}{2}$	PA dec, spect K0	01064s8155
h3435	25.5	360	16	7 - 9	spect F2	01234s5946
h3475	2.6	64	57	7 - 7	neat pair, PA inc slowly, spect F0	01537s6033
h3483	7.2	284	30	$9\frac{1}{2}$- $9\frac{1}{2}$	PA dec, spect G0	02031s7132
h3481	18.2	9	16	$8\frac{1}{2}$- 10	spect A2	02038s5925
h3484	52.7	61	19	$7\frac{1}{2}$- 9	spect F5	02058s5954
h3489	21.9	244	31	7 - 11	spect G5	02111s7111
h3489b	8.0	270	31	11- 14		
R14	10.6	158	19	9 - $9\frac{1}{2}$	(GLI 16)	02113s7255
I 266	3.1	163	33	7 - 12	spect K0	02116s6623
I 458	1.6	274	31	8 - $10\frac{1}{2}$	spect G5	02174s7143
h3522	34.3	290	18	7 - $10\frac{1}{2}$	spect F2	02335s7607
h3522b	11.4	354	18	$10\frac{1}{2}$-$12\frac{1}{2}$		
h3568	15.4	224	39	$5\frac{1}{2}$- 8	cpm pair, relfix, spect F0	03090s7911

LIST OF VARIABLE STARS

NAME	MagVar	PER	NOTES	RA & DEC
RS	9.0--15..	216	LPV. Spect M3e	01091s7706
UX	8.5--10..	600:	Semi-reg; spect M2	03490s7002
VW	8.5--13.4	Irr	Eruptive, U Geminorum type	04095s7125

LIST OF STAR CLUSTERS, NEBULAE, AND GALAXIES

NGC	OTH	TYPE	SUMMARY DESCRIPTION	RA & DEC
602		□	pB,S, 1.5' x 0.7'; in two parts divided by dark rift; possibly member of Nubecula Minor	01284s7349
643		⊕	S,F, Mag 13; outlying cluster of Nubecula Minor	01341s7548
1511		∅	Sa; 12.1; 2.5' x 0.9' pB,pS,mE, gbM	03593s6746

DESCRIPTIVE NOTES

ALPHA   Magnitude 2.84; spectrum F0 V. Position 01572s6149. The star is at a distance of about 30 light years; the actual luminosity is about six times that of the Sun, and the absolute magnitude is +2.9. The annual proper motion is 0.26" in PA 83°; the radial velocity is 4 miles per second in recession.

BETA   Magnitude 2.78; spectrum G1 or G2 IV. Position 00232s7732. This is one of the nearer of the bright stars in space, at a distance of 21 light years. It is a star of the solar type, but about 2.7 times brighter, with an absolute magnitude of +3.7. The annual proper motion is fairly large, at 2.25" in PA 82°; the radial velocity is 13.7 miles per second in recession.

GAMMA   Magnitude 3.27; spectrum M0 II or III. Position 03480s7424. The computed distance is about 300 light years, giving an actual luminosity of about 330 times that of the Sun, and an absolute magnitude of −1.5. The annual proper motion is 0.12"; the radial velocity is 9.6 miles per second in recession.

INDUS

LIST OF DOUBLE AND MULTIPLE STARS

NAME	DIST	PA	YR	MAGS	NOTES	RA & DEC
I 1422	0.6	226	25	8½- 8	dist dec, spect A0	20261s4619
h5204	0.3	136	45	8½- 8½	(Rst 5470)	20287s4532
	6.4	34	36	- 9½	relfix, spect F0	
I 41	1.8	360	26	8 - 8½	spect A3	20331s4544
α	67.4	199	14	3 - 12½	(h5209) spect K0	20341s4728
	62.2	343	14	- 13½	(*)	
I 17	1.0	36	53	8 - 8	7th mag star at 124"; spect A0	20414s5040
I 1429	0.4	140	59	8 - 8	PA slight dec, spect F8	20514s4647
I 129	2.0	351	59	8 - 10	PA dec, spect G0	20526s5928
	35.8	29	27	-13½		
Hu 1332	3.7	86	38	7½- 11	spect A3	20534s4852
I 130	2.9	318	36	7 - 10½	spect K0	21007s4810
Hu 1333	1.7	316	36	7½- 10½	PA slight dec, spect K0	21030s4600
h5246	3.5	129	51	8 - 8	PA inc, spect G0	21067s5447
Hu 1626	1.2	152	59	7½- 8	dist inc, PA dec; spect F8	21078s5233
I 1437	1.2	353	42	7 - 10	spect K0	21116s4918
h5259	27.0	126	59	7 - 10½	PA slight dec, spect G5	21160s4716
θ	6.0	275	57	4½- 7	(h5258) cpm, nice pair, PA dec, dist inc; spect A4	21163s5340
Hu 1536	5.8	170	31	7½- 13	spect F2	21190s5559
h5267	22.8	208	14	7½- 12	(I 1442) spect F8	21233s4617
	44.0	182	14	- 9½		
I 19	1.3	320	47	7 - 9	PA dec, spect F2	21448s6544
Hd 296	0.3	108	45	6½- 7	binary, PA inc, 29 yrs; spect F0	21515s6207
δ	0.2	50	54	5½- 5½	close binary; 12 yrs; PA inc, spect F0 (φ 307)	21545s5514
Hd 297	26.8	89	01	7½- 11½	spect G5	22151s6833
h5325	19.0	267	17	8½- 8½	spect both A0	22196s7303
	32.4	99	17	- 11		
ν	0.1	360	33	6 - 6	(φ285) close binary, spect G0	22204s7230

INDUS

LIST OF VARIABLE STARS

NAME	MagVar	PER	NOTES	RA & DEC
R	8.2--14.6	216	LPV. Spect M2e	22325s6733
S	7.4--14.0	401	LPV. Spect M6e	20527s5431
T	6.0--8.5	320	Semi-reg; spect N	21169s4514
U	9.5--11..	98	Semi-reg; spect M	20387s5116
V	9.2--10.2	.4796	Cl.Var; Spect A0--F2	21082s4517
W	8.3--10..	199	Semi-reg; spect M4e	21108s5314
X	8.0--12...	226	LPV. Spect M4e--M5e	21270s5411
Y	8.7--12..	304	LPV. Spect M6e--M7e	21406s5258
RR	9.0--11..	140	Semi-reg; spect N	21424s6532
RS	9.9--10.3	.6241	Ecl.Bin;	21324s7033
RW	9.6--13..	150:	LPV. Spect M2e	21545s6927
RX	9.5--15..		LPV.	20373s5240
RZ	9.0--14..	250:	LPV. Spect M	21004s4941
SS	9.0--13..	190	LPV.	22043s6723
SU	8.7--9.1	.9863	Ecl.Bin; spect G5	20513s4555
AK	7.8---8.1	Irr	Spect M	21059s4658

LIST OF STAR CLUSTERS, NEBULAE, AND GALAXIES

NGC	OTH	TYPE	SUMMARY DESCRIPTION	RA & DEC
6935		⊖	Sa; 12.9; 1.7' x 1.6' pB,cL,R,1bM	20347s5217
6942		⊖	S0/SB; 12.9; 2.5' x 2.3' pB,pL,R,pslbM	20370s5430
----	I.5063	⊖	E3/S0; 13.0; 1.0' x 0.8' S,F,bM	20482s5716
6970		⊖	SBa; 12.7; 0.7' x 0.6' pB,S,1E,gbM	20486s4859
6984		⊖	SBc; 13.1; 1.7' x 1.1' F,pL,v1E,vgbM	20543s5204
7007		⊖	S0; 12.9; 1.0' x 0.6' pB,S,R, psbM	21019s5245
7014		⊖	E0; 13.2; 0.8' x 0.8' pF,S,R,bM	21054s4724

INDUS

LIST OF STAR CLUSTERS, NEBULAE, AND GALAXIES   (Cont'd)

NGC	OTH	TYPE	SUMMARY DESCRIPTION	RA & DEC
7029			E6/S0; 12.3; 1.1' x 0.7' B,cS,R,pgmbM	21084s4930
7038			Sc; 12.5; 2.8' x 1.5' pB,pL,lE,gbM	21117s4726
7041			S0; 12.2; 2.0' x 0.9' B,cS,cE,psmbM	21130s4835
7049			S0; 11.8; 1.0' x 0.8' vB,pS,E,mbM	21156s4847
----	New 6		Sc; 12.9; 4.0' x 1.0' L,F,bM	21200s4600
7064			SBc; 12.7; 3.5' x 0.5' eF,pL,vmE	21255s5300
7083			Sb/Sc; 12.6; 3.8' x 2.5' pF,cL,vlE,gpmbM	21318s6407
7090			SBc; 11.8; 6.0' x 1.0' pB,L,vmE,pslbM	21329s5447
7096			Sa; 13.1; 1.6' x 1.4' vF,S,R	21374s6408
7124			SBb; 12.9; 2.5' x 1.5' pB,L,pmE,vgbM	21448s5048
7125			SBc; 13.2; 3.0' x 1.9' cF,pL,R; 7126 in field	21456s6056
7126			Sc; 13.2; 0.7' x 0.4' pB,pS,lE,gbM	21457s6050
7155			S0/SB; 12.8; 1.3' x 1.1' pB,S,lE,mbM	21529s4946
7168			E3; 12.7; 0.5' x 0.4' pB,S,R,psbM	21589s5200
7196			E3; 12.3; 0.9' x 0.7' cB,S,R	22026s5022
----	I.5152		I/S?; 12.3; 4.1' x 2.0' F,cL,cE,cbM; 8^m star 1.2' NE	21596s5132
7192			E0; 12.9; 1.1' x 1.1' pB,pS,R,pmbM	22032s6433
7205			Sb/Sc; 11.7; 4.0' x 2.0' pB,pL,cE,pslbM	22051s5740

DESCRIPTIVE NOTES

ALPHA Magnitude 3.11; spectrum K0 III. Position 20341s4728. The computed distance of the star is 85 light years; the actual luminosity is about 33 times that of the Sun, and the absolute magnitude is +1.1. The star shows an annual proper motion of 0.08"; the radial velocity is 0.7 mile per second in approach.

EPSILON Magnitude 4.68; spectrum K5 V. Position 21596s5700. This star is one of the Sun's nearer neighbors in space, at present the 6th nearest of all the naked-eye stars. Only Alpha Centauri, Sirius, Epsilon Eridani, 61 Cygni, and Procyon are known to be closer. (See list on page 640)

Epsilon Indi is located at a distance of 11.4 light years, just a shade more distant than Procyon. The star has an unusually large annual proper motion of 4.69" in PA 123°, among the ten largest motions known. A main sequence K-type star, Epsilon Indi has about 80% of the Sun's diameter and about 13% the luminosity. The computed absolute magnitude is +7.0. Measurements of the spectrum indicate a radial velocity of 25 miles per second in approach; the true space velocity is about 55 miles per second.

This is one of the nearest stars which is fairly similar to our own Sun in size and type, one of the few stars where any planetary companions might be within range of detection. Even an object of the mass of Jupiter could not be seen visually at such a distance, but would produce minute perturbations in the star's motion which could be detected by extremely accurate astrometric plate measurements; this has, in fact, been accomplished in the case of Barnard's Star in Ophiuchus where at least one planetary companion comparable in mass to Jupiter is now known to exist. Smaller planets will probably remain forever undetected by any optical techniques. A search for possible radio signals from this star and a few others has already been made (Project Ozma), so far with entirely negative results. More recently (1972) the "Copernicus" satellite has searched for possible laser signals from these nearby systems, also without positive results. (See also Tau Ceti and Epsilon Eridani)

LACERTA

LIST OF DOUBLE AND MULTIPLE STARS

NAME	DIST	PA	YR	MAGS	NOTES	RA & DEC
Ho 175	1.1	308	57	7 - 10	relfix, spect A5	21589n4324
	32.7	284	00	- 12		
A1451	0.1	352	34	7½ -7½	spect A0	21590n3900
Ho 177	8.2	112	32	6 - 12½	spect M	22001n3644
β694	0.8	4	60	6 - 8½	PA slow inc, spect	22009n4424
	66.8	309	18	-10½		
0Σ462	1.1	320	59	7 - 9	PA slow dec,	22048n3551
	7.5	32	54	- 11	spect A3	
h1735	27.1	110	21	7 - 8½	spect B9	22072n4436
h1735b	22.7	164	13	8½-11		
Es---	22.1	308	18	7 - 9½	spect F5	22086n4740
h1741	27.6	303	23	6 - 11	cpm, PA dec, slow	22092n5035
					dist inc, spect A2	
Σ2876	12.2	68	31	7½- 9	cpm pair, relfix,	22098n3725
					spect F8	
0Σ465	14.8	322	24	7 - 10	relfix, spect F0	22100n4957
	18.0	235	08	- 13		
Σ2882	3.6	328	41	9 - 9	relfix, spect F5	22120n3730
Σ2886	20.3	109	22	7½- 9½	relfix, spect F8	22128n4907
Ho 614	4.7	175	27	7½- 10	spect A0	22131n5114
Σ2890	9.4	11	56	8½- 8½	relfix, spect A0;	22132n4938
	73.0	278	56	- 9½	in cluster NGC 7243	
Ho 180	0.7	233	59	7 - 7	slow PA inc, spect	22137n4339
					B9	
Σ2891	13.0	309	15	8 - 9	relfix, spect A0	22144n4744
Es---	21.2	292	07	7 - 12	spect K0	22154n4853
Σ2894	15.8	194	25	6 - 8	relfix, cpm pair;	22167n3731
	43.7	248	05	- 13	spect dF2	
2	48.1	9	25	4½- 11	(h1755) relfix,	22190n4617
					spect B6	
h1756	22.1	286	24	6½- 10½	spect K2	22197n4025
	22.4	325	25	- 13		
	58.0	76	25	-11½		
Σ2902	6.4	90	42	7 - 8	relfix, spect G5	22214n4506
Σ2906	4.2	3	37	7 - 11	relfix, spect B3	22245n3711
A1464	4.7	98	29	7½- 14	spect A0	22245n5351
β380	36.2	135	22	7½- 8	(0ΣΣ234) spect A	22249n4927
	25.0	323	17	- 12		
Es 1467	7.9	228	15	6½- 13½	spect K5	22260n4351
A187	2.0	128	44	7½- 12½	relfix, spect K0	22264n4817

LIST OF DOUBLE AND MULTIPLE STARS   (Cont'd)

NAME	DIST	PA	YR	MAGS	NOTES	RA & DEC
OΣ472	14.2	6	33	7 - 11$\frac{1}{2}$	spect G5	22279n5209
Σ2917	4.9	71	54	8 - 8	relfix, spect F0	22286n5316
	89.5	186	11	-11$\frac{1}{2}$		
Σ2916	44.8	336	44	7$\frac{1}{2}$- 9	relfix, spect K2	22291n4057
Σ2916b	3.7	31	44	9 - 10		
Σ2918	1.4	246	36	8 - 10	relfix, spect A2	22292n5036
α	36.3	294	25	4 - 12	optical, dist inc, PA dec, spect A2 (β703)	22292n5001
Roe 47	42.5	158	10	6 - 10	spect A3	22302n3931
	32.4	344	10	-10$\frac{1}{2}$		
	106	216	10	-10		
Roe 47d	6.6	175	28	9$\frac{1}{2}$-10	(Mlb 32)	
AG 283	2.8	334	37	8$\frac{1}{2}$- 9	spect F8	22303n5456
Hu 1320	0.2	270	65	8 - 8	binary, 61 yrs; PA inc, spect F5	22308n4908
A1468	0.3	256	53	7$\frac{1}{2}$- 7$\frac{1}{2}$	relfix, spect A2	22322n5350
Arg 44	7.2	169	35	8 - 8	relfix, spect B9	22323n5007
8	22.3	186	58	6 - 6$\frac{1}{2}$	(Σ2922) multiple;	22336n3923
	47.8	169	56	- 10$\frac{1}{2}$	AB cpm, spect B1,B2	
	81.1	145	56	- 9$\frac{1}{2}$		
8c	1.3	254	32	10$\frac{1}{2}$-14	(A1469)	
8d	9.1	226	35	9$\frac{1}{2}$- 13		
A1470	0.2	278	63	8$\frac{1}{2}$- 8$\frac{1}{2}$	binary, 22.5 yrs; PA inc, spect G0	22337n5257
Ho 295	0.3	333	53	7$\frac{1}{2}$- 7$\frac{1}{2}$	binary, 30 yrs; orbit edge-on; spect dG0	22365n4403
OΣ475	15.4	72	32	7 - 11	relfix, spect B3	22368n3707
10	62.1	49	25	5$\frac{1}{2}$- 9	(S813) relfix; spect O9	22370n3848
Ho 187	19.5	286	34	6 - 12	spect G5; A= close pair: 0.1"	22373n3720
β277	0.6	211	59	8 - 8	PA inc, spect A2	22373n4107
	25.6	116	13	-13		
Es 1028	5.7	243	10	7$\frac{1}{2}$- 10	spect B9	22404n5359
Hu 91	0.5	315	58	7 - 8	(OΣ476) close	22409n4654
Hu 91b	0.2	218	58	8 - 10	triple, AB PA dec, spect B9	
R	5.7	50	17	var- 14	LPV, spect M5e	22410n4206

LIST OF DOUBLE AND MULTIPLE STARS   (Cont'd)

NAME	DIST	PA	YR	MAGS	NOTES	RA & DEC
13	14.7	129	58	5½- 11	(0Σ479) relfix	22418n4133
					spect K0	
Σ2942	2.9	273	55	7 - 9	(0Σ478) (β450)	22418n3912
	9.8	238	33	- 12	all relfix, spect K	
Σ2946	5.4	258	55	8 - 8	slow PA inc, F8	22474n4015
h1823	19.2	262	21	6½- 12	multiple group;	22495n4103
	82.2	338	23	- 7½	spect A0	
	119	263	21	- 7½		
h1823c	5.4	136	17	7½ - 9	spect A0	
15	25.7	144	15	5 - 12	(β451) optical,	22498n4303
					PA inc, dist dec,	
					spect gM0	
β382	0.2	251	25	6 - 8	binary, 104 yrs;	22514n4429
	28.0	356	25	- 10½	PA inc, spect A4	
					(h1828)	
0ΣΣ239	51.0	243	23	5½- 8½	(h975) wide pair;	22534n3605
					spect B7	
16	27.5	345	23	5 - 12	(Σ2960) relfix,	22541n4120
	62.7	48	23	- 9	primary variable;	
					spect B2	
β452	6.7	256	33	7 - 11	relfix, spect K5	22552n4245

LACERTA

LIST OF VARIABLE STARS

NAME	MagVar	PER	NOTES	RA & DEC
12	5.0---5.1	.19309	(DD Lacertae) Spect B2; Beta Canis Majoris type	22392n3958
16	5.3---5.4	.16916	(EN Lacertae) Spect B2; Beta Canis Majoris type	22541n4120
R	8.4--14.7	299	LPV. Spect M5e	22410n4206
S	7.8--13..	240	LPV. Spect M5e--M6e	22268n4004
U	8.8--10..	Irr	Spect M4e	22457n5454
V	8.5---9.6	4.9835	Cepheid; spect F2--G5	22466n5603
W	9.4--12..	326	LPV. Spect M7	22054n3729
X	8.4---9.1	5.4444	Cepheid; spect F5--G1	22470n5610
Y	8.8---9.8	4.3238	Cepheid; spect F5--G2	22071n5048
Z	8.3---9.1	10.886	Cepheid; spect F6--G6	22389n5634
RR	8.7---9.5	6.4162	Cepheid; spect F5--K0	22395n5610
RS	9.5--12.4	237	Semi-reg; spect K0	22108n4331
RT	9.0---9.7	5.0740	Ecl.Bin; spect sgG9 + sgK1	21595n4339
RX	8.0---9.8	174	Semi-reg; spect M6	22477n4047
SW	8.5---9.3	.32071	Ecl.Bin; spect both G3; W Ursae Majoris type	22514n3740
SX	8.0---9.0	190	Semi-reg; spect M0	22536n3456
TV	9.9--11..	Irr	Spect N	22540n5357
AR	6.8---7.6	1.9832	Ecl.Bin; spect sgG5 + sgK0	22067n4530
BG	8.6---9.3	5.3319	Cepheid; spect F7--G4	21584n4312
BU	9.4--12..	2000:	Semi-reg; spect M0	22268n5019
CM	8.3---9.2	1.6047	Ecl.Bin; spect A2 + A8	21581n4419
CP	2.2--15.6	---	Nova Lacertae 1936 (*)	22138n5522
CS	9.7--10.0	3.7978	Ecl.Bin; spect B5	21577n4220
CT	9.5--15..	Irr	Spect R	22047n4813
CX	8.8---9.5	130	Semi-reg; spect K5	22056n3952
DI	4.3--14.4	---	Nova 1910	22338n5227
DK	5.0--15.5	---	Nova 1950	22477n5301
EW	5.0---5.3		Class Uncertain, spect B3pe	22549n4825
EV	8.5--11..	Irr	Flare star; spect dM4e	22446n4404
FW	8.6---9.5	Irr	Spect M	22195n4524
FZ	9.3--10.1	60	Semi-reg; spect M	22327n4450
GL	8.7---8.9	9.475	Alpha Canum Venaticorum type; spect A0p	22421n5519

LIST OF STAR CLUSTERS, NEBULAE, AND GALAXIES

NGC	OTH	TYPE	SUMMARY DESCRIPTION	RA & DEC
7209	53[7]		L,cRi,pC, diam 20'; about 50 stars mags 9....12; class D	22018n4616
7243	75[8]		L,1C, mag 8, diam 20'; 40 stars, incl Σ2890; class D	22132n4938
7245	29[6]		S,Ri, diam 3'; 50 vF stars; class D	22136n5405
----	I.5217		Mag 12.5; diam 8" x 6" with 14^m 0-type central star	22219n5043

DESCRIPTIVE NOTES

CP   Nova Lacertae 1936. Position 22138n5522.
A bright and well-observed nova which was first noticed on June 18, 1936, near the Lacerta-Cepheus border, about 1½° south of Epsilon Cephei. The exploding star reached its maximum magnitude of 2.2 on June 20 after an extraordinarily rapid rise from magnitude 15.3. The total light increase was thus about 175,000 times, and was accomplished in three days! This was a typical example of a "fast nova", strongly resembling Nova Aquilae 1918. At the maximum, spectroscopic observations revealed the violent expansion of the outer layers of the star at the incredible velocity of 2200 miles per second, one of the highest velocities ever measured in our Galaxy. Probably the only star surpassing this record was T Coronae Borealis which at its second outburst in 1946 showed an expansion velocity of 2700 miles per second. (Supernova outbursts, of course, show even higher expansion velocities, but these are cataclysms occurring on an enormously greater scale, and should not be compared with the ordinary or "classical"

novae such as CP Lacertae). After passing the maximum, the star began to fade almost immediately. In three weeks it had fallen to 6th magnitude, and within a year had dropped to magnitude 11.   The nova is now a faint white star of magnitude 14.8, and is probably still fading slowly. Novae often require a number of years to reach the pre-outburst brightness, though Nova Aquilae reached that level in only seven years and even the "slow nova" T Aurigae (1891) took only fifteen.

The distance of CP Lacertae has been computed to be about 5400 light years; this implies that the actual luminosity at maximum was about 300,000 times that of the Sun; the absolute magnitude was about -8.9. Before the outburst the star was about ½ magnitude brighter than the Sun, and in its present state is still about one magnitude brighter than the Sun would appear at the same distance.   Nova Aquilae 1918 also seems to be more luminous than the Sun, while other novae at minimum, such as DQ Herculis (1934) appear to be definitely fainter than the Sun. The validity of these conclusions, of course, depends very much upon the accuracy of the derived distances. (See also Nova Aquilae 1918, DQ Herculis, GK Persei, T Coronae Borealis, etc)

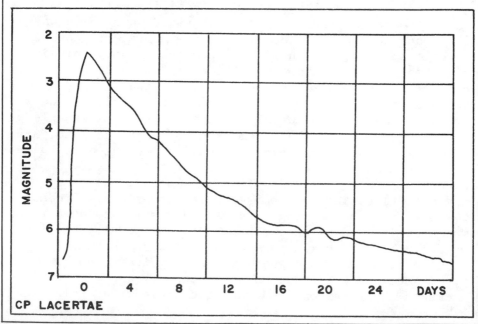

CP LACERTAE

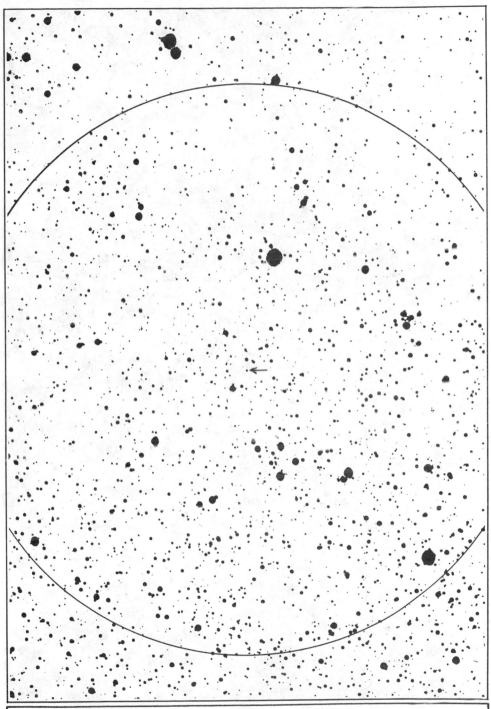

NOVA LACERTAE 1936. Identification field for CP Lacertae, from a plate made with the 13-inch telescope at Lowell Observatory. Circle diameter = 1° with north at the top; limiting magnitude about 15.5.

THE MILKY WAY IN LACERTA. A region of rich star clouds near the Lacerta-Cygnus border, about 2° east of 80 Cygni. The photograph was made with a 5-inch lens. Lowell Observatory.

LEO

LIST OF DOUBLE AND MULTIPLE STARS

NAME	DIST	PA	YR	MAGS	NOTES	RA & DEC
A2477	0.3	301	58	7 - 8½	PA inc, spect G0	09207n1821
κ	2.1	208	59	5- 10½	(β105) cpm, dist dec, spect K2	09218n2624
Σ1353	3.2	309	43	8½- 9	slow PA dec, spect G5	09249n1558
ω	0.4	316	62	6 - 7	(2 Leonis) (Σ1356) binary, 117 yrs; PA inc, spect dF8	09258n0917
3	25.2	80	58	6- 10½	relfix, spect gK0	09258n0824
Σ1360	14.2	242	55	7½- 8	relfix, cpm pair; spect G5	09278n1049
Σ1364	16.9	155	53	7½- 9	(h466) relfix,	09292n2016
	40.0	300	27	-11	spect F5	
6	37.4	75	21	6 - 9	(Sh 107) optical, spect K3	09293n0956
β909	6.0	91	24	7 - 12	spect K0, galaxy NGC 2903 in field	09294n2205
HN 29	30.5	260	58	6 - 10	spect A2	09304n2835
7	41.2	80	46	6½- 8½	relfix, cpm pair, spect A1, F8	09332n1436
Σ1372	0.5	61	59	8 - 8	PA inc, spect F8	09344n1627
A2759	1.8	152	57	7½- 13	spect A2	09357n0756
OΣ204	7.9	102	40	6½- 10	relfix, spect A3	09361n1100
ο	85.4	44	19	4 - 10	optical, dist inc, A= composite spect; A2 + F6	09385n1007
Ho 253	1.1	292	51	7 - 12	relfix, spect G5	09451n1018
20	0.3	214	61	6½- 7	(Kui 44) spect A8	09471n2125
Σ1389	2.3	297	68	8 - 9	PA dec, spect K0	09496n2713
Σ1399	30.8	175	58	6½- 7½	cpm pair, relfix, spect G0	09543n2000
31	7.9	44	34	4½-13½	cpm, spect K4	10053n1015
α	177	307	24	1½- 8	REGULUS; wide cpm pair, spect B7, dK1	10057n1213
αb	2.6	86	43	8 - 13	slight PA dec, spect dK1 (*)	
A2145	0.2	214	61	7½- 7½	binary, 53 yrs; PA dec, spect gF5, A2	10066n2035
34	0.1	304	67	7 - 8	(Hu 874) perhaps slight PA inc, spect F6	10090n1336

LIST OF DOUBLE AND MULTIPLE STARS   (Cont'd)

NAME	DIST	PA	YR	MAGS	NOTES	RA & DEC
Σ1413	2.1	275	46	9 - 9	slight PA & dist dec, spect G0	10096n1636
Σ1417	2.3	259	62	8 -8	relfix, spect F2	10124n1922
OΣ215	1.1	184	66	7 - 7	slow binary, about 450 yrs; dist inc, PA dec, spect dF2	10136n1759
Σ1419	4.5	227	40	8½- 9	relfix, spect A0	10144n1022
39	7.8	299	58	6 - 11	(OΣ523) relfix, cpm, spect dF3, dM1	10145n2322
Σ1421	4.4	331	53	7½- 8½	relfix, cpm, spect F2	10153n2747
Σ1423	1.3	25	63	8½- 9½	binary, about 460 yrs; PA dec, spect dK5	10165n2049
γ	4.4	123	66	2½- 3½	fine binary, PA & dist slow inc, spect K0, G7 (*)	10172n2006
Σ1426	0.8	299	66	8 - 8½	AB PA inc, spect F5	10179n0641
	7.6	8	62	- 9½		
OΣ216	1.0	288	62	7- 10½	(β594) binary, about 250 yrs; PA dec, spect dG2	10200n1536
Σ1429	0.6	205	58	8½- 8½	PA dec, spect G5	10223n2443
Σ1431	3.8	70	34	8 - 9½	relfix, spect F0	10230n0902
Σ1434	6.4	278	56	8½- 8½	cpm, slight PA inc, spect G5	10244n1819
45	37.2	132	24	7 - 12	(h832) spect Ap	10250n1001
OΣ220	0.8	85	60	7 - 9	PA inc, spect F8	10266n1025
Σ1439	1.3	91	67	8 - 8½	PA dec, spect G5	10274n2104
Σ1442	13.3	155	53	7 - 7½	relfix, cpm, spect F0	10292n2218
Σ1446	5.6	252	38	8½- 9½	relfix, spect G0	10309n1528
Σ1447	4.3	125	43	7 - 9	relfix, spect A2	10310n2337
Σ1448	3.6	182	20	7½- 13	(Hu 1338) spect K0	10317n2154
	11.0	259	31	- 9		
49	2.2	158	62	6 - 8½	(Σ1450) relfix, cpm; A= ecl.var. spect A2 (TX Leonis)	10324n0855

LIST OF DOUBLE AND MULTIPLE STARS   (Cont'd)

NAME	DIST	PA	YR	MAGS	NOTES	RA & DEC
Σ1451	8.3	269	40	8½– 9½	relfix, cpm, spect	10326n2632
	41.4	329	26	–13½	G5	
OΣ224	0.5	201	62	7½– 9	binary, about 215	10371n0906
					yrs; PA dec, spect	
					dF6	
OΣ225	0.7	226	66	7½– 9½	PA dec, spect F8;	10374n1930
	6.2	354	58	–10	AC relfix	
OΣ227	0.7	356	67	7½– 8½	PA inc, spect A2	10390n1100
Σ1468	3.9	334	55	8½– 8½	relfix, spect G0	10420n2058
OΣ228	0.7	176	59	7 – 8	PA dec, spect F0	10446n2250
β596	2.4	278	25	6½– 13	spect K0	10468n1725
A2372	0.3	90	58	8 9½	PA inc, spect F8	10484n1619
Σ1482	11.9	305	54	8 – 9	cpm, relfix, spect	10496n0744
					G5	
A2733	1.2	9	60	8 – 9½	cpm, PA dec, spect	10500n0516
					K0	
S617	35.2	178	25	6 – 10	relfix, cpm, spect	10509s0159
					K0, G0	
54	6.5	110	58	4½– 6½	(Σ1487) cpm, PA	10529n2501
					inc, both spect A1	
55	0.8	76	66	6 – 8½	(β1076) binary, 90	10531n0100
					yrs; PA dec, spect	
					dF3	
A2774	1.5	95	37	7½–12½	spect A2	10570n1012
Σ1500	1.7	305	58	7½– 9	slow PA dec,	10575sQ312
					spect F8	
Σ1502	13.1	282	15	8½– 9½	cpm, spect G0	10594n1453
Σ1503	11.3	271	56	8½– 9½	relfix, spect F8;	10596n1010
					A= ecl.var star AM	
					Leonis	
Σ1504	1.2	296	59	7½– 7½	PA inc, spect F0	11014n0355
Σ1506	11.2	219	63	8 – 9½	slow PA inc, spect	11022s0357
					G5, dM0	
Σ1506b	0.2	41	62	10– 10	(A676) binary, 23.5	
					yrs; PA inc	
χ	3.3	262	58	4½– 11	(63 Leonis)(Kui 54)	11024n0737
	76.4	305	24	–9	cpm, PA & dist inc,	
					spect F2	
Σ1507	8.1	165	36	8 – 10	relfix, spect F8	11035n0718

LIST OF DOUBLE AND MULTIPLE STARS   (Cont'd)

NAME	DIST	PA	YR	MAGS	NOTES	RA & DEC
65	2.3	102	58	5½- 11	(β599) cpm, PA inc spect sgG7	11044n0214
Σ1511	7.7	286	43	8½- 8½	relfix, spect K0	11046n1111
67	4.8	244	58	6- 14½	(A677) relfix, spect A3	11061n2457
Σ1517	0.3	159	66	7½- 7½	PA & dist dec, spect G0	11111n2025
Σ1518	104	257	19	7½- 9½	spect K0	11119n0532
Σ1518b	3.4	352	40	10- 10		
Σ1521	3.7	96	57	7 - 7½	relfix, spect A5	11127n2751
Σ1527	1.7	28	67	7 - 8	slow dist dec, PA inc, spect G0	11164n1433
Σ1529	9.4	253	55	7 - 8	cpm, relfix, spect F8	11168s0123
Σ1534	5.1	321	55	8 - 11	cpm, PA dec, spect G5	11192n1828
ι	1.0	205	62	4 - 6½	(Σ1536) binary, about 190 yrs; PA dec, spect F2	11213n1048
Σ1537	2.4	356	55	7½- 8½	relfix, spect G0	11218n2054
Σ3070	9.2	276	38	8½- 9	cpm, slow dist inc	11220s0407
83	28.9	150	58	6½- 7½	(Σ1540) cpm pair; spect dK0, dK5	11243n0317
	90.3	188	37	-10		
Rst4944	0.3	302	67	7 - 8½	PA dec, spect K0	11253s0124
τ	91.1	176	32	5½- 7	wide & easy pair; (ΣI 19) slight PA inc, dist dec, spect G8	11254n0308
Σ3072	9.7	330	09	7½- 10	relfix, spect F8	11283s0627
88	15.4	326	58	6½- 8½	(Σ1547) cpm, PA inc, nice colors; spect dF7, dK6	11292n1439
Σ1548	10.9	128	38	7½- 9	(Rst 4480) relfix; spect K0	11296s0315
Σ1548b	0.5	312	58	9½- 10		
90	3.4	209	58	6 - 7½	(Σ1552) AB cpm, relfix, spect B3, B6, F5	11321n1704
	63.1	235	38	- 9		
A2777	3.6	90	28	7½- 13	spect G5	11337n0335
Σ1555	0.3	315	59	6½- 7	binary, PA inc, spect A3 (h503)	11337n2803
	21.3	149	24	-11		

$$\boxed{\textbf{LEO}}$$

LIST OF DOUBLE AND MULTIPLE STARS   (Cont'd)

NAME	DIST	PA	YR	MAGS	NOTES	RA & DEC
Σ1558	1.2	162	56	8½– 9	PA slow inc,	11341n2145
	43.6	276	56	– 8½	spect G5	
Σ1564	5.2	87	41	8 – 9	relfix, spect A2	11370n2714
Σ1565	21.6	305	55	7 – 8	relfix, cpm, both	11370n1916
					spect F5	
Σ1566	2.7	348	47	8½–10	relfix, cpm	11380n2119
Hu 888	0.7	162	58	8½– 9	PA inc, spect F5	11402n2122
A679	5.0	93	33	7½–14½	spect K0	11412n2417
OΣ239	37.3	24	24	5½– 10	PA slight inc,	11416n2530
					spect K5	
93	74.3	355	25	4½– 8½	wide cpm pair; A–composite spect; A + G5	11454n2030
β603	0.6	14	62	6 – 10	binary, 134 yrs; PA dec, spect dA6 (Refer to note on Beta Leonis)	11461n1434
h1201	14.9	189	34	6½– 11	relfix, cpm, spect A3	11484n1233
Σ1582	12.3	76	20	7½– 9	relfix, cpm, spect F8	11534n2216

LEO

LIST OF VARIABLE STARS

NAME	MagVar	PER	NOTES	RA & DEC
R	5.2--10.5	312	LPV. Spect M7e--M9e (*)	09449n1140
S	9.2--14..	190	LPV. Spect M3e	11083n0544
V	8.5--14.0	273	LPV. Spect M5e	09573n2130
W	8.9--14.5	385	LPV. Spect M7e	10510n1359
X	11.5-15..	Irr	Peculiar, SS Cygni type	09483n1207
Y	9.3--11.5	1.686	Ecl.Bin; Spect A3	09340n2628
Z	9.0--10..	57:	Semi-reg; spect M3	09493n2709
RR	9.7--10.6	.4524	Cl.Var; spect A0--F4	10049n2414
RS	9.5--15..	209	LPV. Spect M5e	09407n2005
RY	8.5--11..	155	Semi-reg; spect M2e	10016n1414
SS	9.9--11.0	.6263	Cl.Var; Spect A7--F6	11313n0015
TX	5.6---5.7	2.455	(49 Leonis) Ecl.Bin; spect A2	10324n0855
UV	7.8---8.3	.6001	Ecl.Bin; spect G0--G2	10357n1432
UW	9.0--10..	Irr	Spect M5	11126s0519
UY	8.5--10..	Irr	Spect gM7	10266n2319
UZ	9.5--9.9	.6184	Ecl.Bin; W Ursae Majoris type, spect A7	10379n1350
VV	8.0--10..	182	Semi-reg; spect M7	10462n0856
VY	6.0---6.4	Irr	(56 Leonis) Spect M5	10534n0627
WX	8.0---8.7	Irr	Spect M5	11536n1601
YY	8.8--10..	41:	Semi-reg; spect M3	10056n2015
ZZ	9.7--11..	148	Semi-reg; spect M0	11343n0948
AC	9.0---9.4	.4895	Ecl.Bin; W Ursae Majoris type, spect K0	09575n1742
AD	9.0---9.4	Irr	Red dwarf flare star; spect M4e	10169n2007
AF	8.0--10..	107	Semi-reg; spect M5	11253n1525
AI	8.7--10.5	Irr	Spect M5	11379n1128
AK	8.4---9.4	Irr	Spect M5	11382n1321
AM	8.2--9.7	.3658	(Σ1503) Ecl.Bin; W Ursa Majoris type, spect F8	10596n1010
AP	9.0---9.7	.43036	Ecl.Bin; W Ursa Majoris type; Spect G0	11025n0526
CX	6.05± .09		α Canum type; spect B8	10250n1001
CZ	8.5--9.0	115:	Semi-reg; spect M5	11288s0401
DF	6.8--7.0	70:	Semi-reg; spect gM4	09208n0756
DG	6.1 ± .02	.0818	δScuti type; spect A8	09470n2125
DE	5.6---5.7		Semi-reg; spect M2	10226n0902

LIST OF STAR CLUSTERS, NEBULAE, AND GALAXIES

NGC	OTH	TYPE	SUMMARY DESCRIPTION	RA & DEC
2903	56[1]	⊖	Sb/Sc; 9.7; 11.0' x 4.7' cB,vL,E,gmbM; many armed spiral (*)	09293n2144
2911	40[2]	⊖	E2p/S0; 13.5; 0.6' x 0.5' F,pL,R,gbM	09310n1022
2964	114[1]	⊖	Sb/Sc; 11.9; 2.3' x 1.1' B,vL,1E,vgbM; 2968 nf 6'	09400n3205
2968	491[2]	⊖	I/Sp; 12.8; 1.2' x 0.7! pB,pL,1E,vglbM	09403n3210
3032		⊖	S0/Sa; 12.8; 0.6' x 0.4' F,S,slbM	09492n2928
3041	98[2]	⊖	Sc; 12.7; 2.7' x 1.6' F,vL,1E,vglbM	09503n1655
3067	492[2]	⊖	Sa/Sb; 12.7; 2.0' x 0.7' pB,pL,E,gbM; 9^m star 4' E	09554n3237
3098		⊖	E7/S0; 13.0; 1.4' x 0.4' pB,S,E,psbMN	09595n2458
3162	43[2]	⊖	Sc; 12.3; 1.9' x 1.5' pF,cL,R,vglbM	10107n2259
3177	25[3]	⊖	Sb; 12.8; 0.8' x 0.7' cF,S,R; psbM	10139n2122
3185		⊖	SBa; 12.7; 1.5' x 0.9' pF,pL,mbM; forms nice group with 3187,3190,3193	10149n2156
3187		⊖	SBc; 13.0; 1.0' x 0.3' vF,E; 3190 in field	10150n2208
3190	44[2]	⊖	Sb; 12.0; 3.0' x 1.0' pB,pL,E, nearly edge-on (*)	10154n2205
3193	45[2]	⊖	E0; 12.0; 0.9' x 0.9' B,S,R, 8½^m star 80" N	10157n2209
3226	28[2]	⊖	E2; 12.7; 1.0' x 0.9' pB,cL,R, 2' pair with 3227	10207n2009
3227		⊖	Sa/Sb; 11.6; 3.0' x 1.2' pB,cL,E; pair with 3226; 50' E of Gamma Leonis	10207n2007
3274	358[2]	⊖	Sc; 13.0; 2.0' x 0.7' F,pL,E,glbM	10296n2756
3287		⊖	SBc; 12.8; 2.2' x 0.7' F,pL,E	10321n2155

LIST OF STAR CLUSTERS, NEBULAE, AND GALAXIES (Cont'd)

NGC	OTH	TYPE	SUMMARY DESCRIPTION	RA & DEC
3300	55[3]		SBa/S0; 13.1; 1.0' x 0.5' cF,cS,1E,pmbM	10340n1426
3301	46[2]		SBa; 12.3; 2.0' x 1.0' cB,S,1E,psbM; nearly edge-on	10343n2208
3338	77[2]		Sb/Sc; 11.6; 4.7' x 2.8' F,cL,E,vgbM, 7^m star 2.4' W	10395n1400
3346	7[5]		Sc/Sd; 12.4; 2.3' x 2.0' cF,vL,R,vgvlbM	10410n1509
3351	M95		SBb; 11.0; 4.0' x 3.0' B,L,R,mbM; θ structure (*)	10413n1158
3367	78[2]		Sc; 12.1; 1.9' x 1.7' pB,cL,R,vglbM	10440n1401
3368	M96		Sb; 10.2; 6.0' x 4.0' vB,vL,1E,vsmbM (*)	10442n1205
3370	81[2]		Sc; 12.4; 2.5' x 1.3' cB,pL,vlE,gbM	10445n1732
3377	99[2]		E5; 11.4; 1.6' x 0.9' vB,cL,1E,svmbM, BN	10451n1415
3379	17[1] M105		E1; 10.6; 2.1' x 2.0' vB,cL,R, group with 3384 & 3389	10452n1251
3384	18[1]		E7/S0; 11.0; 4.0' x 2.0' vB,L,E,psmbM	10457n1254
3389	41[2]		Sc; 12.2; 2.2' x 1.0' F,L,vglbM	10458n1248
3412	27[1]		E5/S0; 11.6; 2.4' x 1.2' B,S,1E,smbMN	10483n1341
3433	20[3]		Sb/Sc; 12.9; 2.2' x 2.0' vF,vL,R,vgbM	10494n1026
3437	47[2]		Sb/Sc; 12.6; 2.0' x 0.6' pB,pL,1E,gbM	10499n2311
3455	82[2]		Sb; 13.1; 2.4' x 1.0' pF,S,E,gbM; 3454 is 3.3' N = pF,E, edge-on spiral	10518n1733
3485	100[2]		SBb; 12.8; 1.4' x 1.2' F,L,R,glbM	10574n1506
3489	101[2]		E6/S0; 11.3; 2.2' x 1.0' vB,pL,1E,smbMN	10577n1410
3495			Sc/Sd; 12.7; 3.8' x 1.0' vF,pL,mE; nearly edge-on	10586n0353

LIST OF STAR CLUSTERS, NEBULAE, AND GALAXIES (Cont'd)

NGC	OTH	TYPE	SUMMARY DESCRIPTION	RA & DEC
3506	22^3	⊶	Sc; 13.2; 0.8' x 0.7' vF,cS,R,vgvlbM	11006n1121
3521	13^1	⊶	Sb; 10.2; 6.0' x 4.0' cB,cL,mE,vsmbMN; multi-arm spiral (*)	11032n0014
3547	42^2	⊶	Sb; 12.9; 1.1' x 0.8' F,S,1E,vlbM	11073n1100
3593	29^1	⊶	S0/Sa; 12.0; 3.0' x 0.9' B,cL,E, psbM	11120n1306
3596	102^2	⊶	Sc; 12.2; 3.6' x 3.0' pF,L,R,glbM	11124n1504
3605	27^3	⊶	E4; 13.5; 0.7' x 0.5' F,S,R,bM; 3607 N foll	11142n1818
3607	50^2	⊶	E1/S0; 11.2; 1.5' x 1.3' vB,L,R,vmbM; 3608 N 6'	11143n1820
3608	51^2	⊶	E3; 12.2; 1.0' x 0.8' B,pL,R,psbM	11144n1826
3611	521^2	⊶	Sa; 12.6; 1.2' x 0.8' pF,cS,R,psmbM; 10^m star 3' north prec.	11149n0450
3623	M65	⊶	Sa/Sb; 10.3; 7.8' x 1.6' B,vL,mE,mbM; M66 in field, also 3628 (*)	11163n1323
3626	52^2	⊶	Sb/S0; 11.4; 1.7' x 1.3' B,S,vlE,sbM	11175n1838
3627	M66	⊶	Sb; 9.7; 8.0' x 2.5' B,vL,mE,mbM; M65 & NGC 3628 in field (*)	11176n1317
3628	8^5	⊶	Sb; 10.3; 12.0' x 2.0' pB,vL,vmE; edge-on	11177n1353
3629	338^2	⊶	Sc; 12.9; 1.7' x 1.3' cF,L,R,vgvlbM	11179n2715
3630		⊶	E7/S0; 12.8; 1.5' x 0.4' pB,S,E,smbMN; edge-on	11177n0315
3640	33^2	⊶	E2; 11.6; 1.1' x 0.9' B,pL,R,psbM	11185n0331
3646	15^3	⊶	Sc; 11.8; 3.5' x 1.8' cF,cL,1E,gbM	11192n2027
3655	5^1	⊶	Sc; 12.3; 1.1' x 0.8' pB,pS,R,bM	11203n1651

LIST OF STAR CLUSTERS, NEBULAE, AND GALAXIES (Cont'd)

NGC	OTH	TYPE	SUMMARY DESCRIPTION	RA & DEC
3659	53[2]	⊖	Sc; 12.9; 1.2' x 0.9' cF,S,1E	11211n1805
3664		⊖	SB; 12.9; 1.4' x 1.4' pF,bM	11217n0335
3666	20[1]	⊖	Sc; 12.6; 3.6' x 0.8' F,L,mE; nearly edge-on	11219n1137
3681	159[2]	⊖	Sb/Sc; 12.6; 1.0' x 0.9' B,pS,R,bM; 3684 nf	11239n1709
3684		⊖	Sc; 12.4; 1.4' x 0.9' pB,pL,E,vgbM; 3686 nf	11245n1718
3686	160[2]	⊖	Sc; 12.0; 2.2' x 1.9' pB,L,v1E,vgbM	11251n1730
3689	339[2]	⊖	Sc; 12.8; 1.0' x 0.8' pB,pL,E,bM	11255n2556
3691	54[2]	⊖	SB; 13.1; 0.9' x 0.7' F,pS,1E	11255n1711
3705	13[2]	⊖	Sb; 12.2; 4.0' x 1.3' pF,pL, E,vsmbM	11276n0933
3720		⊖	E0; 13.0; 0.5' x 0.5' vF,S; 3719 NW 2' =Sb; vF	11298n0105
3773	81[3]	⊖	E0/S0; 13.0; 0.5' x 0.5' cF,cS,R,psbM	11356n1223
3810	21[1]	⊖	Sc; 11.5; 3.6' x 2.5' B,L,v1E	11384n1145
3872	104[2]	⊖	E4; 12.9; 0.7' x 0.4' B,S,R, smbM	11432n1403
3900	82[1]	⊖	SBb; 12.5; 1.7' x 1.0' B,pL,v1E,bMN	11466n2717
3912	342[2]	⊖	Sb/Sc; 13.0; 0.9' x 0.5' F,pL,1E,pgbM	11475n2646
4008	368[2]	⊖	E5/S0; 12.9; 1.0' x 0.6' pB,pS,E,psbM	11557n2828

DESCRIPTIVE NOTES

ALPHA Name- REGULUS, "The Little King"; sometimes called COR LEONIS, "The Lion's Heart". The 21st brightest star in the sky; magnitude 1.36, spectrum B7 V. Position 10057n1213. Opposition date (midnight culmination) is February 19.

The name *Regulus* is the diminutive form of the Latin *Rex*, or "king". According to R.H.Allen the star was known in Arabia as *Malikiyy*, "the Kingly One", while in ancient Greece it was "The Star of the King". Pliny calls it *Regia* or "The Royal One"; in ancient Babylonia it was *Sharru* or "The King"; to the even more ancient Akkadians of Mesopotamia it represented *Amil-gal-ur*, a legendary "King of the Celestial Sphere" who ruled before the Great Flood. The Hindu title *Magha* signifies the "Mighty" or the "Great One" while the Persian name *Miyan* seems to mean "The Central One" or "The Star of the Center". The Latin *Cor Leonis* is the equivalent of the later Arabian *Al Kalb al Asad*, "The Heart of the Royal Lion".

Tycho called the star *Basiliscus*, evidently from the Roman title *Basilica Stella*. The modern name *Regulus*, given by Copernicus, seems to have no certain connection with the famous Roman general Regulus, whose heroism so inspired the Romans during the first of the three great struggles with Carthage. Some connection has indeed been asserted by E.H. Burritt in his *Geography of the Heavens* (1856) but this claim, repeated by a few other writers, has been discounted by R.H.Allen. Regulus was regarded by the ancient Persians as one of the four "Royal Stars" of Heaven, the other three being Aldebaran, Fomalhaut, and Antares.

As the brightest star in Leo, Regulus has been almost universally associated in ancient cultures with the concept of royalty and kingly power. The star lies at the base of the "Sickle of Leo", a pattern formed by the stars Alpha, Eta, Gamma, Zeta, Mu, and Epsilon, and resembling a large reversed question-mark some 16° high. This configuration was in ancient China the *Yellow Dragon* though on some other oriental star-maps it appears as a horse or a *Great Chariot of Heaven*. To modern sky-watchers the Sickle outlines the majestic head and mane of a great westward-facing lion, crouching in the regal pose of the enigmatic Egyptian Sphinx. The origin of the zodiacal lion is somewhat obscure though the Greeks identified it as the famous Nemaean Lion

THE RISING OF THE SICKLE OF LEO is a traditional herald of the coming of Spring. This photograph was made by the author with a 1.3-inch Kodak anastigmat lens.

DESCRIPTIVE NOTES   (Cont'd)

who is said to have originated in the Moon, and whose con-
quest by Hercules constituted one of the Twelve Labors. To
the Egyptians, however, Leo was the *House of the Sun,* and
Pliny stated that the constellation was worshipped because
the annual rise of the Nile coincided with the Sun's
entrance into Leo. On the representation of the Egyptian
zodiac at the Temple of Denderah, Leo is shown as a Lion
standing on an outstretched serpent. Both Leo and the Dog
Star (Sirius) were believed to contribute to the heat and
storms of summer; Aratus refers to this ancient tradition
when he writes:

> *"Most scorching is the chariot of the Sun*
> *.....when he begins to travel with the Lion.*
> *Turbulent north winds then fall on the wide sea*
> *With all their weight; no time is that*
> *For oar-sped barques; broad ships be then my choice;*
> *O helmsman! Keep the stern before the wind! "*

A magnificent head of Leo appears on the fine silver
coinage of the Greek city of Leontini, as the badge and
identifying symbol of the city. The specimen shown here in
Figure 1 was struck about 470 BC. A similar type was coined
at Cyzicum in Mysia (Fig 2) in the 4th Century BC, and at
Cnidus in Caria at approximately the same period (Fig 3).
An impressive figure of a walking lion appears on the
tetradrachms struck by Mazaeus, Satrap of Babylon in the
days of Alexander the Great (Fig 4).

Medieval Christians regarded Leo as a symbol of the
Lion's Den of the Book of Daniel; to earlier Hebrews the
Lion was the traditional tribal sign of Judah, based on the
text of the 49th chapter of *Genesis*. Babylonian tablets
record observations of Regulus dating from about 2100 BC,

and it was through a study of such records, and those of
Spica, that the Greek astronomer Hipparchus detected the
Precession of the Equinoxes, about the year 130 BC. The
longitude of Regulus had changed by some $28\frac{1}{4}°$, or nearly 2
hours of right ascension, since the first observations had
been inscribed on the clay tablets of Babylonia, slightly
over 2000 years before.

Regulus lies only about half a degree from the
Ecliptic, and it is therefore a relatively common sight to
find the Moon passing quite close to the star, and now and
then to occult it. The bright planets also may appear on
occasion very near the star. Regulus was occulted by Venus
on July 7, 1959, an exceedingly rare event. According to
modern computations, no other occultation of a first magnitude star by Venus will occur for several centuries.

Regulus is about 85 light years distant, and has an
actual luminosity of about 160 times that of the Sun. The
computed absolute magnitude is -0.7. Regulus is a late
"Orion" or "helium" type star with a surface temperature
of about 13,000°K and an estimated diameter of about five
times that of the Sun. The star shows an annual proper
motion of 0.25" in PA 270°; the radial velocity is about
2.3 miles per second in recession.

The small companion star to Regulus is an easy object
for the small telescope; it is magnitude 7.9 and 177" from
the primary, but shows the same proper motion. The curious
remark of Miss Agnes Clerke (1905) that the star appears as
"if steeped in indigo" does not appear to be supported by
any modern observer, nor by the known spectral type of dK1.
J.H.Mallas (1969) calls the color "a deep yellow, almost
golden" while E.J.Hartung in his *Astronomical Objects For
Southern Telescopes* (1968) refers to it as "orange-red".
The star is a dwarf with about half the solar luminosity,
and the projected separation from Regulus itself is about
4660 AU.

The faint companion is itself a close double for
larger telescopes, the duplicity having been first detected
by Winlock at Washington in 1867. With magnitudes of 8 and
13 this is a difficult pair because of closeness, the
faintness of the third star, and field glare from Regulus
itself. The separation has decreased from 3.9" to about 2.6
since 1867, and the PA appears to be diminishing at the

rate of about 7° per century. This third star is a dwarf of uncertain type with about 1/250 the luminosity of the Sun. No orbit computation has been made for the B-C pair, but the period must be many thousands of years.

A fourth star, designated "D", is mentioned in R.G. Aitken's *New General Catalogue of Double Stars* (1934) at a distance of 217" from Regulus in PA 274°. This star, of the 13th magnitude, evidently has no real connection with the system, and the separation is slowly decreasing from the proper motion of Regulus itself. The bright primary, in fact, is moving almost directly toward the star D, and the minimum separation of about 14" will occur around the year 2790 AD.

Just 20' from Regulus, almost due north, is the position of the dwarf elliptical galaxy called "Leo I", one of the members of the Local Group of Galaxies, and among the smallest and faintest galaxies known. Discovered by A.G.Wilson on Palomar 48-inch Schmidt plates in 1950, the Leo I System lies some 750,000 light years distant, but is almost impossible to observe visually in any telescope owing to the extremely low surface brightness. On Lowell Observatory 13-inch telescope plates it appears as a barely perceptible smudge some 10' in width, slightly elliptical from east to west, and resembling a minor plate defect rather than a real object. Great reflectors reveal it as a rich swarm of countless tiny star images of magnitude 20 and fainter; the longest dimension measures about 28' and the total photographic magnitude is about 11. With an absolute visual magnitude of about -11, this is one of the least luminous systems known. A "Little Brother" to this system, the Leo II galaxy, lies 97' north from Delta Leonis but is even smaller and fainter than Leo I, with a total photographic magnitude of about 12.8. Studies of the two systems with 200-inch telescope plates have revealed many RR Lyrae type stars among the members, all close to magnitude 21; hence the distance is known with good accuracy.

BETA   Name- DENEBOLA, "The Lion's Tail", from the Arabic *Al Dhanab al Asad*. Other names are Dafirah and Serpha, apparently from *Al Sarfah*, "The Changer" or "Governor of the Weather". According to R.H.Allen the Chinese identified it, with several small neighbor

stars as *Wu Ti Tso*, the "Seat of the Five Emperors". On
Bayer's maps of the early 17th century the star is labeled
Denebalecid while Schickard has *Dhanbol-asadi* and Riccioli
called it *Nebolellesed*. The *Alfonsine Tables* have a similar
title, *Denebalezeth*. In all these names the *Deneb* portion
refers of course to the Lion's tail.

Denebola is magnitude 2.14, spectrum A3 V; position
11465n1451. The star is a main-sequence A-type star, very
similar to Sirius in luminosity and type. At a computed
distance of 43 light years the actual luminosity is about
equal to 20 suns, and the absolute magnitude is about +1.5.
The star shows an annual proper motion of 0.51" in PA 256°;
the radial velocity is very near zero.

The star has several distant optical companions. A
star of magnitude 13 was recorded by S.W.Burnham in 1878,
at 77" in PA 344°. A closer star of magnitude 15.5 is 40"
distant in PA 346°. A brighter star, "D", is 264" distant
in PA 203°, magnitude 8.5, spectrum F8. Finally, at 1134"
in PA 200° is the bright star β603, magnitude 5.9, spectrum dA6. This star, which has no real connection with Beta
itself, is a rather difficult double, discovered by S.W.
Burnham in 1879. The companion of the 10th magnitude, less
than 1.0" away, has shown a decrease of PA of about 1° per
year during the past 50 years. An orbit computation by P.
Baize (1957) gives the following elements: Period = 134
years, Periastron = 1939, semi-major axis = 0.81", eccentricity = 0.68, inclination = 141°.

GAMMA   Name- AL GEIBA, probably from an erroneous
form of *Al Jabbah* or *Juba*, the Lion's Mane".
Magnitude 1.98; spectra K0 III and G7 III. The position is
10172n2006. This is the brightest star in the curve of the
Sickle of Leo, lying some 8° NNE from Regulus.

Gamma Leonis is one of the finest double stars in the
sky, though rather difficult for low power telescopes owing
to the closeness of the pair. Fortunately, as K. McKready
points out in his *Beginner's Star Book*, the pair is slowly
widening and thus becoming increasingly available to small
telescopes. Discovered by Sir William Herschel in 1782, it
is a binary in slow direct motion with a highly eccentric
orbit and a period of some six or seven centuries. Exact
orbital elements are still in doubt. The results of studies

DESCRIPTIVE NOTES   (Cont'd)

and computations by three different authorities are compared here; it will be noticed that the orbital period seems
to be the most uncertain figure in the table:

	PERIOD	SEMI-MAJ AXIS	ECCENT	PERIASTRON
W.Doberck	407.0 yr	1.98"	0.73	1741
Guntzel-Lingner	701.4	2.74"	0.74	1737
W.Rabe	618.6	2.51"	0.84	1743

Individual magnitudes for the two stars have been measured
photometrically as 2.14 and 3.39. The separation of the
components has increased noticeably since the first really
accurate measurements by F.G.W.Struve in 1831, and the PA
has increased about 16° in the last century. Both components are yellow stars, though the fainter has often been
called greenish. Herschel in 1784, however, saw them as
white and pale red, Struve in 1837 as golden yellow and
reddish, and Admiral Smyth as bright orange and greenish-
yellow. C.E.Barns (1929) called them "orange & pale Yellow"
whereas W.T.Olcott in his *Field Book of the Skies* describes
them as "yellow and green". The author of this book has
always seen them as a deep golden yellow, tending toward
orange, and with very little difference between the two
stars. Bright close pairs of this sort are often best seen

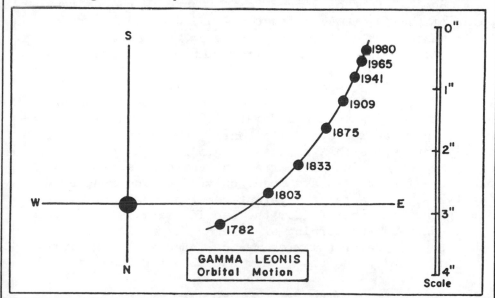

GAMMA LEONIS
Orbital Motion

DESCRIPTIVE NOTES   (Cont'd)

in twilight or moonlight, which helps to reduce the glare
of the stars; Maedler always made his measurements of this
star in complete daylight. At present, the orbital motion
(page 1063) is very slow as the stars continue to widen
toward their maximum separation of about 5" around the year
2100. The expected increase in PA between 1950 and 1980 is
only 3°; the projected separation of the pair at the present time (1975) is about 125 AU, or about 1½ times the
distance across the entire Solar System.

Gamma Leonis is approximately 90 light years distant;
the actual luminosities are about 90 and 30 times that of
the Sun, with computed absolute magnitudes of -0.1 and +1.2.
The annual proper motion is 0.35" in PA 118°; the radial
velocity is 22.6 miles per second in approach.

The ADS Catalogue also lists a third star of the 11th
magnitude, which is not, however, a real member of the
system. It has a large proper motion of 0.50" per year in
a different direction, toward PA 265°; the position (1960)
is about 4.9' from Gamma in PA 290°. Also in the field is
the star 40 Leonis (Mag 4.97, spectrum F5) about 22' south,
a subgiant of luminosity class IV. The reported magnitudes
for this star are quite discrepant in some early catalogs,
and the star has been suspected of variability, though no
modern observations have shown any definite changes.

Some 2° to the NW from Gamma Leonis is the radiant
point of the famous Leonid Meteors which currently reach
their maximum on November 17 each year, and stage more or
less spectacular displays at intervals of 33 years. This
is the swarm that produced the fabulous meteor shower of
November 13, 1833, and the almost equally fine ones of
1799, 1866, and 1966. The Leonids are disintegration products of the Tempel-Tuttle Comet (1866 I) which has a relatively short period of 33.176 years, and whose orbit very
nearly intersects that of the Earth. Although some meteors
appear each year in November as the Earth crosses their
path, the finest displays occur only at third-of-a-century
intervals, just after the parent comet has passed.   The
next great star-shower from the Leonid swarm is expected
before dawn on the morning of November 17 or 18, 1999.

Several faint galaxies may be found in the vicinity
of Gamma Leonis; the close pair NGC 3226 and 3227 lies 50'
to the east.

THE LEONID METEOR SHOWER of November 13, 1833, from a contemporary print. The Leonid Swarm also produced spectacular displays in 1799, 1866, and 1966.

DELTA Name- ZOZMA, apparently from the Greek ζωσμα, "The Girdle". Another name, *Duhr*, is from the Arabic *Al Thahr al Asad*, the "Lion's Back", which appears in the manuscripts of Ulug Beg. The name *Zubra* appears on some star maps, from *Al Zubrah*, "The Mane", but this name would seem more appropriate for Gamma Leonis. The Chinese name, *Shang Siang*, has been translated "The High Minister of State". Delta and Theta together formed the Babylonian *Kakkab Kua*, the "Stars of the Oracle of the God Kua"; the same pair, according to R.H.Allen, were titled *Wadha*, "the Wise", by the star-gazers of ancient Sogdiana. The ancient Egyptian name, *Mes-su*, might signify either "The Heart of the god Su" or "The Son of Su".

Delta Leonis is magnitude 2.55, spectrum A4 V; position 11115n2048. The computed distance is about 80 light years, the actual luminosity about 50 times that of the Sun, and the absolute magnitude +0.6. The star shows an annual proper motion of 0.20" in PA 133°; the radial velocity is 12.4 miles per second in approach. The star shows very nearly the same space motion as Sirius, as is also true of Alpha Ophiuchi, Beta Aurigae, and a number of other bright stars at widely separated locations in the sky. (See page 389)

The close double star Σ1517 lies about 23' to the south, and the faint spiral galaxy NGC 3646 is about 2° to the east and slightly south. (On some editions of the *Skalnate Pleso Atlas of the Heavens*, the number is given by error as 3676)

EPSILON Name- RAS ELASED, from the Arabian *Al Ras al Asad al Janubiyyah*, the "Southern Star of the Lion's Head". Magnitude 2.98, spectrum G0 II, position 09430n2400. The computed distance is about 340 light years; this gives an actual luminosity of 580 times that of the Sun and an absolute magnitude of -2.1. The annual proper motion is 0.05"; the radial velocity is 3 miles per second in recession.

ZETA Name- ALDHAFERA, probably through a mistaken translation of another Arabic name originally applied to Epsilon and Mu Leonis. Zeta Leonis is magnitude 3.44, spectrum F0 III, position 10139n2340. Zeta is the bright star which lies a little more than $3\frac{1}{2}°$ north from Gamma in the curve of the Sickle of Leo. Zeta is about 130 light years distant, and has a true luminosity of some 50 times that of the Sun. The absolute magnitude is about +0.5 The star has a slight annual proper motion of 0.02"; the radial velocity is 9 miles per second in approach.

An optical companion for field glasses, the star 35 Leonis, is located 5.5' distant in PA 338°. It is magnitude 5.87, spectrum dG2. The two stars do not form a true pair as the proper motion of 35 Leonis is about 10 times larger than that of Zeta, and in a different direction.

Near Zeta is the small group of galaxies consisting of NGC 3177, 3185, 3190, 3193, and 3187; these are located about midway between Zeta and Gamma Leonis, and are shown in the photograph on page 1080.

ETA (30 Leonis) Magnitude 3.48; spectrum A0 Ib; position 10046n1700. This is the star which joins the blade of the Sickle to the handle, some 5° north of Regulus. Although occupying an important position in the constellation, it appears to have no proper name in any of the standard lists, nor does R.H.Allen mention any in his exhaustive work *Star Names and Their Meanings*. The star is a highly luminous supergiant of class Ib, probably the most remote of the bright stars of Leo. From the spectral characteristics the absolute magnitude is believed to be about -5.5, or about 13,000 times the luminosity of our Sun. The computed distance is then about 2000 light years; if this star were as close to us as Regulus it would appear nearly three times brighter than Jupiter in our skies, and about six times brighter than Sirius.

THETA Name- CHORT; the versions CHERTAN and COXA are also given. The Medieval Arabian name *Al H´aratan* seems to have referred to Theta and Delta together. Magnitude 3.34; spectrum A2 V; position 11116n1542. Theta Leonis is approximately 90 light years distant; the actual luminosity is some 30 times that of the Sun. The star shows an annual proper motion of 0.10"; the radial velocity is 4.7 miles per second in recession. The fine group of spiral galaxies M65, M66, and NGC 3628 will be found about $2\frac{1}{2}°$ to the SE. (Photograph on page 1074)

R LEONIS. The star is shown relatively faint (top) in March 1930, and near maximum (below) in March 1977. Lowell Observatory photographs made with the 13-inch telescope.

R   Variable star. Position 09449n1140. R Leonis
is one of the brightest of the long-period
variable stars, and the fourth star of the type to be discovered, found by J.A.Koch at Danzig in 1782. In February
of that year he noticed that the star was a naked-eye object whereas it had been listed in his own records as a
7th magnitude star only two years earlier. Keeping watch
on the star, Koch soon found it fading, and by the spring
of 1784 it had declined to 10th magnitude. The following
year Koch sent his observations to Bode at the Berlin
Observatory. The three other stars of the type then known
were Omicron Ceti (Mira), Chi Cygni, and R Hydrae.
        R Leonis is located about 5° west of Regulus and just
half a degree SE from 18 Leonis (Mag 5.6, spect gK4); some
10' north of the variable and slightly west is 19 Leonis
(Mag 6.3, spect A3). R Leonis itself forms a triangle of
about 5' on a side with two stars of magnitudes 9.0 and
9.6 lying to the west; the field is one easily remembered
and quickly located, making this one of the most suitable
variable stars for amateur observers.   Leslie C. Peltier
began his notable variable-star career with an observation
of R Leonis, on the night of March 1, 1918.
        A pulsating red giant of the Mira class, R Leonis
sometimes rises to 5th magnitude at maximum and usually
declines to 10th or fainter at minimum. The average period
is 312 days. This star is noted for the peculiar intensity
of its red light, best described as a rosy scarlet with
often a seeming touch of purple. J.R.Hind called it "one
of the most fiery-looking variables on our list- fiery in

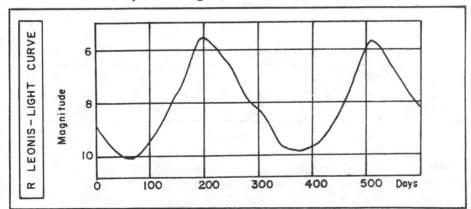

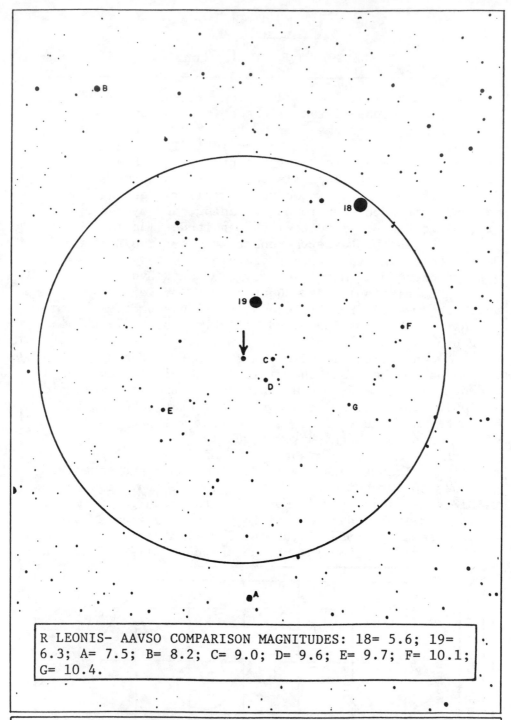

R LEONIS– AAVSO COMPARISON MAGNITUDES: 18= 5.6; 19= 6.3; A= 7.5; B= 8.2; C= 9.0; D= 9.6; E= 9.7; F= 10.1; G= 10.4.

R LEONIS. Identification field, from a plate made with the 13-inch telescope at Lowell Observatory. Circle diameter = 1° with north at the top; limiting magnitude about 15.

every stage, from maximum to minimum, and is really a fine
telescopic object in a dark sky, when its color forms a
striking contrast with the steady white light of the 6th
magnitude star a little to the north". To which T.W.Webb
adds: "One of the finest of its most mysterious class,
which deserves a special investigation, in which amateurs
may do good service without costly instruments."

R Leonis is a late M-type giant, the spectrum varying from M7e to about M9 in the course of the cycle. From
an estimated distance of about 600 light years, the actual
luminosity of the star at maximum would appear to be in
the range of 200- 260 times that of the Sun; the absolute
magnitude is near -1 at peak brightness. The annual proper
motion is 0.05"; the radial velocity is 8 miles per second
in recession. (For a brief account of the long period variables, refer to Omicron Ceti; page 631)

WOLF 359   (LFT 750) Position 10541n0719. This is a
faint but famous star, discovered photographically by Professor Max Wolf at Heidelberg, and at
present the third nearest star known. Only the Alpha Centauri system and Barnard's Star in Ophiuchus are closer to
the Solar System. The distance is 7.75 light years, less
than twice the distance of Alpha Centauri. To locate the
field, center the telescope about 1.4° northwest of the 5th
magnitude star 59 Leonis, then compare the field with the
chart on page 1072 where the circle is 1° in diameter. The
chart was made from a plate obtained in March 1959, so
observers of the future will find it necessary to make a
correction for the large proper motion of the star (4.71"
annually in PA 235°) which amounts to nearly 8' in the
course of a century. It will be nearly 400 years, however,
before the star reaches the edge of the circle drawn on
the chart.

Wolf 359 is an extremely faint red dwarf, one of
the least luminous stars known. It has a spectral type of
dM6e and an apparent visual magnitude of 13.66; the true
luminosity is about 1/63,000 that of the Sun, and the absolute magnitude is +16.8. Stars of such intrinsic faintness
are probably very common in space, but very few are actually known; obviously only the nearby examples can be seen at
all. Other well known examples of such miniature stars are

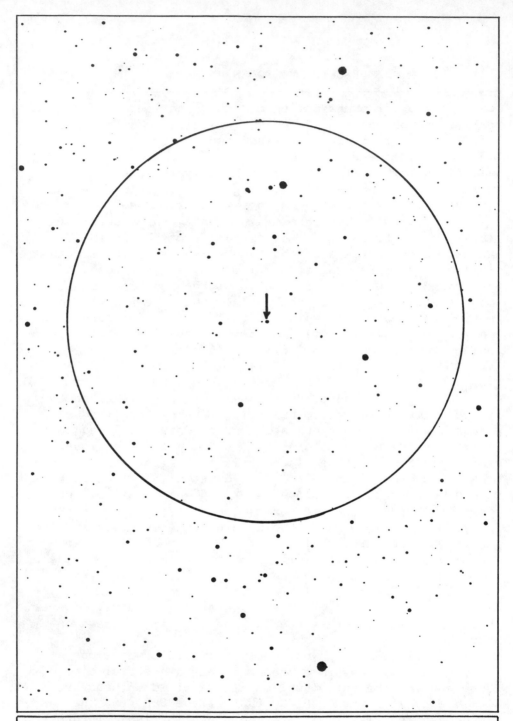

WOLF 359. Identification field for the famous red dwarf, from a Lowell Observatory 13-inch telescope plate. Circle diameter = 1°, north at the top, limiting magnitude 15.

DESCRIPTIVE NOTES (Cont'd)

Ross 614B in Monoceros, Van Biesbroeck's Star in Aquila, and Proxima Centauri. These objects all have masses close to 10% of the solar mass; in the case of Ross 614B, whose mass is accurately known from its orbital elements, the mass is 8% that of the Sun. Wolf 359 is probably quite comparable in size and mass; its diameter is estimated to be roughly the size of Jupiter, or about 10% the diameter of the Sun. Like several other extremely faint red dwarf stars, Wolf 359 is now known to be a flare star. Its sudden outbursts appear to be considerably rarer than those of Proxima Centauri and Krueger 60B in Cepheus, and the amplitude does not appear to exceed one magnitude.

The radial velocity of the star is 7.8 miles per second in recession; the true space velocity is about 33 miles per second. (Refer also to Proxima Centauri, Krueger 60 in Cepheus, and L726-8 in Cetus)

M65 + M66   (NGC 3623 and NGC 3627) Mean position 11170n1320. A fine pair of spiral galaxies lying about midway between the stars Theta and Iota Leonis, about 2½° to the SSE of Theta. They are 21' apart, and can be seen together in the field of a low-power telescope. Both can be detected in good binoculars on a clear night, and form a rather striking pair in the large 20 X 70mm or 20 X 80 binoculars which are becoming increasingly available to modern sky-watchers. M65 and M66 were discovered by P.Mechain in March 1780; it seems that Messier's Comet of 1773 passed directly through the field on November 2, 1773, but the two galaxies were not noted by that diligent observer; Mechain theorized that Messier "no doubt missed them because of the light of the comet". M66 is visually the brighter of the two, with a total magnitude of about 9.7; some observers, however, find M65 actually more conspicuous owing to its more elongated outline.

Both galaxies may be outlying members of the Virgo Cluster of Galaxies which is centered some 15° away, but they seem to be on the near side of the Virgo group, since the measured red shift (about 380 miles per second) is only about half the mean value for members of the Virgo group. This suggests that M65 and M66, together with a few faint field galaxies, possibly constitute a small independent cluster rather than a sub-cluster of the Virgo Cloud.

THE M66 GROUP IN LEO. M66 is at lower left, M65 at lower right, and NGC 3628 at top. This photograph was made with a 12½-inch reflector by Evered Kreimer of Prescott, Arizona

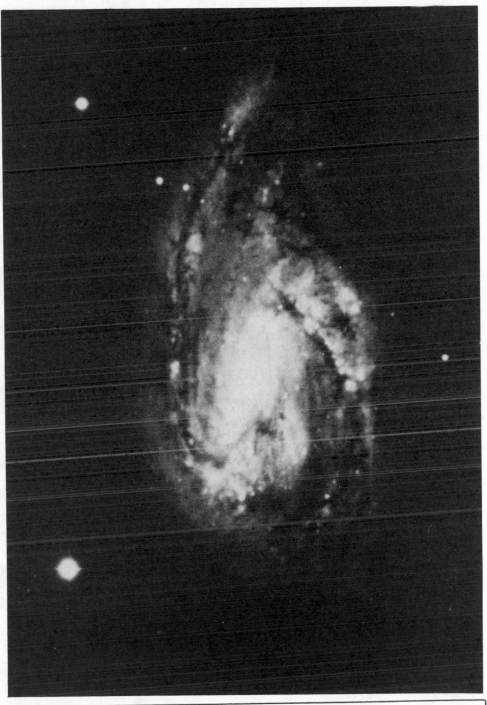

SPIRAL GALAXY M66 in LEO. A galaxy with a rather thick and asymmetric pattern of spiral arms. U.S.Naval Observatory photograph with the 61-inch reflector.

1075

SPIRAL GALAXY M65.   A bright, highly tilted spiral in Leo.
Photograph made with the 61-inch astrometric reflector of
the U.S.Naval Observatory at Flagstaff.

DESCRIPTIVE NOTES (Cont'd)

Both galaxies are spirals of class Sb, though rather different in appearance. M66 is an impressive object with its heavy dust lanes and thick spiral arms studded with coarse masses of star clouds. The largest arm, on the SE side, resembles a huge crab's claw; fainter arms can be traced out for vast distances. Admiral Smyth called this system "a large elongated nebula with a bright nucleus trending N.p. and S.f." Isaac Roberts, on plates made with his 20-inch reflector in the 1890's, described it as spiral "with a well-defined stellar nucleus, forming the pole of the convolutions, in which I have counted 14 nebulous star-like condensations". Lick Publications XIII calls M66 a "very bright, beautiful spiral 8' x 2½' in PA 180°. Bright, slightly elongated nucleus; the whorls are somewhat irregular and show numerous condensations".

The companion galaxy, M65, is a more normal system in appearance, somewhat fainter, and with a smooth and regular pattern of spiral coils. A fine narrow dust lane rims the entire east edge, which is undoubtedly the nearer side. Sir William Herschel found M65 "a very brilliant nebula, extended in the meridian, about 12' long. It has a bright nucleus, the light of which suddenly diminishes on its border, and two opposite faint branches". Lord Rosse in 1848 identified it as a spiral "resolved very well about the nucleus but no other part". D'Arrest reported the impression that the system was "resolved into faint stars" with 147X on his telescope. Modern measurements show that M65 is a spiral tilted about 14° from the edge-on position.

In describing the early photographs made of this pair of galaxies, Miss Agnes Clerke (1905) referred to them as "elliptical spirals, analogues of the colossal structure in Andromeda. Photographed by von Gothard in 1888, the one (M65) showed a bright center with four appendages resembling the sails of a windmill; the other (M66), a complex arrangement of envelopes partially surrounding a nucleus, somewhat like the paraboloidal veils flung around the head of a comet near perihelion. These, however, were incomplete views. On Dr. Roberts' plates, both objects took shape as ovoid formations. composed of closely winding luminous coils thick inlaid, in the case of M66, the chief of the pair, with nebulous condensations." T.W.Webb refers to the two as merely "two rather faint objects, elongated visually

in different directions, in a low-powered field, with several stars".

Just 35' north of M66 lies a third object, the edge-
on galaxy NGC 3628, characterized by a fine narrow dust
lane. Larger but fainter than M65 or M66, and not noticed
by Messier, it measures some 12' in length. These three
galaxies form a noble group for the small telescope.

The currently accepted value for the Hubble constant
(1976) implies a distance of about 38 million light years
for this group, which seems somewhat too large; E.Holmberg
has published a figure of about 29 million light years,
giving about 50,000 light years as the diameter of M66 and
about 60,000 for M65. The absolute magnitude in each case
is close to -21; their separation about 180,000 light years.

M95 + M96   (NGC 3351 and NGC 3368) Positions 10413n
1158 and 10442n1205. A pair of bright
galaxies lying near the center of Leo, about 9° east of
Regulus; their separation is 42', M95 being the western
member of the pair. Both were discovered by P.Mechain in
March 1781, and described as "nebula without star", M95
being recorded as slightly more distinct than M96. Admiral
Smyth found M95 to be a "lucid white nebula.....round and
bright", while M96 was "large and pale white". On modern
photographs, M95 is seen to be a fine barred spiral with
a bright center and a structure resembling the Greek letter
Theta. M96, classed as an Sa system, has a large amorphous
central region and a multiple assortment of faint spiral
arms. A very faint outer ring, 6' in diameter, encircles
the whole system, contacting the inner pattern on the NW
side. The two galaxies appear to be members of the same
sub-cluster, sometimes called the "Leo Galaxy Group", which
includes M65 and M66; the presently accepted distance is
about 29 million light years, and the corrected red shift
about 420 miles per second.

About 48' to the NNE from M96 is the E1 galaxy NGC
3379 (sometimes called M105) and its two companions NGC
3384 and 3389; the three objects form a little triangle
about 8' on a side. These objects are evidently dynamical
members of the M95-M96 group, with a projected separation
from the bright pair of about 400,000 light years. (Refer
also to M65 and M66, page 1073)

SPIRAL GALAXIES IN LEO. Top: The barred system M95. Below:
The Sa-type spiral galaxy M96.

200-inch telescope, Palomar Obsvt.

THE NGC 3190 GROUP IN LEO. The elliptical galaxy 3193 is at
lower left; 3187 at upper left, and 3185 at upper right.
North is at the left. Palomar Observatory photograph.

SPIRAL GALAXY NGC 2903 in LEO. A bright many-armed spiral photographed with the 120-inch reflector at the Lick Observatory.

SPIRAL GALAXY NGC 3521 in LEO. A fine multiple-arm spiral
photographed with the 61-inch astrometric reflector of the
United States Naval Observatory at Flagstaff, Arizona.

OFFICIAL U.S. NAVY PHOTOGRAPH

1082

LEO MINOR

LIST OF DOUBLE AND MULTIPLE STARS

NAME	DIST	PA	YR	MAGS	NOTES	RA & DEC
Σ 1344	3.7	105	41	8½- 9	relfix, spect F2	09203n3921
11	2.0	31	37	5½- 14	(Hu 1128) Dist dec, spect G8	09327n3603
Σ 1374	2.8	298	59	7- 8½	PA slow inc, spect G2	09383n3911
Σ 1375	7.0	306	42	8 - 10	relfix, spect A2	09389n3448
Σ 1382	27.8	107	26	7- 10½	relfix, spect A3	09461n3419
	32.7	255	26	- 11½		
Ho 369	0.3	281	60	7½- 7½	PA inc, spect F2	09482n3643
h2517	43.8	151	26	7 - 12	cpm; spect F5	10009n3816
Hu 631	0.7	260	68	7- 8½	PA dec, spect F8	10011n3253
A2142	0.9	304	60	7½- 8½	Slight PA dec, spect F0	10027n4117
Σ 1406	1.0	227	56	8 -8½	relfix, spect A3	10028n3120
Σ 1405	22.0	251	26	7 - 10		10029n3950
h475	0.2	174	58	6½- 7	(Kui 48) spect F5	10054n3152
	27.6	172	58	- 13		
Hu 634	1.7	169	52	8½- 9	relfix, spect F2	10134n3324
Hu 875	1.0	67	54	7 - 10	PA slight dec, spect A2	10154n3746
Σ 1420	2.5	326	34	8 - 10	relfix, spect G5	10156n3922
β	0.4	223	67	4½- 6½	(Hu 879) binary, 38 yrs, PA inc, spect G8	10250n3659
Σ 1443	5.1	160	56	9 - 9	Slight PA inc, spect G0	10304n3756
Σ 1454	2.1	329	60	7½- 10	cpm; spect K0; PA inc, dist dec.	10354n2652
Σ 1459	5.4	153	56	8 - 8½	relfix, spect K0	10374n3840
40	18.4	112	58	6 - 13	(β913) PA dec,	10403n2635
	32.5	78	23	- 13	dist inc, spect A2	
	54.1	278	19	- 13		
Σ 1492	21.5	164	18	8 - 9½	spect A2	10549n3055

LEO MINOR

LIST OF VARIABLE STARS

NAME	MagVar	PER	NOTES	RA & DEC
R	6.3--13..	372	LPV. Spect M7e- M8e	09426n3445
S	8.0--13.5	234	LPV. Spect M4e	09508n3510
U	9.8--13..	272	Semi-reg; spect M6e	09517n3620
RS	8.1--9..	Irr	Spect M7	09225n3622
RV	8.8---9.2	55	Semi-reg, spect M5	10206n3006

LIST OF STAR CLUSTERS, NEBULAE, AND GALAXIES

NGC	OTH	TYPE	SUMMARY DESCRIPTION	RA & DEC
2859	137[1]	⊘	SBa; 12.1; 4.0' x 3.0' vB,pL,lE,smbM, faint outer ring	09213n3444
2942		⊘	Sb/Sc; 12.9; 1.6' x 1.2' F,pL,vlE, vglbM	09362n3414
2955	541[3]	⊘	Sb; 13.1; 1.2' x 0.6' cF,pS,mE,glbM	09383n3607
3003	26[5]	⊘	SBc; 12.1; 5.0' x 0.9' ! cB,L,vmE, nearly edge-on spiral	09456n3339
3021	115[1]	⊘	Sb; 13,0; 1.2' x 0.6' pB,pS,lE,mbM, 10^m star 1' south following	09480n3347
3158	639[2]	⊘	E2; 12.9; 0.6' x 0.5' cB,cS,R,psbM. Brightest member of small group, incl 3150, 3151, 3152, 3159, 3160, 3161, 3163.	10109n3900

LIST OF STAR CLUSTERS, NEBULAE, AND GALAXIES (Cont'd)

NGC	OTH	TYPE	SUMMARY DESCRIPTION	RA & DEC
3245	86[1]	⊖	E5/S0; 11.8; 1.8' x 0.9' vB,pL,mE, smbMN	10245n2846
3254	72[1]	⊖	Sb; 12.5; 4.3' x 1.0' cB,L,mE,psmbMN; nearly edge-on spiral	10265n2945
3277	359[2]	⊖	Sb; 12.5; 1.0' x 0.9' cB,cS,R,pgmbM	10302n2846
3294	164[1]	⊖	Sc; 11.6; 2.7' x 1.2' cB,L,mE,glbM	10334n3735
3344	81[1]	⊖	Sc; 11.0; 6.0' x 5.1' cB,L,gbM, fine face-on spiral	10407n2511
3395	116[1]	⊖	Sc; 12.4; 1.4' x 0.8' cB,pS,E; pair with 3396	10471n3315
3396	117[1]	⊖	I/SB pec; 12.8; 1.0' x 0.5' pB,pS,lE; interacting pair with 3395, separation 1.7'	10472n3316
3414	362[2]	⊖	S0; 12.1; 1.5' x 1.1' B,pL,lE,mbM, lenticular	10486n2815
3430	118[1]	⊖	Sc; 12.2; 3.2' x 1.8' pB,L,mE,gbM; 5.5' sp is 3424: pF,pL, nearly edge-on spiral	10495n3314
3432	172[1]	⊖	Sc; 12.0; 5.9' x 0.8' pB,pL,vmE, flat streak; nearly edge-on spiral	10497n3654
3486	87[1]	⊖	Sc; 11.2; 5.5' x 4.2' cB,cL,lE,gmbM,BN; fine multiple-arm spiral	10578n2915
3504	88[1]	⊖	Sb; 11.6; 2.0' x 1.8' B,L,E,vmbM; 3512 nf 12'	11005n2815
3510	365[2]	⊖	Sc; 12.8; 3.5' x 0.4' F,L,cE,8^m star 8' np	11010n2909
3512	366[2]	⊖	Sc; 12.8; 1.1' x 1.0' F,pS,R,pgbM; compact spiral	11013n2818

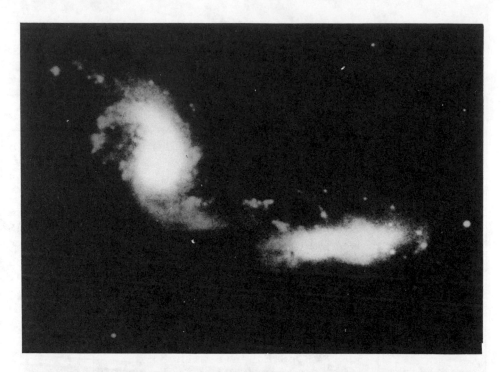

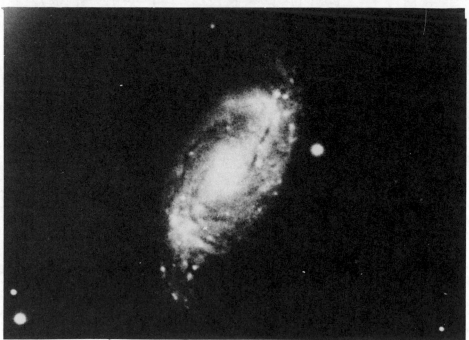

GALAXIES IN LEO MINOR. Top: The interacting pair NGC 3395 and 3396. Below: The barred spiral NGC 3504.

Palomar Observatory

LEPUS

LIST OF DOUBLE AND MULTIPLE STARS

NAME	DIST	PA	YR	MAGS	NOTES	RA & DEC
h3705	22.4	141	20	7½- 9½	spect A0	04546s1613
β 314	0.8	318	61	6½- 7	binary, 56 yrs;	04568s1627
	52.9	34	14	- 8	PA dec, spect dF2	
A2629	0.1	110	60	8 - 8	PA dec	04587s1117
λ44	0.7	336	60	7½- 10	dist dec, spect F2	04592s2347
β884	0.6	197	42	8 - 8	relfix, spect A0	05006s1231
λ47	3.7	60	33	6½- 13	PA slight inc, spect F0	05076s2233
ι	12.8	335	33	4½- 10½	(Σ655) (3 Leporis) relfix, spect B8	05100s1156
I 734	1.4	76	42	8½- 9½	PA slight inc, spect G0	05103s2459
ϰ	2.5	360	59	4½- 8	(Σ661) relfix, spect B8	05109s1300
β317	8.6	12	38	7½- 11	spect G5	05118s2302
	18.1	47	33	-12½		
S473	20.3	307	53	8 - 10	relfix, spect B8	05155s1516
S476	39.2	18	52	6½- 6½	wide cpm pair, relfix, spect B8	05171s1834
h3750	4.1	281	59	5 - 9½	relfix, spect A0	05183s2117
h3752	3.1	97	53	5½- 6½	AB cpm, PA dec,	05198s2449
	58.8	106	00	- 9	spect G7, A3, K0. Globular M79 is 35' to ENE	
Hn 74	6.2	234	48	9½- 9½	relfix	05211s1714
Cor 31	18.1	283	33	7½- 9½	spect A3	05215s2221
Σ710	10.8	195	35	8 - 8½	relfix, spect B9,	05229s1121
	10.0	222	05	- 13½	AB cpm.	
	40.0	88	00	- 13½		
h3759	27.1	318	18	6½- 8	slow dist dec, spect dF4, dF5	05239s1944
β319	4.3	231	39	7½- 10	slight dist inc, spect A3	05243s2045
β	2.5	330	57	3 - 11	(β320) (h3761) (*)	05261s2048
	64.3	145	21	- 11	PA inc, spect G5	
α	35.5	156	20	3 - 11	(h3766) (11 Lep) (*)	05305s1751
	91.4	186	13	- 12	relfix, spect F0	
h3770	4.2	18	59	8 - 12	slight PA inc, spect F8	05314s2422

LIST OF DOUBLE AND MULTIPLE STARS (Cont'd)

NAME	DIST	PA	YR	MAGS	NOTES	RA & DEC
h3780	0.9	147	47	7 - 8	Multiple group =	05371s1753
	89.1	137	16	- 9	cluster NGC 2017;	
	76.0	7	15	- 8	primary spect B9	
	129	299	16	-8½	(*)	
	60.3	49	16	-10		
h3780c	1.6	357	46	9- 9½		
β322	2.5	104	42	8 - 9½	relfix, spect B9	05375s2510
Hd 78	11.8	124	61	7½- 8½	relfix, spect B8	05376s2028
	32.2	83	45	- 11½		
h3788	25.9	154	29	8- 9½	spect F5	05393s2622
h3791	10.9	58	40	8 - 9	relfix, spect A2	05411s2041
γ	94.9	351	59	3½- 6	(S498) wide color-contrast pair (*)	05424s2228
γᵇ	45.0	345	00	6 - 11		
Ho 336	19.4	238	45	7 - 12	relfix, spect B3	05427s2141
h3798	15.4	69	18	8½- 9	spect F2	05452s2431
β405	14.1	126	01	8½- 11	cpm; spect K2	05456s1332
h3799	3.9	151	47	8- 8½	relfix, spect A0	05462s1843
β94	2.3	173	43	6- 9½	PA dec, spect G5	05473s1430
A2512	0.9	278	59	7 - 8½	relfix, spect B9	05490s1139
A3019	44.8	42	33	7½- 11	spect B9	05522s1544
A3019b	3.2	4	33	11- 13		
GAn2	9.8	20	41	7½-10½	spect G5	05523s1943
S504	3.5	68	57	8½- 8½	(Lal 43) PA slow dec, spect F0	05563s2010
h3821	18.0	213	20	8 - 9½	spect A2	05587s2100
	19.8	87	00	-12		
B96	2.2	131	59	5 - 13	spect gK3	06012s2617
h3833	44.9	72	20	6 - 10	spect A2	06044s2306
h3835	30.2	83	35	7½ -11	spect A0	06048s2305
β17	3.2	184	35	6½-10½	(4 Monoc) spect A2	06061s1108
	9.0	249	33	- 11½		
Σ875	5.2	327	58	8½- 10	PA dec, spect K0	06098s1308

LIST OF VARIABLE STARS

NAME	MagVar	PER	NOTES	RA & DEC
R	5.9--11..	432	LPV. Spect N6e; very red "Crimson Star" (*)	04573s1453
S	6.4--7.7	90	Semi-reg; spect M6	06037s2411
T	7.4--13.5	368	LPV. Spect M6e--M8e	05027s2158
U	9.5--10.4	.5815	Cl.Var; Spect A7--F6	04541s2117
V	9.8--10.5	1.070	Ecl.Bin; lyrid, spect A5	06088s2012
W	9.9--11..	485	Semi-reg	06077s2231
X	9.5--14..	279	LPV. Spect M7e	05161s1624
Y	9.0--11..	109	Semi-regl; Spect M4	05068s2429
RS	9.3--11.5	1.288	Ecl.Bin; spect A0	05572s2013
RT	9.5--11..	399	LPV.	05405s2344
RX	5.9--7.0	Irr	Spect gM6	05090s1154
RY	8.2--9.1		Ecl.Bin? Spect F0	05460s2002
SS	4.8--5.1	Irr	Symbiotic, spect A0e+gM1	06027s1629

LIST OF STAR CLUSTERS, NEBULAE, AND GALAXIES

NGC	OTH	TYPE	SUMMARY DESCRIPTION	RA & DEC
1744		⊖	SBc; 12.3; 7.0' x 3.3' F,vL,vmE,vglbM	04579s2606
1784		⊖	SBc; 12.6; 3.8' x 2.2' pB,pL,cE,vgbM	05032s1156
1832	292²	⊖	Sc; 12.3; 2.1' x 1.1' pB,lE,mbM	05100s1547
1904	M79	⊕	Mag 8.4, diam 7½', cRi,eC, class V, stars vF (*)	05222s2434
----	I.418	◎	vS,B, 11^m star in nebulous disc, diam 14" x 11"	05254s1244
1964	21⁴	⊖	Sb; 11.8; 5.0' x 1.6' F,pL,vsmbM, mE	05312s2159
2017		⠐	L, scattered group = multiple star h3780 (*)	05371s1753
2139	264²	⊖	Sc pec; 12.2; 1.8' x 1.5' F,S,1E	05590s2349
2179		⊖	Sa; 13.0; 0.9' x 0.7' F,pS,1E,glbM	06059s2144
2196	265²	⊖	Sb; 12.6; 1.8' x 1.4' pF,pS,vlE, pmbM	06101s2147

ALPHA   Name- ARNEB or ARSH. Magnitude 2.58;
             spectrum F0 Ib. Position 05305s1751. The
computed distance of the star is about 900 light years,
giving an actual luminosity of about 5700 times that of the
Sun. The annual proper motion is less than 0.01"; the
radial velocity is 15 miles per second in recession.
        A faint companion star, of the 11th magnitude, is
probably not a true physical companion; the separation of
35.5" has not changed since the early measurements of John
Herschel in 1835. R.H.Allen calls the colors of this pair
"pale yellow and gray". About 1.7° east of Alpha lies the
multiple star h3780, also listed as an open cluster under
the designation NGC 2017. (See photograph, page 1100)

BETA   Name- NIHAL. Magnitude 2.85; spectrum G5 III
             Position 05261s2048. The distance of the
star is about 115 light years; the actual luminosity about
70 times that of the Sun. The spectral characteristics
suggest an absolute magnitude of +0.1. Beta Leporis shows
an annual proper motion of 0.08"; the radial velocity is 8
miles per second in approach.
        The star is a close and unequal double, difficult
for any but large telescopes. The companion is an 11th
magnitude star at 2.5" distance; owing perhaps to the
observational difficulties there are large discrepancies
in the reported magnitudes of the small star. The Lick
Observatory "Index Catalogue of Visual Double Stars" gives
the figure as 7.5. A slow decrease in the separation of
this pair has been measured since discovery by S.W.Burnham
in 1875, and the P.A. has been increasing at the rate of
about 40° in the last 55 years. The orbital period remains
uncertain, but must be at least several centuries. The
color of the small star is described as blue by R.H.Allen,
possibly from the effect of contrast with the fine deep
yellow tint of the primary. Allen also makes the curious
remark, attributed to Holden, that the companion is "suspected to be a planet", a suggestion which could never have
been taken seriously unless the star was thought to be
much closer than Alpha Centauri! At the accepted distance
the computed luminosity is about 1/25 that of the Sun, a
normal luminosity for a dwarf star of about type K.

GAMMA   Magnitude 3.60; spectrum F6 V, Position
05424s2228. Gamma Leporis is a wide double
star with a pleasing color contrast, an easy and appealing
object for even the smallest telescopes. The components
show the same proper motion of 0.47" per year in PA 219°,
and are therefore known to form a true physical pair though
there is no evidence for orbital motion. The space motion
of the pair closely matches that of Sirius, as is also true
of a number of other bright stars in various parts of the
sky.

The individual magnitudes of the stars are 3.59 and
6.18; spectra F6 V and dK2. Parallax measurements lead to
a distance of 29 light years; the apparent separation of
95" then corresponds to about 900 AU. The primary is a main
sequence star with about 3 times the solar luminosity; the
radial velocity of the system is 6 miles per second in
approach.

As with many color-contrast pairs, descriptions
vary considerably. Smyth, strangely enough, reported the
companion to be pale green; T.W.Webb called them pale
yellow and garnet, while E.J.Hartung in his "Astronomical
Objects for Southern Telescopes" (1968) describes them as
yellow and orange. "This brilliant wide pair", says Hartung
"makes a beautiful low power object in a field sprinkled
with many stars".

A possible third component to the Gamma Leporis
system was discovered by Van Biesbroeck in 1956; it is a
faint red star of the 16th magnitude, 18.8' distant in PA
67°. This star, however, has a distinctly larger motion
(about 0.63" annually) than the bright pair, and a PA about
10° lower; at a separation of at least 9000 AU these differences would give the faint star a relative velocity well
in excess of the escape velocity of the system. Evidently
the third star cannot be a true member of the Gamma Leporis
system, despite the general similarity of the proper
motion.

Another component, an 11th magnitude star at 45"
from Gamma B, is also found to be merely an optical companion which the bright motion pair is presently passing.

EPSILON   Magnitude 3.20; spectrum K5 III. Position
05034s2226.   The distance of the star is
about 170 light years; the actual luminosity about 120
times that of the Sun.   The annual proper motion is 0.08";
the radial velocity is less than 1 mile per second   in
recession.

MU   Magnitude 3.29; spectrum B9 or A0p;
Position 05107s1616.   The computed dis-
tance is about 390 light years, giving an actual luminosity
of about 580 times that of the Sun.   The star shows an
annual proper motion of 0.05"; the radial velocity is 16
miles per second in recession.

17   Magnitude 5.04 (slightly variable).
Position 06028s1629.   Spectrum given by
various authorities as B9, A0, or A2p with superimposed
features of gM1.   An emission type star or "shell star"
related apparently to the erratic nova-like variables or
the symbiotic stars.   17 Leporis differs from these classes
however, in showing no large changes in light. Variations
of about 0.07 magnitude have been measured. Spectroscopic
studies reveal a turbulent atmosphere whose outer layers
appear to be in a continual state of violent expansion with
a velocity of 35- 40 miles per second. The spectrum shows
wide bright lines and peculiar variations, at times resembling the spectrum of a nova outburst.   Several times a
year there may be periods of unusual activity, and it is
estimated that the star loses about one-millionth of its
mass each year in this manner.

   Spectral features characteristic of a red giant
companion were first noted by A.Slettebak (1950) and it now
appears that 17 Leporis is actually a binary system strongly resembling such giant pairs as T Coronae Borealis and
Z Andromedae.   From the studies of K.G.Widing (1965) and
A.P.Cowley (1966) the period of the system appears to be
about 260 days, with probable masses of 1.4 and 4.6 suns
for the red and blue components, respectively. Accepting a
diameter of about 75 suns for the red giant, the computed
orbit places the components so near each other that some
form of mass exchange must occur; thus it seems that the

DESCRIPTIVE NOTES   (Cont'd)

gaseous shell of the blue star is composed of material ejected by the red companion. A rough periodicity in the system's major outbursts appear to be correlated with the orbital period of the pair.

The distance of the star is uncertain, but the best parallax measurements seem to place it at less than 300 light years. The value of 0.023" reported in the "Yale Catalogue of Bright Stars" (1964) is equivalent to 141 light years, but the spectroscopic features suggest an absolute magnitude of about +0.5 which would imply a some- what greater distance of about 260 light years. The true luminosity is thus in the range of 60- 100 times that of the Sun, fairly comparable to the minimal magnitude of the T Coronae Borealis system. These two binary systems, in fact, seem so similar in type, mass, and period, that some speculation concerning the future career of 17 Leporis seems justified.   T Coronae is, of course, the brightest of the recurrent novae, with violent outbursts in 1866 and 1946. We might wonder whether 17 Leporis will eventually develop into a true recurrent nova (with an expected maximum apparent magnitude of about -5!) or whether the star has perhaps already passed through such a stage at some time in the past.

The annual proper motion of 17 Leporis is quite small, measured at 0.006"; the radial velocity is about 12 miles per second in recession. (Refer also to T Coronae Borealis, Z Andromedae, R Aquarii, and RS Ophiuchi)

R   Position 04573s1453. Hind's "Crimson Star", a famous long period pulsating red variable. It was first seen by J.R.Hind of London in October 1845, and the variability was detected by Schmidt from observations made between 1852 and 1855. The light range is about 100 times in a period averaging 432 days. At maximum the star may rise to nearly naked-eye visibility, though these periods of greatest brilliancy have been scarce in recent years. The light curve is somewhat abnormal, showing a temporary standstill on the rising branch, about 100 days after minimum.

The color of this remarkable star is an intense smoky red, described by various observers as resembling a glowing coal, a ruby, or an illuminated drop of blood.

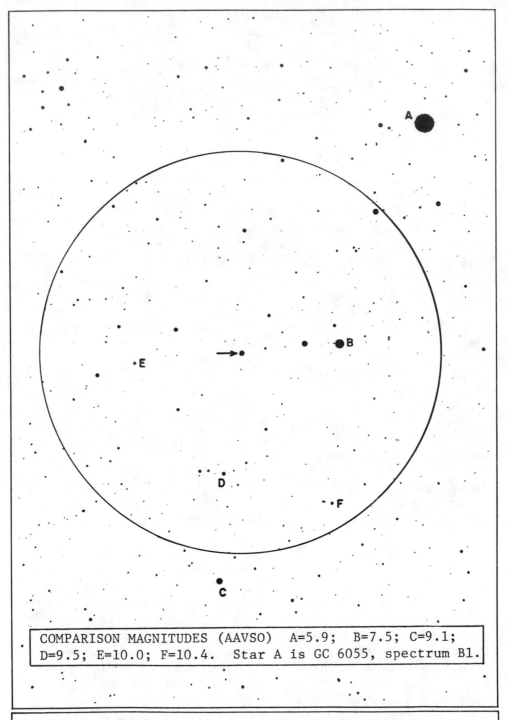

COMPARISON MAGNITUDES (AAVSO) A=5.9; B=7.5; C=9.1;
D=9.5; E=10.0; F=10.4. Star A is GC 6055, spectrum B1.

R LEPORIS IDENTIFICATION FIELD. The circle is one degree in
diameter with north at the top. Limiting magnitude about 14
Chart made from a Lowell Observatory 13-inch telescope
plate.

Hind himself found it "of the most intense crimson, resembling a blood-drop on the background of the sky; as regards
depth of color, no other star visible in these latitudes
could be compared with it." E.J.Hartung (1968) mentions it
as gleaming "like a crimson jewel in a field well sprinkled
with stars" while Miss Agnes Clerke (1905) remarked that
even the colors of Antares and Betelgeuse were "mere pale
shades" when compared with the wine-red hue of R Leporis.
"As with most other variables, however, increase of light
brings with it a paling of colour. Near maximum, intense
redness gives place, partly through a well-known physio-
logical effect, to a coppery hue. Its spectrum is of the
fourth type, with particularly strong absorption of the
blue rays, a small proportion of which penetrate its dense
veil of carbonaceous vapors."

The phrase "spectrum of the fourth type" as used
by Miss Clerke, refers to the rare objects later assigned
to spectral class N, the "Carbon Stars". Another class,
"R" was created to cover a few stars of rather similar type
but having higher temperatures, weaker colors, and a lower
carbon content. R Leporis itself is classed as N6, while
the star S Camelopardi is a typical example of type R. In
recent years, a combining of the two classes has been

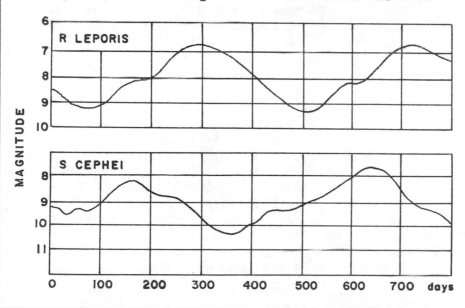

suggested, into a single class "C", the Carbon stars, with a sub-script added to indicate the carbon abundance.   R Leporis, on this system, is classed as $C7_4e$.

In all the stars of the type, the spectra show strong bands caused by carbon compounds, indicating temperatures of 2600° K and cooler. The extreme absorption in the blue and violet portion of the spectrum appears to be due to the abundance of the tri-atomic molecule Carbon-3. The opinion was formerly held that such stars were "dying suns, smothered in absorbing vapors", but they are now known to be pulsating red giants of the same general type as Mira. However, the view that an N-type star is merely an extreme example of an M-type star, differing only in temperature, is probably incorrect. Current studies seem to confirm real differences in chemical composition and probably indicate a somewhat different evolutionary history.

As none of these stars are sufficiently near to provide accurate parallaxes, the true luminosities are not precisely known. From statistical studies, the maximum absolute magnitude of an N-type star is thought to be in the range of -1.5 to possibly -2.0; this suggests a distance of about 1500 light years for R Leporis. The star shows an annual proper motion of 0.03"; the radial velocity is 20 miles per second in recession. A very similar star in the northern sky, S Cephei, appears to be somewhat more remote, with a computed distance of about 2000 light years. Possibly the nearest star of the type is Y Canum Venaticorum (page 361) whose distance is estimated to be somewhere in the range of 400 to 600 light years.

At the 1963 Spring meeting of the AAVSO, evidence was presented by T.A.Cragg that R Leporis possibly oscillates in two superimposed periods. The similar star V Hydrae, for example, has a period of about 530 days, but this cycle is superimposed upon a longer wave with a period of about 18 years. R Leporis itself appears to have had a number of faint maxima before 1910, then a series of brighter maxima until the late 1940's, followed by another faint series from 1950 to 1963. From 1956 to 1963 the star did not rise above magnitude 9.3, and went to 11.3 on at least two occasions. In 1970 it seemed evident that the maxima

R LEPORIS. The exceptionally red color of this star is here illustrated by a comparison of red and blue exposures. The red photograph is at the top.   Palomar Observatory.

DEEP SKY OBJECTS IN LEPUS. Top: The globular cluster M79.
Below: The spiral galaxy NGC 1832.
                    Mt. Wilson and Palomar Observatories

DESCRIPTIVE NOTES (Cont'd)

were once again increasing in brightness, and the star
should now be observed very carefully in all its subsequent
cycles. From all the observations compiled to date, it
would appear that the longer cycle, if it is real at all,
must be at least 20 years, and possibly more than 40.

(Refer also to S Cephei, V Hydrae, TX Piscium, and
Y Canum Venaticorum. The classic long-period variable star
Mira is described on page 631)

M79   (NGC 1904) Position 05222s2434. Globular star
cluster, discovered by Mechain in October 1780
and reobserved by Messier in December of the same year. It
lies about 34' ENE of the double star h3752, which in turn
is easily located by drawing an imaginary line from Alpha
through Beta Leporis and extending it out 4.4° to the
south. M79 is not one of the more brilliant globulars, and
becomes a truly impressive object only in rather large
telescopes. Sir William Herschel found it "beautiful and
extremely rich" in his 20-foot telescope, while both Smyth
and Webb observed it as a round "nebula, blazing toward the
center". Some appearance of mottling becomes evident with
higher powers, while an 8" to 10" instrument begins to
resolve the outer edges into a hazy sprinkling of 14th
magnitude stars. Large apertures reveal a rich crowding of
images toward the center; the compact central mass measures
about 2' in diameter.

According to Sawyer's "Bibliography of Individual
Globular Star Clusters" (First Supplement, 1963) the full
angular diameter of M79 is about 7.8', equivalent to about
110 light years at the adopted distance. The total integrated magnitude is 8.39, and the combined spectral type is
F3. From several color-magnitude studies, the distance
modulus appears to be close to 16 magnitudes, giving a
true distance of about 50,000 light years; the resulting
absolute magnitude is -7.6. This is equivalent to the
light of 90,000 suns. Only five variable stars have been
detected in M79, a small total compared to the nearly 200
now recognized in the cluster M3 in Canes Venatici. M79,
like many globulars, shows a very high radial velocity,
118 miles per second in recession. (For a general account
of the globular clusters, refer to M13 in Hercules)

DEEP SKY OBJECTS IN LEPUS. Top: The multiple star h3780 as photographed with the 13-inch telescope at Lowell Observatory. Below: Galaxy NGC 1964, photographed at Palomar.

LIBRA

LIST OF DOUBLE AND MULTIPLE STARS

NAME	DIST	PA	YR	MAGS	NOTES	RA & DEC
h2714	30.7	283	59	7½- 12	PA & dist inc, spect K0	14212s1935
Σ1837	1.2	285	68	6½- 8½	cpm, PA dec, spect both dF1	14220s1127
β225	35.0	295	55	6 - 6	spect A8	14226s1944
β225b	1.2	93	55	6½- 8	BC PA dec	
h4679	16.6	306	33	8 - 9	relfix, spect F2	14231s2154
β117	2.2	84	60	7 - 9	PA dec, spect G5	14286s1524
	107	335	11	- 12		
Hu 140	1.2	190	59	8½- 9	PA inc, spect G5	14298s1247
5	3.1	249	55	6½- 11	(Hn 20) cpm pair, spect gK2	14432s1515
β346	2.0	269	60	7 - 8	cpm, PA & dist inc, spect G0	14457s1708
Hu 141	0.1	258	58	8 - 8½	PA dec, spect F8	14465s1037
μ	1.8	355	58	5½- 6½	(β106) slight dist inc, PA inc, spect A2p	14466s1357
α	231	314	13	3 - 5	(Shj 186) wide cpm pair, spect A?, F5	14481s1550
IIh 457	21.8	302	59	6 - 7½	(P212) (Sh 190)	14545s2111
	45.4	192	11	- 13	AB cpm, dist inc, spect K5, dM2	
h2757	12.0	95	32	7½- 10	relfix, spect F5	14558s2212
18	19.6	39	35	6 - 9½	(Σ1894) relfix, color contrast pair spect K3, G8	14562s1057
β1085	9.3	23	34	6 - 13	cpm, spect dF6	14563s0447
Σ1899	28.2	67	37	7 - 9½	relfix, spect K2	14590s0258
β119	1.9	287	60	8 - 8½	PA dec, spect G0	15029s0649
Σ3090	1.1	281	60	8½- 9	PA slight inc, cpm,	15061s0047
	95.5	130	22	- 10	dist dec, spect G0	
ι	58.6	111	43	4½- 9½	cpm, spect B9 or A0	15094s1936
ι a	0.1	16	59	5 - 5½	(B2351) PA inc	
ι b	1.9	17	43	10- 10½	(β618)	
B288	1.8	76	59	6 - 11	spect gG5	15109s2600
Shj 195	47.4	140	16	7 - 8	spect F5	15116s1815
β350	0.7	141	58	7 -8	PA dec, dist slow dec, spect F2	15127s2725

LIBRA

LIST OF DOUBLE AND MULTIPLE STARS (Cont'd)

NAME	DIST	PA	YR	MAGS	NOTES	RA & DEC
Σ 3091	0.4	234	68	8 - 8	Binary, about 144 yrs; PA dec, spect F8, eccentric orbit	15134s0443
β 352	14.2	66	31	7½ -9½	spect K2	15149s2649
β 227	2.1	168	59	7½ -9½	cpm, PA dec, spect A2	15162s2405
h4756	0.7	290	59	7½ -8	(β228) dist & PA dec, spect F2	15167s2405
h4769	9.6	192	40	8 - 9	relfix, spect K0	15225s2145
Σ 1939	9.2	130	37	8 - 9	PA slight dec, spect G0	15248s1048
β 33	3.0	40	59	8 - 10	Spect F8	15258s1249
	23.3	132	59	- 12		
β 34	6.5	56	59	11 - 11	4' from β 33	15259s1251
β 1114	1.1	317	59	7 - 7½	(h4774) PA dec, spect F8	15259s2842
	9.2	6	31	-10½		
h4783	10.7	282	58	6 - 9	cpm, relfix, spect A4, dF5	15288s2000
LI 123	9.3	301	51	7 - 7	(S673) all cpm;	15302s2419
LI 123b	0.3	153	62	8 - 8½	(λ238) binary, 57 yrs, PA inc, A&B spect A3, K5	
υ	3.3	160	44	4 - 11½	(39 Librae) (I1271) spect K5	15340s2758
Σ 1962	11.8	189	56	6½ -6½	neat relfix pair; cpm, both spect dF6	15360s0838
β 121	1.6	280	47	7½- 8	relfix, spect A0	15366s2729
β 122	1.7	221	66	7 - 7	cpm, relfix, spect F2	15370s1936
β 35	2.3	104	46	7 - 8	cpm, relfix, spect G4	15400s1551
β 354	5.6	289	38	7 - 9	relfix, spect F0	15402s2515
β 260	0.4	165	59	7 - 7	(h4803) spect A5,	15432s2754
	50.7	214	14	- 9	C spect = G0	
Σ 3096	3.8	80	50	9 - 9	slight PA dec, spect G5	15452s0510
Σ 3097	4.0	183	37	9 - 9	relfix, spect G0	15482s0852
Hu 153	0.4	78	59	8 - 8	relfix, spect F2	15492s1223
47	0.5	129	58	6 - 8	(Hu 1274) binary, PA dec, spect B5	15521s1914

LIBRA

LIST OF VARIABLE STARS

NAME	MagVar	PER	NOTES	RA & DEC
δ	4.8--5.9	2.327	Ecl.Bin; Spect A0 (*)	14583s0819
σ	3.2--3.35	Irr	Spect M3	15011s2505
48	4.85± 0.1	Irr	(FX) Shell star (*)	15554s1408
R	9.7--15..	242	LPV. Spect M5e	15508s1605
S	8.0--13..	193	LPV. Spect M2e	15185s2013
T	9.6--15.5	238	LPV. Spect M4e	15079s1950
U	9.1--14..	226	LPV. Spect M3e	15391s2101
V	9.0--14.5	255	LPV. Spect M5e	14376s1727
W	9.9--15..	203	LPV.	15350s1600
X	9.9--14..	165	LPV. Spect M3e	15333s2059
Y	7.8--14..	275	LPV. Spect M5e	15091s0549
RR	8.0--14..	277	LPV. Spect M4e--M5e	15535s1810
RS	7.0--12.5	217	LPV. Spect M7e--M8e	15214s2244
RT	8.3--14..	252	LPV. Spect M3e--M5e	15036s1832
RU	7.4--13..	317	LPV. Spect M5e--M6e	15305s1509
RV	8.8--9.4	10.722	Ecl.Bin; Spect G5	14330s1749
RW	9.0--14..	203	LPV. Spect M5e	15202s2354
SV	9.0--11..	403	LPV. Spect M	15304s2701
SX	9.0--14..	334	LPV. Spect M6e	14399s2000
TT	9.5--15..	278	LPV. Spect M3e	15061s1519
UW	9.4--10.0	85	Semi-reg; Spect G2e--K0	14281s1635
UZ	9.3--9.6	9.500	Ecl.Bin; Spect K pec	15297s0822
YY	8.0--11..	230	LPV. Spect M	15053s2059
EE	9.0--12..	209	LPV. Spect M6e	15409s0913
EF	9.8--10.5	Irr	Spect M3	15186s2830
EH	9.5--10.0	.0884	Cl.Var; Spect A5--F3	14564s0045
EI	9.4--10.4	1.987	Ecl.Bin; Spect A2	15314s2250
EP	9.7--13..	186	LPV.	14371s2222
ES	7.0--7.4	.883	Ecl.Bin; Spect A2	15141s1251
FW	9.8--10.2	1.495	Ecl.Bin; Lyrid	15076s2059
FY	7.1--7.5	45:	Semi-reg; spect gM5	14550s1214
FZ	6.9--7.1		Semi-reg; spect gM4	15167s0858
GG	6.8--6.9		Semi-reg; spect M5	15296s2343

LIBRA

LIST OF STAR CLUSTERS, NEBULAE, AND GALAXIES

NGC	OTH	TYPE	SUMMARY DESCRIPTION	RA & DEC
5595	121[3]	⊖	Sc; 12.4; 1.5' x 0.8' F,pL,1E,vgbM, 5597 sf 4'	14215s1630
5597	122[3]	⊖	Sb; 12.6; 1.4' x 1.2' vF,L,v1E,vg1bM	14217s1633
5605	120[3]	⊖	Sc; 13.1; 1.3' x 1.0' vF,pL,R,vgbM	14223s1257
5728	184[1]	⊖	SBb; 12.4; 2.0' x 0.8' pF,pL,pmE,mbM, θ shape	14396s1703
5756		⊖	Sb/Sc; 13.1; 1.3' x 0.6' pB,pL,pmE,gpmbM	14449s1439
5757	690[3]	⊖	SBb; 12.6; 1.2' x 1.0' vF,S,R,1bM	14450s1853
5768	373[3]	⊖	I/Sd; 12.9; 1.2' x 0.7' F,pR,bM	14496s0220
5792	683[2]	⊖	Sb; 12.9; 6.0' x 1.2' pB,pL,mE,mbM	14558s0054
5791	691[3]	⊖	Sa; 13.0; 1.0' x 0.6' pF,cE,E	14560s1904
5796		⊖	E1; 12.8; 0.7' x 0.6' F,pS,1E,bM	14566s1626
5812	71[1]	⊖	E1; 12.6; 0.7' x 0.6' cB,S,R,vsmbM	14582s0716
5861	192[2]	⊖	Sc pec; 12.4; 2.7' x 1.2' F,L,E,bM, distorted arms on north side	15064s1108
5878	736[3]	⊖	Sb; 12.6; 2.6' x 0.9' pB,pL,pmE,psmbM	15110s1405
----	F703	⊖	SBc; 12.8; 2.2' x 1.9' F,S,R,bM	15110s1518
5885	116[3]	⊖	Sc; 12.4; 2.2' x 2.2' F,cL,R,bM, with inner ring	15124s0953
5897	19[6]	⊕	Mag 10, diam 8½', class XI; pF,L,Ri, stars faint (*)	15145s2050
5898	138[3]	⊖	E1; 12.6; 0.6' x 0.5' F,S,R,gbM; 5903 nf 5.5'	15152s2355
5903	139[3]	⊖	E2; 12.8; 0.8' x 0.7' cF,pS,R,pmbM	15156s2351
5915		⊖	Sb; 12.5; 1.0' x 0.7' B,S,R, compact spiral	15188s1255
----	Me2-1	◎	(VV 72) Mag 12, diam 6"; S,F	15195s2327

ALPHA   Name- ZUBEN EL GENUBI, "The Southern Claw";
Magnitude 2.75; spectrum A3 with strong
metallic lines. Position 14481s1550. Alpha Librae is about
65 light years distant; the actual luminosity is about 25
times that of the Sun, and the absolute magnitude close to
+1.2.   The star shows an annual proper motion of 0.13";
the radial velocity is 6 miles per second in approach.

The field glass companion at 231" is the star 8
Librae (GC 19970); magnitude 5.16, spectrum F5 IV. Despite
the wide separation, the two stars show the same proper
motion, and undoubtedly form a true physical pair with a
projected separation of about 4800 AU.

In addition, the primary star is itself a spectroscopic binary of uncertain period. On May 31, 1966   the
duplicity of the star was verified photoelectrically by
observations made during an occultation of the star by the
Moon. Using the 36-inch reflector at Kitt Peak National
Observatory in Arizona, H.L.Poss and T.R.Kremser obtained
a light curve which revealed the star to be an extremely
close double differing by 0.4 magnitude in blue light and
separated by about 0.01". From the estimated masses and
apparent separation, the orbital period of the system is
expected to be on the order of 20 days.

BETA   Name- ZUBEN ESCHAMALI, "The Northern Claw";
Magnitude 2.61; spectrum B8 V.   Position
15143s0912. The computed distance is about 140 light years;
the actual luminosity about 145 times that of the Sun. The
annual proper motion is 0.10"; the radial velocity is 21
miles per second in approach.

There is a classic controversy over the supposed
change in brightness of Beta Librae since ancient times.
It is said that Eratosthenes had assigned the star first
place among all the stars in Scorpius and Libra.   Ptolemy,
several centuries later, found it the equal of Antares.
R.H.Allen suggests, however, that it may be Antares which
has changed.   Another mystery concerns the fact that this
white star has so often been described as "greenish" or
"pale emerald".   Olcott refers to it as "the only naked-eye
star that is green in color", while T.W.Webb refers to its
"beautiful pale green hue".   Star colors are strangely

elusive, of course, and there are many such discrepancies in the guidebooks, but modern observers generally agree that the only stars which appear definitely green are the close companions to red stars such as Antares itself. If Beta Librae appears truly greenish- which each observer must decide for himself- it is the only bright single star which does so.

DELTA   Position 14583s0819. A bright eclipsing variable star of the Algol type, discovered by Schmidt in 1859. It has a range of just over one magnitude (4.79 to 5.93 photographic) and a period of 2.32735 days. The fall to minimum requires nearly 6 hours, and the star at primary minimum is about 66% obscured by its larger but fainter companion. Only one spectrum, of type A0 or A1 is seen, but the computed surface brightness of the companion corresponds to class G or possibly early K. From a photometric study of the system, the following facts have been derived:

	Spect	Diam	Mass	Lum	Abs.Mag.
A	A0	3.4	2.7	46	+0.7
B	G?	3.7	1.2	3	+3.8

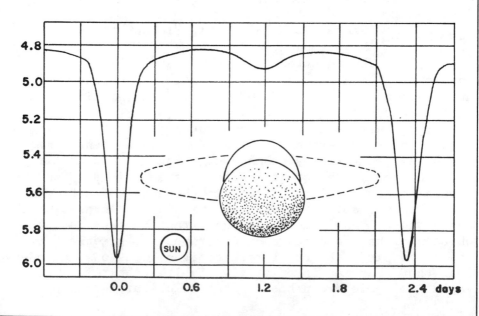

DESCRIPTIVE NOTES (Cont'd)

The two stars are about 4.5 million miles apart, and the bright component is about 1.5 million miles from the center of mass of the system. In the light curve we find the usual secondary minimum (about 0.1 magnitude) which occurs when the brighter star partially occults the faint one, and demonstrating that the "dark" star is not truly dark at all, but has a luminosity at least several times that of the Sun. There are also the common ellipticity and reflection effects which prevent the light curve from being entirely flat outside of eclipse. The surface brightness of the fainter star is nearly doubled on the side facing the primary, due to reflected light from the more brilliant component. This effect, combined with the somewhat ovoid shapes of the stars when seen near elongation, causes the total light of the system to increase long after each eclipse is ended.

Trigonometric parallaxes suggest a distance for Delta Librae of 160- 200 light years; this is in fair agreement with the distance of about 225 light years derived from the spectroscopic luminosity criteria. The star shows an annual proper motion of about 0.06"; the radial velocity is 24 miles per second in approach.

IOTA   Magnitude 4.66; spectrum A0p; Position 15094s1936. Iota Librae has an unusual spectrum showing strong lines of silicon. The computed distance is about 250 light years, giving an actual luminosity of about 60 times that of the Sun. The annual proper motion is 0.06"; the radial velocity is variable, with a mean value of 7 miles per second in approach.

Iota Librae is a close visual binary, discovered as recently as 1940 by Van den Bos; he obtained a separation of 0.2" in PA 342° and estimated magnitudes of 5.2 and 5.8. By 1959 the separation had decreased to 0.1" in PA 16°, a change of about 1.5° per year. Needless to say, this star is always an extremely difficult pair, resolvable only in large telescopes. Iota Librae, however, has another more distant companion star, 58.6" distant, first measured by Sir William Herschel in 1782. Showing common motion with the primary, this star, of magnitude 9.7, is itself a close double with nearly equal magnitudes and a separation of about 1.9". The position angle between the components is decreasing slowly and the star is undoubtedly a binary of very long period. Iota Librae, then, is actually a system of four stars, arranged as two close pairs with a vast distance from one pair to the other. The projected separations are: AB= 8 AU; CD= 150 AU; AB- CD= 4600 AU. There appear to be no spectral types available for the CD pair, but the derived luminosity of each star is about half that of the Sun, suggesting main sequence stars of late type G.

In the vicinity of Iota Librae, about 1.7° to the southeast, lies the large, dim, and very loose structured globular star cluster NGC 5897 (photograph on facing page)

SIGMA Magnitude 3.31; spectrum M4 III. Position 15011s2505. The parallactic distance is about 60 light years; the actual luminosity about 15 times that of the Sun, and the absolute magnitude +1.9. Sigma Librae shows an annual proper motion of 0.09"; the radial velocity is 2.5 miles per second in approach.

This star, on some older atlases, was assigned to the constellation Scorpius, and given the designation Gamma Scorpii.

48 Magnitude 4.85; position 15554s1408. Spectral type given by various authorities as B3, late B or early A, but peculiar with emission lines and evidently variable. This is the classic example of a "shell star", first noted during a spectroscopic survey at Harvard, and described in the Henry Draper Catalogue as "peculiar in showing sharply defined hydrogen lines with very wide and ill-defined helium lines". A suggestion that the star might be a close binary was ruled out by further research. 48 Librae is a blue-white star about 450 times the luminosity of the Sun and about 5 times the diameter. A member of the extensive Scorpio-Centaurus moving group, the star is some 640 light years distant, shows an annual proper motion of 0.03" and a mean radial velocity of about 3.5 miles per second in approach.

The star is encircled by an outer shell or ring of turbulent gases which occasionally show violent expansion velocities up to 60 miles per second. Activity of the

NGC 5897. An unusually loose-structured globular cluster
with slight central condensation. Palomar Observatory
200-inch telescope photograph.

DESCRIPTIVE NOTES   (Cont'd)

shell causes large changes in the measured radial velocity
which varies from about 15 miles per second in recession to
48 miles per second in approach. Major periods of shell
activity have occurred in 1935, 1945, 1952, and 1967. A.B.
Underhill (1966) finds that the apparent period of the
radial velocity variations has been close to 10 years but
"the range of variations was not closely the same in each
cycle which has been observed in the years since 1935".
Spectral lines of the shell change not only in position and
intensity during periods of strong activity, but also in
shape, the asymmetry being especially evident in the lines
of hydrogen and the Fe II lines of iron. In 1967 and 1968,
when the star again showed strong activity, many spectral
lines were distinctly asymmetric. Large differences in the
radial velocities of different lines also appeared to imply
the presence of numerous shells or rings of gas, expanding
at different rates.

According to a study by R.Faraggiana (1969) most of
the material comprising the shell is probably confined to a
thick equatorial ring. In relatively quiescent times the
shell appears to be about twice the diameter of the star,
and may be connected in some way with the abnormally large
rotational velocity of the star, some 240 miles per second
at the equator. The shell is thus a symptom of some sort
of stellar instability, but exact details of the process of
shell formation are still lacking. As in the case of some
other shell stars, there is evidence for chemical strati-
fication in the shell; the Fe II lines, Ti II lines, and
the hydrogen absorption features all appear to originate
at different levels above the surface. It is also interesting to note that 48 Librae shows at least one distinct
emission line, at 6320 angstroms, which remains unidentified; the similar star Zeta Tauri also shows this same
mysterious feature and two others as well. Attempts to
compare this star with other examples, however, are complicated by the fact that a considerable variety of objects
show some type of "shell" activity. 48 Librae appears to be
a single star and shows only slight light variations;
Omicron Andromedae is probably a close binary, 17 Leporis
has a red giant companion, Gamma Cassiopeiae shows large
irregular light variations, and P Cygni is a high-luminos-
ity nova-like variable! (See also Zeta Tauri)

LIST OF DOUBLE AND MULTIPLE STARS

NAME	DIST	PA	YR	MAGS	NOTES	RA & DEC
h4672	3.8	301	34	6 - 8	cpm, relfix, spect G5	14170s4250
R244	4.3	121	36	6½- 9½	relfix, spect B2	14194s4806
τ^2	0.3	181	60	5½- 5½	(I 402) relfix, spect dF7	14229s4509
I 426	10.6	310	32	5½- 11	spect B9	14269s4506
a	19.4	25	33	5 - 9	(not Alpha)(ϕ 318) (h4690) relfix,	14340s4555
a a	0.1	35	59	6 - 6	spect K0	
△163	60.7	106	38	8 -8½	probably optical, PA dec slowly, spect F0, B8	14347s5418
α	27.6	232	01	2½- 14	spect B2 (*)	14386s4711
△168	5.4	203	51	7½- 7½	cpm, relfix, spect F5	14392s5458
h4696	35.3	206	13	7- 12½	spect B9	14427s4439
h4698	8.9	260	33	5 - 13	(b) cpm; spect K0	14435s5210
Hd 240	39.1	290	29	6½- 11	(I.6074) spect A0	14437s5200
△171	17.5	226	32	7½- 9½	relfix, spect B8	14500s4539
	17.0	332	32	-12½		
h4715	2.5	278	52	6 - 7½	(△174) relfix, spect B9	14531s4741
I 1261	3.9	78	60	7 - 12	spect F0	14560s4316
h4723	5.5	169	37	7½- 11	relfix, cpm, spect K0	14584s5143
π	1.4	73	56	5 - 5	(h4728) dist inc, PA dec, spect B5,B5	15017s4651
λ	0.2	178	60	5 - 5½	(λ219) binary, 73 yrs; PA inc, spect B3	15055s4505
CPO--	50.6	22	02	7½- 8	cpm, spect both G0	15073s4332
△ 178	1.1	312	59	6½- 10½	AC optical, PA & dist dec, spect gK0	15082s4505
	32.2	264	59	- 7		
$\varkappa$	27.0	144	51	4 - 6	(△177) easy pair cpm, relfix, spect B9, A0	15085s4833
I 228	1.3	25	56	8 - 8	PA dec, spect A2	15106s4336
△179	10.5	46	59	7½- 8½	relfix, spect A2	15112s4312
Hd 244	13.9	39	34	7 - 10	spect B9	15120s4358

LIST OF DOUBLE AND MULTIPLE STARS   (Cont'd)

NAME	DIST	PA	YR	MAGS	NOTES	RA & DEC
h4750	13.3	19	33	7 - 10	cpm, relfix, spect A2	15124s4753
μ	1.2	142	55	5 - 5½	(h4753) (△180) all cpm, AB PA dec, spect B8	15150s4742
	23.7	130	55	- 7		
Hwe 76	5.6	124	33	6½- 8½	slight PA inc, cpm, spect A0	15182s3802
ε	0.6	247	60	4 - 6	(△182) all cpm, AB PA dec, spect B3	15193s4431
	26.6	171	55	- 9		
B 1288	0.3	309	59	8½- 8½	slow PA inc, spect B9	15211s4824
	8.7	343	30	- 12		
△185	13.2	32	34	6 - 9	(λ234) relfix, spect gG2	15248s5126
I 239	0.3	325	60	7½- 8½	PA dec, spect A2	15257s3118
h4772	7.7	280	21	7½- 11	Lupus-Norma border, PA inc, dist dec, spect K2	15259s5113
h4776	6.0	228	51	6½- 8	relfix cpm, spect A0	15270s4145
Hwe 78	0.3	185	60	7 - 9	(B2036) spect A2 AC PA dec	15281s3339
	1.6	138	60	- 9		
△187	28.3	229	22	7 - 10	(h4784) dist & PA dec, spect F0, G0	15300s4722
λ239	13.5	7	13	8 - 12	spect K0	15310s4704
γ	0.1	12	27	3½ - 4	(h4786) binary, 104 yrs, spect B2	15318s4100
h4788	2.4	4	55	5 - 7½	(d Lupi) PA inc, spect B5	15324s4448
I 243	0.7	340	59	8 - 8½	PA dec, spect F5	15348s3105
I 89	1.2	158	53	7 - 8	PA inc, spect F5	15378s3949
Hwe 79	3.7	348	52	6½- 8	(Hrg 121) cpm, PA dec, spect A0	15410s4140
λ248	0.2	136	26	8½- 9	spect A0	15419s3457
I 245	0.7	331	41	8 - 9	PA inc, spect K0	15435s4405
B847	0.1	36	37	6 - 6½	spect B6	15435s3432
△192	0.3	42	60	7 - 8	AB PA inc, spect B9	15439s3522
	34.8	144	38	- 7½		
λ249	15.2	131	53	6 - 13	(GC 21205) cpm; spect G6, DA; NGC 5986 in field	15442s3745

LIST OF DOUBLE AND MULTIPLE STARS   (Cont'd)

NAME	DIST	PA	YR	MAGS	NOTES	RA & DEC
ξ	10.6	49	51	5½– 5½	(△196) fine relfix pair; spect A1, A0	15537s3349
h4821	19.3	145	19	8½– 8½	relfix	15547s3149
η	15.2	21	34	3½– 7½	(△197) (Rmk 21) cpm, relfix, spect B2 (*)	15568s3815
Cor 190	0.8	240	25	6½– 10	(I 1280) Spect A0	15575s4018
	8.1	159	33	– 10		
Hwe 82	2.6	346	54	7½– 7½	relfix, spect F0	16006s3256
I 373	8.6	217	30	7 – 11	spect B9	16013s3918
I 1284	0.2	92	31	8½– 9	spect F8	16014s4036
Cor 193	2.5	102	59	8½– 8½	PA inc, dist dec, spect G5	16044s3753

LIST OF VARIABLE STARS

NAME	MagVar	PER	NOTES	RA & DEC
α	2.23--2.31	.2599	β Canis type, spect B1 (*)	14386s4710
δ	3.21--3.24	.1655	β Canis type, spect B2	15181s4028
τ¹	4.10--4.14	.1774	β Canis type, spect B2	14229s4500
R	9.4--14.0	236	LPV. Spect M5e	15502s3609
S	8.0--13.5	343	LPV. Spect Se	14501s4625
T	9.2--10..	Irr	Spect N3..	14190s4937
U	9.6--11.5	87	Semi-reg; spect G2--K0e	15576s2947
W	9.5--12..	237	LPV.	15120s5037
Y	8.2--15..	402	LPV. Spect M7e	14559s5445
Z	9.0--11..	Irr	Spect N0	14328s4309
RS	9.5--10..	Irr	Spect N	14202s4718
RT	9.5--15..	364	LPV.	14274s4828
RU	9.0--13.0		RW Aurigae type, spect G5	15534s3740
RW	8.5--12..	198	LPV. Spect M3	14232s4356
RX	9.0--13..	237	LPV. Spect M3e--M6e	15429s4754
RY	9.5--12..		RW Aurigae type, spect G0	15561s4014
SW	9.0--12..	377	LPV.	15336s3737

LIST OF VARIABLE STARS   (Cont'd)

NAME	MagVar	PER	NOTES	RA & DEC
EV	9.8--12.5	15.31	Ecl.Bin.	15129s4408
FH	9.4--14..		LPV.	15122s3537
FQ	8.4--10..		LPV.	15403s3701
FT	8.8--9.2	.4701	Ecl.Bin; spect F5	14566s4247
FW	8.6--9.0	16.73	Cepheid; spect G0	15191s4044
FZ	9.4--9.8	2.267	Ecl.Bin? spect A0	14290s5328
GG	5.5---6.0	1.85	Ecl.Bin; spect B5+A0	15156s4037
GH	7.2---7.6	9.285	Cepheid; spect G5	15209s5240
GO	7.0---7.2		Semi-reg? Spect M4	15250s3711

LIST OF STAR CLUSTERS, NEBULAE, AND GALAXIES

NGC	OTH	TYPE	SUMMARY DESCRIPTION	RA & DEC
5530		⊘ (galaxy)	Sc; 12.3; 3.5' x 2.0' vF,pmE,svmbM, BN	14154s4309
----	I.4406	◎ (planetary)	Mag 11, diam 100" x 35"	14193s4355
5593	△357	⁙ (open cluster)	pL,1C, diam 8', class E, group of 10 stars mags 10...	14224s5435
----	I.4444	⊘ (galaxy)	Sb/Sc; 12.2; 1.5' x 1.3' F,pS,1E	14285s4312
5643	△469	⊘ (galaxy)	SB; 11.4; 2.5' x 2.3' pB,L,R,vg1bM	14294s4359
5749	△356	⁙ (open cluster)	pL,pRi,1C, diam 10', about 20 stars mags 10... class E	14453s5419
5824		⊕ (globular)	Mag 9½, diam 3', class I; pB,S,bM, stars eF	15009s3253
5822		⁙ (open cluster)	vL,Ri,1C, diam 40', about 100 stars mags 9...12; class D	15016s5409
5873		◎ (planetary)	Mag 12, diam 3"; stellar	15094s3754
5882		◎ (planetary)	Mag 10½, diam 7", vB,S, disc	15133s4527
5927	△389	⊕ (globular)	Mag 9, diam 6', class VIII; cB,L,R,vRi, mags 15...	15244s5029
5986	△552	⊕ (globular)	Mag 8, diam 5', class VII; vB,L,R,gbM, stars mags 13...	15428s3737
6026		◎ (planetary)	Mag 12½, diam 50", ring-shape (Erroneously identified as galaxy in Shapley-Ames list and Skalnate-Pleso Atlas)	15581s3425

DESCRIPTIVE NOTES

ALPHA   Magnitude 2.28 (slightly variable).
Spect B1 or B2 II; some authorities, how-
ever, have assigned the star to luminosity class IV. The
position is 14386s4710. This star, like many of the other
bright stars in this part of the sky, is a member of the
large Scorpio-Centaurus moving group. It is also a giant
variable of the Beta Canis Majoris type with a period of
0.2599 days. The 14th magnitude companion at 27.6" is
probably only a background star, but a modern measurement
of the pair should make it possible to accept or reject the
possibility of common motion. The most recent observations
appear to be those made in 1901!
Alpha Lupi has a computed distance of about 430
light years; the actual luminosity is about 1700 times that
of the Sun. The annual proper motion is 0.03"; the radial
velocity is about 4.5 miles per second in recession.

BETA   Magnitude 2.69; spectrum B2 IV. Position
14552s4256. Another member of the great
Scorpio-Centaurus moving group. The distance of the star
is about 540 light years; the actual luminosity about 1900
times that of the Sun. Beta Lupi shows an annual proper
motion of 0.07"; the radial velocity is close to zero.

GAMMA   Magnitude 2.80; spectrum B2 V. Position
15318s4100. Gamma Lupi is a close and
difficult binary star which reaches a separation of about
1" at widest; the orbit is oriented only 5.5° from the
edge-on position which causes the images to fuse together
when the components are passing each other, as in 1926.
Orbit elements are still somewhat uncertain; the results
of two different computations are compared here:

	Period	Semi-major axis	Eccen.	Periastron
Dawson	104 yr	0.78"	0.31	1905
Heintz	147 yr	0.59"	0.49	1887

The components of this system appear to be of very similar
type, with individual magnitudes of 3.55 and 3.75. At a
computed distance of about 570 light years, the total

luminosity of the pair is about 1900 times the light of
the Sun, and the mean separation corresponds to about 140
AU, or nearly twice the distance across the orbit of Pluto.
Gamma Lupi shows an annual proper motion of 0.04"; the
radial velocity is 3.5 miles per second in recession.

DELTA   Magnitude 3.24; spectrum B2 IV. Position
15181s4028.   Another member of the widely
scattered Scorpio-Centaurus moving group. The computed
distance is about 680 light years; the actual luminosity
close to 1900 times that of the Sun. Delta Lupi shows an
annual proper motion of 0.03"; the radial velocity is one
mile per second in recession.

EPSILON   Magnitude 3.36; spectrum B3 IV. Position
15193s4431.   Epsilon Lupi is a close
binary star in slow retrograde revolution with a period
estimated to be at least 700 years.   Since the early
measurements of 1896, the separation has decreased slightly
and the PA has diminished from 282° to 247° in 1960.   The
third member of the system, at 26.6", shares the motion of
the bright pair. In addition to the three visible stars,
the primary is known to be a spectroscopic binary with a
period of 0.9014 day; Epsilon Lupi is thus a quadruple
system.
        The parallactic distance is about 300 light years,
giving the bright pair a total luminosity of about 240
suns, and a true separation of some 60 AU.   The A-C span
of 26.6" corresponds to a projected separation of about
2470 AU.   Epsilon Lupi shows an annual proper motion of
0.03"; the radial velocity is about 2.5 miles per second
in recession.

ZETA   Magnitude 3.42; spectrum G8 III. Position
15087s5155.   Zeta Lupi is about 90 light
years distant; the actual luminosity is 28 times that of
the Sun; the annual proper motion is 0.13"; the radial
velocity is 6 miles per second in approach.
        Recent measurements indicate that a nearby star of
magnitude 7.8 shares the proper motion of the primary. It
lies 71.9" distant toward PA 249°; the projected separation
of this wide pair is about 2000 AU.

ETA   Magnitude 3.45; spectrum B2 V. Position
15568s3815. Eta Lupi is another member
of the Scorpio-Centaurus moving group, at a computed distance of about 570 light years. The actual luminosity of
the star is close to 1000 times that of the Sun; the annual
proper motion is 0".04; the radial velocity is 4 miles per
second in recession.

For the small telescope the star is an attractive
double with a separation of 15.2", magnitudes 3.5 and 7.5.
No change has been measured in this pair since the early
measurements of John Herschel in 1834. According to E.J.
Hartung (1968) the colors are "white and ashy". A common
motion pair, the projected separation of the components is
about 2600 AU.

NOVA 1006   Approximate position 14596s4142. The
appearance of this brilliant "new star"
is chronicled in a number of medieval records from Europe,
China, Japan, and Islamic Egypt. Although the exact position of the object is uncertain, the numerous accounts seem
to leave no doubt that a very brilliant object - most
probably a supernova - appeared in the vicinity of the star
Beta Lupi in the Spring of 1006 A.D. A study of all the
records has been made by Bernard R. Goldstein of Yale
University, and the reality of the object now seems to be
established beyond any reasonable doubt. This star, then,
appears to be the earliest historically authenticated
supernova outburst observed in our Galaxy, preceding by 48
years the appearance of the famous Taurus supernova of
1054 A.D. which produced the Crab Nebula.

Among the surviving records of the object is a
very interesting eye-witness account by the Islamic astrologer Ali ibn Ridwan of Cairo, who first noted the new star
on May 1 or possibly the preceding evening:

".....a spectacle that I saw at the beginning of
my education. This....appeared in the zodiacal sign Scorpio
in opposition to the Sun....it was a large star, round in
shape, and its size 2½ to 3 times the magnitude of Venus.
Its light illuminated the horizon and it twinkled very much
....its brightness was a little more than a quarter of the
brightness of the Moon. It continued to appear.... until

DESCRIPTIVE NOTES   (Cont'd)

the Sun arrived at the zodiacal sign Virgo.....and it
ceased appearing suddenly. This apparition was also seen
at the time by other scholars just as I have recorded it".

The word used by the Egyptian astrologer to designate the object is "nayzak", a rare Arabic term of Persian
derivation which might refer either to stars, meteors, or
comets. The accounts, however, appear to leave no doubt
that the unknown object had a fixed position with respect
to the stars; neither is there any mention of a tail which
should have been extremely conspicuous on a comet of such
extraordinary brightness.

The object is also described in the Chronicles of
Ibn al-Athir (13th Century) as follows:

"In that year (396 A.H.) at the new moon of the
month Shu'ban, a large star similar to Venus appeared to
the left side of the qibla (the direction to Mecca) of
Iraq. Its rays on the earth were like the rays of the Moon
and it remained until the middle of the month Dhu al-Qa'da
when it disappeared..."

Another 13th Century account is that of Bar Hebrae-
us, which reads as follows in a translation by Professor
E.A.W.Budge:

"In A.H. 396 there appeared a star resembling
Aphrodite in greatness and splendor in the zodiacal sign
Scorpio, its rays revolved and gave out light like that of
the moon - it remained four months and disappeared..."

In the annals of Hepidannus, a monk who lived at
St.Gall in Switzerland, the new star was described as being
of unusual brightness and visible for three months in the
extreme south. In another Arabian record, the "Annales
Regum Mauritaniae" of the 14th Century, we read of the
appearance of "a wondrous star, excessively brilliant...
it lasted for a period of six months and then disappeared.
In that year there were strong winds, bursts of lightning,
and shattering thunder without rain...." The year of this
alarming apparition is again given as A.H. 396.

Far Eastern accounts supply independent confirma-
tion of the new star. In a report presented to the Sung
Dynasty emperor at Kaifeng in 1044 A.D. it is stated that
"....at the first watch of the night, on the second day of
the fourth month (May 1, 1006) a large star, yellow in
color, appeared to the east of K'u Lou in the west of

DESCRIPTIVE NOTES   (Cont'd)

Ch'i Kuan.   Its brightness had gradually increased.......
According to the star manuals there are four types of
auspicious stars, one of which is called 'Chou-po', yellow
and brilliant....forboding great prosperity for the State
over which it appears....."
           The nova is mentioned in several places in the
official history of the Sung Dynasty, compiled by Toktaga
and Ouyang Hsuan, and completed about 1345 A.D.   In the
chapter   dealing with astronomical phenomena it is stated:
"...the fourth month of the third year of the Ching-Te
reign period (May 6, 1006) a Chou-po star appeared to the
south of the Ti lunar mansion, and one degree to the west
of Ch'i Kuan...its shape was like that of a half-moon and
it shone so brightly that objects could be seen by its
light....it appeared at the east of K'u Lou.   During the
eighth month (late August to late September) it went below
the horizon following the rotation of the heavens.   During
the eleventh month it was again sighted..." The yellow
color of the nova appears to be confirmed by the account of
the Buddhist monk Weng Ying (about 1078) who stated that
"its bright rays resembled a golden disc...a huge star at
the west of the Ti lunar mansion. No one could identify its

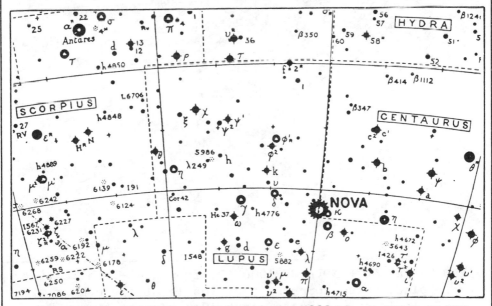

THE PROBABLE POSITION OF THE NOVA of 1006 A.D. IS SHOWN ON THIS
PORTION OF A CHART FROM NORTON'S STAR ATLAS.

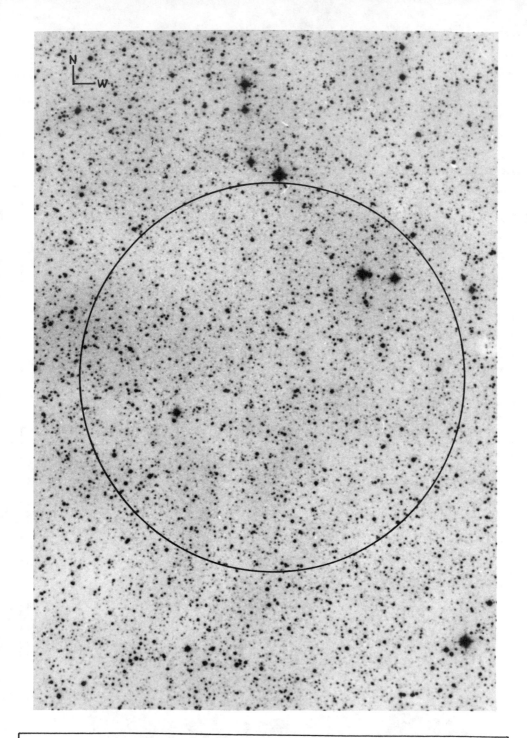

FIELD OF THE SUPERNOVA OF 1006 AD, from a 48-inch Schmidt Camera plate. The computed position is at the center of the ½° circle and stars to about 19th magnitude are shown.

'omen category'. Chou K'o-ming, the Chief Official of the
Spring agency said that the star was a Chou-po.....it is
said that the color of the Chou-po is golden, its rays
brilliant, and peace and prosperity to the State are fore-
told by its appearance..." This view is at least pleasingly
optimistic in contrast to the notions of 14th Century
Arabian astrologers, who seemed to have regarded the nova
as an alarming portent of divine wrath and diverse
calamities.

Of the Japanese accounts, which confirm the date and
approximate position, the following passage from the
"Meigetsuki" of Fujiwara (about 1235 A.D.) has been chosen:
"The second day of the fourth month in the third year
of the Kanko reign period of Ichijo-In (May 1, 1006)   a
large 'guest star' resembling Mars, bright and sparkling,
appeared after nightfall for successive nights directly
south within the Kikan constellation". Another account
mentions that a report on the new star was presented to the
Emperor on July 21, 1006, and that offerings were then made
to the various shrines in accordance with the Emperor's
decree.

For the modern observer, the most interesting feature
of the medieval records lies in the information concerning
the exact position, which offers the possibility of locating the remnant of the star. From the great brilliance
and the long period of conspicuous visibility, it appears
virtually certain that the star was a supernova, probably
of type II, and evidently much closer to the Solar System
than either Tycho's Star of 1572 or Kepler's Nova of 1604.
From the studies of oriental scholars, it seems possible
to narrow the position down to an area within a few degrees
of the star which we now call Beta Lupi.

A search of this area by W.Baade with 48-inch Schmidt
plates has not revealed any certain trace of a nebulous
remnant. According to R.Minkowski (1965) a few isolated
features that might be very short faint filaments were
found, but judged to be more likely blends of several faint
stars. A faint ring-shaped nebula surrounding a 17th magnitude star has been reported near the predicted area by
C.Jackson (1965) but has been identified as a planetary
nebula, and does not appear to be a logical suspect for a
supernova remnant. Another planetary nebula, NGC 5882,

DESCRIPTIVE NOTES   (Cont'd)

lies about 4° from the assumed position, also seems
disqualified for the same reasons. The distinguishing
feature of a supernova cloud is, of course, the enormous
expansion velocity, which makes any connection with the
planetary nebulae seem highly unlikely. The Crab Nebula
in Taurus, the most conspicuous known supernova remnant, is
expanding at a rate of about 600 miles per second, while
some of the filaments in the Cassiopeia A cloud have been
measured at more than 3500 miles per second. Nothing
resembling these objects has been identified near the Lupus
position, but at the same time it should be remembered that
only the faintest shreds of nebulosity mark the positions
of Tycho's Star, Cassiopeia A, and Kepler's Star. The vast
cloud of the Crab Nebula must then be regarded as a unique
type of object, or possibly the illuminating conditions
are unusually favorable.

         At present, the most promising suspect seems to
be the prominent radio-source detected with the 210-foot
radio telescope at Parkes, Australia in 1964. According to
F.F.Gardner, J.G.Bolton, D.K.Milne, and M.B.Mackey, it is
located at 14596s4142, near the Lupus-Centaurus border,
about 1½° northeast of Beta Lupi, and less than 1° ENE of
Kappa Centauri. The southwest portion of this source was
listed previously in the catalogue of Bolton, Gardner, and
Mackey (1964) as Source #1459-41. Radio-intensity studies
show the source to be roughly round in outline, about 40'
in diameter, with two inner lobes and a definite concentric
"shell structure". Very faint nebulous filaments have now
been detected (1976) in the area, on plates made with the
4-meter reflector at Cerro Tololo in Chile. (Refer also to
Tycho's Star in Cassiopeia, Kepler's Star in Ophiuchus, and
the "Crab Nebula" M1 in Taurus)

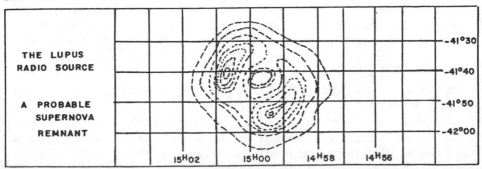

THE LUPUS
RADIO SOURCE

A PROBABLE
SUPERNOVA
REMNANT

-41°30
-41°40
-41°50
-42°00

15H02   15H00   14H58   14H56

LYNX

LIST OF DOUBLE AND MULTIPLE STARS

NAME	DIST	PA	YR	MAGS	NOTES	RA & DEC
4	0.8	124	59	$6\frac{1}{2}$- 8	(Σ881) PA inc,	06176n5924
	26.2	96	35	-13	spect A2	
	100	356	08	-11		
5	31.4	139	24	6 - 10	(S514) spect K4	06224n5827
	96.0	272	24	- $8\frac{1}{2}$		
Σ936	1.3	274	59	7 - $8\frac{1}{2}$	PA inc, spect G5	06354n5809
Σ946	4.1	130	54	$7\frac{1}{2}$- $9\frac{1}{2}$	relfix, spect G0	06404n5930
12	1.8	101	59	$5\frac{1}{2}$- 6	(Σ948) binary,	06418n5930
	8.5	308	59	- $7\frac{1}{2}$	about 700 yrs, PA	
					dec, spect A2, all	
					cpm	
Σ958	4.9	257	56	6 - 6	relfix, cpm, spect	06440n5546
					dF5, dF6	
Σ960	22.0	67	24	$7\frac{1}{2}$- 9	relfix, spect F0	06456n5305
14	0.3	117	25	6 - 7	(Σ963) binary,	06487n5931
					about 600 yrs, PA	
					inc, spect gG0, A2	
Σ968	20.7	288	25	8 - 9	spect A3	06489n5245
OΣ158	16.8	304	25	7- $11\frac{1}{2}$	relfix, spect F5	06495n5135
	56.0	63	25	- 11		
15	0.8	33	59	5 - 6	(OΣ159) PA inc,	06530n5829
	29.0	346	24	- $12\frac{1}{2}$	spect G5, AB cpm,	
					AC optical	
Σ1001	9.0	66	56	7 - 9	relfix, spect G0	06591n5415
Σ1001b	1.7	3	39	9 - $9\frac{1}{2}$	BC PA inc.	
Σ1002	0.3	2	60	9 - 9	(A1324) PA inc,	07000n5631
	30.3	317	17	9 -$9\frac{1}{2}$	spect G	
Σ1009	3.9	150	55	7 - 7	PA dec, dist inc,	07017n5250
					spect A2	
Σ1025	25.6	133	55	$7\frac{1}{2}$- $7\frac{1}{2}$	PA dec, dist inc,	07088n5553
					spect K0	
Σ1032	2.6	110	35	7- $10\frac{1}{2}$	PA inc, spect A0	07101n4835
	32.7	319	08	- 10		
Σ1033	1.5	277	59	$7\frac{1}{2}$- 8	relfix, spect F0	07108n5238
Σ1044	12.2	168	30	$8\frac{1}{2}$- $8\frac{1}{2}$	relfix, spect A2	07127n4744
Σ1050	20.2	20	24	$7\frac{1}{2}$- 8	relfix, cpm pair,	07158n5501
					spect A0	
20	15.1	254	50	7 - 7	(Σ1065) cpm pair,	07184n5015
					relfix, spect F0,F0	

LIST OF DOUBLE AND MULTIPLE STARS   (Cont'd)

NAME	DIST	PA	YR	MAGS	NOTES	RA & DEC
19	14.7	315	56	$5\frac{1}{2}$- $6\frac{1}{2}$	(Σ1062) easy pair,	07188n5523
19b	74.2	287	08	- 11	relfix, spect B8, AO	
β758	1.0	93	51	6 - 10	(Kui 30) all cpm,	07251n4817
	17.1	94	34	- 10	spect B9	
A2047	2.0	257	21	$7\frac{1}{2}$-$13\frac{1}{2}$	spect K5	07305n4616
OΣ174	2.0	86	57	$6\frac{1}{2}$- 8	relfix, spect FO	07324n4309
O$\Sigma\Sigma$87	64.4	178	24	$7\frac{1}{2}$- 8	both spect F5	07351n4235
24	54.8	320	11	5 - 9	(h2405) spect A3	07388n5850
Es---	11.1	278	13	8 - $11\frac{1}{2}$	spect KO	07408n4529
Σ1172	1.6	243	54	$7\frac{1}{2}$- $9\frac{1}{2}$	relfix, spect AO	08007n5453
Σ1174	0.2	318	59	$8\frac{1}{2}$- $8\frac{1}{2}$	(A2050) PA inc,	08011n4726
	5.8	213	43	- $8\frac{1}{2}$	spect F5	
27	45.6	266	26	$4\frac{1}{2}$- $12\frac{1}{2}$	(Es 70) spect A2	08047n5139
27b	7.3	348	26	- 13		
OΣ189	4.2	293	34	7 - 10	relfix, spect AO	08114n4311
Σ1200	8.2	360	19	$8\frac{1}{2}$- $8\frac{1}{2}$	relfix, spect GO	08123n4956
Σ1204	12.0	102	29	8 - 9	relfix, spect F8	08126n3838
Σ1211	0.4	309	67	$8\frac{1}{2}$- 9	PA dec	08150n3909
	27.1	1	00	- $12\frac{1}{2}$		
	100	131	25	- $9\frac{1}{2}$		
β576	1.5	142	41	7 - $11\frac{1}{2}$	relfix, spect KO	08184n3406
Σ1217	29.0	241	20	7 - $8\frac{1}{2}$	cpm, spect GO, KO	08208n4507
Σ1222	10.3	49	30	8 - 9	relfix, spect A3	08231n3743
OΣ193	14.1	295	16	7 - 11	relfix, spect KO	08250n3342
Σ1225	3.7	192	55	$8\frac{1}{2}$- $8\frac{1}{2}$	relfix, spect A2	08261n5122
Hu 716	0.5	122	62	$7\frac{1}{2}$- 9	binary, 33 yrs, PA dec, spect G5	08284n3508
Σ1234	23.5	67	57	7 - $8\frac{1}{2}$	cpm, dist inc, PA dec, spect KO	08292n5532
Σ1239	12.9	289	29	$8\frac{1}{2}$- 10	relfix	08292n3740
Σ1240	26.8	78	58	7 - 10	dist & PA slow inc,	08300n3336
	51.1	246	09	- 10	spect AO	
Σ1259	5.0	342	55	$8\frac{1}{2}$- 9	relfix, spect GO	08434n3840
Σ1274	8.9	42	38	7 - $8\frac{1}{2}$	relfix, spect A2	08458n3832
Σ1279	1.3	268	58	$8\frac{1}{2}$- $8\frac{1}{2}$	relfix, spect GO	08467n3947
Σ1282	3.7	278	56	7 - 7	cpm pair, relfix, spect F8	08476n3515
Σ1289	3.7	5	67	$7\frac{1}{2}$- $8\frac{1}{2}$	cpm pair, relfix, spect F8	08514n4347

LIST OF DOUBLE AND MULTIPLE STARS (Cont'd)

NAME	DIST	PA	YR	MAGS	NOTES	RA & DEC
A2132	0.1	242	58	$7\frac{1}{2}$- $7\frac{1}{2}$	PA inc, spect F0	08524n4152
Σ3120	1.1	356	67	$7\frac{1}{2}$- $8\frac{1}{2}$	PA inc, spect G0	08528n4352
Σ1296	2.3	74	46	$8\frac{1}{2}$- 9	slight PA inc, spect G5	08562n3508
10 UMa	0.6	350	62	4 - 6	rapid binary, 22 yrs, PA dec, spect F5 (BS 3579)	08574n4159
Σ1333	1.6	49	66	$6\frac{1}{2}$- 7	dist & PA inc, spect A4	09154n3535
38	2.9	228	67	4 - $6\frac{1}{2}$	(Σ1334) PA slow dec, spect A3	09158n3701
	87.7	212	09	- 11		
Σ1338	0.9	232	67	$6\frac{1}{2}$- 7	binary, 220 yrs, PA inc, spect dF2, dF3	09179n3824
Σ1342	17.1	320	23	$8\frac{1}{2}$- 11	slow PA dec, spect A2	09182n3439
Σ1369	24.7	149	56	7 - 8	relfix, AC spect= F2, G0	09323n4011
	118	325	56	- 8		

LYNX

LIST OF VARIABLE STARS

NAME	MagVar	PER	NOTES	RA & DEC
R	7.4--13.9	379	LPV. Spect S3e--S6e	06572n5524
S	9.0--14..	298	LPV. Spect M6e	06403n5758
T	8.0--12..	419	LPV. Spect N0e	08195n3341
U	8.9--15..	436	LPV. Spect M8e	06363n5955
V	8.5--11..	---	Irr or semi-reg; spect M5	06251n6135
W	8.8--13..	295	LPV. Spect M6	08134n4017
X	9.5--16.0	321	LPV. Spect M5e	08223n3534
Y	6.9---7.5	110:	Semi-reg; spect M5	07246n4606
RR	5.6---6.0	9.945	Ecl.Bin; spect A5	06222n5619
RS	9.0--10.5	285:	Semi-reg; spect M7	07145n4837
RT	9.0--12..	395	LPV. Spect M6e	08116n3750
RU	9.5--14..	243	LPV. Spect M3	07367n3646
SS	9.0--10..	Irr	Spect M5	08022n5150
SU	8.6--10..	126	Semi-reg; spect M4	06388n5532
SV	7.0--8..	Irr	Spect M5	08004n3629
SW	9.2--9.8	.6441	Ecl.Bin; spect F2	08043n4157
SX	9.9--11.3	2.023	Ecl.Bin; spect A2	08099n5725
SY	9.5--11..	79	Semi-reg; spect M	07456n5316
SZ	8.7--9.5	.1205	Cl.Var; period variable; spect F2	08061n4437
TT	9.3--10..	.5974	Cl.Var; spect A6--F4	08598n4447
UV	9.4--9.8	.4150	Ecl.bin; W U.Maj type; Spect F8	09003n3818
UW	4.9---5.0		Semi-reg? Spect M3	06133n6132
UX	6.6---6.8		Semi-reg; Spect M6	09006n3857

LIST OF STAR CLUSTERS, NEBULAE, AND GALAXIES

NGC	OTH	TYPE	SUMMARY DESCRIPTION	RA & DEC
2419	218[1]	⊕	Mag 11½, diam 2', vRi,eC, 8^m star prec 4'. Most distant globular in the Milky Way Galaxy (*)	07348n3900
2474	830[3]	◎	vF,L, annular, mag 13½, with 17^m central star. Diam 6' x 5½'	07538n5333
2500	709[3]	⊘	SBc; 12.2; 2.0' x 1.6' F,L,R, vgbM	07582n5054

LIST OF STAR CLUSTERS, NEBULAE, AND GALAXIES (Cont'd)

NGC	OTH	TYPE	SUMMARY DESCRIPTION	RA & DEC
2537	55[4]	⊖	Sd; 12.3; 1.0' x 0.9' pB,pL,R	08097n4609
2541	710[3]	⊖	Sc; 12.0; 4.5' x 2.0' F,L,E, vgbM	08111n4915
2549		⊖	E6/S0; 12.1; 1.6' x 0.7' pB,S,mE,psbM	08149n5758
2552	711[3]	⊖	I/Sd; 12.5; 2.6' x 2.0' eF,cL,1E	08154n5011
2683	200[1]	⊖	Sb; 10.6; 9.0' x 1.3' vB,vL,vmE,gmbM, nearly edge-on spiral	08496n3338
2712		⊖	SBb; 12.7; 2.0' x 1.0' pB,pL,E,vgbM	08562n4507
2776		⊖	Sc; 12.0; 2.1' x 1.9' pB,pL,R,vgbM	09089n4511
2782	167[1]	⊖	Sb; 12.4; 1.8' x 1.6' cB,R,mbM,BN, distorted structure, extending fila- ment on east side	09109n4019
2793		⊖	Sc; 12.9; 0.9' x 0.9' vF,S,R	09137n3439
2798	708[2]	⊖	SB; 13.0; 2.0' x 0.5' pB,S,E; edge-on system 2799 lies 1' east	09144n4210
2832		⊖	E4; 13.1; 0.6' x 0.6' F,vS,R; lens-shape system 2831 lies 0.8' southwest	09168n3359
2844	628[3]	⊖	Sa; 13.0; 0.9' x 0.5' cF,cS,E	09186n4022

DESCRIPTIVE NOTES

ALPHA   Magnitude 3.14; spectrum M0 III. Position
              09180n3436. From parallactic measurements
the distance appears to be about 155 light years, giving an
actual luminosity of 110 times that of the Sun, and an
absolute magnitude of -0.3. The annual proper motion of
the star is 0.22"; the radial velocity is 24 miles per
second in recession.

NGC 2419   Position 07348n3900, about 7° north of
                 Castor in Gemini. This is the famous
"Intergalactic Wanderer", the most distant of the Milky
Way's globular star clusters, tentatively classed by some
authorities as an extra-galactic object. It was discovered
in December 1788 by Sir William Herschel and reobserved in
1833 by John Herschel. A description of the object appears
in a list published by Lord Rosse in 1861; although no
definite resolution was achieved even in his giant reflector, Rosse suggested that NGC 2419 was probably a globular
cluster. This surmise was finally verified at Lowell
Observatory in 1922 when C.O.Lampland obtained excellent
photographs of the object with the 42" reflector.
        NGC 2419 is located nearly opposite the galactic
center where most of the other globulars are found; in fact
there is no other globular cluster within 60°. A great
distance is suggested also by the small apparent size and
the unusual faintness of the individual stars, even the
brightest members not quite reaching 17th magnitude. On
this basis, H.Shapley (1922) estimated the distance as
50,000 parsecs. A later study of the cluster with the 100-
inch telescope at Mt.Wilson identified 36 variable stars in
the group, 31 of which are typical short period cluster
cepheids of the RR Lyrae class. The apparent magnitude of
these stars averages about 19.2; since their true luminosity is known to be in the range of 50- 60 suns, the distance can be easily computed. After a slight correction for
galactic absorption the resulting values are: 182,000 light
years from the Sun, and 210,000 light years from the center
of the Galaxy. The distance is comparable to that of the
Magellanic Clouds and suggests the possibility that NGC
2419 is an independent intergalactic object. The fact that
such clusters do exist is indicated by the discovery of

DESCRIPTIVE NOTES   (Cont'd)

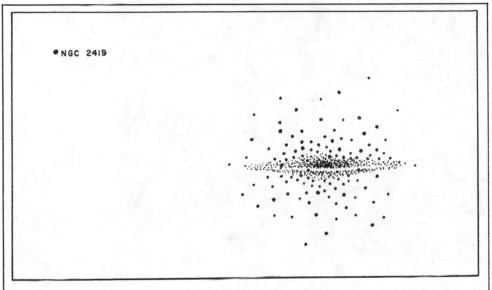

NGC 2419

a number of very faint and distant globulars on the 48-inch
Palomar Survey plates; some of these objects are definitely
more remote than the Magellanic Clouds.

Despite its vast distance, the cluster can be seen
in small telescopes as a small fuzzy spot of the 11th
magnitude. This indicates a very high actual luminosity
which is suggested also by the unusual richness of the
cluster as revealed by photographs made with the 200-inch
telescope.   The computed absolute magnitude of -8.2 is
equivalent to 175,000 suns. The brightest individual stars
are red and yellow giants about 360 times the brightness of
the Sun.

Approximately 100 globular clusters are now known,
distributed around the Galaxy on all sides, and virtually
all contained within a sphere 65,000 light years in radius
centered on the Galactic Center (diagram above). The two
chief exceptions are NGC 2419 and NGC 7006 in Delphinus,
the latter being located on the far side of the Galaxy,
some 150,000 light years from the Galactic Center.

Summary of data on NGC 2419: Total magnitude (pg)
10.9; apparent diameter 2.0' visual; 7.2' photographic
maximum; linear diameter 380 light years (maximum from
photographs); distance modulus about 18.75 magnitudes;
integrated spectral type F5; radial velocity= 12 miles per
second in recession.

NGC 2419 in LYNX. The very remote globular cluster as it appears on a plate obtained with the 42-inch reflector at Lowell Observatory.

LIST OF DOUBLE AND MULTIPLE STARS

NAME	DIST	PA	YR	MAGS	NOTES	RA & DEC
Ho 432	16.8	287	34	6½- 13	spect K2	18223n3843
β134	1.1	134	55	8 - 10	relfix, spect A0	18239n4651
Es2173	5.6	300	25	7½- 12	spect K0	18249n3608
Ho 434	13.7	186	25	7½- 12	dist inc, spect G5	18272n2931
Σ2328	3.6	73	56	8 - 8½	relfix, spect A2	18275n2953
Σ2333	6.4	334	52	7½- 8	relfix, spect A0	18292n3213
β1253	7.2	157	58	6 - 13	relfix, spect B8	18309n3031
Hu1293	0.9	79	22	7½- 14	spect K0	18328n3605
Σ2351	5.1	340	58	7½- 7½	relfix, spect A0	18346n4114
Σ2349	7.3	203	58	5½- 11	relfix, spect A0p	18348n3326
Σ2352	15.8	284	10	7½- 10	relfix, spect K0	18352n3449
α	62.8	173	46	0 - 10	VEGA. Optical, PA	18352n3844
	54.4	285	50	- 11	& dist inc, spect A0 (*)	
Σ2356	1.0	61	68	8 - 9	PA inc, spect F0	18364n2839
Σ2358	2.5	220	42	8½- 9	relfix, spect F8	18367n3041
Σ2362	4.2	184	50	7 - 8½	slow PA inc, spect A5	18367n3601
Σ2367	0.3	63	62	7 - 7½	binary, 90 yrs; PA	18394n3015
	14.3	193	39	- 8½	dec, spect gG5;	
	22.9	95	00	- 12	ABC all cpm.	
Σ2371	9.7	56	51	8½- 8½	relfix, spect A0	18402n2736
Σ2372	0.2	254	58	6½- 7½	(B2546) relfix,	18403n3442
	25.1	83	58	- 8	spect B5	
	91	65	10	- 11		
Σ2380	25.6	9	51	6½- 8	relfix, spect G0	18414n4452
Σ2376	22.3	64	49	7½- 8½	relfix, spect A0	18418n3021
	81.7	207	34	- 11		
ε	208	173	24	5 - 5	(ΣI37) famous "double-double" (*)	18427n3937
ε¹	2.8	359	66	5½- 6½	(Σ2382) PA dec, spect A2, A4 (*)	
ε²	2.2	98	66	5 - 5½	(Σ2383) PA dec, spect A3, A5 (*)	
ζ	43.7	150	55	4½- 5½	(ΣI38) (β968)	18430n3733
	46.1	272	23	- 13	fine easy pair, AB	
	61.8	301	60	-11½	cpm, relfix, spect A0, F0	
Σ2393	14.8	24	33	7½- 10	optical, dist inc, spect K0	18435n3816

LIST OF DOUBLE AND MULTIPLE STARS   (Cont'd)

NAME	DIST	PA	YR	MAGS	NOTES	RA & DEC
Σ2392	3.2	317	55	8 - 10	relfix, spect A2	18435n3910
	23.4	175	55	- 9½		
Σ2394	7.2	201	40	8½- 9	slight dist dec.	18439n4159
Σ2390	4.2	158	51	7½- 8½	relfix, spect A5	18440n3428
Σ2397	3.8	267	51	7 - 9½	relfix, spect G5	18453n3121
Hu 936	1.8	104	60	9 - 9	relfix	18469n3358
Σ2406	4.8	2	41	7 - 11	relfix, spect A2	18479n2622
8	34.8	73	24	6 - 11½	(ν') spect B2	18479n3245
	58.2	122	23	- 10½		
9	19.0	177	34	5 - 13	(ν^2) (Ho 440)	18480n3230
					spect A3	
h1352	11.9	243	18	7 - 11	spect B8	18481n2945
β	46.6	149	55	3½- 7	(ΣI39) AB spect	18482n3318
	68.6	318	29	- 9	B8,B7; primary	
	85.7	19	25	- 9	typical "lyrid"	
	46.4	247	13	- 13	type "Ecl.Bin. (*)	
Es2026	19.7	108	23	7½- 13	spect B3	18503n3727
Es2026b	5.0	340	23	13- 13½		
β137	1.4	150	59	8 - 8½	PA inc, spect G0	18522n3719
	20.7	143	25	- 11½		
δ^2	86.2	349	23	4½- 11	spect M4 (*)	18528n3650
δ^2 b	2.2	138	23	11- 11½		
0Σ525	1.7	128	58	6- 10	relfix, spect G5	18530n3354
	45.2	350	35	- 7½	color contrast	
Ho 270	8.4	306	34	6 - 13	relfix, spect gG8	18533n4132
	24.0	40	34	- 12		
Σ2419	3.3	174	39	8½- 9	PA dec	18535n2910
Σ3130	0.5	168	00	7½- 8½	(0Σ365) spect A2	18544n4410
	2.7	262	58	- 11	AB uncertain, not	
					seen in recent	
					years.	
Σ2422	0.7	82	66	7½- 7½	PA dec, spect A0	18551n2602
β648	1.0	223	62	5½- 7½	binary, 61 yrs; PA	18552n3250
	55.1	287	21	- 12	dec, spect G0	
Σ2427	51.6	60	50	8½- 9	PA dec, dist inc,	18564n3809
Σ2427b	7.1	80	50	10 -10	spect K0	
γ	13.8	304	58	3 - 12	(AGC 9) optical, PA	18571n3237
					slow inc, spect B9	
Σ2431	19.1	236	35	7 - 9	relfix, spect B5	18572n4037

LIST OF DOUBLE AND MULTIPLE STARS (Cont'd)

NAME	DIST	PA	YR	MAGS	NOTES	RA & DEC
Σ2430	1.9	1	46	8½– 8½	relfix, spect G0	18574n2932
Σ2441	5.5	275	44	7½– 9½	PA dec, spect A3	19008n3120
	96.3	141	08	–11½		
Σ2448	2.7	191	45	8 – 8	relfix, spect A3	19019n3540
β1285	11.1	295	00	7 – 13	spect B5	19021n3405
	39.8	208	00	–10½		
Da 9	2.0	178	35	7½– 11	relfix, spect B9	19027n4348
Σ2454	0.2	33	55	8 – 8	PA inc, spect K0	19042n3022
	1.1	276	66	– 9		
Σ2458	11.9	227	14	8⅓– 9	spect A0	19050n2741
	70.2	63	11	– 9		
Σ2459	13.7	233	50	8½– 9	relfix, spect A2,A2	19053n2554
17	3.7	308	58	6 – 10	(Σ2461) PA dec, spect dA7	19055n3225
Σ2466	2.4	105	55	8 – 8½	PA slow dec, spect A0	19060n2943
Σ2469	1.6	122	56	7½– 8½	relfix, cpm pair, spect A3	19061n3851
Σ2467	10.3	264	18	8½– 9	relfix, spect F8	19062n3043
Σ2472	20.9	338	58	7½– 9	dist inc, spect K0	19068n3750
	75.4	346	25	– 9	C= Σ2473	
Σ2473	6.3	293	56	10– 10	relfix, spect F5	19068n3751
Σ2470	13.5	271	33	6½– 8	relfix, spect B3	19069n3441
Σ2474	16.7	262	49	6½– 8	slight PA inc, spect dG1, dG5. Σ2470 in field.	19072n3431
Ho 572	18.6	314	02	6½– 12	spect B9	19074n3029
Σ2481	4.5	208	61	8 – 8	PA dec, dist inc, spect G5	19095n3842
Σ2481b	0.3	116	67	8 – 8½	binary, PA dec, 61 yrs; spect dK0	
J1263	34.2	254	16	7½– 12	spect K0	19096n2848
J1263b	5.0	72	16	12 – 13		
Σ2480	15.3	24	35	7– 10½	spect A3	19097n2610
Hu 941	1.2	144	58	7½– 13	relfix, spect A0	19104n3210
Σ2483	9.7	318	51	7 – 8	relfix, spect A0	19104n3016
	71.0	236	34	– 9		
η	28.2	83	55	4 – 8	(Σ2487) optical, spect B2	19120n3903

LIST OF DOUBLE AND MULTIPLE STARS (Cont'd)

NAME	DIST	PA	YR	MAGS	NOTES	RA & DEC
0Σ366	21.8	230	46	7 - 10	spect B9	19124n3408
β975	33.4	228	35	7 - 9½	(0Σ367) (Ho 648)	19126n3428
	21.4	74	13	- 12½	spect F5	
β975b	1.3	244	57	9½- 9½		
0Σ371	0.9	161	58	7 - 7	PA inc, spect B5	19139n2722
	47.4	268	33	- 9		
Σ2491	1.2	226	59	8 - 9	PA inc, spect A2	19142n2811
θ	99.8	71	24	5 - 10	Spect K0, color contrast	19146n3803
Ho 102	86.1	345	00	7 - 10	spect B5	19148n3302
Ho 102b	1.9	57	02	10 -11		
Σ2502	1.7	205	35	8 - 10	relfix, spect A2	19173n3910
Σ2505	10.7	316	35	8 - 9	spect A0	19181n3527
	19.3	224	05	- 13		

LIST OF VARIABLE STARS

NAME	MagVar	PER	NOTES	RA & DEC
β	3.4---4.3	12.91	Ecl.Bin; typical "lyrid"; spect B8p (*)	18482n3318
δ²	4.5---5.0	?	Spect M4 (*)	18528n3650
R	4.1---5.0	46	Semi-reg; spect M6	18538n4353
S	9.8--15.3	438	LPV. Spect S	19111n2555
T	7.5---9.3	Irr	Spect R6; very red star	18306n3658
U	8.3--13..	457	LPV. Spect N0e	19184n3747
V	8.8--15.0	374	LPV. Spect M7e	19071n2935
W	7.5--13.0	196	LPV. Spect M3e--M6e	18132n3639
X	8.6--10..	Irr	Spect gM3	19110n2641
Z	9.2--15.0	288	LPV. Spect M5e	18578n3453
RR	7.1---8.0	.5668	Standard "cluster-type" variable (*)	19239n4241
RS	9.2--15.6	305	LPV. Spect M5e	19112n3320
RT	9.1--15.2	251	LPV. Spect M5e	18595n3727
RU	9.9--15..	370	LPV. Spect M6e--M8e	19107n4113
RW	9.9--15..	504	LPV. Spect M7e	18437n4335

LIST OF VARIABLE STARS   (Cont'd)

NAME	MagVar	PER	NOTES	RA & DEC
RY	9.1--15..	326	LPV. Spect M5e--M6e	18431n3437
RZ	9.9--11.2	.5112	Cl.Var; Spect A9--F4	18418n3245
SS	8.5--13.0	349	LPV. Spect M5e	19119n4654
ST	9.5--13..	299	LPV. Spect M4e	19082n4332
SY	9.9--11..	Irr	Spect M6	18395n2846
TT	9.1--10.6	5.244	Ecl.Bin; spect A0	19260n4136
TU	9.5--11..	Irr	Spect M6	18186n3144
TV	9.9--13.8	263	LPV. Spect M4e	18197n3026
TW	9.5--13.5	376	LPV.	18223n3933
TY	9.3--12.5	332	LPV. Spect M8e	19078n2759
TZ	9.8--10.8	.5288	Ecl.Bin; spect K0	18142n4105
UZ	9.9--11.0	1.891	Ecl.Bin; spect A2	19194n3750
WZ	9.5--14..	377	LPV. Spect M9e	19009n4708
XY	6.1---6.7	Irr	Spect M4--M5	18365n3937
EY	8.7--10.7	66	Semi-reg; spect M5	18390n3137
FI	9.5--10.5	146	Semi-reg; spect M	19401n2854
FL	9.7--10.5	2.178	Ecl.Bin; spect G5	19106n4614
HK	8.5--10.6	Irr	Spect N	18411n3654
HM	8.7---9.5	Irr	Spect M6	18505n3328
HN	9.8--13..	406	LPV. Spect gM7e	19158n4243
HR	6.5--15..	---	Nova 1919	18515n2910
KP	9.3--10.3	146	Semi-reg; spect M9	18292n3836
OQ	8.9---9.6	Irr	Spect G6	18549n3108
OU	9.0---9.6	90	Semi-reg; spect M3	19098n3231

LIST OF STAR CLUSTERS, NEBULAE, AND GALAXIES

NGC	OTH	TYPE	SUMMARY DESCRIPTION	RA & DEC
6703			E0/S0; 13.0; 1.1' x 1.1' B,S,R,bM	18459n4530
6720	M57		!! Mag 9, diam 80" x 60" B,pL,lE, annular "Ring Neb" typical planetary neb (*)	18517n3258
6779	M56		Mag 8, diam 5', class X, B,L,R, mCM, stars mags 11...	19146n3005
6791			L,F, diam 20', mag 11, class E; vRi, several hundred faint stars (*)	19190n3740

VEGA. The region of the 5th brightest star, photographed
with the Lowell Observatory 5-inch camera. The bright
double star at lower left is Epsilon Lyrae.

DESCRIPTIVE NOTES

ALPHA   Name- VEGA, sometimes WEGA, the "Falling Eagle" or the "Harp Star". The fifth brightest star in the sky, formerly assigned 4th place, but shown by modern photometric measurements to be slightly fainter than Arcturus. Vega is the brightest of the three stars forming the large "Summer Triangle" consisting of Vega, Deneb, and Altair. Vega is magnitude 0.04; spectrum A0 V, position 18352n3844. Opposition date (midnight culmination) is about July 1.

The name is derived from the Arabic *Al Nasr al Waki*, "The Swooping Eagle"; the alternate forms of *Waghi, Vagieh,* and *Veka* also appear on Medieval charts, where the star and its constellation are depicted as an eagle, vulture, or falcon, often shown bearing a harp or lyre in its beak or talons. The Babylonian *Dilgun*, the "Messenger of Light", may be a reference to Vega. Pliny's title, usually translated "The Harp Star", is a reference to the legendary 7-stringed lyre of Hermes, later the property of Orpheus, but associated also with a veritable galaxy of gods and heroes including Apollo, Mercury, King Arthur, the Biblical David, and the Greek poet Arion. This is the Lyre whose strings, according to James Russell Lowell *"give music audible to holy ears"*. For obvious reasons, this celestial music goes virtually unheard anywhere in the modern world. Longfellow in his *"Occultation of Orion"*speaks of the heavenly lyre:

> "............*with its celestial keys,*
> *Its chords of air, its frets of fire,*
> *The Samian's great Aeolian lyre,*
> *Rising through all its sevenfold bars,*
> *From Earth unto the fixéd stars..."*

Hafiz of Persia calls it the *Lyre of Zurah;* to the Medieval Arabs it was *Nablon* or *Nablium,* the "Phoenician Harp", or *Al Lura* which became the *Allore* or *Alohore* of the *Alphonsine Tables*. The Greek title *Cithara* is from κιθαρα , the harp or lyre of the ancient Greek bards; the word "zither" evidently comes from the same root. Aratus, however, titles it the "Little Tortoise" or "Tortoise Shell", from the legend that Hermes created the first harp from the empty shell of a tortoise. The design of a turtle or tortoise shell on the early silver coins of Aegina very probably relates to this tradition. The Lyre of Orpheus is shown in classic style on the silver tetradrachms of the Chalcidian League, minted at Olynthus in Macedonia about 400 BC. The design was adopted by other Greek cities during the next century; the obverse of these coins usually shows a fine head of Apollo, god of music, poetry, and the arts.

The music of the Lyre, in Greek legend, cast such a spell that Orpheus charmed every living creature with it, even persuading the grim guardians of the Underworld to allow him to rescue his beautiful wife Eurydice from the Land of the Dead. Having been warned to cast no glance upon her until the couple had safely reached the upper world, Orpheus unfortunately lost Eurydice at the last moment by disobeying the fateful order. The story is one of the most popular of the Greek legends, and was the subject of the opera *Orpheo ed Euridice* by Gluck in 1762, and a ballet by Stravinsky in 1947.

Vega plays a prominent part in one of the few star-legends which have come down to us from ancient China, the appealing tale of the "Herd-Boy and the Weaving-Girl". The origin of the story is unknown, though it is mentioned in the *Shih Ching* or "Book of Songs", the ancient anthology of poetry of the Chou Dynasty of the 6th century BC; the *Shih Ching* was regarded as a classic in the days of Kung-fu-tze (Confucius) who, according to some scholars, may have had a part in compiling and editing it. This was one of the great classics, incidentally, whose destruction was ordered by the megalomaniac emperor Shih Huang Ti (221-210 BC) who hoped to be immortalized as the builder of the Great Wall, but who is remembered instead by Chinese scholars as *"He who burned the books"*.

Vega, in the legend is the "Weaving-Girl", while the "Herd-Boy" is Altair and the two stars on either side. The young lovers, lost in "amorous dalliance", neglected their duties to Heaven, and are now eternally separated by the

"On the seventh night of the seventh moon
  Vega glows in radiant splendor
   On the edge of the River of Stars....."

Celestial River, the impassable barrier of the Milky Way.
In China, however, there was always compassion. Once a
year, on the seventh night of the seventh moon, the lovers
are allowed to meet, when a bridge of birds temporarily
spans the River of Stars.

The most revered poet of Old China, the incomparable
Tu Fu of the T'ang Dynasty, mentions the Heavenly lovers
in more than one poem. Praised by the perceptive modern
critic and poet Kenneth Rexroth for his "sensibility acute
past belief", Tu Fu has an almost supernatural power of
evoking an entire landscape in a single line. In the de-
ceptively simple four lines of his *Autumn Night*, he not
only conjures up a perfect mood-picture of the silent night
but seems to evoke, through some indefinable magic, the
very essence of all Chinese civilization:

"Silver candles, autumn night, a cool screen,
Soft silks, a tiny fan to catch the fireflies...
On the stone stairs the night breathes cool as water..
I sit and watch the Herd-Boy and the Weaving-Girl.."

Astronomical allusions occur frequently in the poems
of Tu Fu, and in the almost equally evocative poems of the
4th century hermit-poet Tao Yuan Ming and the T'ang Dynasty
painter-poet Wang Wei. *"In this sapphire night.....I play*
my flute to the summer moon" writes Tu Fu. *"Over the Triple*
Gorge the Milky Way pulsates between the stars..." And in
another poem he speaks of his visit to the Monastery of
Ta-Yun:

"...the courtyard lies silent in the deep spring night;
The Northern Crown
Glitters on the edge of the temple roof
Where the iron phoenix soars
Toward the stars. The murmur of prayers
Floats from the temple hall....
The echo of ancient bells
Fades to silence
Beside my bed..."

The great friend and friendly rival of Tu Fu was the
other immortal T'ang Dynasty poet Li Po. In his thoroughly
readable survey of ancient cultures *"The Living Past"*,

Ivar Lissner compares them to *"two comets which drift in the Universe and only come into contact every few million years"*. Here is Li Po, expressing the ancient Chinese reverence for Nature and that *"other heaven and earth beyond the world of men"* :

> *"Here it is night...I stand before the Summit Temple;*
> *The stars are almost within reach of my hand...*
> *In the awesome silence I speak no word*
> *That I disturb not the dwellers of Heaven.."*

To observers in the Earth's Northern Hemisphere, Vega reigns as the leader of the "stars of a summer night" and dominates the heavens from its position virtually at the zenith in the evening hours of late July and August. From its sharp blue-white glitter, it has often been called by popular writers "The Arc-light of the Sky"; to the author of this book the color has always seemed distinctly bluer than Sirius. The color is, of course, intensified in the standard visually corrected refractor; the out-of-focus blue light produces a dazzling azure corona around the image. Manilius, writing in the days of Augustus, seems to anticipate the effect of Vega in a modern telescope when he notes that the star "displays a vigorous light and darts surprising rays". Mary Proctor, in 1924, wrote of seeing Vega through the great refractor at Yerkes Observatory where *"its vivid blue blazes and the twirlings of the diffraction rings which surround the great star make it appear a marvel of beauty"*. To T.W.Webb, the great star was "a pale sapphire...a lovely gem!" E.J.Hartung, however, calls the color of Vega "a very brilliant pale yellow", a curious impression which is probably explainable by the very low altitude of the star as seen from his observing station in Australia.

Perhaps the most striking reference to Vega in modern literature appears in the bizarre and colorful Dunsanian fantasy *"The Dream-Quest of Unknown Kadath"* by the American master of the strange and macabre, Howard P. Lovecraft. In the climactic scene of the tale, the dream-adventurer receives the ultimate revelation from one of the Elder Gods:

> *"Look! Through that window shine the stars of eternal night. Even now they are shining above the scenes you have known and cherished, drinking of their charm that they may*

*shine more lovely over the gardens of dream.....Go now....
the casement is open, and the stars wait outside...Steer
for Vega through the night.....into space toward the cold
blue glare of boreal Vega.....*

*The stars danced mockingly, almost shifting now and
then to form pale signs of doom that one might wonder one
had not seen and feared before; and ever the winds of
nether howled of vague blackness and loneliness beyond the
cosmos...."*

Vega is approximately 27 light years distant and has
an actual luminosity of about 58 times that of our Sun
(absolute magnitude +0.5). Vega is a star of the "Sirian"
type, with a surface temperature of about 9200°K, nearly
twice as hot as the Sun. Its diameter is estimated to be a
little more than 2.7 million miles. Vega has a mass of
about 3 solar masses and a density of about 0.2 the solar
density. The annual proper motion of the star is 0.35" in
PA 36°; the radial velocity is 8.5 miles per second in
approach.

During the course of the slowly changing orientation
of the Earth's axis in space (the Precession of the Equinoxes) Vega was the Pole Star some 12,000 years ago and will
occupy the position again about the year 12,000 AD. At its
closest, Vega is some 4½° from the true Pole, whereas it
is now some 51° distant.

It is toward a spot in the general direction of Vega
that the Sun- and the entire Solar System- moves in the
depths of space at a velocity of 12 miles per second. This
position is known as the "Apex of the Sun's Way", or simply
the "Solar Apex". Some idea of the vastness of space may
be gained from the fact that it would take the Sun over
450,000 years to reach Vega at this speed, even if it were
moving directly toward it.

The exact position of the Solar Apex cannot be very
precisely determined, however, since noticeably different
results are obtained by using different varieties of stars
as reference objects. Using a mere handful of the brighter
stars, William Herschel made the first estimate of its
position in 1783; his result placed the Apex near Lambda
Herculis, about 17° SW of Vega. A.Eddington in 1910, by
using some 5000 stars from the Boss *Preliminary General*

Catalogue, placed the Apex near Theta Herculis, about 7°
west of Vega. An analysis of nearly 20,000 stars in the
Cape Catalogue in 1954 placed the position about 10° more
to the south, while W.J.Luyten's study of over 92,000 stars
in the Bruce Proper Motion Catalogues (1941) indicated a
position near R Lyrae, about 6° NE of Vega. A preliminary
analysis of some 9000 stars in the Lowell Observatory
Proper Motion Catalogues again gave a somewhat different
result, placing the Apex about midway between Vega and Eta
Cygni. Another determination by Luyten (1967) used 25,800
proper motion stars measured on 48-inch Palomar Schmidt
plates, and going to a photographic magnitude of 21; these
results agreed closely with the Lowell analysis and placed
the Apex very near the star 4 Cygni, about midway between
Vega and Eta Cygni. The Luyten and Lowell surveys, of
course, concentrate chiefly on faint stars of large proper
motion; these lists therefore contain a large proportion of
high-velocity dwarfs and subdwarfs. If bright naked-eye
stars are used- which are virtually all giants- different
results must be expected simply because the two classes
of objects have different space motions.

P.Guthnick and R.Prager (1918) found slight variations in the light of Vega, and observations by E.A.Fath
(1935) at Lick Observatory confirmed a range of about 0.4
magnitude with suggestions of a periodicity of some 2 to
3 hours. His measurements were made in blue light with the
Lick 12-inch refractor, using Zeta-1 Lyrae as a comparison
star. Recent attempts to verify these results have not
been successful. E.G.Ebbighausen (1966) found no certain
variations as great as 0.01 magnitude, nor any suggestion
of short-period variability. Ebbighausen suggests that the
error may have resulted from the choice of Zeta-1 Lyrae as
the comparison star; this A-type star has a peculiar
metallic-line spectrum and has been suspected of variability.

The angular diameter of Vega has been measured with
high accuracy with the new stellar intensity interferometer
at Narrabri, Australia in July and August 1963. This new
instrument employs two large arrays of mirrors, each 22
feet in diameter and composed of some 250 small hexagonal
mirrors. The two mirror arrays may be separated by up to
618 feet on a circular track. The plotted results of many

measurements gave an angular diameter of 0.0037" for Vega, with a probable error of about 5%. This corresponds to about 3.2 times the solar diameter, or about 2.8 million miles.

Vega was the first star to be photographed, on the night of July 16-17, 1850. The historic photograph was made by the daguerreotype process at Harvard Observatory with the 15-inch refractor. The exposure was 100 seconds.

Vega has a small bluish companion star of the 10th magnitude, about 1' distant. There is no real connection between the two stars, and the separation is gradually increasing from the proper motion of Vega itself. It was 43" in 1836, 56" in 1925, and 63" in 1950. According to Webb, the faint star has been detected with a 8½-inch mirror only 30 minutes after sunset. Modern observers will find it not at all difficult with a good 6-inch glass. Vega has another faint optical companion of the 12th magnitude at about 54"; the separation of this A-C pair is also increasing from the motion of Vega itself.

BETA   Name- SHELIAK, from *Al Shilyak*, one of the many Arabian names for the whole constellation and signifying "The Tortoise". Magnitude 3.38 (variable); Spectrum B8e p;   Position 18482n3318. This star is the typical example of the "lyrid" or "bright-eclipsing" type of variable. The magnitude varies continually in a sinusoidal curve; the maxima are equal (magnitude 3.4) and the minima alternate between 3.8 and 4.1. These variations are due to the mutual eclipses of two unequally bright stars which revolve around their common center of gravity in a period of 12.9079 days. The light variations were first noticed by John Goodricke in 1784.

The light changes require only slight optical aid or none at all, and are most easily followed by comparing the star with the nearby Gamma Lyrae which is magnitude 3.25. For a few days the two stars will appear of nearly equal brightness, but about every 13 days Beta will fade away to only half the brightness of Gamma. The secondary minimum occurs some six days later; during this phase the star is about 65% as bright as Gamma.

Beta Lyrae is one of the most frustratingly interesting stars in the sky to the astrophysicist, and has

STAR FIELD IN LYRA. Beta Lyrae is the brightest object at lower center; Gamma is near the lower left corner. Vega and Epsilon Lyrae are at the top of the print.

LOWELL OBSERVATORY

DESCRIPTIVE NOTES   (Cont'd)

probably been the subject of more studies and investigations than any other star in the heavens with the single
exception of our own Sun.   In his review of the Beta Lyrae
problem, J.Sahade (1966) has stated that Beta Lyrae is an
object, if not THE object, which merited more thought and
effort from Prof. Otto Struve than any other of the several
hundred stars he investigated during his lifetime.   The
spectrum presents many peculiar features, some of which are
not entirely explained even today.   Struve in 1957 even
suggested the possibility that the spectral type of the
primary is not quite the same on opposite sides of the
star! The attribution of supergiant status to the star now
seems to be definitely an error, as were the unusually
large masses of about 80 and 50 suns derived from the
radial velocity curves.   It is now recognized that the
velocity measurements are severely distorted by the moving
streams of gas in the system. An accurate distance for the
star would at least settle the question of the true luminosity, but Beta Lyrae is too remote for a direct parallax
measurement. From observations of the two companion stars,
however, a modulus of 7 - 8 magnitudes is indicated; this
gives a distance in the range of 800- 1300 light years with
present evidence favoring the lower value. The star, from
these results, has an absolute magnitude of about -4,
rather than -7 as was formerly assumed.   H.A.Abt,   H.M.
Jeffers, J.Gibson, and A.R.Sandage (1962) found a distance
modulus of 7.1, giving the distance as about 860 light
years and the absolute magnitude as -3.9.

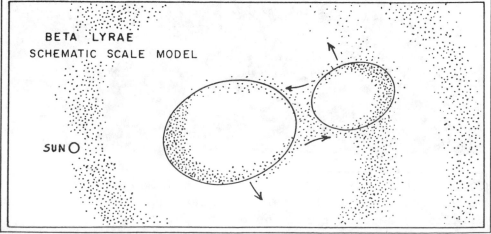

BETA LYRAE
SCHEMATIC SCALE MODEL

SUN ○

DESCRIPTIVE NOTES   (Cont'd)

From these results, our present picture of the
system may be summarized about as follows:

The primary star is a brilliant B8 type star of
luminosity class II or III, about 19 times the diameter of
the Sun and some 3000 times brighter. The companion star
has not been detected by the spectroscope, but its computed
surface brightness corresponds to spectral type late A or
early F. From six-color photometric observations of the
system, M.J.S.Belton and N.J.Woolf (1965) have derived a
probable spectral type of about A7. From the observed light
curve, the diameter of the secondary appears to be about 15
times the size of the Sun. The orbit of the pair is nearly
circular, with an eccentricity of 0.024. With a center-to-
center separation of about 22 million miles the two stars
are so close together that their atmospheres are intermingled, and both components are ellipsoidal in shape as a
result of rapid rotation and gravitational distortion. It
is not quite certain whether the primary eclipse is annular
or partial; secondary eclipse is probably total. According
to spectroscopic analysis, the two stars are connected by
a great filament of gases along which matter streams at the
high velocity of 180 miles per second, from the larger star
to the smaller. This flow of matter will gradually tend to
equalize the masses of the two stars, and also produce a

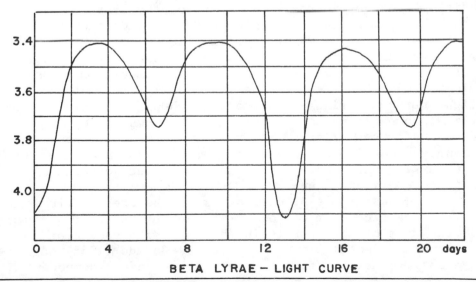

BETA LYRAE — LIGHT CURVE

DESCRIPTIVE NOTES (Cont'd)

retarding effect on both bodies which results in a slow
lengthening of the period. At present the period is growing longer at the rate of 9.4 seconds annually.
        The gas stream between the two stars produces
other effects in the spectrum which are difficult to interpret. Material which is not regained by either star appears
to be expanding into space in the form of a vast ever-
widening spiral of gas. Thus there is a continual mass loss
from the system, a fact of great interest to astronomers
studying the problems of stellar evolution. Perhaps the
most startling discovery about Beta Lyrae is the finding
(1962) that the components do not obey the mass-luminosity
relation, the fainter star having at least twice the mass
of the bright primary. The exact individual masses are
still in doubt. J.Sahade (1966) found probable masses of
9.7 and 19.5 suns; Abt, Jeffers, Gibson and Sandage found
11 and 20.2; Woolf (1965) has a much large mass difference
with 2.0 and 11.5. The fainter star, in any case, appears
to be clearly underluminous by several magnitudes.
        During 1959, Beta Lyrae was the subject of an
international program of spectroscopic and photometric
observations. Variations in the period were confirmed; it
was found that the mean orbital period for the year was
12.9355 days, whereas the Moscow General Catalogue (1958)
has 12.9079 days, revised to 12.9081 in the newer (1970)
edition. Changes in the total light of the system- aside
from the eclipse variations- were found, and an over-all
fading of about 0.1 magnitude (blue) between 1958 and 1959
was measured. Definite changes in the shape of the light
curve were also discovered; in 1959 the rise from primary
minimum was much steeper than in 1958. The form of the
light curve at primary eclipse also shows clear dissimilar-
ities; some minima show nearly flat bottoms while slight
dips and irregularities are evident in others. These
changes and differences are undoubtedly caused chiefly by
the varying shape and size of the gas cloud surrounding
the secondary component.
        A possible explanation for the severe underluminos-
ity of the secondary component is the suggestion that much
of the mass of the star is in the form of a vast cloud
ring, or disc encircling the star. J.Sahade (1958) suggested the probability that the massive secondary component

has already passed through the giant stage, having evolved
more rapidly than the less massive primary. The present
transfer of mass is causing the components to move closer
to each other, which in turn accelerates the mass transfer.
S.Huang (1963) has pointed out that "the time scale of a
close binary in this mode of mass exchange is necessarily
short; this explains why Beta Lyrae is an unusual object
in the Galaxy. The shortening of the separation will stop
only when the masses.....become equal. After this point, a
further transfer of mass from the primary to the secondary
will reverse the trend and widen the separation. This is
the present situation of Beta Lyrae.....According to our
interpretation, the phase of drastic exchange of mass.....
is now over".

     The future history of the star, in any case, seems
bound to be a curious one. The secondary, presently growing in mass and strongly underluminous, will eventually
evolve again to the giant state and start a new cycle of
mass-transfer - this time from the secondary to the primary
or just the reverse of the present situation. By that time
of course, the secondary will outshine the "primary" so
that the designations will need to be reversed!

     Even stranger than this picture of periodically
reversing evolutionary cycles is the suggestion (1972) that
the odd secondary component may be a "collapsar" or a so-
called "black hole" rather than a real luminous star. A
black hole is, of course, the next theoretical step beyond
the white dwarf and the neutron star; it is a degenerate
mass of such density that not even light can escape the
gravitational field. A star of one solar mass, to achieve
this state, would need to be compressed into a sphere about
$3\frac{1}{2}$ miles in diameter. Although such a body could not be observed directly, there are fairly convincing reasons for
accepting their theoretical probability; the odd object
Cygnus X-1 is the most convincing candidate at present. In
the case of such binary systems as Beta Lyrae and Epsilon
Aurigae, the black hole hypothesis can be admitted as only
one of several possibilities. (See pages 413 and 793)

     Of the visual companions to Beta Lyrae (page 1132)
only the 7th magnitude star at 46.6" and the 9th magnitude
star at 86" seem to be physical members of the system; the
other stars listed are evidently optical companions only.

The brighter of these two stars is a spectroscopic binary
with a period of 4.348 days; only one spectrum, of type B7
V is seen.   The absolute magnitude of this star is about
0.0 (luminosity= 80 suns) and the projected separation from
Beta Lyrae itself corresponds to about 12,000 AU.

The annual proper motion of Beta Lyrae is less than
0.01"; the mean radial velocity is 11.5 miles per second in
approach.

GAMMA   Magnitude 3.25; spectrum B9 III. Position
18571n3237.   This is the most suitable star
to use as a comparison object for the nearby variable Beta
Lyrae, which lies about 1.7° to the west and slightly
north.   The famous Ring Nebula M57 lies between the two
stars, but somewhat closer to Beta.

Gamma itself has a companion star of the 12th
magnitude, discovered by A.G.Clark in 1868, but evidently
an optical companion only.   There is also a 10th magnitude
star at 177" in PA 21°. The computed distance of Gamma is
about 370 light years; the actual luminosity about 525
times that of the Sun. The annual proper motion is less
than 0.01"; the radial velocity is 13 miles per second in
approach.

DELTA   Delta 1= Magnitude 5.51; spectrum B3; position
18520n3654.   Delta-2= Magnitude 4.52; spectrum
M4 II; position 18528n3650.   These two stars are separated
by 10½' and form a wide field-glass pair with noticeable
color contrast.   They show the same radial velocity of 15½
miles per second in approach, and may form a true physical
pair despite their wide separation.   The annual proper
motion of about 0.01" is too slight to serve as a reliable
indicator of common motion. Several other stars ranging
from magnitude 7½ to 10 lie in the field and create the
appearance of a sparse and scattered cluster. The group has
been studied only enough to make its reality fairly certain
though the distance of about 800 light years places it
among the nearest star clusters.

O.J.Eggen (1964) has observed about 100 stars in
the Delta Lyrae region, and finds a color-magnitude array
which suggests a fairly young cluster, comparable in age
to the Pleiades.   Curiously, Eggen also finds that the

space motion of the Delta Lyrae group is identical to that
of the Pleiades. This is not a unique circumstance, however, as it has long been known that the Hyades Cluster in
Taurus and Praesepe in Cancer share a common space motion.
The debate concerning the possible common origin of such
clusters has never been definitely resolved.

The brightest star of the group- Delta-2 -is an
M-type variable of small range and uncertain class with an
absolute magnitude of about -3 (luminosity= 1300 suns). At
a distance of 86" will be found an 11th magnitude companion
which is probably not a true physical member of the group.
This faint star is itself a close double of 2.2" separation
and approximately equal magnitudes. The surrounding region
is very attractive for rich-field telescopes. "Glorious
field" says Webb.

EPSILON   Position 18427n3937. (ADS 11635) This is
                       the renowned "Double-double" star, one of
the most famous examples of a multiple star in the sky. It
is easily located about 1½° northeast of Vega. Appearing
to the average eye as a single star, it may be seen double
in an opera glass or by a very keen eye unaided; the wide
pair has a separation of about 3.5'. In a good 3-inch glass
each of the two stars is found to be a close double itself.
The close pairs appear to have been first noticed by Sir
William Herschel in August 1779. "A very curious double-
double star. At first sight it appears double at some
considerable distance, and by attending a little we see
that each of the stars is a very delicate double star..."
Smyth saw the whole system as "an irregular looking star
near Wega, which separates into two pretty wide ones under
the slightest optical aid. Each of these two will be found
to be a fine binary pair". The first accurate measurements
of the close pairs were made by Struve in 1831 and a fine
series of observations has been compiled since that time.

The northern component, called Epsilon 1, has a
separation of 2.8", decreasing from 3.2" during the last
century. The General Catalogue magnitudes for the two
stars are 5.06 and 6.02; the spectra are A2 and A4, and the
brighter star is a spectroscopic binary of uncertain
period. A change in angle between the components of the
visual pair has been evident since the time of Herschel;

DESCRIPTIVE NOTES   (Cont'd)

the decrease in PA has amounted to 21° in the last century.
According to a preliminary orbit computation by Guntzel-
Lingner (1956) the period of this pair may be about 1165
years with periastron in 2318 AD. The computed orbit has a
semi-major axis of 2.78"; an inclination of 138°, and the
true separation of the two stars averages about 155 AU.

The southern pair, Epsilon 2, has a present separation of 2.6", also with slight decrease since discovery.
Since the time of Herschel the PA has decreased by about
74°, or 40° in the last century. The Catalogue magnitudes
for the two stars are 5.14 and 5.37; spectra A3 and A5. For
this pair, the computed orbit gives a period of 585 years
with periastron in 2229 AD; the semimajor axis is 2.95" and
the inclination is 120°. The true separation of the two
stars averages about 165 AU.

There is very little color contrast in this system
and modern observers generally find all four stars to be
simply "white". Struve, however, found the components of
Epsilon 1 to be "greenish-white and bluish-white" while
Herschel thought the fainter star inclined toward a red
tint; this was also the impression of Admiral Smyth who
reported them as "yellow and ruddy".

All the stars share a common proper motion of
about 0.06" in PA 14° though there is no sign of orbital
revolution in the wide pair (208") which probably has a
period, as Smyth estimated, of "something less than one
million years". Parallax measurements indicate a distance
of about 180 light years for the system, giving absolute
magnitudes of +1.4 and +2.5 for the components of Epsilon
1, and +1.4 and +1.7 for the stars of Epsilon 2. The true
separation of the wide pair is close to 13,000 AU, or 0.2
light year; and the whole gigantic system shows a radial
velocity of 17½ miles per second in approach, Epsilon 1
having a slightly higher velocity than Epsilon 2.

There is a curious discrepancy in the masses of the
four stars obtained from the computed orbital elements:
For Epsilon 1 a total mass of 2.71 suns is found, with
individual values of 1.58 and 1.13; this appears completely
normal. For Epsilon 2, however, the result is a total mass
of 12.86, with individual values of 6.74 and 6.12. As the
four stars are all very similar in type and luminosity,

DESCRIPTIVE NOTES (Cont'd)

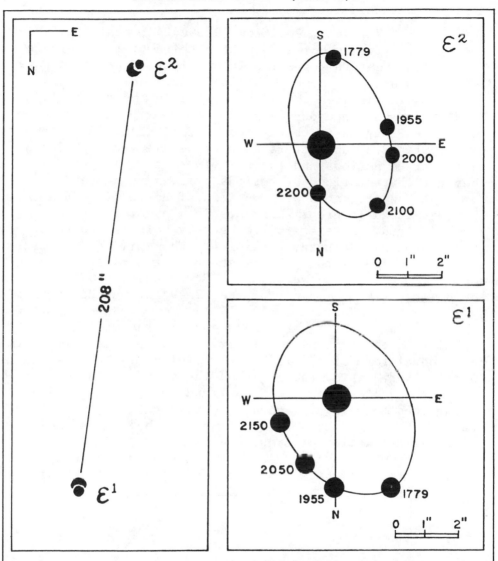

it appears impossible to accept the abnormally high masses derived for Epsilon 2. The most likely cause for this error is erroneous orbital elements, always to be expected in systems where only a fraction of the full orbit has actually been traversed by the components. The period of Epsilon 2 is probably significantly longer, and the orbit rather larger, than the best present computations suggest. (Orbit diagrams above are based on the text, page 1152)

RR   Variable. Position 19239n4241, near the Lyra-
              Cygnus border, about a third of the distance
from Delta Cygni to Vega. The identification chart on page
1155 shows stars to about 9th magnitude, and includes all
the stars in the area which are listed in the BD Catalogue.
Grid squares are one degree on a side; north is at the top.
        RR Lyrae is the typical example of a large number
of pulsating variable stars which resemble the cepheids but
have shorter periods and lower luminosities. In general,
they are characterized by periods of less than one day,
though a few are known with periods of up to 1.35 days. As
they are found in great abundance in many of the globular
star clusters, these stars are usually referred to under
the designation "cluster variables". They are not especially rare outside of the globulars, but not one is close
enough to be a naked-eye object. RR Lyrae itself is still
the brightest known member of the class; it was discovered
by W.Fleming at Harvard in 1901.
        The cluster variables resemble the well known
cepheids in the precise regularity of the pulsations and in
the general form of the light curve. The rise to maximum
is often very rapid, the light of the star increasing to
more than double in less than half an hour; the subsequent
fading then follows at a more leisurely pace. A typical
star of the class has a period of around half a day and a
brightness variation of about one magnitude. RR Lyrae has
a period of 0.566837 day, and a magnitude variation of 7.1
to 8.0.   It is a white star whose spectral class changes

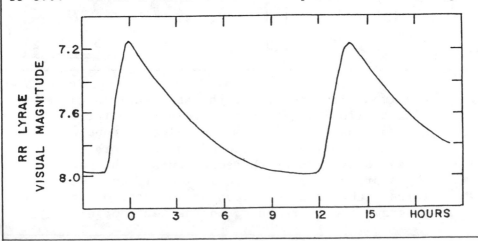

DESCRIPTIVE NOTES   (Cont'd)

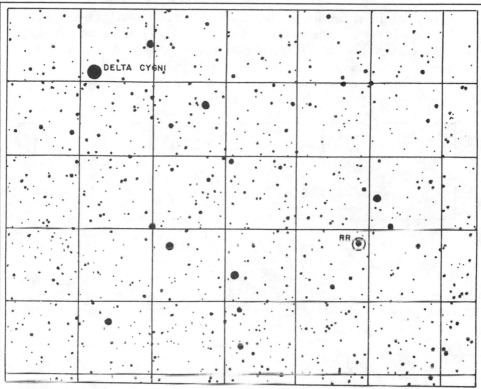

from A to F during the light cycle. Owing to certain
spectral peculiarities, the exact classification is somewhat arbitrary. In the Moscow General Catalogue (1958) the
range was given as A2 to F1; but in the newer (1970) edition it has been revised to A8---F7. Another peculiarity is
that the star appears to be oscillating in at least two
superimposed periods; the exact period and shape of the
light curve change over successive cycles in an interval of
40.8 days. The maximum magnitude (photographic) may thus
vary from 7.2 to 7.8, but variations in the minimum do not
exceed 0.13 magnitude. Color measurements show that the
star becomes somewhat bluer as it brightens.
        When S.I.Bailey published the results of his
studies of the variable stars in the great globular cluster
Omega Centauri, he reported (1902) that three separate
types of light curves seemed to be displayed by the stars.
Stars of the shortest periods, under 0.45 day, had a smooth
and symmetrical curve of small amplitude which he called
type "c". At about 0.48 day the light curve becomes much

steeper, with a sudden rise and a greater amplitude (type "a"). Finally, as the period increases again to about 1.3 day, the light curve becomes less steep and shows a more moderate amplitude. Bailey's types "a" and "b" appear to form a continuous sequence and are usually considered as a single group by modern astronomers.

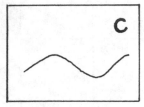
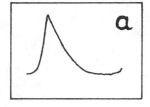
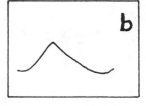

The RR Lyrae stars, then, appear to fall into two well marked groups, when arranged in order of period. Stars of Group I contain the short-period stars of Bailey's type "c"; while Group II includes his types "a" and "b". Some typical light curves for the stars of the two groups are shown here, with their periods in days.

GROUP I

RV Arie	.09
AQ CorA	.12
RW Arie	.26
ω Cent v	.30
ω Cent v	.34
ω Cent v	.41
ω Cent v	.47

GROUP II

V467 Sgtr	.43
ω Cent v	.50
ω Cent v	.61
ω Cent v	.73
ω Cent v	.90
DE CorA	1.01
V527 Sgtr	1.26
ω Cent v	1.35

The interesting and valuable fact about the cluster variables is that they all seem to be nearly equal in actual luminosity, regardless of the length of the period. In the Omega Centauri cluster, for example, they all have apparent magnitudes near 14.6, while in the cluster M3 in

Canes Venatici they appear in great numbers at magnitude
15.7.   Stars of the RR Lyrae type thus seem remarkably
identical, all having luminosities of 50 to 65 times that
of the Sun, diameters of 6 or 7 times the Sun, and spectral
types of A or F. The fact that these stars are very nearly
identical in actual luminosity makes them important as
distance indicators; a comparison of the apparent and the
absolute magnitude (m-M) gives the distance immediately
(Table IV, page 67). To use this principle, of course, the
absolute magnitude of standard variables must be determined
accurately. The formerly accepted value of 0.0 has been
found by recent studies to be about half a magnitude too
high; present evidence favors a range of about +0.3 to
+0.6. Using the value of +0.3 for RR Lyrae itself, the distance is found to be close to 900 light years. These stars,
like the better known cepheids, can be used as "measuring
sticks" in determining cosmic distances.

        In connection with this problem, it is interesting
to note that although some of the globular clusters contain
numerous RR Lyrae stars, others seem to contain none at
all; and while they occur in moderate numbers in our own
Galaxy, they appear to be relatively scarce in the nearby
Magellanic Clouds. In our own Galaxy they are definitely
associated with the "halo" component of our stellar system
rather than with the spiral arms, and their numbers show a
rapid increase toward the central hub of the Galaxy. From
these facts, and from their unusually high space velocities
ranging up to more than 200 miles per second, the cluster
variables may be classed definitely as Population II stars;
their distribution in space is very similar to that of the
planetary nebulae.

        Unfortunately, the RR Lyrae stars are not suffic-
iently brilliant to permit their use as distance indicators
for the external galaxies. RR Lyrae, for instance, would
appear as an object of magnitude 24½ if removed to the
distance of the Andromeda Galaxy M31; the identification of
such an object could not be accomplished with any telescope
in the world today. It is in our own Galaxy, and particularly in the globular star clusters, that the RR Lyrae
stars make themselves valuable as distance indicators. As
a matter of fact, it was Baade's failure to identify RR
Lyrae stars in M31 with the 200-inch telescope that first

suggested that the accepted distance of 750,000 light years must be a serious under-estimation. Studies of the RR Lyrae stars thus led, though in a negative way, to a major revision of the distance scale of the Universe.

The high space velocity of a typical RR Lyrae type star is an interesting feature which demands a word of explanation. The Sun is a Population I member, lying in the plane of the Galaxy, and sharing the rotation of the spiral pattern. Cluster variables are members of the great spherical "halo" component of the Galaxy; they may travel in highly inclined orbits around the galactic center and can cut through the galactic plane at all angles. The high relative velocity is thus a result of the two different motions; to an observer on RR Lyrae it would be the Sun which would be identified as a "high-velocity" star. One of the cluster variables, RZ Cephei, has the highest accurate space velocity measured to date, probably exceeding 400 miles per second. RR Lyrae itself, although about 900 light years distant, shows an annual proper motion of 0.23" in PA 213°.

DWARF CEPHEIDS. In 1952, the star SX Phoenicis was found by O.J.Eggen to have the abnormally short pulsation period of 79 minutes. Another similar object, CY Aquarii, has a period only slightly longer, 88 minutes. These stars, and a few others, are now known to be definitely inferior to RR Lyrae in size and luminosity. Their position on the H-R diagram (opposite page) also seems to make it plain that they should not be classed among the true RR Lyrae stars. The Harvard astronomer H.J.Smith has grouped them in a separate category which he calls "dwarf cepheids". These are remarkable objects, smaller and probably denser than the true RR Lyrae stars, several magnitudes fainter in actual luminosity, and with periods all under 0.2 day. Absolute magnitudes range from about +1 to about +5 and spectral classes again seem restricted to types A and F. Evidently related to the dwarf cepheids are a third class of object, the so-called "Delta Scuti" stars. These are also A and F-type stars with lower luminosities than the RR Lyrae stars, but with light curves of very low amplitude and periods of less than 0.2 day. The five best known examples all have absolute magnitudes in the range of +1.4

DESCRIPTIVE NOTES   (Cont'd)

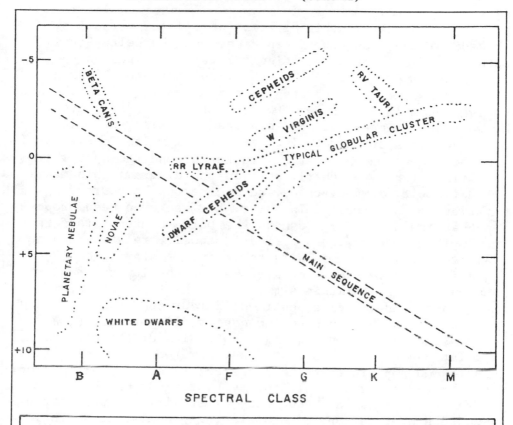

THE CLASSIFICATION OF THE SHORT PERIOD VARIABLE STARS ON
THE H-R DIAGRAM. ABSOLUTE MAGNITUDE SCALE IS AT THE LEFT.
POSITIONS OF PRE-NOVAE STARS AND THE CENTRAL STARS OF THE
PLANETARY NEBULAE ARE ALSO SHOWN.

to +2.2; periods range from 1.3 to 4.7 hours but the
amplitudes of the light curves are generally much less
than the RR Lyrae stars, seldom reaching 0.3 magnitude. In
some of the brightest known examples, the amplitude is less
than 0.1 magnitude. Stars of the class include Rho Puppis,
Delta Scuti, Delta Delphini, DQ Cephei, and CC Andromedae.
The bright stars Beta Cassiopeiae and Epsilon Cephei have
recently been identified as probable members of the class;
the latter star is especially remarkable for its ultra-
short period of 61 minutes; the amplitude of the light
curve, however, is only 0.03 magnitude.

DESCRIPTIVE NOTES   (Cont'd)

The relationship between the cluster variables, classical cepheids, dwarf cepheids, and the Delta Scuti stars is best seen by plotting the various types on the H-R diagram (page 1159). The RR Lyrae stars then form a well-marked group of spectral types A and F, lying somewhat above the Main Sequence. Dwarf cepheids average several magnitudes fainter and lie in a band crossing the Main Sequence at about spectral type A8. The Delta Scuti stars, on the other hand, appear to cluster around the spot where the first two types nearly meet on the diagram; this makes it difficult, in many cases, to decide on the correct attribution of a particular star. It has even been proposed that the dwarf cepheids and the Delta Scuti stars form a single sequence with a period-luminosity relation and an inverse amplitude-luminosity relation. A star like CY Aquarii, in this view, would represent the lower end of the sequence, with low luminosity, ultra-short period, but large amplitude; Delta Scuti would represent the upper end of the sequence, with high luminosity, longer period, and only slight variations in light. Studies of the stellar populations, however, seem to cast some doubt on this view and it now appears that the different types of short-period stars have different evolutionary histories. RR Lyrae is known to be in an advanced state of stellar evolution; the abundance of this type of star in the globular clusters is strong evidence for extreme age. This appears to be true also for the shorter period stars we call dwarf cepheids. The Delta Scuti objects, on the other hand, seem to be sub-giants which are beginning their evolution away from the Main Sequence, and probably have a long future history in the sub-giant and giant stage.

In connection with these questions, it may be of some importance to realize, from theoretical studies of stellar structure, that the mass of a typical RR Lyrae star appears to be well below the mass of the Sun. According to an investigation by R.F.Christy (1966) a satisfactory theoretical model for an RR Lyrae star requires a mass of about 0.5 sun. In the entire lifetime of the Galaxy it would seem that there has not been sufficient time for such low mass stars to evolve significantly. Yet, the RR Lyrae stars are known to be in an advanced state of stellar evolution. This flat contradiction seems to require the hypothesis of extensive mass-loss at some time in the past, presumably during an unstable giant stage. A.J.Deutsch (1968) considers that the RR Lyrae stars are descended from main sequence stars of about 1.2 solar mass, suggesting that more than half of the original mass has been ejected into space. According to A.R.Sandage, a star of the type possibly spends about 80 million years in the RR Lyrae stage, undergoing a gradual change in luminosity and a slow decrease in period during this interval. A study of the RR Lyrae stars in the globular clusters also confirms the suspected great age of these stars. In a typical globular cluster H-R diagram, they occur in the so-called "horizontal branch" (refer to pages 366-367 and 990-992) where the stars are known to have evolved past both the main sequence and red giant stages. (Refer also to Delta Cephei, Delta Scuti, RZ Cephei, CY Aquarii, and SX Phoenicis)

M56   (NGC 6779) Position 19146n3005. Globular star cluster, located in the southeast part of the constellation, slightly less than midway between Beta Cygni and Gamma Lyrae. This is one of Messier's discoveries, found in fact on the same night on which he discovered one of his comets, January 19, 1779. Although Messier was unable to resolve the object with his modest telescope, the lack of motion by January 23 demonstrated its non-cometary nature, and it was marked on Messier's chart as a "nebula without stars and having little light". William Herschel in 1784 resolved it into a mass of 11th to 14th magnitude stars; modern observers will find it readily resolvable around the edges with a good 6-inch mirror; the central mass requires a somewhat larger aperture. T.W.Webb thought it "perhaps resolvable with a 3.7 inch aperture" which may seem slightly over-optimistic. The field is richly sprinkled with vast numbers of faint and distant stars.

     M56 shows a fairly uniform structure with no very evident central condensation. According to Sawyer's "A Bibliography of Individual Globular Star Clusters" (First Supplement, 1963) the over-all diameter is about 5' and the integrated photographic magnitude about 9.55; the integrated spectral type is F5. H.Shapley in 1930 reported only one variable star in M56, but ten additional ones have since been discovered. Color measurements of the stars of

DEEP SKY OBJECTS IN LYRA. Top: The globular cluster M56, photographed with the 13-inch telescope at Lowell Observatory. Below: The Ring Nebula M57, a classic "planetary".

DESCRIPTIVE NOTES (Cont'd)

M56 suggest an obscuration of about 0.6 magnitude due to interstellar dust; allowing for this effect the true modulus of the cluster seems to be about 15.7 magnitudes; this gives a distance of about 46,000 light years. Not one of the richer clusters, M56 appears to have a true diameter of some 60 light years and a total luminosity of about 90,000 times the Sun. The cluster shows the high radial velocity of 88 miles per second in approach.

M57   (NGC 6720)   Position 18517n3258. The famous
             Ring Nebula in Lyra, probably the best known example of a planetary nebula though not the largest nor the nearest object of the type. The term "planetary nebula" is purely descriptive, implying no connection with planets. A typical nebula of this class appears as a glowing disc resembling Uranus or Neptune as seen in a large telescope. M57 is the classic example of the type and the first one to be discovered, found by the French astronomer Antoine Darquier of Toulouse in 1779. Using a telescope of about 3-inch aperture, he described it as a perfectly outlined disc as large as Jupiter, but dull in light and looking like a fading planet. Messier, in observing the comet of 1779, found it a short time later and thought it possibly "composed of very small stars...but with the best telescope it is impossible to distinguish them; they are merely suspected..." Sir William Herschel in 1785 referred to the nebula as "among the curiosities of the heavens.... a nebula that has a regular concentric dark spot in the middle...and is probably a ring of stars. It is of an oval shape, the shorter axis being to the longer as about 83 to 100..." John Herschel found the interior "filled with a feeble but very evident nebulous light: like gauze stretched over a hoop". Father Secchi, curiously enough, thought he had resolved the nebula into "minute stars, glittering like silver dust". T.W.Webb found the light apparently "fluctuating and unsteady, like that of some other planetary nebulae; an illusion arising probably from an aperture too small for the object...It is somewhat oval and bears magnifying well."
        The Ring Nebula is easily located, about 45% of the distance from Beta to Gamma Lyrae, and slightly south of a line joining them. It may be found with certainty in

DESCRIPTIVE NOTES   (Cont'd)

a 3-inch or 4-inch telescope, the total brightness being about 9th magnitude. A 6-inch telescope is usually needed to see the "ring" appearance clearly, and fairly high powers may be used since the surface brightness is high. With a good 6-inch or 8-inch glass the nebula strikingly resembles a "tiny ghostly doughnut"; the apparent size is about 80" X 60". The peculiar color, a soft bluish-green, becomes increasingly evident with larger apertures.

A faint outer envelope with a wreath-like structure was detected on photographs made by J.C.Duncan with the 100-inch telescope at Mt.Wilson in 1936. The diameter of this outer loop is about 170". Faint streaks traversing the ring were seen visually by Lord Rosse as early as 1844, and were first photographed with the Crossley reflector at Lick Observatory in 1899. F. Von Hahn in Germany appears to have been the first observer to report the central star of the nebula, about the year 1800. Lick Publications XIII gives the central star a magnitude of 15.4 visual; W.Liller finds a value of 14.2 photovisual while L.Berman reports a figure of 14.4 photographic. However, there is some suspicion that the star may be variable, since it is often a difficult object even in large telescopes. In their list of "The Finest Deep Sky Objects", J.Mullaney and W.McCall report having glimpsed the central star with the 13-inch Allegheny refractor at 600X. In the summer of 1959, on the other hand, the star did not appear to be as bright as 16th magnitude visually, and was seen only with difficulty through the 40-inch reflector at Flagstaff, Arizona. As the visibility of a faint star embedded in nebulosity is very critically dependent on the seeing conditions, it is still uncertain whether the star is truly variable. The planetary nebula NGC 7662 in Andromeda is another object which contains a suspected variable as its central star.

In the Ring Nebula, the faint central star is a peculiar bluish dwarf or sub-dwarf with a seemingly continuous spectrum; the estimated surface temperature is about 100,000°K and the computed density is several thousand times that of the Sun. The position on the H-R diagram (page 1159) seems to indicate that such a star is approaching the end of its history, and its evolution is nearing the white dwarf state. Stars of this type are the hottest known and produce strong ultraviolet radiation

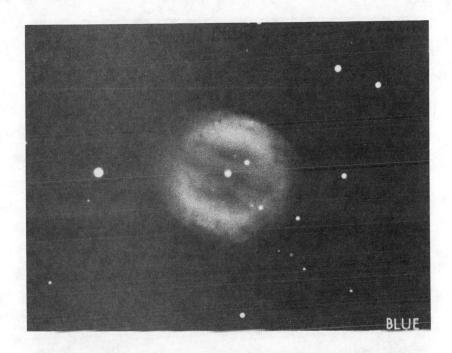

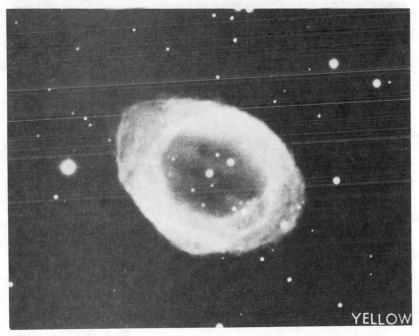

THE RING NEBULA M57 in LYRA, photographed in blue light
(top) and in yellow (below) with the 200-inch reflector at
Palomar Observatory.

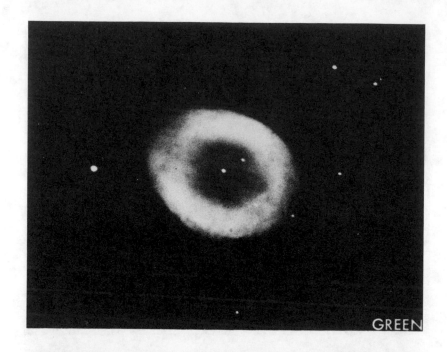

GREEN

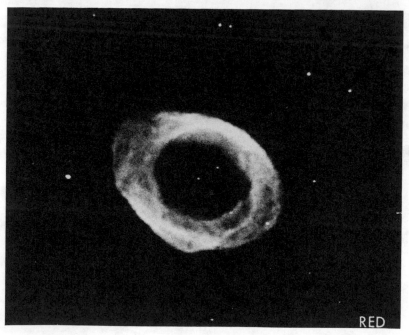

RED

THE RING NEBULA M57 in LYRA, photographed in green light
(top) and red (below) with the 200-inch reflector at
Palomar Observatory.

DESCRIPTIVE NOTES (Cont'd)

which causes bright fluorescence in the rarified gases of
the nebula. The typical bluish-green tint of the nebulae,
however, is the result of two strong emission lines (5007
and 4959 angstroms) of doubly ionized oxygen, the strange
"forbidden lines" at first attributed to a hypothetical
element called "nebulium". As a result of the work of I.S.
Bowen (1927) the atomic transitions which produce this
radiation are now well understood. In the Bohr model of the
atom, the electrons may be imagined as occupying various
possible orbits around the atomic nucleus; these orbits
represent different energy levels and an electron which has
absorbed energy will be raised to a higher level. In the
same way, an electron emits energy as it falls to a lower
level; since the structure of each type of atom limits the
number of possible "transitions" the emitted radiation must
appear as certain definite wavelengths in the spectrum. In
a gaseous nebula, however, the atoms are absorbing high-
energy photons coming from the ultra-hot central stars; in
many cases the energy is sufficient to allow electrons to
escape the parent atoms entirely. These atoms are then said
to be "ionized". Stray electrons captured by such atoms may
land in "metastable states", energy levels intermediate
between the highest possible levels and the lowest or
"ground" state. Transitions from the metastable levels are
several million times rarer than the common transitions
which produce the normal radiations observed in laborator-
ies on Earth. Under the conditions on Earth, the atoms of
any gas are in a constant state of mutual collision, and
the electrons are ejected from their metastable states
before they have a chance to make the transition to the
ground state. In the extremely rarified gases of a great
nebula, however, atoms may remain undisturbed for so long
a time (minutes instead of millionths of a second) that the
very rare transitions from the metastable state eventually
occur.

The forbidden lines, then, are not truly "forbidden"
at all; they are merely, so to speak, frowned upon severely
by the well-behaved atoms in an earthly laboratory. To the
eye, in fact, these strange radiations outshine all others
in the Ring Nebula. Other colors are also present, but are
not seen directly through the telescope because the eye
loses sensitivity to color at low light levels. The human

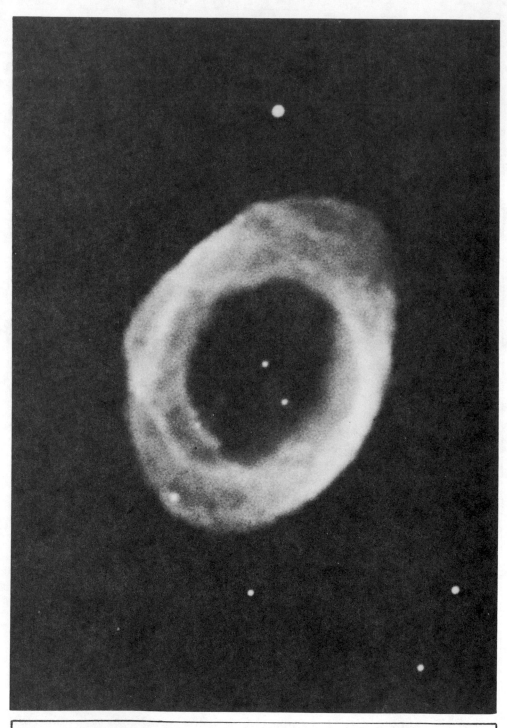

M57 in LYRA. Probably the best known planetary nebula in the heavens. Palomar Observatory photograph with the 200-inch reflector.

DESCRIPTIVE NOTES   (Cont'd)

eye is most sensitive to wavelengths in the yellow-green
portion of the spectrum, and the greenish glow of a bright
planetary nebula obliterates any other faint tints that may
be present.   Some remarkable color photographs of the Ring,
the Dumbbell (M27 in Vulpecula) and other nebulae have been
obtained by W.Miller with the 200-inch telescope; these
show spectacular colors which the eye would see if the
intensity of the illumination could be sufficiently increased.   The dark central portion of the Ring appears to be the
bluest portion of the Nebula; the annulus itself grows more
yellowish as one proceeds outward, becoming orange and
finally reddish near the outer rim.   The central portion is
dark only by contrast with the bright ring, but shows much
nebular detail on long exposure photographs.
       The "ring" appearance itself has usually been
interpreted to mean that the nebula has the structure of a
hollow shell or bubble of gas; the amount of material lying
in the line of sight would be greatest when looking through
the edge of the shell, and the center would appear comparatively blank.   NGC 7293 in Aquarius shows a very similar
appearance, as do many other planetaries. However, it has
been pointed out by Minkowski and Osterbrock (1960) that
the "shell model" is in many cases inadequate to account
for the great difference in brightness between the ring and
the center, which in M57 is a ratio of about 20 to 1. They
suggest that the ring shape is actually the true form of
some of these nebulae, and that some odd-shaped planetaries
(as M76 in Perseus) are actually toroidal rings seen edgewise.
       The distance of the Ring Nebula is not known with
any degree of precision, but reasonable calculations have
been made from studies of the nature of the illuminating
process and the probable absolute brightness of the central
star. The distance is currently believed to be well under
2000 light years, and probably closer to 1500. The Russian
astronomer I.S.Shklovsky (1957) obtained a distance of 1270
light years, while the value accepted by K.G.Jones as the
best modern estimate in his book "Messier's Nebulae and
Star Clusters" was 430 parsecs or about 1400 light years.
This distance indicates an actual diameter for the Ring of
just 0.5 light year, or about 30,000 AU.   The resulting
total luminosity of the Ring is about 50 times that of the

DESCRIPTIVE NOTES   (Cont'd)

Sun, and the central star appears to be about a magnitude
fainter than the Sun in true luminosity, with an absolute
magnitude of about +6.

The total amount of material in M57 is well under one
solar mass. The density of a typical planetary nebula is
fairly well known from theoretical calculations; according
to L.H.Aller (1956) the value is about 10,000 ions per
cubic centimeter. A density 1000 times greater would still
be considered an excellent vacuum by usual earthly standards. Aller has also studied the chemical abundances of
a typical planetary; he finds that for each 10,000 atoms of
oxygen, the following elements are present in the propor-
tions given:

Hydrogen-	17,000,000	Sulfur-	900
Helium-	1,350,000	Argon-	130
Oxygen-	10,000	Chlorine-	34
Nitrogen-	5,000	Fluorine-	4
Neon-	1,500		

According to R.Minkowski (1962) a total of 672 planetary
nebulae have been identified in our Galaxy, including a
number of very faint objects discovered during the 48-inch
Schmidt Sky Survey at Palomar Observatory. L.Perek and L.
Kohoutek listed a total of 704 known in 1963, while the
total number listed in their "Catalog of Galactic Planetary
Nebulae" (1967) was 1036. Probably at least 10,000 of the
objects exist in our Galaxy. The nearest appear as great
ghostly shells like NGC 7293 in Aquarius, 15' in diameter;
the most remote appear as tiny stellar points which can be
identified only by their peculiar spectra. The smaller
planetaries often have a remarkably high surface brightness
(NGC 7027 in Cygnus and NGC 6572 in Ophiuchus) and in a
number of cases no central star can be detected, apparently
because of the "glare" of the nebula!

All planetary nebulae do not display the neat disc
or ring shape of M57; many are more or less amorphous and
some are quite irregular. Considerable filamentary detail
is discernable in all the larger examples, and even the
irregular ones usually show some indication of symmetry.
Concentric rings separated by dark zones are frequently
present; another common appearance is the "zeta" pattern,
a Z-shaped structure resembling the form of a barred spiral

DESCRIPTIVE NOTES   (Cont'd)

galaxy. Magnetic fields may also have some connection with these structural patterns. NGC 7293 in Aquarius shows a number of radial streaks on the inner edge of the annulus, each pointing outward from the central star; the appearance suggests a slow expansion of the nebula, which is confirmed also by radial velocity measurements.

What part does a planetary nebula play in the life history of a star? The problem is still being studied and no complete answer can yet be given. The gaseous shell obviously consists of material ejected from the central star, but the exact details are uncertain. Various investigators have connected them with a wide variety of objects; novae, shell stars, B-type emission stars, Wolf-Rayet stars R Coronae Borealis stars, P Cygni stars, red giants, and white dwarfs. Some of the central stars of the planetaries show continuous spectra; others are class 0 or W, but all are hot and bluish and resemble the post-nova stars.

The distribution of the planetaries in our Galaxy definitely classes them as Population II objects. They are found occasionally at great distances above the galactic plane, there is a marked concentration toward the galactic center, and the distribution is very similar to that of the RR Lyrae stars and the globular star clusters. A planetary nebula, in fact, is known to be a member of one globular cluster, M15 in Pegasus. Thus the planetaries appear to be older objects whose central stars are nearing the white dwarf state and shedding excess mass, but quietly rather than explosively. According to one view, the planetary nebula stage may follow the red giant stage, the core of the star contracting suddenly into a hot dense sub-dwarf following the exhaustion of nuclear fuel, and allowing the tenuous outer layers to expand into space. A connection with the peculiar Wolf-Rayet stars has also been postulated since these active objects are known to be ejecting mass and must be surrounded by nebulae resembling the planetaries, although such nebulae may be too small to be detected visually. Some of the central stars of the planetaries do have spectra resembling Wolf-Rayet stars, but the actual luminosities appear to be very much lower. Absolute magnitudes for the central stars range from about +1 down to about +9; the most luminous ones thus appear to have about the same brilliance as a typical RR Lyrae star, while the

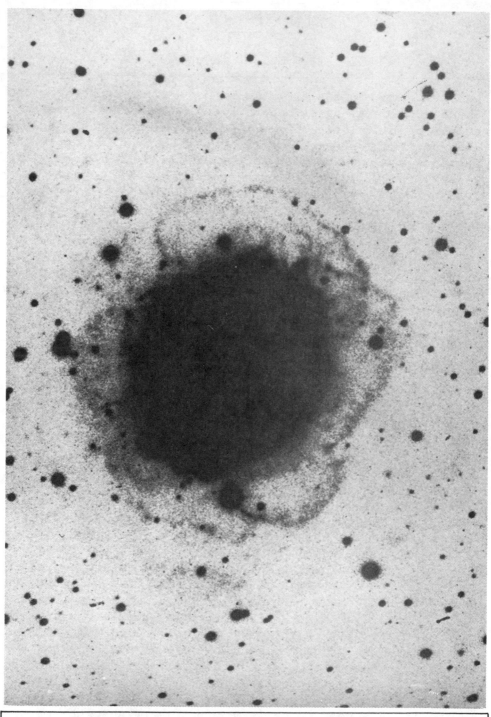

OUTER DETAILS OF THE RING NEBULA. A negative print from the photograph made by J.C.Duncan with the 100-inch telescope at Mt. Wilson.

DESCRIPTIVE NOTES   (Cont'd)

faintest ones appear to be very near the white dwarf state.
The planetary nebula, then, would seem to be the result of
the final period of mass ejection, before the star becomes
a true white dwarf.

Spectroscopic studies show that the planetary
nebulae are slowly expanding. Plates of M57 made some 60
years apart have shown a barely measurable increase in size
of about 0.3". Several such attempts to measure the rate of
expansion by direct comparison of photographs have been
made, but the results are often inconclusive. The large
planetary M27 in Vulpecula has expanded by about 0.6" in
62 years, but the predicted increase in size from radial
velocity measurements is three times greater. In NGC 2392
in Gemini, the doppler shift gives a predicted expansion
of 1.0" in only 30 years; yet no definite increase in size
has been detected in a 60-year interval. The explanation
may lie in the fact that the apparent edge of a planetary
does not represent the real boundary, but simply marks the
zone where the distance from the central star has become
too great to allow the illuminating process to operate
efficiently. The material lying inside this zone may be
said to be inside an "ionization sphere"; the size of which
may itself be changing since it depends upon such factors
as the density of the gas and the temperature of the illuminating star. Thus the apparent expansion of a planetary
nebula depends upon the relative sizes and expansion rates
of both the gas sphere and the ionization sphere.

The spectroscopic expansion velocity found for M57
is some 12 miles per second, implying that the Ring has
required only about 20,000 years to attain its present
diameter. This is on the assumption that the expansion has
proceeded at a fairly constant rate since the beginning.
Similar results are obtained for other planetaries, and do
not support the suggestion that these objects are simply
the remains of ancient novae. It is true that gaseous
shells have been observed around a number of bright novae
but these displayed enormous expansion velocities and
vanished within a few years. The shell of Nova Persei is
still visible after 70 years, and the famous Crab Nebula,
a supernova remnant, has lasted nearly a thousand years,
but these objects do not resemble typical planetary nebulae.
Compared with a nova shell, the planetaries seem to be

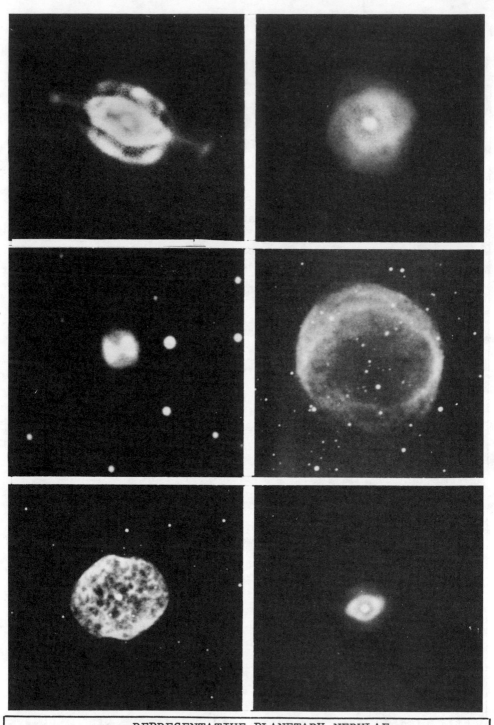

REPRESENTATIVE PLANETARY NEBULAE

NGC 7009, Aquarius	NGC 6543, Draco
NGC 3195, Chamaeleon	NGC 6781, Aquila
NGC 1501, Camelopardalis	NGC 3242, Hydra

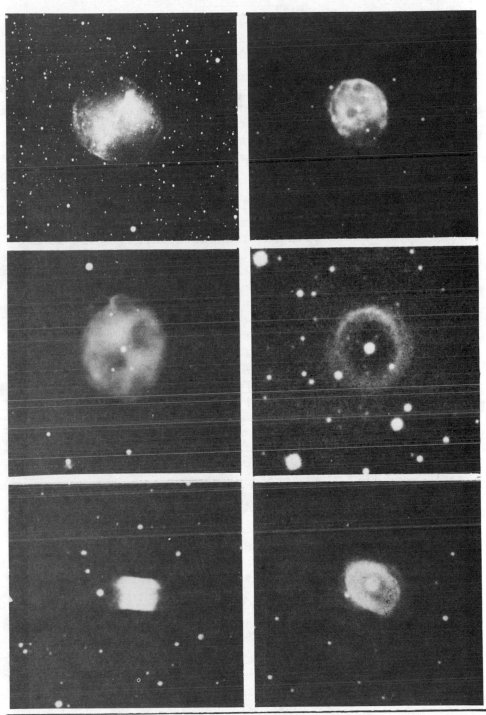

REPRESENTATIVE PLANETARY NEBULAE

M27, Vulpecula	NGC 246, Cetus
M97, Ursa Major	Sp 1, Norma
IC 4406, Lupus	NGC 3132, Vela

relatively permanent structures; if they are related to
novae at all they must be considered exceptionally sluggish
examples of nova activity. A further objection is the comparatively large mass of a planetary nebula, some 10% of
the mass of the Sun. A nova shell is believed to contain
only a few millionths of the mass of the exploding star.

In a planetary nebula, then, we have a fairly symmetrical ring-shaped or bubble-shaped gas cloud, about
half a light year in diameter, formed by mass-ejection from
a hot bluish dwarf or subdwarf, but apparently not in a
nova-like explosion. These stars are seemingly intermediate
in size and density between the normal stars and the true
white dwarfs; they are the hottest stars known and are
believed to be nearing the final white dwarf state. In some
of them the ejection of material still seems to be taking
place, while others show no sign of present activity.

In a study of 65 central stars of planetary nebulae
G.O.Abell (1965) found photographic absolute magnitudes of
from 0 to about +10, with a mean value near +5; temperatures ranged from 30,000°K up to about 400,000°K, and the
computed diameters of the stars varied from less than 0.01
the solar diameter up to nearly the size of the Sun. These
results support the conclusion of C.R.O'Dell (1964) that
some of the central stars have the dimensions of true white
dwarfs, and that the planetary nebula phenomenon probably
represents a stage in the evolution of certain stars to
the white dwarf state.

It is interesting to reflect upon the planetary nebulae which we see today, speculating that they must be
ephemeral objects by astronomical standards, and are fated
to expand outwards into invisibility in a mere 30 or 40
thousand years. This implies that new ones are forming at
the rate of several per year, in order to account for the
present estimated total of 10,000 in our Galaxy. The birth
of a planetary nebula, then, appears to be about as frequent as a nova outburst, but it is not certain that there
is any connection between the two classes of objects, other
than the fact that both are the results of some sort of
stellar instability. There are, however, a few cases in
which a star may be in the process of "growing" a planetary
nebula at the present time; the curious star FG Sagittae
is perhaps the most noteworthy suspect. Since about 1890

NGC 6791. An unusually rich galactic star cluster in the
Lyra Milky Way, photographed with the 48-inch Schmidt
telescope at Palomar Observatory.

this star has slowly brightened from magnitude 13.7 to about 10; modern observations show that the star is nicely centered in a round nebulous disc about 30" in diameter which, from spectroscopic studies, appears to be a true planetary. The nebula, however, is much too large to have originated in the 1890's; evidently the star had consider- able emission activity at some time in the distant past. Yet, spectroscopic analysis indicates that the eruptive activity is in progress at the present time, so it would seem that the formation of a second gaseous shell is cur- rently underway. The bubble-like shell around the strange nova-like variable Eta Carinae is another possible example though the star itself does not resemble the known nuclei of any of the classic planetaries.

For the modern observer, a considerable number of interesting planetary nebulae are within range of a good amateur instrument, ranging from tiny stellar-appearing objects up to great phantom rings such as NGC 7293 in Aquarius. To identify the more stellar planetaries, the interested observer may experiment with a simple technique that impressively demonstrates their peculiar radiation: merely obtain a small piece of replica diffraction grating and place this between the eyepiece and the observer's eye. Seen in this way, all the images of stars will be drawn out into narrow colored streaks, but a planetary nebula will appear as a series of discrete individual images, each one indicating a definite wavelength in which the object is radiating. The observer should try this unusual technique on some of the smaller and brighter planetaries, such as NGC 6572 in Ophiuchus or NGC 6210 in Hercules, before attempting to identify more distant nearly stellar examples.

To discern the central star of a planetary nebula, a "blinking" technique is sometimes helpful in objects where the total light of star and nebula are about equal. If one gazes directly at the central star, the nebula may seem to vanish; with averted vision the nebula reappears but the star is then lost in the nebulosity. Alternating rapidly from direct to averted vision gives a "blinking" effect which is quite striking in a few objects, as the "Blinking Planetary" NGC 6826 in Cygnus. (For a list of bright planetary nebulae, refer to M27 in Vulpecula)

LIST OF DOUBLE AND MULTIPLE STARS

NAME	DIST	PA	YR	MAGS	NOTES	RA & DEC
h3607	37.2	127	39	8 - 8½	perhaps slight PA inc, spect K0, A5, color contrast	03403s8102
h3612	19.4	162	19	8 - 9	spect F8	03451s8011
h3673	10.2	67	39	7½- 7½	slight PA inc, spect F8	04271s7748
I 472	0.9	266	27	7½- 9½	PA dec, spect A2	04336s7844
	18.6	78	27	- 11		
h3741	46.5	122	19	6 - 10	spect K0	05026s7822
Hrg 2	1.1	172	47	7½- 8	relfix, spect A0	05026s7425
γ	38.2	107	18	5 - 11	(h3795) optical, dist inc, spect K4	05338s7623
I 277	3.9	190	00	7½- 11	spect K0	05359s7110
h3911	22.0	47	18	7 - 11	PA inc, spect G5	06496s7648
h3932	8.3	284	18	7½- 10	spect A2	06565s7743
h3996	16.6	255	19	7½- 11	spect B9	07199s8423
I 312	1.1	161	44	8½- 8½	PA dec, spect F5	07202s7547

LIST OF VARIABLE STARS

NAME	MagVar	PER	NOTES	RA & DEC
R	8.5--11..		Semi-reg, spect M6	05444s7516
S	9.0--10...	Irr	Spect M4	06592s7554
U	7.0--10..	410	semi-reg	04139s8159
X	9.9--14..	380	LPV, Spect M3e	03335s7637
TY	7.8--8.0	.1875	Cl.Var., spect A3	05317s8137
TZ	6.2---7.0	8.569	Ecl.Bin.; spect A0	05398s8449
UX	9.2--9.6	2.091	Ecl.Bin.; spect F8	05319s7617

MICROSCOPIUM

LIST OF DOUBLE AND MULTIPLE STARS

NAME	DIST	PA	YR	MAGS	NOTES	RA & DEC
Gls259	3.8	156	59	8 - 8	AB cpm, PA inc,	20285s4104
	10.3	356	59	- 13	spect G0; AC dist	
					inc, optical	
h5206	16.4	193	18	7½- 12	spect A2	20296s3133
I 1425	0.4	150	25	8½- 8½	spect G0	20360s2836
h5211	20.5	300	16	6½- 11	spect G5	20376s4235
h5211b	6.3	248	16	11- 13		
λ428	4.5	185	36	7- 11½	spect G0	20420s3521
h5218	9.9	191	35	7 - 12	relfix, spect F0	20423s3040
Stn 53	17.3	177	00	7½- 11	spect G5	20449s2756
α	20.6	166	33	5 - 9½	(h5224) relfix,	20469s3358
					spect gG7	
h5228	32.3	104	19	7½- 10	spect K0	20484s4106
β765	1.2	120	53	7 - 12	PA & dist dec,	20575s3529
					spect K0	
△236	57.8	73	51	6½- 7	cpm pair, spect	20589s4312
					G3, K0	
λ435	0.3	292	59	6½- 7½	relfix, spect K0	21002s2756
Daw153	3.5	280	41	7- 11½	spect G5	21060s2841
β251	2.6	233	43	7½- 9½	relfix, spect F5	21091s3048
θ^2	0.5	267	59	6½- 7	(β766) PA dec,	21212s4113
	78.4	66	00	-10½	spect A0	
B528	6.9	170	33	7 - 13	spect K0	21216s3511
β767	2.8	148	58	5½- 8½	PA slight inc,	21238s4246
					spect A3	

MICROSCOPIUM

LIST OF VARIABLE STARS

NAME	MagVar	PER	NOTES	RA & DEC
θ^1	4.8± .05	2.1219	α Canum type; spect A2p	21176s4101
R	8.4--13..	138	LPV. Spect M4e	20370s2858
S	7.8--13..	209	LPV. Spect M3e	21238s3004
T	6.0--8..	347	Semi-reg; spect M6e	20249s2826
U	8.5--13..	334	LPV. Spect M6e	20259s4035
V	9.4--12..	381	LPV. Spect M6e	21206s4055
W	8.8--13..	199	LPV. Spect M4e	21198s4211
X	8.5--13..	240	LPV. Spect M4e--M5e	21015s3328
Y	9.0--10.5	182	Semi-reg	21040s3429
RR	9.6--10.8	31.782	Cepheid, W Virginis type	21125s4325
RS	8.6--13..		LPV.	20260s2841
RU	9.5--10..	165:	Semi-reg	20355s3651
RY	8.6--12..	190	LPV.	20552s4025
VY	8.4--8.7	4.436	Ecl.Bin; spect A3	20460s3355
ZZ	9.4---9.7	.0672	Cl.Var; spect A4--A8	20570s4251
AU	8.6--9.0	4.865	Ecl.Bin + flare star?	20421s3131

LIST OF STAR CLUSTERS, NEBULAE, AND GALAXIES

NGC	OTH	TYPE	SUMMARY DESCRIPTION	RA & DEC
6923		⊘	Sb; 12.9; 2.0' x 1.0' pF,cS,1E,gbM	20286s3101
6925		⊘	Sb; 12.1; 3.0' x 1.0' cB,L,mE,pslbM	20312s3209
	I.5039	⊘	Sb; 13.1; 1.0' x 0.5' eF,pS,vmE	20402s3003
6958		⊘	E0/E1; 12.5; 0.7' x 0.6' B,cS,R,vmbM	20454s3811
----	I.5105	⊘	E4; 13.0; 1.0' x 0.6' vF,vS,E	21212s4050

MONOCEROS

LIST OF DOUBLE AND MULTIPLE STARS

NAME	DIST	PA	YR	MAGS	NOTES	RA & DEC
A322	4.1	359	32	7½–13½	cpm, spect G0	05559s0439
3	1.8	353	35	5½– 9	(β16) relfix, spect B5	05595s1036
AC 3	0.8	193	58	6½– 9	PA inc, spect B9	06093s0439
β566	1.7	209	38	6½– 12	PA dec, spect A0	06121s0433
β323	2.3	101	21	8½– 10	spect A3	06123s0142
A505	0.7	254	17	7½– 13	spect K0	06123s0426
β567	4.2	242	58	7 – 11	PA dec, spect A2	06130s0454
A668	0.3	29	60	6½– 6½	PA inc, spect B9	06130s0901
A323	1.1	218	43	7 – 10½	spect A0	06158s0538
A2667	0.4	129	62	7½– 8	binary, 90 yrs; PA inc, spect A5	06188n0218
8	13.2	27	34	4½– 6½	(ε) (Σ900)	06211n0437
	93.7	254	11	–12½	relfix, spect A5, dF4; fine field.	
J53	1.8	127	57	7 –10½	spect K0	06212n0242
	33.9	103	43	– 10		
β97	0.7	269	66	7 – 9	PA inc, spect B8	06220s0123
β569	1.4	119	43	8 – 10	slight PA dec, spect K0	06229s1054
Σ910	66.6	151	32	7 – 8½	spect G5, G5	06242n0029
Σ910b	0.6	158	43	8½– 9	PA dec	
Σ911	13.0	156	42	8½– 8½	slight dist dec, spect K0	06242n0406
Σ914	20.9	298	38	6½– 9	relfix, spect A0	06243s0729
Σ915	5.8	37	21	8 – 9	relfix, spect B9	06256n0518
	38.8	126	12	–12		
11	7.4	132	55	5 – 5½	(β) (Σ919) fine	06264s0700
	25.9	56	32	– 12	multiple star, ABC	
11b	2.8	105	55	5½– 6	cpm, relfix, spect B3, B3 (*)	
0Σ142	8.6	353	01	7½– 11	relfix, spect B3	06272n0709
Σ920	9.3	212	28	8 – 11	relfix, spect B3	06278n0422
Σ921	16.3	3	58	6 – 8	relfix, spect B2, A5	06284n1117
0Σ144	22.5	145	28	7½– 12	PA dec, spect G0	06288n0257
Σ926	11.7	289	48	7½– 8½	slight dist inc, spect A0	06290n0548

LIST OF DOUBLE AND MULTIPLE STARS (Cont'd)

NAME	DIST	PA	YR	MAGS	NOTES	RA & DEC
GAn3	3.2	284	16	7 - 12	(A2927) spect B2,	06293n0459
	6.9	319	17	- 12	in cluster NGC	
	12.5	288	16	- 12	2244	
	13.2	197	16	- 10		
OΣ146	31.7	140	57	5½- 9½	optical, slight	06296n1143
					dist dec, spect K0	
β98	0.7	152	66	8 - 8	PA inc, spect B8	06302s0518
A2673	1.3	304	43	7½- 10	spect K0	06317n0321
	28.7	107	20	- 12		
14	10.2	209	58	7 - 11	(Σ938) relfix,	06321n0737
					spect A0	
A2821	3.3	213	22	7½- 13	spect A2	06344n0534
A509	1.4	141	43	7½- 10	spect K0	06355s0844
	9.6	72	16	- 14		
Rst4820	6.0	298	41	7½- 11	spect K5	06378s0425
15	3.0	213	57	5 - 8½	(S Mon) (Σ950)	06382n0957
	16.6	14	57	- 11	slight PA inc.	
	41.1	309	38	- 11	Chief star of the	
	74.0	140	57	- 9	cluster NGC 2264;	
					spect O7, B7 (*)	
Σ949	3.3	292	21	8½- 9	relfix, spect A0	06382n0545
D11	3.7	45	43	9 - 9	cpm; in cluster	06383n0957
	40.8	222	24	- 9	NGC 2264	
Σ953	7.0	330	32	7 - 7½	relfix, cpm pair;	06384n0902
					spect F5	
Σ954	13.3	152	25	7½- 10	relfix, spect B3	06384n0931
Σ3117	0.8	88	43	9 - 9½	PA dec, spect B9,	06386n0947
					in NGC 2264	
Σ3118	2.6	172	36	9 - 9½	relfix, in cluster	06388n0953
					NGC 2264	
A1055	2.7	281	27	7 - 14	spect B8	06390s0707
A2677	0.8	16	23	7½- 12	spect B9	06398n0255
	21.4	312	23	- 14		
OΣ157	0.4	254	66	7½- 8	binary, about 265	06452n0024
					yrs; PA dec, spect	
					A2	
A58	4.5	154	50	7½- 8½	PA inc, spect F8	06461s0401
β897	6.2	28	33	6½- 12	relfix, spect F2	06483s0029
Σ987	1.2	172	66	7 - 7½	PA inc, spect A3	06517s0547
Σ986	5.7	165	31	8½- 9	relfix, spect A2	06522n0934

LIST OF DOUBLE AND MULTIPLE STARS   (Cont'd)

NAME	DIST	PA	YR	MAGS	NOTES	RA & DEC
Σ 992	14.2	299	38	8 - 9½	relfix, spect G0	06533s0925
β 326	1.2	51	43	8 - 9½	PA dec, spect G5	06536n0223
Σ 998	3.5	208	54	8 - 8½	relfix, spect A5	06544s0525
β 327	0.6	101	66	7½- 8	relfix, spect B3	06560s0257
	13.5	100	30	-11½		
Σ1003	3.9	320	38	9 - 9	relfix, spect F2	06562s0906
β1060	3.6	59	53	7 - 12	perhaps slight dist inc, spect G8	06563n0340
Σ1010	23.1	6	31	8 - 9	(A518) spect B3	06590s0303
	2.0	187	29	-15		
β573	0.8	287	66	7½- 8	PA inc, spect F8	06595s1049
Ho 241	8.8	183	00	8 - 13	spect B9	07017n0439
Σ1015	4.8	197	49	8½-8½	relfix, spect A0	07025s0542
A1741	1.0	18	58	8½-8½	PA inc, spect F5	07026s0047
Σ1019	6.5	278	33	7 - 11	(D12) relfix,	07035s1035
	37.9	295	20	-9½	spect B3	
Σ1029	2.0	25	53	7½- 8	relfix, spect F0	07055s0436
Σ1030	15.8	43	38	8 - 9	relfix, primary spect composite= K + A0	07064s0836
Σ1034	2.3	16	38	8½- 9	relfix	07070s0814
δ	32.0	170	10	4 - 13	(22 Mon) spect A0	07093s0025
Σ1043	2.5	249	37	9 - 9	relfix, spect F5	07101s0036
Σ1045	5.6	239	56	8 - 9	PA inc, spect F5	07102s0305
Σ1049	3.6	40	46	8 - 10	relfix, spect A0	07113s0850
A524	2.7	157	58	6½- 11	PA inc, spect K5	07117s0349
24	3.8	313	46	6 - 12	(β1268) relfix, spect G5	07128s0004
Σ1056	4.0	299	33	7½- 8½	relfix, spect G0	07130s0146
Ho 242	4.4	64	24	7 - 12	spect A5	07176s0454
Σ1084	14.1	286	19	7 -9½	slight dist inc, spect K0	07215s0353
A1967	1.3	358	66	7½- 9½	spect G5	07285s0204
Σ1109	3.2	17	55	9 - 9	relfix, spect A0	07294s0025
	27.0	307	07	- 11		
Σ1111	20.0	220	38	8 - 8½	relfix, primary spect composite= A + G	07296s0835
Σ1112	23.6	113	38	7½- 10	spect F5	07297s0846

MONOCEROS

LIST OF DOUBLE AND MULTIPLE STARS   (Cont'd)

NAME	DIST	PA	YR	MAGS	NOTES	RA & DEC
A534	0.1	148	37	7½- 10	(B2525) both PA	07353s0229
	1.1	303	43	- 10	inc, spect A5	
Σ1128	15.8	169	38	8 - 10	spect G5	07374s0608
Σ1132	19.7	235	22	8 - 8½	relfix, cpm pair;	07397s0324
					spect K5	
β1195	0.3	91	50	7½- 7½	PA inc, spect A0	07489s0917
Σ1154	2.6	353	66	7½- 10	relfix, spect A5	07496s0255
Σ1157	0.9	214	66	8 - 8	PA & dist dec,	07520s0240
					spect F0	
A1580	0.1	188	51	7½- 9	PA inc, spect F5	07593s0827
Σ1183	30.9	326	35	5½- 7½	(A543) spect A0	08040s0906
Σ1183b	1.2	323	36	-12		
ζ	32.0	105	36	5- 10½	(29 Mon) (Σ1190)	08061s0250
	66.4	245	36	- 8½	all optical, spect	
					G2, G8, K2	

LIST OF VARIABLE STARS

NAME	MagVar	PER	NOTES	RA & DEC
R	10--12	Irr	Nucleus of Nebula NGC 2261	06364n0847
			RW Aurigae type ? (*)	
S	4.5--5..	Irr	Spect O7; chief star of	06382n0957
			cluster NGC 2264 (*)	
T	5.6---6.6	27.02	Cepheid; spect F7--K1	06225n0707
U	5.8---7.7	92.26	RV Tauri type, spect F8e-	07284s0940
			K0p	
V	6.3--13.7	334	LPV. Spect M5e--M8e	06202s0210
W	9.0--13..	---	Irr or Semi-reg; spect N	06499s0705
X	6.9--10.2	156	Semi-reg; spect gM3e--M4	06548s0900
Y	8.5--14..	231	LPV. Spect M4e--M5e	06541n1119
Z	9.0--10.1	Irr	Spect K5	06304s0850
RR	8.5--15..	393	LPV. Spect S7e	07150n0111
RS	9.6--12.7	264	LPV. Spect M3e--M6e	07048n0504
RT	8.2--10.3	115	Semi-reg; spect M3	08064s1038

LIST OF VARIABLE STARS (Cont'd)

NAME	MagVar	PER	NOTES	RA & DEC
RV	7.0--8.9	132	Semi-reg; spect N	06557n0614
RW	9.1--11.0	1.906	Ecl.Bin; spect A0	06320n0852
RX	9.0--12..	343	LPV. Spect M6e	07269s0410
RY	7.7--9.2	466	Semi-reg; spect R	07045s0729
SU	8.1---9.4	Irr	Spect S3	07399s1046
SV	8.0--9.6	15.23	Cepheid; spect F8--K5	06188n0630
SW	9.2--10.7	112	Semi-reg; spect M4	06244n0525
SX	8.5--10..	100:	Semi-reg; spect M6	06493n0450
SY	8.6--14..	422	LPV. Spect M6e	06350s0121
TT	8.0--12..	323	LPV. Spect M6e	07232s0544
TU	9.0--10.9	5.049	Ecl.Bin; spect B5 + A5	07508s0255
UX	7.9--8.8	5.905	Ecl.Bin; spect A6p + G2p	07568s0722
UY	9.0--9.7	1.261	Ecl.Bin; spect F8	06561n0942
VV	9.6--10.4	6.051	Ecl.Bin; spect G0	07009s0540
AO	9.3--9.9	1.885	Ecl.Bin; spect B3 + B5	07041s0433
AP	9.0---9.5	Irr	Spect M3	07071s0639
AR	9.0--9.8	21.21	Ecl.Bin; spect K0	07183s0510
AU	8.3---9.2	11.11	Ecl.Bin; spect B5 + F0	06524s0119
AX	7.0---7.3	Irr	Spect B1e + gK0	06279n0554
BB	9.9--10.4	.7327	Ecl.Bin; spect A0	06589s0837
BC	9.0--12..	272	LPV. Spect M3e	07565s0421
BG	9.2--10.4	30:	Semi-reg; spect N	06537n0708
BI	9.5--14..	433	LPV.	06583n1049
BS	9.5--10.4	Irr	Spect M0p	07216s0327
BT	4.5----16	---	Nova 1939	06412s0158
BX	8.5--12..	1374	LPV. Spect Me/pec	07229s0330
DI	9.5--11..	Irr	Spect M6	06471n0314
FW	9.5--10.5	3.874	Ecl.Bin; spect B5 + F2	07552s0703
GI	5.2----15	---	Nova 1918	07243s0635
GL	9.0--10..	104:	Semi-reg; spect M6	06333s0520
GY	7.7---8.9	Irr	Spect N3	06507s0431
HH	9.0---9.6	Irr	Spect F5	06528s0722
IM	6.5---6.6	1.190	Ecl.Bin; lyrid, spect B5 + B8	06205s0315
KS	8.5--10..	Irr	Spect N	06173s0516
KT	9.8----15	---	Nova 1942	06226n0528
MR	9.5--10..	120:	Semi-reg; spect M5	06446n0135
NR	9.0--10..		Semi-reg; spect M6	07241s0849
PS	9.0--10..		Semi-reg ? Spect M	08065s1015

MONOCEROS

LIST OF VARIABLE STARS   (Cont'd)

NAME	MagVar	PER	NOTES	RA & DEC
PZ	8.8---9.7		Flare star? spect dK2e	06459n0116
QR	9.3--10.4		Semi-reg; spect M5	06515n0051
V448	9.4---9.9	1.118	Ecl.Bin; spect A2	06451n0126
V474	6.0--6.36	.1352	Delta Scuti type; spect F2	05566s0923
V505	7.5---8.0		Ecl.Bin? Spect B5	06432n0233
V519	8.2--9..		LPV? Spect gM5	06489n0002
V523	7.0---7.2	45:	Semi-reg; spect M3	06557s0857
V526	8.0-- 8.4	2.675	Cepheid; spect G0	06594s0055
V569	6.5± .06	.267	β Canis type; spect B0	07035s1035
V571	5.5± .03	.0999	δ Scuti type; spect A8	07098s0013
V592	6.15--6.3	2.976	α Canum type; spect A2p	06483s0759
V613	7.7± .07		Semi-reg? Spect M2	06457n0536
V614	7.2--7.6	60:	Semi-reg; Spect R5	06595s0311

LIST OF STAR CLUSTERS, NEBULAE, AND GALAXIES

NGC	OTH	TYPE	SUMMARY DESCRIPTION	RA & DEC
2149		☐	Mag 12.5, diam 2.0' x 1.3' S,F, stellar nucleus (erroneously classed as a galaxy in Shapley-Ames list and Skalnate-Pleso Atlas)	06011s0944
2170	19[4]	☐	9^m star in F neby 1' diam	06052s0623
2182	38[4]	☐	9^m star in F neby 3' diam	06071s0619
2185	20[4]	☐	vF neby 2' x 2' with 12^m star	06087s0612
2215	20[7]	⋰	L,pRi,pC, diam 8', class D; about 25 stars mags 11....	06184s0716
2232	25[8]	⋰	5^m star (10 Mon) with small group mags 6---8	06241s0443
2236	5[7]	⋰	Diam 5', pRi,pC, class F, about 50 stars mags 11.....	06270n0653
2237		☐	eL,F,great diffuse halo diam 80' x 60' surrounding cluster NGC 2244. "Rosette Nebula"	06296n0440
2244	2[7]	⋰	Mag 5.5, diam 40', vL,B,!C, about 15 stars mags 6....9 incl 6^m star (12 Mon) and surrounding neby NGC 2237 (*)	06297n0454

LIST OF STAR CLUSTERS, NEBULAE, AND GALAXIES (Cont'd)

NGC	OTH	TYPE	SUMMARY DESCRIPTION	RA & DEC
2245	3^4	□	pL,F, 3' x 5' with 11^m star	06299n1012
2251	3^8	⋯	vL,E, diam 10', about 25 stars mags 10....	06320n0824
2259	28^6	⋯	Diam 4', cRi,C, class D, about 25 stars mags 12....	06358n1055
2262	37^7	⋯	S,C, diam 3', about 15 faint stars	06358n0114
2261	2^4	□	! B,mE, Mag 10, diam 2'; Comet-like shape with variable star R Mon at tip. "Hubble's Variable Nebula" (*)	06364n0846
2264	5^8	⋯	5^m star (S Mon) and L,B,lC group of 20 stars mags 6...10 with F neby inv; incl dark "Cone Nebula" (*)	06384n0956
2282		□	10^m star in R,F,pL neby diam 3'; star spect B5	06443n0123
2302	39^8	⋯	L,P,lC, about 15 faint stars	06485s0710
2301	27^6	⋯	Mag 6, diam 15', class D; L,Ri, about 60 stars mags 8..	06492n0031
2309	18^6	⋯	Diam 4', pL,pRi,mC; about 25 faint stars mags 12....	06538s0708
2324	38^7	⋯	Diam 9', L,Ri,cC; class E about 50 stars mags 11....	07004n0108
2323	M50	⋯	Mag 6, diam 10', class E;1 vL,Ri,pC, about 100 stars mags 9---14 (*)	07005s0816
----	I.2177	□	vL,eF, neby 85' x 25'	07031s1029
2343	33^8	⋯	Diam 4', about 15 stars mags 10....	07059s1034
2353	34^8	⋯	Diam 15', lC, one 6^m star + 20 stars mags 10... class D	07123s1012
----	Mel 72	⋯	pS,F,Ri, diam 5', class D about 40 stars mags 12....	07357s1033
2506	37^6	⋯	pL,vRi,C, mag 11, diam 10'; class G; about 75 stars mags 11..... (*)	07577s1029

DESCRIPTIVE NOTES

BETA   (11 Monocerotis) Position 06264s0700. Magnitude 4.60; spectrum B3 Ve. This is a fine triple star, one of the best examples of a true ternary system in the sky for the small telescope. It was discovered by Sir William Herschel in 1781, and was described by him as "one of the most beautiful sights in the heavens". This is one of the unusual cases where three stars of a system have nearly equal magnitudes; bright pairs with one or more faint and distant companions appear to be more the usual rule in the heavens.

The AB pair of this system consists of two 5th magnitude stars 7.4" apart; the third "C" component is 2.8" distant from B, and the three stars form a narrow triangle with only the slightest change (if any) since the accurate measurements of F.G.W.Struve in 1831. E.J.Hartung, however, thinks that both separations and angles seem to be increasing very slowly. The Yale "Catalogue of Bright Stars" gives the magnitudes of the three components as 4.50, 5.22, and 5.60; all three stars have spectra of class B3 with emission lines. There is no color contrast in this otherwise beautiful system; all three stars appear brilliantly white.

Beta Monocerotis shows an annual proper motion of about 0.025"and the radial velocity of the system is about 12 miles per second in recession; variations in the radial velocity of the A component suggest that the star may be a close binary. The distance, from trigonometrical parallax measurements, appears to be in the range of 150 to 200 light years, suggesting absolute magnitudes of about +1.0, +1.6, and +2.0 for the three stars. The projected separation of the AB pair is about 400 AU.

ROSS 614   Position 06269s0246, about 2° north of the coarse cluster NGC 2232. Ross 614 is a faint but notable binary system, containing two of the smallest stellar masses yet discovered. The star was first detected by F.E.Ross at Yerkes Observatory in 1927, through a comparison of plates made in 1907 and 1926. The star has an annual proper motion of 1.00" in PA 131°and is a red dwarf of type dM6e with a visual magnitude of 11.1. At a distance of 13.1 light years this is one of the 25 nearest stars known. The absolute magnitude is 13.1; about 2100

DESCRIPTIVE NOTES (Cont'd)

stars of this brightness would be required to equal the luminosity of our Sun. Ross 614 has an estimated diameter of about a third the solar diameter.

Photographs of Ross 614 have been obtained at the McCormick Observatory since 1932 and at Sproul Observatory since 1938. Careful measurements of the proper motion show a periodic variation, evidently indicating the presence of an unseen companion. In 1950 the period was determined to be 16.5 years. All visual attempts to detect the companion were unsuccessful until the star was observed with the 200-inch telescope at Palomar in March 1955. Dr.W.Baade was able to detect the companion visually and also obtained photographs; the pair then being, as predicted, near its greatest apparent separation of 1.2".

The elusive companion is a star of visual magnitude 14.8, revolving about the primary at an average distance of 3.9 AU. Both stars are among the smallest stellar masses yet determined, with 0.14 and 0.08 the mass of the Sun. Ross 614b is very exceptional in this respect, being at discovery the smallest stellar mass known up to that time. At present, only one visible star is known to have a still smaller mass: the binary L726-8 (UV Ceti System) where each star has a mass of 0.04 sun.

With a mass of 0.08 that of the Sun, Ross 614B has about 80 times the mass of Jupiter, and a luminosity of 1/63,000 that of the Sun. The absolute magnitude is about +16.8, placing the star among the faintest known. The true diameter may be about 0.1 that of the Sun, or about a third the size of the primary. The radial velocity of the system is 14.4 miles per second in recession; the true space velocity is about 18 miles per second.

Ross 614 is one of the classic cases where the presence of a close companion was detected before it was actually seen; Sirius and Procyon are other well known examples. At the present time there are a number of nearby stars which show slight periodic perturbations, and which must be attended by unseen companions; some of the computed masses are smaller than any known star. Outstanding among such cases are Lalande 21185 in Ursa Major, whose companion must have about 0.01 the solar mass; 61 Cygni C with a computed mass of 0.008, and the "planetary" companion to Barnard's Star in Ophiuchus with a mass of 0.0015.

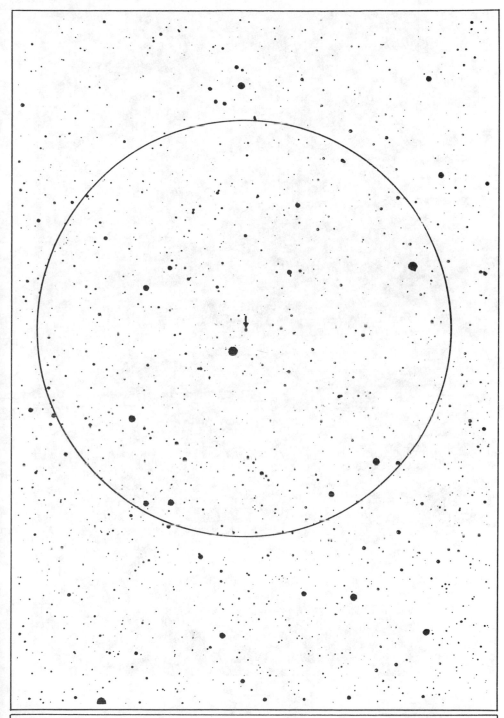

ROSS 614 Identification chart, made from a Lowell Observatory 13-inch telescope plate. The circle is one degree in diameter with north at the top; limiting magnitude 15.

PLASKETT'S STAR. The famous high-mass star is encircled on this Lowell Observatory photograph. NGC 2244 is in the lower right corner; 13 Monocerotis is at upper right.

DESCRIPTIVE NOTES   (Cont'd)

PLASKETT'S STAR   (HD 47129) (BD+6°1309) (GC 8631)
Magnitude 6.06, spectrum O8e.
Position 06348n0611.   Plaskett's Star is a giant binary
system, usually regarded as the most massive pair of stars
yet identified in our Galaxy.   It was first investigated by
J.S.Plaskett of the Victoria Observatory in 1922. Located
about 1½°southeast of 13 Monocerotis, the star is very
probably a member of the huge aggregation NGC 2244 and its
associated nebula NGC 2237, less than 2° distant.
Plaskett's Star consists of two giant O-type stars
orbiting their common center of gravity in a period of
14.414 days, with a separation estimated to be about 50
million miles.   The inclination of the system is uncertain
but as no eclipses have been detected the orientation must
be far from the edge-on position.   This pair resembles Beta
Lyrae in the presence of turbulent masses of gas surrounding the components and flowing between them. It is now
recognized that such gas streams affect spectroscopic
radial velocity measurements, leading to erroneous orbital
elements and incorrect masses.   Thus the original masses of
90 suns each, derived by Plaskett, are now believed to be
spuriously high, According to recent studies (1959) the
total mass of the system is probably about 100 solar masses
but it is not definitely known whether the stars are nearly
equal in mass.   J.Sahade in 1961 reported probable masses
of 40 and 60 suns for the primary and secondary, respectively.   This apparent violation of the mass-luminosity
relation is of great interest; Beta Lyrae is another system
where the visually brighter star appears to have a smaller
mass than the secondary.   It is thought that the mutual
evolution of the members of such a pair is being strongly
affected by the exchange of mass between the stars, leading
to the odd situation that the less massive star is at present more advanced in its evolution than the other. In ages
to come the situation will gradually reverse; the fainter
star will evolve until it outshines the primary, and the
designations "primary" and "secondary" will then need to be
interchanged!
The distance of Plaskett's Star is not well determined, but is assumed to be at about the same distance as
the NGC 2244 group, probably about 2700 light years according to recent studies by W.W.Morgan, W.A.Hiltner, and their

associates in 1965. Allowing for some light loss through
interstellar absorption, the total luminosity of the star
appears to be about 3000 times that of the Sun; the true
absolute magnitude is near −4.0. The measured annual proper
motion of the star is very slight, less than 0.01"; the
radial velocity is about 15 miles per second in recession.

Binary systems of this type are extremely rare in
space, but can be identified at vast distances because of
the very high luminosity. Sir James Jeans spoke of this
system as "most massive and absolutely the brightest star
whose elements are known with fair certainty. Temperature
28,000°C. Every square centimeter of surface emits sufficient energy to run a locomotive at full speed for millions
of years..." (Refer also to Beta Lyrae, AO Cassiopeiae,
UW Canis Majoris, and Y Cygni)

M50   (NGC 2323) Position 07005s0816. A fine open
           star cluster, located a little more than a
third of the way along a line drawn from Sirius to Procyon.
This is probably the object discovered by G.D.Cassini some
time before 1711 "in the area between Canis Major and Canis
Minor". Charles Messier rediscovered it in April 1772 while
observing the comet of that year, and J.E.Bode found it
again in 1774 while searching for the reported Cassini
object. M50 is easily found in binoculars, and the total
integrated magnitude of about 6.3 suggests naked-eye visibility under the best conditions.

The cluster is a moderately compressed group about
10' in diameter with many outlying stragglers increasing
the apparent size to about 20' x 15'. Curving arcs of stars
give the perimeter a rather heart-shaped outline, nicely
high-lighted by a single reddish star some 7' south of the
cluster center. This star, a red M-giant, is not quite
the "blood-ruby star" that C.E.Barns described, but still
stands out rather prominently against a background of blue
and white stellar points. T.W.Webb called the whole group
"a brilliant cluster, straggling, J.Herschel says, to 30'
and containing a red star....in a superb neighborhood where
the Creator has 'sowed with stars the heaven thick as a
field'." J.E.Gore, from star counts on plates obtained by
Isaac Roberts in 1893, estimated that the main body of the
cluster contained about 200 stars.

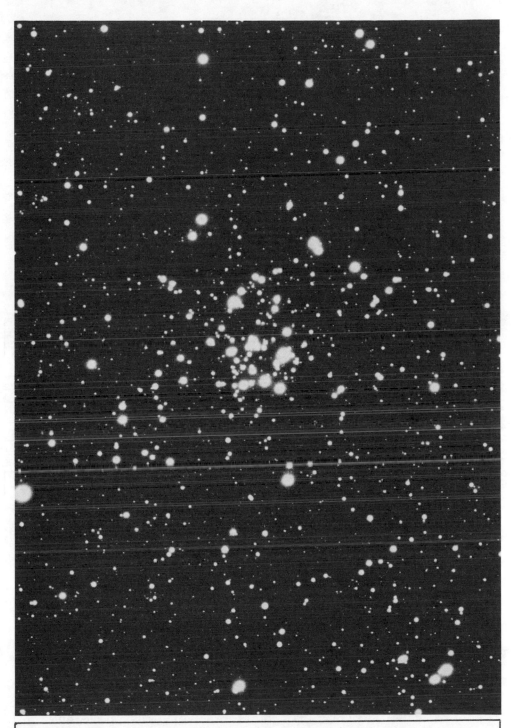

STAR CLUSTER M50 in MONOCEROS. An attractive group for the small telescope. This photograph was made with the 13-inch camera at Lowell Observatory.

DESCRIPTIVE NOTES   (Cont'd)

       In a study of M50, J.Cuffey (1941) found that a
color-magnitude plot of the cluster showed a well-defined
main sequence whose brightest members had spectral types of
B8, B9, and A0; a few yellow giants of absolute magnitude
about -1 were also present. His results indicated that more
than 70% of the stars brighter than (red magnitude) 13 are
true cluster members, and more than 50% of the fainter
stars are cluster members. A.Wallenquist (1959) counted 50
stars as probable members within 7' of the cluster center.
       The distance of the group is not precisely known;
published values range from about 700 parsecs up to 1100;
perhaps the best present compromise is the figure of 910
parsecs accepted by R.H.Garstang in the list of the Messier
objects in the British Astronomical Association "Handbook"
for 1964.   This corresponds to about 2900 light years; the
true diameter of the cluster (the 10' central mass) must
then be about 9 light years and the total luminosity about
1600 times the light of the Sun.

NGC 2237 + 2244   Position 06297n0454. A very large and
                      complex wreath-like nebula usually
called the "Rosette Nebula", surrounding a bright galactic
star cluster. NGC 2244 is the cluster; it appears in the
lower right corner of the photograph on page 1192. The four
brightest portions of the nebula were recorded separately
in the NGC and given the numbers 2237, 2238, 2239, and 2246
though the entire object is often mentioned in astronomical
literature simply as NGC 2237. The nebula is a huge and
irregular annulus some 80' in diameter, and is a rather
difficult object for small telescopes. With good binoculars it may usually be detected as a formless aura of soft
light encircling the star cluster. Visible to the naked
eye under good conditions, the cluster itself is a large
roughly rectangular aggregation about 40' in size containing five O-type stars, sixteen B-type stars, and an indet-
erminate number of fainter members. The brightest member,
12 Monocerotis, is a yellow giant of magnitude 5.85, of
spectral type K0 III. This star, however, must have an
abnormally high luminosity (about 2500 suns) if it is at
the same distance as the cluster; either it is a foreground
object or the attribution to luminosity class III is very

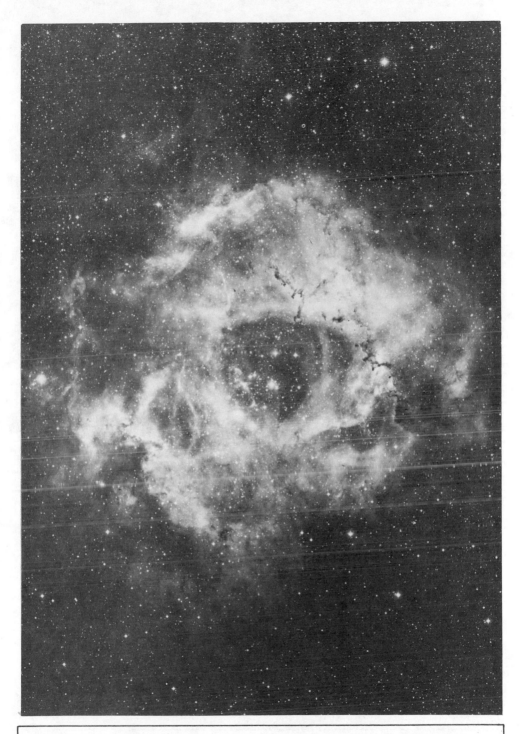

THE ROSETTE NEBULA IN MONOCEROS. The complete annulus is
shown in this Palomar Observatory 48-inch telescope plate.
The central cluster is NGC 2244.

DETAILS IN THE ROSETTE NEBULA. An enlargement of a portion of the photograph on page 1197, showing tendrils of dark nebulosity and some of the much-discussed "globules".

seriously in error.   The normal luminosity of a type K0 III
star is about 30 times that of the Sun.
       According to a study by W.W.Morgan and W.A.Hiltner
(1965) the distance of the Rosette is probably about 2600
light years, giving the cloud an actual diameter of some 55
light years.   This appears to be one of the most massive
nebulae known. T.K.Menon (1962) derived a mass of 11,000
suns, agreeing well with an earlier computation   (about
10,000 suns) by R.Minkowski in 1949.   The nebula shows a
wealth of wonderful detail on long-exposure photographs.
Scattered throughout the shining nebulosity are numerous
slender dark lanes and irregular tendrils of obscuring
matter; also prominent are many·tiny dark spots averaging
a few seconds of arc in diameter and known as "globules".
These are present in a number of other dark nebulae, and
are suspected to be new stars in the process of formation.
B.J.Bok (1970) finds that some of these spherical masses of
dark matter have diameters as small as 0.04 of a parsec, or
in the range of 7,000 to 10,000 AU.   Such an object, slowly
contracting under the influence of gravitation, will eventually become a self-luminous star, or perhaps, we may
conjecture, an entire Solar System resembling ours. Most of
the modern theories of the origin of our planetary system
begin with a primeval dust cloud, strongly resembling the
globules which we see in the Rosette.
       The enclosed star cluster, NGC 2244, lies in a
relatively clear area in the center of the wreath;   this
clear region may be the result of the exhaustion of the
nebulous matter through the formation of the cluster, but
Fred Hoyle has also suggested that intense radiation from
the newly-formed hot stars would blow away the remaining
gases in their vicinity.   The spectrum of the Rosette is of
the emission type, showing lines of hydrogen, singly-
ionized oxygen, and the famous "forbidden lines" of doubly
ionized oxygen which are also prominent in the spectra of
the planetary nebulae.   The whole vast cloud is probably
deeply embedded in obscuring material, of which the bright
nebulosity represents the ionized region excited by the hot
stars of the cluster.   Also the cluster NGC 2264, about 6°
distant, may be physically associated with the Rosette
complex; the two fields seem to be linked by a continuous
faint nebulosity.   (Refer also to NGC 2264, page 1206.)

NGC 2244. The central cluster of the Rosette Nebula in
Monoceros. The photograph was made with a 12½-inch reflector by Evered Kreimer of Prescott, Arizona.

STAR FIELD IN MONOCEROS. The large cluster NGC 2264 is at
upper left; the Rosette Nebula complex NGC 2237-2244 is at
lower right. The circle indicates Hubble's Variable Nebula
NGC 2261.   Lowell Observatory 5-inch camera photograph.

NGC 2261   Position 06364n0846. This is "Hubble's Variable Nebula", a peculiar gaseous nebula enveloping the variable star R Monocerotis. Discovered by Sir William Herschel in 1783, the nebula is rather comet-like in form, with the star located near the southern tip and the "tail" extending northward. The variability of the associated star was discovered by Schmidt at Athens in 1861 and a range of about 2 magnitudes has been established by observations over the last century.

The total brightness of the nebula is usually about 10th magnitude; the brightest portion being about 1' in length. Not a difficult object for modern telescopes, the nebula can be studied under fairly high powers since the surface brightness is high. A bluish color is evident in large instruments. R Monocerotis itself, however, is very often a difficult object to observe, and is frequently lost in the bright glow of the nebula.

In the course of comparing photographs in 1916, Dr. E.Hubble found that the nebula itself was variable, not only in size and brightness, but also that various nebular details showed perceptible changes from month to month and even from one night to the next. The nebula was studied for more than 30 years by C.O.Lampland at Lowell Observatory and a valuable series of over 900 photographs was obtained. In some cases the apparent displacement of details in the cloud was found to be as high as 1.0" in four days.   No regular periodicity was detected, nor did the changes seem to follow the variations of the star. However, it soon became evident that many of the observed changes could not represent actual motion of nebulous material, since the velocities required would be impossibly high. Photographs show that the same details reappear again and again, after having been obscured or dimmed for an unpredictable period of time. Evidently the observed changes are due to changing light conditions and moving shadows cast on the cloud by dark masses drifting near the illuminating star. A similar phenomenon is known in at least two other nebulae: NGC 1555 in Taurus, and NGC 6729 in Corona Australis. In both cases a peculiar variable star illuminates the cloud. In the case of NGC 2261 itself, Lampland concluded that at the accepted distance, the displacement of 1" in four days almost exactly equalled the velocity of light!

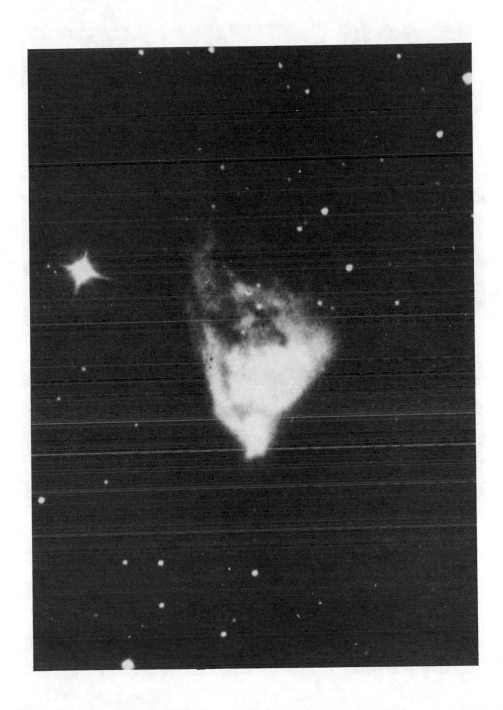

HUBBLE'S VARIABLE NEBULA. NGC 2261 in Monoceros, photographed with the 100-inch reflector at Mt. Wilson. The
star R Monocerotis is located at the southern tip.

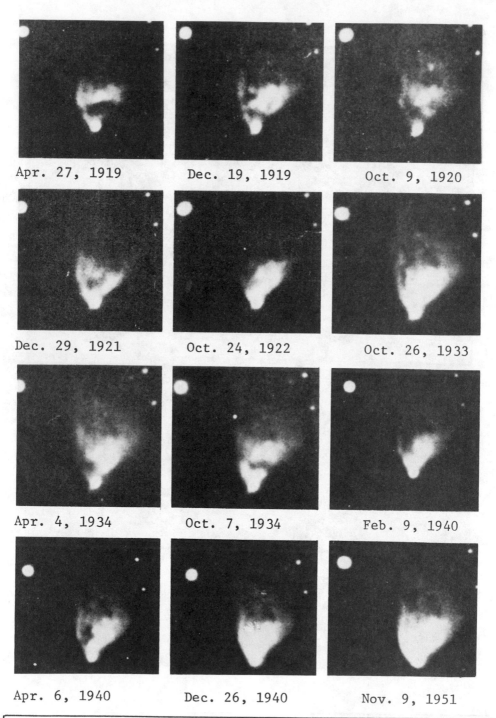

NGC 2261. Selected photographs from the Lowell Observatory collection, illustrating the apparent variations in the nebula over a period of more than thirty years.

DESCRIPTIVE NOTES   (Cont'd)

The illuminating star, like the nebula, is an unusual object whose exact classification remains highly uncertain. In some catalogs it has been classified as one of the RW Aurigae type stars; others call it a T Tauri type star, and occasionally it is referred to merely as a "very unusual nebular variable". As early as 1917, V.M.Slipher at Lowell Observatory had found that the object showed a seemingly continuous spectrum with some superimposed bright lines, but not those typical of diffuse nebulae. Spectra of the nucleus and the nebula were virtually identical, and the emission lines resembled those of a nova in the early stages of its development. Slipher suggested that " the nebula may be shining by reflected light of the pulsating nucleus, and the flow of light-pulses over the nebula causes its observed variations..." In the Moscow General Catalogue (1958) the star is given a spectral range of A to Fe peculiar while in more recent years the observed color suggests a type of late F or early G. To further confuse the situation, it is not at all certain that the various observers have been measuring a "star" in the true sense of the word. The object called R Monocerotis, as seen in great telescopes, is a bright nebulous condensation with perceptible apparent size. Using the Lick Observatory 120-inch reflector in 1968, G.Herbig found this nuclear mass triangular in shape with blunt corners, about 5" in width, and with a tiny stellar point at the southern tip. "At no time during these observations has a normal stellar absorption spectrum been detectable, much less that of a G or K-type star....the absorption lines present always resemble those of a shell..."
        C.O.Lampland and J.Ashbrook (1948) reported that the "star" always appeared embedded in a small dense knot of nebulosity,distinctly elongated NW to SE. Their study showed that the object was varying in at least two cycles: a longer cycle of about 1 magnitude in the course of a year and shorter fluctuations of a few tenths of a magnitude over an interval of several weeks. The light was near the minimum in 1919, 1930, and 1946; near maximum in 1947 and 1948. The extreme range from 1916 to 1948 was 11.3 to 13.8 (photographic). Measurements by E.E.Mendoza (1966) revealed that the star is a strong emitter of infrared radiation; it is in fact significantly brighter in the infrared than

in visual light. According to a theoretical model proposed
by F.J.Low and B.J.Smith (1966) R Monocerotis may be a
protoplanetary system— a star embedded in a dense cloud of
coarse (10 micron) dust grains. More than 99% of the star's
light, on this model, is absorbed by the grains and then
re-emitted at infrared wavelengths. The light of the cloud
shows strong but variable polarization. R.C.Hall (1964)
found the eastern portion of the nebula to be more highly
polarized than the western part; the polarization increases
with the distance from R Monocerotis, and reaches a maximum
of about 34% in the outer regions. According to B.Zellner
(1970) the changes in the degree of polarization do not
appear to be correlated with changes in the light.

The distance of NGC 2261 is not known by direct
measurements; A.Van Maanen in 1928 found no detectable
parallax, and the associated star is so odd an object that
it cannot be used as a reliable distance indicator. It
seems very likely, however, that the nebula is an outlying
member of the huge NGC 2264 group which is centered only a
degree away, and whose distance is fairly well determined
at about 2600 light years. If this distance is accepted,
the 1' length of the nebula corresponds to about 0.7 light
years, and the central star, at maximum, has an actual
luminosity of about 15 times that of the Sun. The emission
lines give a radial velocity of 12.5 miles per second in
recession.

Hubble's Variable Nebula was the first object to
be photographed with the 200-inch reflector at Palomar
Observatory; January 26, 1949.

NGC 2264   Position 06384n0956. A very large and
scattered cluster of about 20 bright stars
and over a hundred fainter members, called from its shape
the "Christmas Tree" by L.S.Copeland. The group forms a
pattern resembling an arrowhead, about 26' in length,
pointing nearly due south. The "tree" is thus seen upside-
down in the sky, but is reinverted to the usual orientation
in the astronomical telescope. The cluster is surrounded
by a vast but faint nebulosity, and has been the subject of
many modern investigations. According to a study by M.F.
Walker (1955) the distance of the group is about 2600 light
years, giving the true distance across the "arrowhead" as

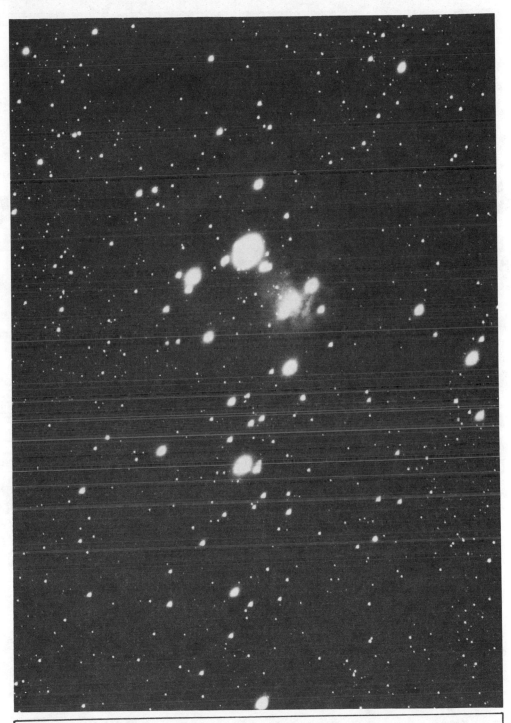

NGC 2264. This large scattered group is shown here with north at the top; invert the photograph to see the "Christmas Tree" outline. Lowell Observatory 13-inch telescope.

FIELD OF NGC 2264. The bright star S Monocerotis is at the top; the dark Cone Nebula stands out prominently against the southern portion of the nebulosity. The photograph was made with the 48-inch Schmidt camera at Palomar Observatory

DESCRIPTIVE NOTES   (Cont'd)

about 20 light years.   In a comprehensive study of the
cluster in 1965, W.W.Morgan, W.A.Hiltner and their co-
workers derived a distance modulus of 9.5 magnitudes, or
about 2600 light years, in excellent agreement with the
earlier determination.   NGC 2264 thus appears to be at very
nearly the same distance as the Rosette Nebula; studies of
the two H-R diagrams suggest that NGC 2264 is probably at
a slightly closer distance. Both objects are evidently
members of a vast complex of nebulosity which extends from
the Monoceros-Gemini border toward the southwest into the
brilliant regions of Orion.   R.D.Davies at Jodrell Bank in
England has found the entire NGC 2237-2244-2264 complex to
be a source of radio radiation.

In NGC 2264, the twenty brightest stars have the
following magnitudes and spectra:

1.	Mag	4.62;	spectrum	O7	11.	Mag 8.96;	spectrum	B?
2.	"	7.14	"	B2	12.	" 8.97	"	B7
3.	"	7.47	"	B2	13.	" 9.02	"	B5
4.	"	7.93	"	B3	14.	" 9.08	"	B7
5.	"	8.11	"	B3	15.	" 9.08	"	B6
6.	"	8.26	"	K3	16.	" 9.10	"	
7.	"	8.44	"	B5	17.	" 9.19	"	A5
8.	"	8.52	"	K2	18.	" 9.21	"	B8
9.	"	8.70	"	B8	19.	" 9.29	"	A0
10.	"	8.81	"	B8	20.	" 9.32	"	G5

Stars known to be non-members have been excluded from the
list.   Star numbers are keyed to the identification chart
on page 1210.

The brightest star of the cluster is the giant 5th
magnitude star S Monocerotis, spectrum O7 IV, a super-hot
sun which has been found slightly variable. The greatest
observed magnitude range appears to be about 4.2 to 4.6
photographic. This star has a computed absolute magnitude
of about -5.0 and an actual luminosity of about 8500 suns.
Also designated "15 Monocerotis", the star is a visual
double with a companion of magnitude 8.5 at a distance of
3.0".   Information on more distant companions is given in
the list on page 1183.

DESCRIPTIVE NOTES   (Cont'd)

The other members of the cluster are chiefly B and A stars, but the group also includes several yellow giants of types G and K, the brightest of which have luminosities of about 300 suns. M.F.Walker (1956) listed five yellow giants as members of the cluster, but S.Vasilevskis (1965) finds that at least two of these are probably field stars.

The nebulosity which surrounds NGC 2264 is not seen in small telescopes, but is dramatically revealed on long-exposure photographs. The brightest portion is a nebulous condensation illuminated by star #7, but the most awesome and spectacularly beautiful section is the famous "Cone Nebula", located just south of star #2. It appears as a great dark pinnacle some six light years high from north to south, wonderfully outlined against glowing nebulosity, and brilliantly illuminated at its summit; the whole structure forming a picture of such strangeness and splendor that it scarcely seems natural. One must wonder what the ancients would have seen in such a celestial marvel; they who gave us our constellations and the rich mythology of the skies. Had they known of such a wonder could they have called it by any lesser name than "The Throne of God"? Here, as in the Great Orion Nebula, even the modern observer is touched by a strange sensation of having been present at the drama

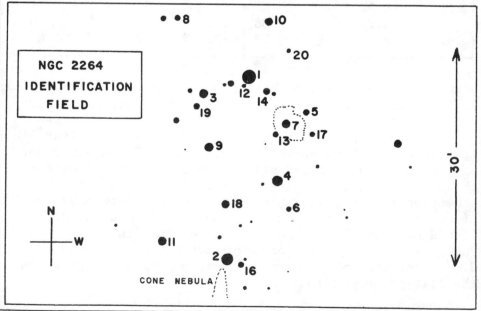

NGC 2264
IDENTIFICATION
FIELD

CONE NEBULA

THE CONE NEBULA. This spectacular formation is located in the south-outer regions of NGC 2264. Palomar Observatory photograph in red light with the 200-inch telescope.

DESCRIPTIVE NOTES   (Cont'd)

of creation. And, as in the case of the Orion Nebula, this
instinctive impression is supported by modern knowledge.
The great diffuse nebulae are undoubtedly the regions where
star formation is still in progress, and NGC 2264 itself is
recognized as one of the youngest clusters known.
        During a study of the cluster by M.F.Walker (1954) it
was found that the color-magnitude diagram of the cluster
is distinctly unusual. Normal hydrogen-burning stars, as
mentioned in various places throughout this book, fall on
a well-defined band called the Main Sequence, but as a star
group ages, the evolution of the more massive stars carries
them away from the main sequence toward the upper right
portions of the diagram. In describing the Hercules (M13)
Cluster, an explanation of cluster age-dating by this
method was given. In NGC 2264, however, a very different
effect appears. It is the fainter stars which are displaced
from the main sequence, the members later than class A0 and
fainter than 10th magnitude. Also, a great many of the
fainter stars are erratic variables of the T Tauri class.
Both of these findings imply extreme youth; the stars still
appear to be contracting gravitationally, and have not yet
reached the main sequence state. Current studies indicate
that the majority of the members of this cluster may not be
more than 1 or 2 million years old. (Refer also to M16 in
Serpens, NGC 2362 in Canis Major, and the Double Cluster
NGC 869+884 in Perseus)

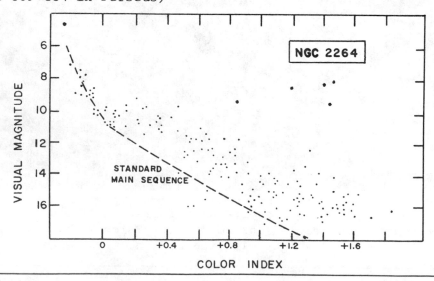

NEBULOSITY NEAR S MONOCEROTIS. The star is the brightest
member of the cluster NGC 2264, and is slightly variable.
Yerkes Observatory photograph

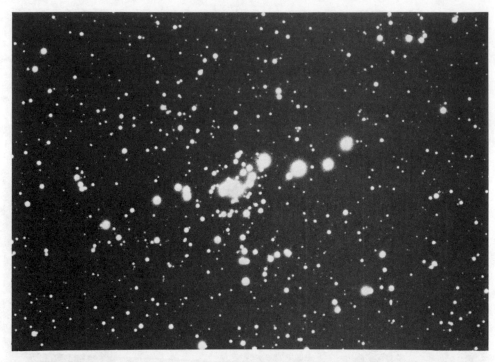

STAR CLUSTERS IN MONOCEROS. Top: The bright elongated
group NGC 2301. Below: The more compact cluster NGC 2506.
Lowell Observatory photographs with the 13-inch telescope.

MUSCA

LIST OF DOUBLE AND MULTIPLE STARS

NAME	DIST	PA	YR	MAGS	NOTES	RA & DEC
h4432	2.5	303	48	5½- 7½	PA slow inc, spect B5	11212s6441
B1699	0.3	356	59	7 - 7	spect B3	11262s7211
NZ 23	1.1	238	39	8 - 8½	spect B9	11303s6536
I 34	0.2	178	54	5½- 6½	(B1705) PA inc,	11371s6508
	38.7	234	01	- 12	spect G0, A0	
	41.8	32	01	-12½		
L4920	1.8	159	40	5 - 7½	(Cor 130) relfix, spect B4	11494s6456
h4498	0.1	119	60	7 - 7	(φ 367) relfix,	12038s6526
	8.7	61	40	- 8	all cpm, spect dF8, C spect= A3	
h4522	12.8	67	18	7½ - 8	relfix, spect B8	12226s6912
h4450	21.0	40	17	7- 10½	spect B9	12297s7338
I 296	1.7	276	32	6½- 9	cpm, spect B9	12359s7506
h4535	17.2	339	18	6½-11½	spect B0	12359s6655
h4545	9.2	192	54	8 - 8½	relfix, spect A0	12424s7454
β	1.6	14	55	4 - 4	(R207) binary, PA inc, spect B2 (*)	12432s6750
h4550	13.6	98	18	8 - 9	PA slow inc, spect A2	12452s6651
G1s185	8.7	9	20	7½- 9	(Hrg 77) spect A0	12460s6519
θ	5.3	186	52	5½- 8	(Rmk 16) relfix, spect B0 + WC6	13049s6502
Hd 224	0.5	212	36	7 - 8	spect B9	13144s6814
B1736	0.3	316	30	6½- 7½	spect B5	13191s7153
h4586	3.1	144	29	7 - 9	PA slow dec, spect A3	13248s6736
I 298	0.9	187	47	7 - 9	PA dec, spect F2	13288s6859
h4596	1.6	282	18	8 - 8	relfix, spect B9	13339s6441
h4598	13.1	46	18	7 - 11	spect A0	13374s7452

MUSCA

LIST OF VARIABLE STARS

NAME	MagVar	PER	NOTES	RA & DEC
R	6.4---7.3	7.510	Cepheid, spect F9--G6	12390s6908
S	6.4---7.2	9.659	Cepheid, spect F8--G4	12101s6952
T	7.0--9..	93:	Semi-reg; spect N	13173s7411
U	9.2--13..	356	LPV. Spect M6e	13216s6424
Y	9.8--11..	---	R Coronae Bor. type	13026s6515
RT	8.5---9.6	3.086	Cepheid, spect F8	11422s6702
TU	8.4---8.8	1.387	Ecl.Bin; lyrid, spect both B3	11289s6528
UU	9.2--10.2	11.636	Cepheid, spect G0	11498s6507
BO	6.0---6.7	Irr	Spect M3	12320s6729
BP	9.6--12.0	3.321	Ecl.Bin.	12474s7130
BR	9.9--10.3	.7982	Ecl.Bin; lyrid	12022s7236
CX	8.7---9.3	5.903	Ecl.Bin; spect B9	11453s6842

LIST OF STAR CLUSTERS, NEBULAE, AND GALAXIES

NGC	OTH	TYPE	SUMMARY DESCRIPTION	RA & DEC
4372		⊕	Mag 8, diam 18', class XII, pF,L,R, stars mags 12...	12230s7224
4463		⋰	diam 3', v1C, about 20 stars class F	12271s6430
----	H6	⋰	L,pRi, diam 7', about 70 stars, class E	12350s6810
4815		⋰	Diam 4', pL,pRi, about 40 stars mags 10... class F	12549s6441
4833	△164	⊕	Mag 8½, diam 6', class VIII, B,L,R,mbM, stars mags 12....	12560s7036
----	I.4191	◎	mag 12, diam 5", stellar	13055s6722
----	H8	⋰	S,1C, diam 4', about 25 stars, class F	13150s6649
5189	△252	□	! B,pL,pE, 185' x 130' with 10^m star of type B9	13299s6543

MUSCA

DESCRIPTIVE NOTES

ALPHA Magnitude 2.70; spectrum B3 IV. Position 12342s6852. The computed distance of the star is about 430 light years; the actual luminosity about 1200 times that of the Sun. Alpha Muscae shows an annual proper motion of 0.04"; the radial velocity is 11 miles per second in recession. This star is one of the many members of the extensive Scorpio-Centaurus moving group.

Slight variations of small amplitude have been recorded, both in magnitude and radial velocity. The maximum recorded variation in light appears to be from magnitude 2.66 to 2.73. It is not certain to which class this star should be assigned.

BETA Magnitude 3.06; spectrum B2 V. Position 12432s6750. The star is a fine double of nearly equal magnitudes forming a slowly rotating binary of uncertain period. The separation has been increasing from about 0.6" in 1880 to 1.6" in 1955; the position angle has changed from 317° to 14° in the same interval, in a counter-clockwise direction. Orbital motion during the last century suggests a period of four or five centuries.

The computed distance is about 470 light years; the total luminosity about 580 times that of the Sun. Beta Muscae shows an annual proper motion of 0.04"; the radial velocity (slightly variable) is about 26 miles per second in recession.

LIST OF DOUBLE AND MULTIPLE STARS

NAME	DIST	PA	YR	MAGS	NOTES	RA & DEC
I 967	4.5	326	25	8½- 11½	spect K2	15280s5148
h4777	5.6	298	33	7 - 9	relfix, spect F0	15288s5714
△186	39.5	116	13	8½- 8½	cpm, relfix, spect both F2	15291s5801
△190	5.0	96	34	7½- 9½	(I 372) spect M;	15390s5758
	33.7	49	00	- 11½	relfix, color contrast pair	
I 543	1.3	343	44	8½- 10½	spect A3	15393s5840
I 1099	0.4	359	49	8 - 8	PA inc, spect A3	15404s5410
h4797	22.4	255	13	7 - 11	spect F8	15405s5004
h4795	7.5	223	33	8- 10½	relfix, spect A2	15408s5858
	20.9	136	17	- 13		
	45.3	177	17	- 10		
△191	0.3	173	28	8 - 8½	(φ 234) (h4796)	15412s5832
	32.6	267	17	- 8½	spect F8	
H1d 124	2.4	201	53	6½- 8½	PA dec, spect A2	15413s5038
I 546	0.8	260	43	8½- 9	spect A0	15417s5550
I 974	0.1	46	59	7½- 7½	dist dec, PA inc, spect A2	15428s5759
B1792	1.2	292	31	6- 10½	spect K0	15450s5217
I 548	2.7	178	34	6 - 11	spect A5	15468s4515
I 548b	0.6	298	48	11½- 12	BC PA inc	
△193	18.0	15	60	6 - 8½	dist & PA dec, cpm, spect B2, K	15472s5454
△195	11.9	10	54	7½- 8	relfix, spect A2	15511s5011
	27.8	293	42	- 11½		
h4813	3.8	99	33	6 - 10	relfix, spect A3	15513s6002
λ254	1.5	204	59	8 - 9½	PA inc, spect F8	15576s4624
ι'	0.6	139	69	6 - 6	(λ258) (h4825) AB	15595s5738
	10.8	246	46	- 8	binary, 27 yrs; PA dec, spect A5; all cpm.	
λ	0.3	130	60	6 - 7	(λ271) PA inc, spect A3	16158s4233
△200	40.7	196	14	6 - 9½	slow dist dec, cpm, spect G5	16188s4346
I 1291	5.3	38	40	6½- 12	spect A2	16213s4514
Cor 197	1.7	131	51	8 - 8	both PA dec, spect G5, AC optical	16216s4902
	15.7	95	59	- 11		
I 93	1.1	286	35	7½- 10½	relfix, spect B8	16231s4756

NORMA

LIST OF DOUBLE AND MULTIPLE STARS   (Cont'd)

NAME	DIST	PA	YR	MAGS	NOTES	RA & DEC
ε	22.0	335	13	4½– 7½	(h4853) relfix, cpm spect B3, A	16235s4727
h4857	4.7	233	47	8– 13½	(Rst 5413) PA slow	16274s4623
	6.6	70	13	– 9½	dec, spect B3	

LIST OF VARIABLE STARS

NAME	MagVar	PER	NOTES	RA & DEC
R	6.7--13..	490	LPV. Spect M3e--M6e	15324s4921
S	6.1--6.8	9.755	Cepheid; in cluster NGC 6087; spect F8--G2	16147s5747
T	6.4--13..	243	LPV. Spect M3e--M6e	15402s5450
U	8.6---9.7	12,641	Cepheid; spect G0--K5	15384s5509
V	8.8--10..	156	Semi-reg; spect M4	16063s4907
W	9.5--10.4	135	Semi-reg; spect M	16129s5229
Z	9.4--10.2	2.557	Ecl.Bin; spect B5	16014s4610
RR	9.2--10.1	3.028	Ecl.Bin; spect A0	15088s5508
RS	9.6--10.4	6.198	Cepheid	16013s5347
SY	9.0--9.9	12.65	Cepheid	15508s5425
TV	8.7--9.3	8.524	Ecl.Bin; spect A0	16004s5124
IL	7.0---16	---	Nova 1893; 1.4^m east of globular cluster NGC 5927 in Lupus	15258s5025
IM	9.0--16.	---	Nova 1920	15357s5210
IQ	9.9--10.4	8.232	Cepheid	15091s5434
IT	9.4--9.8	3.509	Ecl.Bin; spect A0	16136s4451

LIST OF STAR CLUSTERS, NEBULAE, AND GALAXIES

NGC	OTH	TYPE	SUMMARY DESCRIPTION	RA & DEC
----	H9	⊙	S,F, diam 3', about 30 stars, class E. Very remote cluster	15300s5326
5946		⊕	Mag 11, diam 2', class IX; cB,pL,R,1bM; stars mags 16...	15318s5030
----	Sp 1	◎	8½^m ring, diam 72" with 13½^m central star	15474s5121
5999	△343	⊙	L,pRi, diam 4', class F; about 100 stars mags 12....	15482s5620
6005	△334	⊙	Diam 3', pS,pRi,mC; about 30 faint stars, class F	15518s5718
6031	△359	⊙	S,mC, diam 3', about 10 stars mags 11....14	16037s5356
6067	△360	⊙	vB,vL,vRi, diam 15', over 100 stars mags 10..... class F; fine field	16093s5405
----	VV78	◎	vF, grayish disc, diam 25"	16106s5450
6087	△326	⊙	B,L, diam 20', about 40 stars mags 7....10, class D; contains cepheid variable star S Normae	16147s5747
----	H10	⊙	L,1C, about 30 stars over 30' diam field; class D	16158s5452
6134	△412	⊙	cL,pRi, diam 10', about 60 stars mags 11... class F	16240s4904
6152		⊙	L,1C, diam 30', about 60 stars mags 9... Class E	16288s5231
6164 6165		□	F neby surrounding 7^m star, spect B0	16300s4759

OCTANS

LIST OF DOUBLE AND MULTIPLE STARS

NAME	DIST	PA	YR	MAGS	NOTES	RA & DEC
I 177	1.0	17	00	8½- 9½	spect F5	00255s8325
Gls 14	5.5	55	39	7½- 8½	relfix, spect G5	01383s8232
h3582	19.5	297	19	7½- 11	spect F0	03144s8343
R38	2.2	246	45	6½- 8	relfix, spect B9	03508s8526
△82	15.7	271	40	7 - 7	relfix, spect F2	09399s8547
h4310	3.9	269	41	8 - 8½	PA slight dec, spect G5	10082s8351
h4468	22.2	147	19	6 - 10	dist inc, spect K0, slight PA dec	11392s8250
h4490	25.0	145	40	6 - 9	relfix, spect K2	11595s8521
h4798	20.8	134	19	7½- 11	spect K0	15586s8406
h4912	25.1	122	19	7 - 12	spect B8	17186s8244
h5182	26.7	357	18	6- 11½	spect K0	20251s8108
h5186	7.9	96	40	8 - 8½	relfix, spect F5	20255s7723
h5199	28.9	210	18	7½- 12	spect G5	20318s7704
μ^2	17.4	17	40	7 - 7½	(△232) cpm pair; spect both G5	20358s7531
Gls 263	4.8	248	18	7½- 10	cpm, no definite change, spect A5	20593s8054
I 257	1.6	300	32	8 - 8½	PA inc, spect F8	21028s7611
h5233	11.7	270	18	7½- 12	spect K0	21069s8328
h5235	3.3	264	18	8 - 8½	cpm, no certain change, spect F0	21113s8431
I 670	0.5	11	33	8 - 8½	PA dec, spect B9	21137s8032
h5262	24.5	94	18	6½- 11½	spect A0	21269s8016
h5261	5.1	201	18	8½- 8½	relfix, spect F5	21409s8604
λ	3.1	70	46	5½- 7½	(h5278) slow PA dec, spect G0, A3	21435s8257
h5301	10.6	204	17	8 - 10	relfix, spect A5	21575s7734
h5306	34.6	72	17	6 - 10	spect F2	21583s7622
△238	20.1	81	17	6 - 9	spect G0	22213s7516

OCTANS

LIST OF VARIABLE STARS

NAME	MagVar	PER	NOTES	RA & DEC
R	6.6--13..	405	LPV. Spect M6e	05411s8626
S	7.5--13.7	259	LPV. Spect M4e--M5e	17460s8648
T	9.2--14..	218	LPV. Spect M2e--M4e	21061s8219
U	7.4--13.9	303	LPV. Spect M4e--M6e	13180s8358
X	7.4--11..	205	Semi-reg; spect M3e--M6e	10281s8405
RR	8.8--12..	273	LPV.	20501s7510
RT	9.3--12..	180	LPV.	22528s8719
RU	9.0--12..	373	LPV.	00069s8627
TW	9.0--11..	132	Semi-reg; spect M1	19377s7723
UV	8.9--9.8	.5426	Cl.Var; spect A6--F6	16202s8348
UZ	9.4---9.8	1.149	Ecl.Bin.	04527s8454

DESCRIPTIVE NOTES

SIGMA   Magnitude 5.46; spectrum F0. Position 20151s8908. This small star is mentioned here only because of its position within 1° of the South Celestial Pole. It is, so to speak, the "Polaris" of the Southern sky, although a much less conspicuous object. The distance from the true pole was 45' in the year 1900. In 1950 this had increased to 58', and will be 1°03' by the year 2000.

No parallax attempts are reported for the star in the standard catalogs. If the luminosity is normal for a main sequence F0 star, however, the absolute magnitude should be about +2.7, implying a distance of about 120 light years. The annual proper motion is 0.03"; the radial velocity is about 7 miles per second in recession.

OPHIUCHUS

LIST OF DOUBLE AND MULTIPLE STARS

NAME	DIST	PA	YR	MAGS	NOTES	RA & DEC
β948	1.1	118	66	7- 9½	(Σ2005) PA dec,	16031s0609
	28.7	233	58	-10	spect F5	
	52.6	195	24	-11		
A692	3.1	223	25	7½- 15	spect A5	16205s0044
ρ	3.1	344	59	5 - 6	(5 Oph) PA dec,	16226s2320
	152	360	25	- 8	In faint nebula IC	
	156	253	25	- 8	4604; spect B3, B2	
ρ d	0.6	358	59	8 - 9	(β1115) PA dec	
Σ2048	5.3	298	51	6½- 9	slow dist inc,	16261s0801
					spect dF3	
λ	1.0	355	67	4 - 6	(10 Oph) (Σ2055)	16284n0206
	119	170	60	- 11	AB binary, PA inc,	
					129 yrs; spect A1,	
					AC= cpm	
Ho 407	14.2	217	10	7 -12	spect A5	16302s1027
Rst5414	5.9	22	46	7 -13	spect K0	16345n0021
J448	2.4	145	49	9- 10	spect G0	16413s0026
Σ2086	13.9	157	14	7½- 10	relfix, spect A0	16417s0028
Σ2088	20.0	334	10	8 - 12	spect A2	16422n0226
Σ2088b	8.2	105	10	12- 13		
19	23.2	89	58	6- 9½	(Σ2096) PA slight	16446n0209
					dec, spect A2	
Σ2106	0.4	185	67	6½- 8½	binary, about 2000	16487n0930
					yrs; PA dec, spect	
					dF4	
21	0.4	115	66	6 - 8	(OΣ315) PA dec,	16489n0118
					spect A0	
β241	0.4	1	59	7 - 7	PA inc, spect B8	16526s2129
Σ3106	2.1	253	57	8½- 8½	PA inc, spect G0	16530s0506
24	0.8	294	59	6½- 8½	(β1117) PA inc,	16538s2305
					spect A0	
Pi 236	4.5	232	51	6 - 8	(Sh 240) relfix,	16541s1928
h4902	11.2	31	20	8 - 9	spect A0	16547s2732
Σ3107	1.4	83	66	8½- 8½	PA dec, spect G0	16564n0402
	43.2	322	00	- 11		
Ho 554	9.8	357	04	8- 12½	spect G5	16581s2937
	36.0	352	04	- 10		
Σ2114	1.2	180	66	6½- 7½	PA inc, spect A0	16596n0831
Hu 164	1.9	345	37	6½- 12	slow PA inc, spect	17017s1236
					K0	

LIST OF DOUBLE AND MULTIPLE STARS (Cont'd)

NAME	DIST	PA	YR	MAGS	NOTES	RA & DEC
Σ2119	2.3	11	49	8 - 8	slow PA dec, spect F8	17027s1352
β823	0.9	90	61	8 - 9	binary, about 500 yrs; PA inc, spect G0	17040n0043
Σ2122	20.4	280	58	6½- 8½	relfix, cpm; spect A2	17043s0135
h4922	21.8	310	19	7½- 10½	spect A0	17055s2009
A1145	0.3	155	52	6 - 8	cpm, PA dec, spect A0	17056s0101
I 246	1.3	44	42	7½- 10	PA inc, spect F2	17070s2743
η	0.2	325	63	3½- 4	(β1118) binary,	17075s1540
	94.6	143	11	- 12	84 yrs; PA dec, spect A2 (*)	
h589	10.0	304	40	7½- 8	relfix, spect A3	17077s2453
Hu 169	0.2	140	58	8 - 8	PA dec, spect A3	17086s1626
β125	1.7	67	42	8 - 11	relfix, spect G5	17091s2659
Σ2132	1.5	108	42	8½- 9	relfix, spect F2	17101s0400
OΣ325	0.5	231	67	7 - 9	PA inc, dist dec, spect F0	17106n0748
β1247	1.5	343	44	7½- 10	relfix, spect A3	17109s0914
36	4.4	163	62	5½- 5½	(Sh 243) binary,	17123s2630
	732	74	60	- 7½	about 550 yrs; PA dec; all cpm, spect K0, K2, K5	
β282	4.2	147	58	7 - 12	relfix, spect K0	17125s1432
β957	0.3	208	62	8 - 8	binary, 106 yrs; PA dec, spect F5	17128s1015
A2592	0.4	273	59	7½- 8	PA dec, spect F5	17129s0945
41	1.0	346	59	4½- 7½	(A2984) PA inc, spect K2	17140s0023
U	20.4	358	58	6 - 13	(h854) relfix, A= Ecl.Bin; spect B5	17140n0116
HI 35	5.8	335	37	6½- 9	(S385) perhaps PA slight inc. Labeled "38" in Norton's Atlas. spect A0	17146s2635
39	10.8	355	51	6 - 7	relfix, cpm pair; color contrast, spect K1, F6	17150s2414

LIST OF DOUBLE AND MULTIPLE STARS (Cont'd)

NAME	DIST	PA	YR	MAGS	NOTES	RA & DEC
B337	0.6	203	33	7½– 10½	spect A2	17152s2950
β126	2.0	261	67	6½– 7½	relfix, spect A0	17170s1742
	11.4	141	58	–11½		
A2241	5.1	76	26	6½– 14	spect G0	17176s1917
β127	5.2	92	51	8 – 9	relfix, spect F5	17178s2718
ξ	3.7	50	59	4½– 9	(40 Oph) PA dec, dist slight inc; spect F2	17180s2103
A28	1.7	38	52	8½– 8½	(Ho 631) relfix, spect F0	17181s0859
β959	3.4	257	24	7 – 12	spect G5	17196n0503
β46	2.0	204	59	7½– 10½	relfix, spect K5	17213n1328
Σ2156	3.5	33	41	8½ 9	relfix, spect F2	17214s0047
Iln 134	4.3	144	33	6 – 12	(Ho 266) spect G7	17217s2124
I 1069	5.5	350	27	7½– 11½	spect B9	17229s2734
β128	4.0	324	34	7½– 10	relfix, spect B9	17237s2618
Σ2166	27.4	283	49	7½– 8	relfix, spect A0,A3	17255n1126
Rst5085	10.8	133	43	6½– 11½	spect F0	17256s0420
β129	1.0	115	24	7½– 8	PA inc, spect F0	17256s2528
Σ2170	3.6	69	45	8½– 9	PA dec, spect K5	17264n1032
Σ2171	1.3	62	67	8 – 8	PA dec, spect F2	17265s0958
β1089	1.3	336	60	7 – 11	PA dec, spect G5	17271s0552
Σ2173	0.7	146	68	5½– 6	binary, 46 yrs; PA dec, spect G8	17278s0101
A2386	0.1	3	55	6½– 6½	PA inc, spect gG3	17288n0246
OΣ331	1.4	349	59	7½– 9	PA & dist inc, spect B8	17295n0252
	51.0	241	02	– 11		
54	22.0	71	24	6½– 11	(Σ2184) relfix; spect G5	17321n1312
53	41.3	191	49	5½– 7½	(ΣI34) relfix, spect A2	17322n0937
	94.0	345	12	– 11		
	91.4	223	12	– 11		
Σ2185	27.7	5	34	7 – 10	spect F8; AC PA inc, dist dec	17324n0603
	80.8	225	49	– 8		
	89.6	202	12	– 9		
Σ2187	0.3	332	45	9 – 9	(A2985) spect F2	17332n0410
	3.3	177	37	–9½		
A1879	0.3	86	58	7½– 10	PA inc, spect A3	17332n1324
Σ2186	3.0	81	53	7½– 7½	relfix, spect B8	17333n0102
Σ2188	5.7	203	37	8½– 9	relfix, spect A5	17338n0639

LIST OF DOUBLE AND MULTIPLE STARS   (Cont'd)

NAME	DIST	PA	YR	MAGS	NOTES	RA & DEC
β961	8.0	141	25	7- 11½	spect K0	17370n0325
	30.9	152	13	- 13		
Σ2191	26.4	267	34	7 - 8	relfix, spect both F2	17371s0457
Σ2191b	8.1	32	02	8- 12½		
β631	0.2	29	58	7 - 7	PA dec, spect A0	17374s0037
Σ2200	1.6	161	43	8 - 9	PA slight dec, spect A0	17414n0552
61	20.6	93	55	6 - 6½	(Σ2202) relfix, cpm, spect A0, A0	17420n0236
A32	0.6	237	41	7½- 9½	(h4977) spect A0	17429s0329
	22.0	143	10	-13½		
Σ2212	3.1	340	37	8½- 9	relfix, spect A0 in cluster IC 4665	17440n0543
Σ2211	10.2	115	33	8 - 9	relfix, spect G0	17441s0112
Σ2223	18.3	211	24	7½- 8½	spect F0	17465n0459
0Σ337	0.3	197	60	8 - 8½	PA dec, spect F2	17482n0715
S694	81.8	237	23	6½- 7	spect K2, A0	17495n0108
Σ3128	0.4	2	38	7- 10½	PA & dist dec, spect G5	17503s0754
Σ2235	19.5	124	24	7½- 9	slight dist inc, spect K0	17504s0215
Σ2244	0.9	285	59	7 - 7	PA inc, dist dec, spect A2	17545n0004
β1299	0.1	2	63	8½- 8½	binary, 130 yrs;	17551n1058
	26.8	63	20	-11½	PA inc, spect K5	
Σ2252	3.8	22	54	8 - 8½	relfix, spect A2	17565n0202
	94.2	164	31	- 8½		
Σ2254	3.0	263	35	8½- 9	relfix, spect A2	17567n1226
0Σ161	62.8	77	23	7 - 9	spect G5, K0	17578n0852
	15.0	125	11	- 11		
67	54.6	142	25	4 - 9	(β1124) spect B5,	17581n0256
67a	6.6	196	25	4- 13½	B3	
67b	8.4	128	25	9- 12½		
68	0.7	68	59	5 - 10	(β1125) PA inc, spect A1	17592n0118
τ	1.8	271	67	5 - 5½	(Σ2262) binary, about 280 yrs; PA inc, spect dF2, dF2	18004s0811

LIST OF DOUBLE AND MULTIPLE STARS   (Cont'd)

NAME	DIST	PA	YR	MAGS	NOTES	RA & DEC
70	2.8	72	67	4 – 6	(Σ2272) binary, 88 yrs; PA dec, spect K0, dK6 (*)	18029n0232
Σ2276	7.1	258	58	6 – 7	relfix, spect A0; cpm pair	18034n1200
HV 74	42.3	138	60	6 – 9	PA & dist inc, spect A2	18055n1304
β636	4.6	127	14	7 – 12	spect A2	18056n0213
	13.6	100	20	– 13		
	41.5	349	23	– 11½		
73	0.7	61	40	5½– 7	(Σ2281) binary, PA dec, about 390 yrs; spect dF0	18071n0359
	67.9	193	60	– 12½		
Σ2283	0.6	71	67	7 – 7½	PA & dist dec, spect F5	18071n0608
β637	7.3	191	34	6½– 12	relfix, cpm; spect F5	18074n0307
74	28.1	285	35	5– 11½	(h5495) spect G8	18184n0321
	57.8	80	25	– 13		
A582	2.8	45	27	7½– 13½	cpm, spect G0	18270n0719
Σ2322	20.1	169	35	6 – 11	relfix, spect B2	18276n0402
	66.9	70	12	– 13		
Σ2329	4.6	48	56	8 – 9	relfix, spect B9	18290n0625
OΣ354	0.6	188	66	7½–8½	PA inc, spect F5	18296n0645
OΣ355	38.7	248	46	6– 9½	relfix, spect B8 or A pec.	18310n0814
OΣ357	0.3	128	62	7½– 7½	binary, about 255 yrs; PA dec, spect A2	18336n1141
Σ2346	23.6	292	25	7½– 9	optical, dist & PA inc, spect G0	18348n0729
X	0.4	150	60	var–8½	(Hu 198) PA dec, A= LPV, spect M5, K0 (*)	18360n0847
Σ2355	23.1	133	60	6½–11½	optical, PA & dist dec, spect G8	18374n0719

OPHIUCHUS

LIST OF VARIABLE STARS

NAME	MagVar	PER	NOTES	RA & DEC
θ	3.25±0.02	.1405	Beta Canis Majoris type, Spect B2 (*)	17189s2457
χ	4.2--5.0	Irr	Spect B2e pec	16241s1821
66	4.55- 4.8	Irr	(V2048) γ Cass type? Spect B2--B6ne	17578n0422
R	7.0--13.6	302	LPV. Spect M4e--M6e	17049s1602
S	8.9--14.6	233	LPV. Spect M5e	16314s1704
T	8.9--14.2	367	LPV. Spect M6e	16309s1602
U	5.7--6.7	1.677	Ecl.Bin; Spect B5+B5 (*)	17140n0116
V	7.3--11.5	298	LPV. Spect N3e	16239s1219
W	9.2--14.5	331	LPV. Spect M6e	16187s0735
X	6.0--9.3	334	LPV. Spect M5e--M7e (*)	18360n0847
Y	6.5---7.3	17.123	Cepheid; spect F8--G3	17500s0608
Z	7.6--13.2	348	LPV. Spect K5e--M5e	17170n0134
RR	8.1--14.8	293	LPV. Spect M3e	16461s1923
RS	4.0--12..	---	Recurrent Nova; 1898, 1933 1958, 1967 (*)	17475s0642
RT	8.6--15.5	425	LPV. Spect M7e	17542n1110
RU	8.5--14.2	202	LPV. Spect M3e	17305n0927
RX	9.8--13..	322	LPV.	16503n0529
RY	7.6--13.8	150	LPV. Spect M3e--M6e	18141n0340
SS	7.8--14.5	180	LPV. Spect M5e	16552s0241
SU	9.3--11.3	110	Semi-reg; spect M5	17369n0138
SV	9.4--13..	216	LPV. Spect M2e	17539n0323
SY	8.3--10.1	132	Semi-reg; spect M5	16521s2153
TT	9.5--11.3	61	RV Tauri type; spect G2--K0	16471n0343
TX	9.3--11.2	135	RV Tauri type; spect F5--G6	17015n0503
TY	9.5--11.0	Irr	Spect N	18289n0420
UX	9.4--13.5	117	LPV. Spect M4e	17022s1208
UZ	9.4--11.6	87	RV Tauri type; spect G2--G8	17196n0658
VW	9.5--13..	284	LPV. Spect M5e	17249n0416
WZ	9.1--9.8	4.184	Ecl.Bin; spect dG0+ dG0	17042n0751
XX	9.1--11.1	Irr	Spect Be pec (*)	17413s0615
BF	7.3--8.2	4.068	Cepheid; spect F8--K2	17030s2631
V438	8.1--10.3	154	Semi-reg; spect M7e	17123n1107
V447	8.5--12.5	Irr	Spect gM4e	17096n0823
V451	7.9--8.5	2.197	Ecl.Bin; spect A0 + A2	18269n1051

OPHIUCHUS

LIST OF VARIABLE STARS   (Cont'd)

NAME	MagVar	PER	NOTES	RA & DEC
V502	8.4--9.0	.4534	Ecl.Bin; W Ursae Majoris type; spect G2 + F9	16388n0036
V508	8.8--9.7	.3448	Ecl.Bin; W Ursae Majoris type; spect G5	17569n1331
V513	8.7--10.1	58:	Semi-reg; spect M5	18115n0517
V521	9.6--11.0		Irr or semi-reg; spect M4	17207s2826
V533	7.5---9.5	30:	Semi-reg; spect M6	17504s0234
V551	8.0--8.3		LPV. Spect M2	17377s2722
V566	7.2--7.7	.4096	Ecl.Bin; W Ursae Majoris type; spect F4	17544n0459
V574	8.7--10..	72:	Semi-reg; spect M4	18037n0324
V679	9.0--9.3	Irr	Spect S4	18395n0646
V839	8.7--9.3	.4090	Ecl.Bin; W Ursae Majoris type; spect G0	18070n0908
V840	6.2--17..	---	Nova 1917	16516s2933
V841	4.3--13..	---	Nova 1848	16567s1249
V843	-2.5.....	---	Supernova 1604 (*) "Kepler's Star"	17276s2126
V849	7.3--15..	---	Nova 1919	18118n1136
V851	9.3--10.9	188	Semi-reg; spect M5	18324n0701
V906	8.4--13..	---	Nova 1952	17234s2150
V972	8.0--16..	---	Nova 1957	17316s2808
V986	6.1--6.13	.2847	Beta Canis Majoris type; spect B0	18021n0155
V988	7.8--8.6	63	Semi-reg; spect M7	18244n0353
V1010	6.2---7.0	.661	Ecl.Bin; spect A3	16466s1535
V2052	5.82±0.01	.1399	β Canis type; spect B2	17538n0040
---	5......?	---	Nova 1976	18011n1148

LIST OF STAR CLUSTERS, NEBULAE, AND GALAXIES

NGC	OTH	TYPE	SUMMARY DESCRIPTION	RA & DEC
----	I.4604	▢	eL,vF, Irr neby, 140' x 70' surrounding 4^m star Rho Oph	16226s2320
6171	40^6	⊕	Mag 9, diam 4', class X; L,vRi,mC,R, stars mags 14....	16297s1257
6218	M12	⊕	Mag 8, Diam 10', class IX; !! vB,vL,R,gmbM, stars mags 11..... (*)	16446s0152
6235	584^2	⊕	Mag 10, diam 2', class X; pB,L,R, stars mags 14.... Class uncertain, possibly not a globular cluster	16504s2205
6254	M10	⊕	Mag 7, Diam 8', Class VII; ! vB,vL,R,gvmbM, stars mags 11....15 (*)	16545s0402
----	I.4634	◎	Mag 12, diam 20" x 10"; eS,B, with 17^m central star	16585s2144
6273	M19	⊕	Mag 7, Diam 6', class VIII; vB,L,R,vmCM, stars mags 16... (*)	16595s2611
6284	11^6	⊕	Mag 10, diam 2', class IX; B,L,R,mCM, stars mags 15....	17015s2441
6287	195^2	⊕	Mag 10, diam 2', Class VII; cB,L,R,pmCM; stars mags 16...	17021s2238
6293	12^6	⊕	Mag 8, Diam 2', Class IV; vB,L,R,psbM; stars mags 16...	17071s2630
6304	147^1	⊕	Mag 9, diam 2', Class VI; B,cL,R,lbM; stars mags 16...	17114s2924
6309		◎	Mag 11.5; Diam 20" x 10" B,S, with 14^m central star	17112s1251
6316	45^1	⊕	Mag 10, Diam 1', Class III; cB,pS,R,gvmbM, stars eF	17134s2805
6325		⊕	Mag 12, Diam 1', Class IV; pF,L,R, stars vF	17150s2342
6333	M9	⊕	Mag 8, Diam 4', Class VIII; B,L,R,eCM; stars mags 14.. (*)	17162s1828
6342	149^1	⊕	Mag 11, Diam 1', class IV; cB,pS,1E	17182s1932
6355	46^1	⊕	Mag 10, Diam 1'; class uncertain; cF,S,R, stars eF	17209s2619

LIST OF STAR CLUSTERS, NEBULAE, AND GALAXIES (Cont'd)

NGC	OTH	TYPE	SUMMARY DESCRIPTION	RA & DEC
6356	48[1]	⊕	Mag 8.5; diam 2', class II; vB,cL,vmbM, stars eF	17207s1746
----	B72	▣	S-shape dark nebula, diam 30'; 1.5° NNE from Theta Oph.	17210s2335
----	I.1257	⬡	Diam 1', vS,F,vC; possibly a distant globular cluster	17245s0703
6366		⊕	Mag 12, Diam 4', Class XI; F,L,vlbM; stars vF	17251s0502
----	H15	⬡	Mag 9, Diam 10', about 15 faint stars; class E	17260s2929
6369	11[4]	◎	Mag 11, Diam 28"; pB,S,R, perfect ring with 16^m central star	17263s2344
6384		⊘	Sb; 12.3; 4.0' x 3.0' pB,S,E (*)	17299n0706
6401	44[1]	⊕	Mag 11, Diam 1'; pB,pL,R; Class uncertain	17356s2353
6402	M14	⊕	Mag 9, Diam 6', Class VIII; B,vL,R,eRi, stars mags 15.... (*)	17350s0313
6426	587[2]	⊕	Mag 12, Diam 2', Class IX; F,cL,E,vlbM; stars vF	17424n0312
----	I.4665	⬡	vvL sparse group; Diam 55'; Mag 6, about 20 stars mags 7.. Class C; incl Σ2212	17438n0544
6517	199[2]	⊕	Mag 12, diam 1', Class IV; B,pL,R; stars eF	17591s0857
6572	Σ6	◎	Mag 9, diam 15" x 12"; vB,vS, 1E; bluish-green disc with 12^m central O-type star	18097n0650
6633	72[8]	⬡	Mag 5, Diam 20', B,L,mE,1C, about 65 stars mags 7.... Class D (*)	18251n0632

DESCRIPTIVE NOTES

ALPHA   Name— RAS ALHAGUE, "The Head of the Serpent-
Charmer". Magnitude 2.09; spectrum A5 III.
Position 17326n1236. Distance about 60 light years; actual
luminosity about 40 times the Sun. Alpha Ophiuchi shows an
annual proper motion of 0.26" in PA 153°; the radial velocity is 7.5 miles per second in recession.

Alpha Ophiuchi shows the same space motion as the
stars of the Ursa Major group; this is also true of a number of other bright stars at widely different positions in
the sky. The star is also one of the few nearby stars which
shows interstellar absorption lines in its spectrum; this
is a very remarkable feature for a star only 60 light years
distant.

Careful astrometric measurements show a periodic
oscillation in the proper motion of the star, indicating
the presence of an unseen companion with a period of about
8.5 years. Computed masses for the two components are 2.4
and 0.6 times the solar mass; the expected magnitude of the
companion is about 7, and the true separation of the pair
must be close to 6 AU. Widest apparent separation of 0.4"
occurs in 1973-1974.

BETA   Name— CHELEB, "The Shepherd's Dog". Magnitude
2.77, spectrum K2 III; Position 17410n0435.
The computed distance is about 125 light years, giving a
true luminosity of about 100 suns. The annual proper motion
is 0.16"; the radial velocity is 7 miles per second in
approach.

The large sparse galactic star cluster IC 4665
is centered about 1.4° to the northeast.

DELTA   Magnitude 2.72; spectrum M1 III; position
16117s0334. The computed distance is about
140 light years; the actual luminosity about 130 times that
of the Sun. Delta Ophiuchi shows an annual proper motion of
0.16"; the radial velocity is 12 miles per second in
approach.

EPSILON   Magnitude 3.22; spectrum G9 III. Position
16157s0434. The distance is about 90 light
years; the actual luminosity about 35 times that of the
Sun. Epsilon Ophiuchi shows an annual proper motion of

DESCRIPTIVE NOTES   (Cont'd)

0.09"; the radial velocity is 6 miles per second in
approach. Epsilon and Delta form a wide naked-eye pair
with a separation of slightly under 1.5°; Epsilon is the
south-east star of the two.

ZETA   Magnitude 2.57; spectrum 09 or B0 V; position
16344s1028.   Zeta Ophiuchi has a computed
distance of 520 light years; the actual luminosity is close
to 4000 times that of the Sun after a correction for light
loss through space absorption. The star shows an annual
proper motion of 0.02"; the radial velocity is 11.5 miles
per second in approach.
     The remote globular star cluster NGC 6171 lies
2.8° distant, toward the SSW. This object has been added to
the original Messier list in some modern catalogs, and
given the number M107.

ETA   Name- SABIK. Magnitude 2.46; spectrum A2 V;
Position 17075s1540. Eta Ophiuchi is about
70 light years distant; the actual luminosity is about 40
times that of the Sun. The annual proper motion is 0.10";
the radial velocity is 0.5 mile per second in approach.
     The star is a close and difficult binary, but an
excellent test object for larger telescopes. It was first
resolved by S.W.Burnham with the 36-inch refractor at the
Lick Observatory in 1889. According to recent computations
by Van Biesbroeck (1960) the orbital period is 84 years,
and the apparent orbit is a very eccentric ellipse oriented
about 7° from the edge-on position. Periastron occurs in
2020 AD. Other orbital elements are: Semi-major axis =
1.06"; eccentricity = 0.89; inclination = 97°. The two
stars differ by about half a magnitude in brightness, and
are probably of very similar spectral types. Individual
catalog magnitudes for the components are 2.9 and 3.4. The
mean true separation of the components is about 20 AU,
about equal to the separation of Uranus and the Sun.
     The long-period variable star R Ophiuchi lies 0.8°
to the SW.

THETA   Magnitude 3.26; spectrum B2 IV; position
17189s2457. The computed distance is about
700 light years; the actual luminosity is about 1900 times

that of the Sun. This star, like many other bright stars in
this portion of the sky, is a member of the large Scorpio-
Centaurus moving cluster. The annual proper motion is
0.02"; the radial velocity is 2 miles per second in
approach.

Theta Ophiuchi is one of the spectral variables of
the Beta Canis Majoris type, with a period of 3.37 hours
or 0.1405 day. The variations consist of a periodic shift
of the spectral lines with an accompanying slight change in
the light of 0.02 magnitude. Stars of this type, sometimes
called "quasi-cepheids", are believed to be rather young,
massive stars which are just beginning to evolve away from
the main sequence. Beta Canis Majoris and Beta Cephei are
the two standard examples of the type.

Theta Ophiuchi is located in one of the most awesome regions of the Milky Way amid a vast complex of star
clouds. The huge dark mass of the "Pipe Nebula" may be seen
with the naked eye a few degrees to the south. Barnard's
peculiar "S-Nebula" lies 1.5° to the north. (Refer to page
1268)

KAPPA   Magnitude 3.20; spectrum K2 III. Position
16553n0927.   This star was suspected of
irregular variability by Kopff, with a range of more than
half a magnitude. The variability has never been confirmed
and the star is listed as "constant" in the Moscow General
Catalogue (1970).

The computed distance is about 150 light years; the
actual luminosity about 90 times that of the Sun. The star
shows an annual proper motion of 0.29" in PA 267°; the
radial velocity is 33 miles per second in approach.

NU   Magnitude 3.34; spectrum G9 III. Position
17563s0946. Nu Ophiuchi is approximately 140
light years distant, and has an actual luminosity of about
70 times that of the Sun. The annual proper motion is 0.12"
and the radial velocity is 7 miles per second in recession.

The star is located in a heavily obscured dust-
laden region of the Milky Way, near the NE edge of a large
obscuring cloud. The faint globular cluster NGC 6517 lies
one degree to the NE.

70   Position 18029n0232. Magnitude 4.01, spectra
KO V and dK6. This is one of four stars which
form the obsolete V-shaped asterism known as the "Bull of
Poniatowski". The other three stars are 66, 67, and 68.

70 Ophiuchi is one of the best known binary stars,
discovered by Sir William Herschel in 1779, and probably
among the most thoroughly studied dozen binaries in the
heavens. The two components are magnitudes 4.2 and 5.9 and
have a noticeable color contrast, described as yellow and
red by some observers, gold and violet by others. As with
most colored doubles, there are disagreements as to the
precise colors, and even suggestions that the tints have
changed from time to time. Herschel recorded no color in
either component, but Admiral Smyth saw them as "pale topaz
and violet" whereas Flammarion in 1879 thought that the
fainter star was "rose-colored". E.J.Hartung (1968) calls
them "bright yellow and orange" whereas the author of this
book has always seen them as golden and rusty-orange, with
a persistent hint of violet in the fainter star.

The apparent separation of the pair varies from
6.7" (1933) to 1.7" (1982) in a period of 87.85 years. The

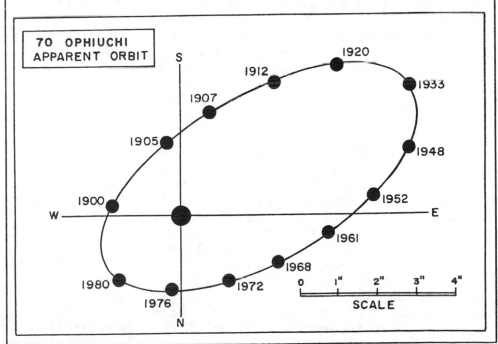

70 OPHIUCHI APPARENT ORBIT

true distance between the stars averages about 23 AU, comparable to the separation of Uranus and the Sun. Since the first measurements were made the stars have completed two orbital revolutions. Other orbital elements, according to K.Strand, are as follows: Semi-major axis= 4.55"; eccentricity= 0.50; inclination= 121°; motion retrograde, and periastron in the year 1895. The chief facts about the two stars are given in the table below:

	Mag.	Spect.	Lum.	Mass.	Diam.	Abs.Mag.
A	4.2	K0 V	0.40	0.82	0.9	+5.8
B	5.9	dK6	0.08	0.60	0.7	+7.5

Both components are dwarf stars, with a total light output of about half that of our Sun. The system is only 16.5 light years distant, and shows a large annual proper motion of 1.13" in PA 167°. The radial velocity is 4.5 miles per second in approach.

70 Ophiuchi is one of the few binary systems which is close enough to permit detailed investigations of its orbital motion. For some years it has been claimed that the star shows unexplained irregularities in the motions of the visible pair, indicating the presence of an unseen companion having about 1% the mass of the Sun. Since there are no visible stars known of such a small mass, it has been suggested that the unseen body may be a very massive planet, about 10 times the mass of Jupiter. The nature of such an object, and even the evidence for its existence, has long been a subject for controversy.

J.H.Madler (1842), W.S.Jacob (1855), T.J.J.See (1896), E.Doolittle (1897) and T.Lewis (1906) are among the earlier observers who found that 70 Ophiuchi showed clear deviations from a Keplerian motion. Computed periods for the unseen body ranged from 6 years to 36 years. Miss Agnes Clerke (1905) stated that "The stars have hitherto so persistently refused to keep to their predicted places that Madler, Jacob, and Sir John Herschel suspected disturbance by an invisible member of their system calculated by Dr.See in 1895 to revolve in a period of thirty-six years.....yet its vagaries of movement are not even thus completely explicable..." She concludes wryly "The mechanism of 70 Ophiuchi has still obscure springs". In 1937 a detailed analysis of the observations was made by Dr.Strand, who found no real evidence for a third body in the system. He attributed the supposed deviations to systematic errors in measurement. In 1943, however, the question was re-opened by D.Reuyl and E.Holmberg at the McCormick Observatory. In an analysis of astrometric plates made between 1914 and 1942 they found indications of a 17-year perturbation with an amplitude of about 0.015"; the mass of the assumed third body would be about 1% of the solar mass. At present it can only be said that more observations are needed to settle the question of the hypothetical third body. It is highly interesting to note, however, that the computed mass of this object is very near the mass derived for the unseen planet-like body in the 61 Cygni system. (Refer to page 768)

U   (38 Ophiuchi) Position 17140n0116. Magnitude 5.70 (variable), spectrum B5 V. This is a short period eclipsing binary star, discovered by B.Gould in 1871. It consists of two bright B-type stars of nearly identical size, type, and luminosity, orbiting in a period of 1.677347 day, with a center-to-center separation of about 5.5 million miles. The chief facts about the two components are given in the table:

	Spect.	Lum.	Diam.	Mass	Abs.Mag.
A	B5 V	480	3.2	5.4	-1.9
B	B5 V	480	3.2	4.7	-1.9

The combined mass of the system is about 10 solar masses, and the stars revolve in almost perfectly circular orbits with velocities of 120 and 110 miles per second. The combined magnitude is normally 5.7 so that the star is usually a naked-eye object. During eclipse the light falls to magnitude 6.7, a light decrease of 2.5 times. There are two minima in the course of each orbit revolution. The distance of the system, from the computed luminosities, appears to be a little more than 1000 light years. The mean radial velocity is 6.5 miles per second in approach; the annual proper motion is 0.02".

John Herschel, in 1820, found a faint companion star of the 13th magnitude, at a distance of 20" in PA 358°.

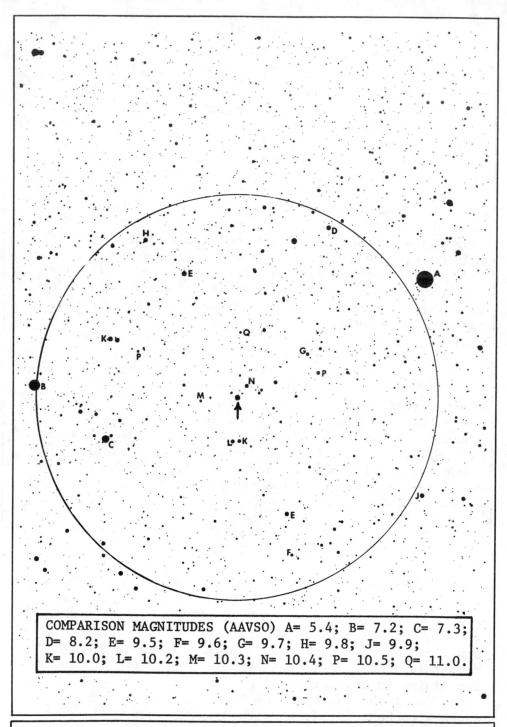

COMPARISON MAGNITUDES (AAVSO) A= 5.4; B= 7.2; C= 7.3; D= 8.2; E= 9.5; F= 9.6; G= 9.7; H= 9.8; J= 9.9; K= 10.0; L= 10.2; M= 10.3; N= 10.4; P= 10.5; Q= 11.0.

X OPHIUCHI. Identification field from a 13-inch telescope plate made at Lowell Observatory. The circle is 1° diameter with north at the top. Limiting magnitude about 15.

The star may not be a true physical member of the system. There has been no relative change in separation or angle since discovery.

The long-period variable star Z Ophiuchi lies 0.8° distant, toward the ENE.

X — Variable. Position 18360n0847. A long-period Mira-type variable star, first observed by T.E.Espin in 1886. It is a pulsating red giant with an average period of 334 days and a magnitude variation of 6 to about 9, which keeps it always within range of a small telescope. The light curve has the peculiarity of a fairly flat and shallow minimum at about 9th magnitude, while the maxima may vary considerably in height. As far back as 1900, this circumstance suggested the presence of a close companion, which would remain visible at a constant magnitude even though the variable itself had faded from sight. In the same year (1900) this expectation was fulfilled at Lick Observatory when W.J.Hussey detected the companion with the 36-inch refractor.

The companion is a 9th magnitude star located 0.3" from the primary; the position angle was 195° in 1900 but appears to be decreasing at the rate of about 30° in 25 years; the separation has remained fairly constant. The two stars undoubtedly form a binary system, one of the few cases known where a long-period variable is a member of a double star system. Omicron Ceti itself (Mira) is perhaps the only other well established case. Such binary systems are of immense importance to the astrophysicist, since they

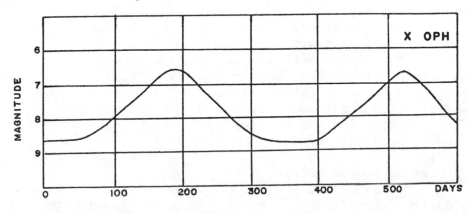

furnish our only opportunity to make a direct determination
of the mass of a Mira-type star. P.W.Merrill (1923) found
the spectral class of the companion to be about K0. In an
analysis of the system, J.D.Fernie (1959) derived an orbit
period of about 560 years, and a distance of about 780
light years. The true separation of the two stars is on
the order of 75 AU, and the total mass of the system is
about twice a solar mass. Absolute magnitudes found for
the components were -0.3 for the M-star at maximum, and
+1.1 for the K-star. It is not known definitely which star
has the greater mass, but the variable, in any case, cannot
be significantly more massive than the Sun. A rather similar result was found for the famous star Mira, and it now
appears certain that these stars are not very massive
objects as was formerly assumed. Fernie suggests that the
long-period variables originate from main sequence stars of
types A8 to F5, stars which are only slightly more massive
than the Sun. Such pulsating red supergiants as Betelgeuse
and Alpha Herculis do indeed appear to have much larger
masses, and must be a fundamentally different type of star.
        X Ophiuchi shows an annual proper motion of 0.02";
the radial velocity is 43 miles per second in approach.
(Refer also to Omicron Ceti, Page 631)

RS   Position 17475s0642. RS Ophiuchi is a very
                  remarkable recurrent nova, one of seven known
stars of the type, and with the exception of T Coronae
Borealis, the only one which reaches naked-eye visibility
at maximum. Four outbursts have been recorded, in 1898,
1933, 1958, and 1967.
        The variations of the star were first detected in
1901, and the star was then listed as an irregular 11th
magnitude variable with a total range of about half a
magnitude. A check of Harvard plates revealed that a high
maximum had been recorded on June 30, 1898, when the star
had attained magnitude 7.7; the brightness was then falling, however, and the true maximum probably had occurred
several days earlier at about 5th magnitude. Spectroscopic
studies showed a typical nova-type spectrum resembling that
of Nova Geminorum 1903.
        Except for a minor outburst of about a magnitude in
1900, the variations of the star were small and irregular

until the second recorded maximum in 1933. On the night of August 10, 1933, the star began to brighten, and the next evening had attained magnitude 5.8. A further increase to magnitude 4.3 occurred during the next 24 hours. The total range of the nova was thus slightly over 500 times; much less than the observed ranges of T Coronae (2500 times) or WZ Sagittae (4000 times). After passing the maximum the nova began to fade almost immediately, and had decreased by three magnitudes in the first ten days. Within two months it had fallen to 10th magnitude. About 130 days after the outburst, the star began to brighten again, and within 20 days had increased by 0.3 magnitude; after about a month it continued its normal fading to minimum. This slight but definite "secondary maximum" occurred also, with even higher amplitude in the light curve of T Coronae, but has not been observed in any other recurrent nova.

Spectra obtained by Adams and Joy at Mt.Wilson showed characteristic nova features; strong bright bands of hydrogen, fainter emission bands attributed to helium and ionized iron, and a rather strong continuous spectrum which began to fade after the first week. The color of the star was a bright orange-red shortly after maximum due to strong hydrogen (H-alpha) emission. A few days later the first nebular lines had begun to appear. A discovery of great interest was made some six weeks later, when a spectrogram

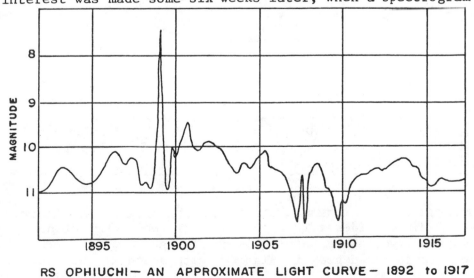

RS OPHIUCHI— AN APPROXIMATE LIGHT CURVE — 1892 to 1917

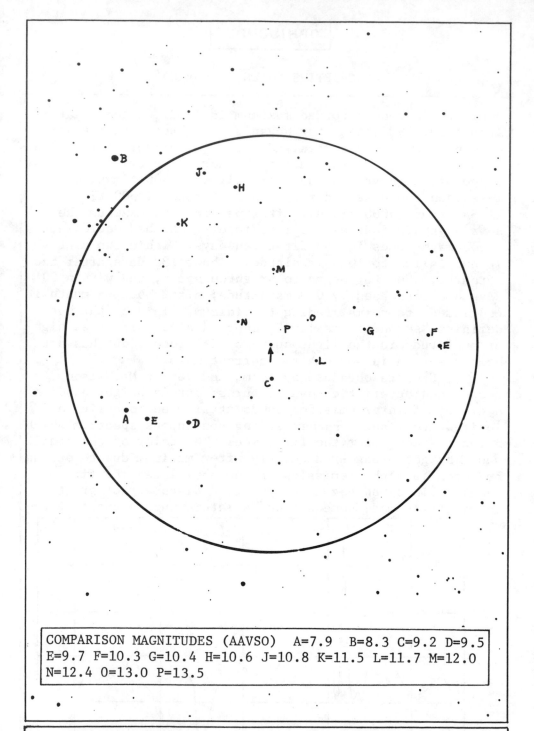

COMPARISON MAGNITUDES (AAVSO) A=7.9 B=8.3 C=9.2 D=9.5
E=9.7 F=10.3 G=10.4 H=10.6 J=10.8 K=11.5 L=11.7 M=12.0
N=12.4 O=13.0 P=13.5

RS OPHIUCHI. Identification field for the recurrent nova,
from a Lowell Observatory 13-inch telescope plate. The
circle is one degree in diameter with north at the top.

DESCRIPTIVE NOTES (Cont'd)

revealed several lines characteristic of the corona of the
Sun, and previously seen nowhere else.

   The third recorded maximum of RS Ophiuchi occurred
in the summer of 1958. On July 12 the star was in its
normal 11th magnitude state, but on the next evening it had
brightened 100 times and had risen to magnitude 6. The
maximum was reached on July 14 at magnitude 5.0, and the
star then began to fade rapidly at the rate of a third of
a magnitude per day. The color was white at maximum but
turned strongly red a few days later as H-alpha emission
dominated the spectrum. During the first few nights the
measured expansion velocity of the nova shell was about
1800 miles per second. There is spectroscopic evidence that
the star was surrounded by a nebulous envelope before the
outburst occurred, but it is not certain whether this
cloud originated in previous outbursts, or whether there is
a continuous mass-loss from the star.

   As one of the two brightest recurrent novae, RS
Ophiuchi is an object of great interest, and has probably
been the most thoroughly studied star of its type. From

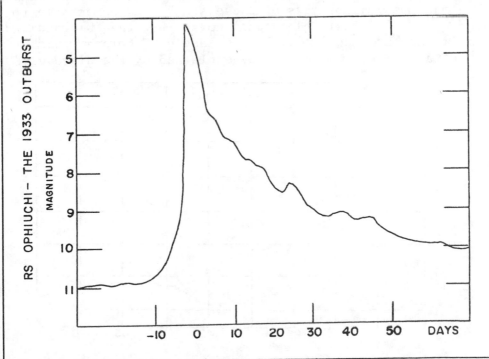

DESCRIPTIVE NOTES   (Cont'd)

the moderate range and the period of about 30 years, the
star seems to bridge the gap between the erratic SS Cygni
variables and the true full-scale novae. It remains to be
determined whether this circumstance is accidental or
whether the three types actually form some sort of evolutionary sequence.

From the strength of the interstellar lines in 1933
the distance of RS Ophiuchi was estimated to be at least
3000 light years. At the 1958 outburst a more detailed
study of the spectrum indicated a probable distance of
about 12,000 light years. The resulting figures for the
true luminosity appear in the table below:

	Apparent Mag.	Absolute	Lum.
At normal minimum	11.1	-1.7	440
Maximum 1933	4.3	-8.5	230,000
Maximum 1958	5.0	-7.8	120,000

These results are rather similar to those obtained
for the only other naked-eye recurrent nova, T Coronae.
The maximum luminosities of these stars seem comparable to
those of the normal "classical" novae, but the observed
amplitudes are much less. In the case of T Coronae this
peculiarity has been partially explained by the discovery

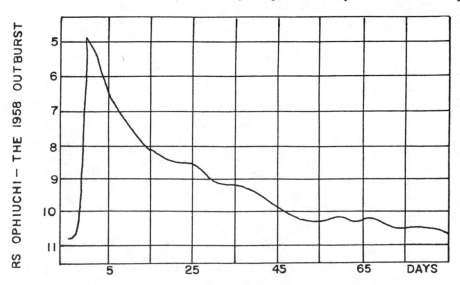

RS OPHIUCHI. The star is shown here on the night of its fourth recorded maximum, October 27, 1967. The comparison plate was made in July 1932. Lowell Observatory 13-inch camera plates.

DESCRIPTIVE NOTES   (Cont'd)

that the star is a close binary; the object we see at minimum is not the actual nova-star, but a larger, cooler companion. Thus the actual range of the nova-component may exceed ten magnitudes. Recent studies indicate that RS Ophiuchi is a very similar system, and that the companion is possibly a larger G-type star. This suggests the possibility that the outbursts of these stars are triggered by some process of interaction or interchange of material between the close components. Since the nova-star itself appears to be a hot subdwarf, the accretion of material from the companion might produce an explosive reaction. In the SS Cygni stars and in the related object AE Aquarii it is almost certain that we are witnessing some process of this sort on a smaller scale. A study of the remarkable star WZ Sagittae may provide additional support of this theory. The star has the normal luminosity of a white dwarf and attains a luminosity of only 25 suns even at maximum; the resemblance to such objects as RS Ophiuchi thus seems to be only superficial. Yet evidence obtained in 1961 clearly indicates that the star is a close binary with a period of only 81.6 minutes; it strongly resembles the SS Cygni stars and the classical novae DQ Herculis and GK Persei as well. The likely conclusion seems to be that the same fundamental process operates in all these stars, and that the nova phenomenon is the result of some type of interaction between two very close stars, at least one of which may be partially degenerate.

     The seven known recurrent novae are given here with the dates of their outbursts: T Coronae Borealis (1866, 1946). RS Ophiuchi (1898, 1933, 1958, 1967). U Scorpii (1863, 1906, 1936, 1979). T Pyxidis (1890, 1902, 1920, 1944, 1967). WZ Sagittae (1913, 1946, 1978). VY Aquarii (1907, 1962). V1017 Sagittarii (1901, 1919, 1973).

XX   Position 17413s0615. An irregular variable
                star of uncertain type, located about 1.5° WNW
from the recurrent nova RS Ophiuchi. It was first studied at Harvard Observatory in 1908, and later christened the "Iron Star" by P.W.Merrill because of the unusual strength of the lines of ionized iron in its spectrum. XX Ophiuchi is normally an object of about 9th magnitude, but fades at unpredictable intervals to about 11th, sometimes remaining

DESCRIPTIVE NOTES (Cont'd)

faint for as long as two years. Deep minima occurred in 1900, 1921, and 1931. Between 1931 and 1946 the star was observed to remain near peak brightness, showing only small fluctuations which are characteristic of this star when at maximum. The light curve resembles the R Coronae Borealis type stars but has a much smaller amplitude; the spectral peculiarities also appear to place it in a class by itself.

The spectrum is of type B with bright hydrogen and metallic lines; the color index at maximum is about equal to spectral type A0. The spectral features, however, show large changes from year to year, one remarkable case being the development of a strong absorption spectrum of ionized titanium in 1925. In the following years the famous "iron spectrum" slowly returned to its original appearance.

P.W.Merrill in 1960 found that the absorption lines showed approach radial velocities of about 195 miles per second, indicating the presence of a diffuse expanding shell of rarified gases surrounding the star. XX Ophiuchi is possibly related to the erratic "shell stars" such as 48 Librae and 17 Leporis although the large light changes constitute a unique feature. The cause of the occasional periods of fading is still unknown. (Refer also to R Coronae Borealis)

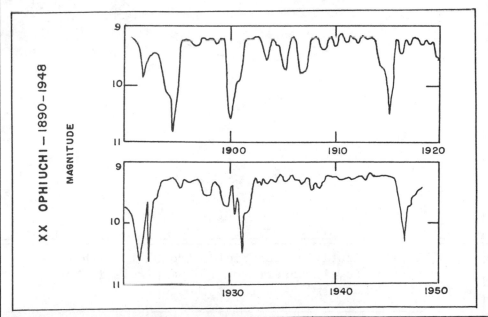

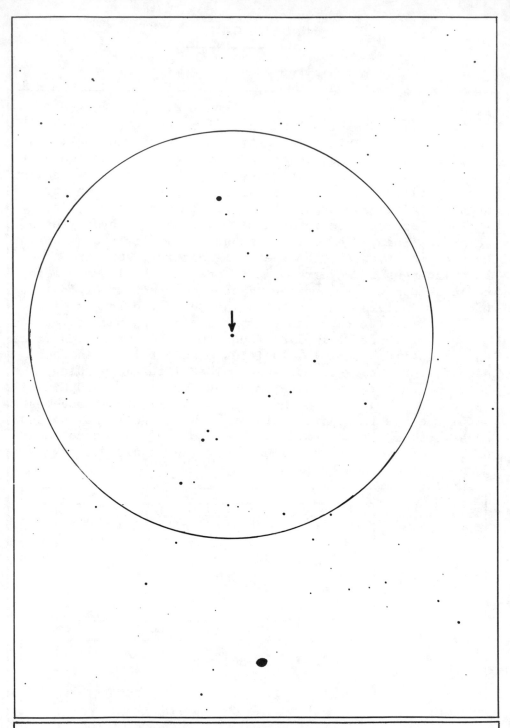

XX OPHIUCHI. Identification field, from a 13-inch camera plate made at Lowell Observatory. The circle is one degree in diameter with north at the top.

DESCRIPTIVE NOTES (Cont'd)

NOVA 1604 AD (V843 Ophiuchi) Position 17276s2126. The last example of a supernova in our Galaxy; the fourth to be observed in the past 1000 years. This brilliant new star was first observed on October 9, 1604 when it had already reached a brightness exceeding any other star in the sky. It is interesting to note that the two planets Jupiter and Mars were in conjunction on the same night, only a few degrees from the nova position! Thus the nova was detected immediately. Among the earliest observers were Altobelli in Verona, Clavius in Rome, Capra and Marius in Padua, and Brunowski in Prague. Brunowski notified Kepler, who saw the nova for the first time on October 17, after a spell of cloudy weather. Kepler made a special study of the phenomenon which so closely resembled Tycho's famous nova of 1572 in Cassiopeia, and the nova is often called Kepler's Star in his honor.

The nova was as bright as Mars when first seen, but within a few days it had surpassed Jupiter in brilliance and remained for several weeks the outstanding object in the heavens. An analysis of all the observations has been made by W.Baade (1943); he finds that the maximum occurred on October 17 and the peak magnitude was about -2.25. The star was still comparable to Jupiter when it was lost in the twilight sky in November, and at its reappearance in the morning sky in January 1605, Kepler still found it brighter than Antares. Finally, in March 1606 the star vanished from sight, having been visible to the naked eye for some 18 months. A study of the light curve shows that the star was a supernova of Type I, as was Tycho's Star of 1572. The light curves of the stars are compared on page 506.

In 1941 an attempt was made at Mt. Wilson to detect the remnant of the supernova with the 100-inch telescope. Red-sensitive plates were used since the star appeared in a heavily obscured region of the sky, deeply embedded in the dust clouds of the southern Milky Way. A faint fan-shaped nebulosity, about 40" in extent, was discovered at the expected position, appearing as an irregular mass of filaments and bright condensations; the total integrated magnitude is about 19. Radial velocity measurements on some of the brighter filaments show an approach velocity of 120

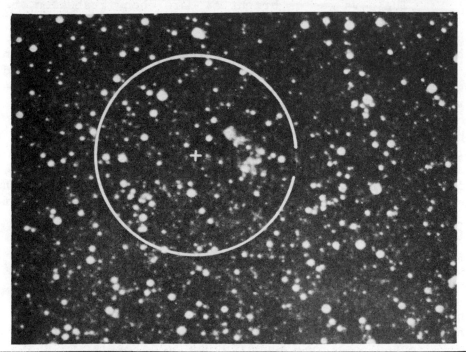

FIELD OF KEPLER'S STAR. Top: A field 40' wide centered on the nova position. Below: A 2' circle centered on the nova position, showing some faint remnants of the star to the right of the cross. Mt.Wilson 100-inch telescope photo.

DESCRIPTIVE NOTES (Cont'd)

miles per second. Fainter outlying wisps may increase the
full size of the nebulosity to about 100". The actual post-
nova star has not been identified, and must be fainter
than 18th magnitude, assuming that a true stellar remnant
still exists. A star of about 18.6 magnitude is located
near the tip of the fan-shaped nebulosity, but its identification as the actual nova-star is still not certain. The
spectrum of the nebulosity, obtained at Mt.Wilson, shows
features resembling the Crab Nebula in Taurus.

The distance of Kepler's Star is not accurately
known, but if the absolute magnitude was about -16, the
distance was about 20,000 light years, making no correction
for absorption. The true distance may have been much less.
(Refer also to Tycho's Star in Cassiopeia, the Nova of
1006 AD in Lupus, and the "Crab Nebula" M1 in Taurus)

BARNARD'S STAR   Position 17554n0424. Barnard's
"Runaway Star" (Munich 15040) (LFT
1385). This is a faint red dwarf star, famous for having
the greatest known apparent motion ("proper motion") of any
known star. It was discovered by E.Barnard in 1916 through
a comparison of plates made in 1894 and 1916. After the
discovery, images of the star were detected by E.Pickering
on Harvard plates made as early as 1888. The annual motion
of the star is 10.29" in a direction nearly due north (PA
356°) and the star requires 351 years to change its place
by one degree in the sky. The positions for various years
and predictions for the next century are indicated on the
chart on page 1252.

Barnard's Star is a dM5 red dwarf of apparent
magnitude 9.53 and absolute magnitude +13.4. The actual
luminosity is about 1/2500 that of the Sun. J.C.Duncan has
found the mass to be about 16% that of the Sun and the
diameter to be about 140,000 miles; the resulting density
is on the order of 40 times the solar density. This is a
fairly "cool" star, with a surface temperature of about
3200°K.

At a distance of 6.0 light years, Barnard's Star
is the second nearest star to our Solar System; only the
Alpha Centauri triple system is closer. The large apparent
motion is thus a result of the nearness of the star combined with an unusually high space velocity of about 103

DESCRIPTIVE NOTES   (Cont'd)

miles per second. The radial velocity is rather large also, amounting to about 87 miles per second in approach. Owing to the steady approach, the proper motion of Barnard's Star is gradually increasing, and will reach 25.6" per year some 8000 years from now. At that time the star will be less than 4 light years distant, and will have brightened to magnitude 8.6.

P.van de Kamp has announced a periodic variation in the motion of this star, indicating the presence of at least one companion of planetary mass with a period of 24 years and a mean separation of 2.4" or close to 4 AU. From the orbital elements, a mass of 0.0015 solar mass, or about 1½ times the mass of Jupiter, is derived; the orbit has an eccentricity of 0.6; periastron occurred in 1974.

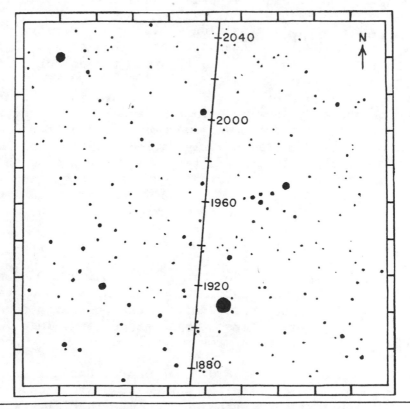

THE PATH OF BARNARD'S STAR — 1880 to 2040 AD.   THE CHART IS 30' ON A SIDE, MARGINAL MARKS ARE 3' APART

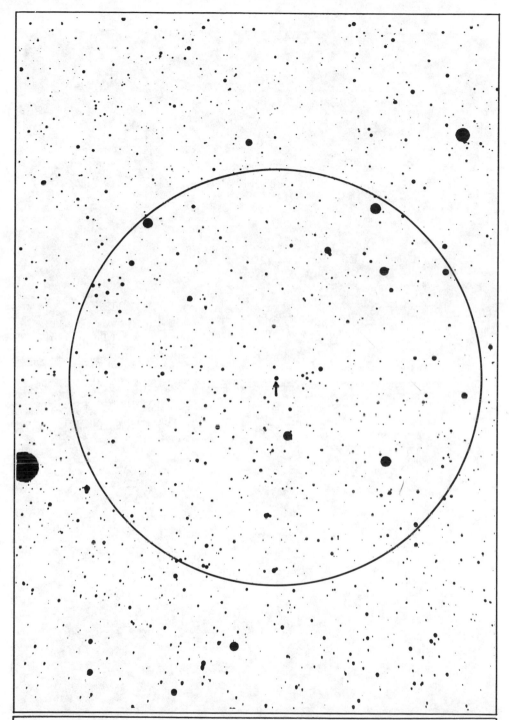

BARNARD'S STAR. Identification field from a Lowell Observatory 13-inch telescope plate. The circle is 1° in diameter, north is at the top. The bright star at the extreme left edge is 66 Ophiuchi, magnitude 4.8. Plate made July 1960.

BARNARD'S STAR, photographed at Lowell Observatory in 1937 (top) and 1960. To identify the two images, refer to the chart on page 1252. The change in position in 23 years amounts to about 4'.

DESCRIPTIVE NOTES   (Cont'd)

If the majority of the stars in the sky had motions
as great as Barnard's Star, the familiar constellation
patterns would be altered noticeably in a few centuries.
It may indeed seem that such changes should be expected
since the entire Galaxy is rotating, the Solar System is
drifting through space, and all the stars are moving in
various directions with velocities averaging many miles per
second.   The answer to this puzzle lies in the vast spaces
between the stars.   A star may be moving at 30 miles a
second, but if we must observe it at a distance of several
hundred trillion miles, we shall be unable to detect any
direct evidence of motion except with the most sensitive
measuring instruments used over a period of years. And the
star's naked-eye position will not be noticeably changed,
even after centuries.

A large proper motion thus definitely indicates
one thing- that the star is relatively near. Evidently a
search for proper motion stars is the simplest way of
making a census of the Solar neighborhood and discovering
very faint nearby stars which would otherwise go entirely
unnoticed.   Most of the known white dwarfs and many other
faint and unusual nearby stars have been found in this way.
Typical examples are UV Ceti, Lalande 21185 in Ursa Major,
Wolf 359 in Leo, Ross 614 in Monoceros, Van Biesbroeck's
Star in Aquila, etc.

The "discovery" of proper motion itself was made by
the English astronomer Edmund Halley and announced in 1718.
He noticed that the positions of the bright stars Sirius,
Arcturus, and Aldebaran were different from the positions
given in ancient star catalogs, Arcturus having moved over
a degree since the time of Ptolemy. In announcing this
discovery and its significance Halley wrote:

"It is scarce credible that the Ancients could be
deceived in so plain a matter, three observers confirming
each other. Again these stars being the most conspicuous in
Heaven, are in all probability nearest to the Earth, and if
they have any particular motion of their own, it is most
likely to be perceived in them, which in so long a time as
1800 years may show itself by the alteration of their
places, though it be utterly imperceptible in the space of
a single century of years. Yet as to Sirius, it may be
observed that Tycho Brahe makes him 2 minutes more north-

DESCRIPTIVE NOTES   (Cont'd)

erly than we now find him." Halley also mentioned an
observation of an occultation of Aldebaran in March 509 AD
which was seen at Athens, and pointed out that the phenomenon was an impossibility unless the declination of the
star had been much less at that time.

Modern observations show that the proper motion of
Arcturus in 2000 years amounts to about 76'. Sirius moves
about 41' in this interval, and Aldebaran about 7'.

The overthrow of the "fixed star" idea created a
need for increasingly accurate position determinations so
that stellar motions could be detected and measured. A
pioneer in this field was Tobias Mayer who in 1760 determined the motions of several stars by comparing his positions with those obtained by O.Romer some 50 years before.
In 1792 the first really large motion was discovered by
Piazzi; his "Flying Star" was the now famous double star
61 Cygni, with a motion of 5".22 annually. In 1842 a star
of the 6th magnitude in Ursa Major was found by Argelander
to have a motion of 7".04 annually; this is the star now
usually called "Groombridge 1830". The discovery of a
motion of 8".70 was announced in 1897 by Kapteyn. This star
(Cordoba 5h 243) is a 9th magnitude object in the southern
constellation Pictor, and is still the second largest
motion known. It is usually called in honor of the finder
"Kapteyn's Star".

Modern proper motion surveys have been made entirely by photography, one of the earliest systematic programs
being that begun by Max Wolf in 1906. In more recent years
numerous proper motion stars have been discovered by F.Ross
and W.J.Luyten, and a proper motion survey of the entire
northern sky was begun at Lowell Observatory in 1958. A
catalog published by Luyten in 1955 listed all known proper
motions exceeding 0.5" per year, a total of 1849 stars. Of
these, 369 had motions greater than 1.0" per year, and 62
had motions exceeding 2.0" per year. The list of proper
motions exceeding 3.0" annually is quite small, and is
given in the table opposite.

Proper motion catalogs contain "basic data"; facts
which may be used in a variety of different studies and
investigations, ranging from the discovery of intrinsically
faint dwarf stars to the study of the dynamics of the Local
Star Cloud and the rotation of the Galaxy. Among the other

DESCRIPTIVE NOTES (Cont'd)

LIST OF STARS OF LARGE PROPER MOTION								
STAR	LFT	CON	MU	PA	MAG	SPECT	RA & DEC	
Barnard's Star	1385	Ophi	10.29"	356	11.3	M5	17554n0424	1
Kapteyn's Star	395	Pict	8.70	131	10.0	M0	05097s4500	
Groombridge 1830	855	UMaj	7.04	145	7.0	G5	11501n3805	1
Lacaille 9352	1758	PscA	6.90	79	8.6	M2	23026s3609	
Cordoba 32416	5	Scul	6.11	113	10.0	M3	00025s3736	
Ross 619	569	Canc	5.40	167	14.2	M6	08092n0902	
61 Cygni A	1604	Cygn	5.22	52	6.2	K5	21044n3828	
B	1605	Cygn	5.22	52	7.2	K7		1
Lalande 21185	756	UMaj	4.78	187	8.9	M2	11007n3618	1
Wolf 359	750	Leo	4.71	235	15.7	M6	10541n0719	3
Epsilon Indi	1677	Indi	4.69	123	5.9	K5	21596s5700	
Lalande 21258 A	757	UMaj	4.53	282	10.2	M2	11030n4347	
B	758	UMaj	4.53	282	16.0	M8		3
Omicron 2 Eridani	338	Erid	4.08	213	5.3	K1	04130s0744	
B	339	Erid	4.08	213	9.8	DA		2
C	340	Erid	4.08	213	12.3	M5		
Wolf 489	1023	Virg	3.87	253	15.5	DK	13344n0358	2
Alpha Centauri C	1110	Cent	3.85	282	13.2	M5	14263s6228	3
BD+5°1668	527	CMin	3.76	171	11.7	M5	07247n0529	1
Mu Cassiopeiae	107	Cass	3.75	115	5.7	G5	01049n5441	4
Wash. 5583	1178	Libr	3.68	196	10.5	K0	15075s1613	
Wash. 5584	1177	Libr	3.68	196	9.8	G6		
Alpha Centauri A	1127	Cent	3.68	281	0.8	G2	14362s6038	
B	1128	Cent	3.68	281	2.9	K1		
LP9-231	---	Drac	3.59	337	15.4	DA	17570n8244	2
Cordoba 29191	1617	Micr	3.46	251	7.9	M1	21143s3904	
L726-8 A	144	Ceti	3.35	80	14.2	M6	01364s1813	
L726-8 B= UV Ceti	145	Ceti	3.35	80	14.7	M6		3
L789-6	1729	Aqar	3.25	46	14.3	M7	22357s1536	
Ross 451	834	Drac	3.20	174	13.7	K4	11376n6736	
e Eridani	277	Erid	3.14	76	5.0	G5	03179s4316	
Ross 578	298	Erid	3.06	152	14.6	M2	03358s1137	
Van Maanen's Star	76	Pisc	2.98	155	12.9	DF	00465n0509	2

Magnitudes in the table are photographic. Notes in the last column: 1= Invisible companion. 2= White dwarf star. 3= Flare star. 4= Close visual companion.

applications are: the identification of moving star groups
such as the Ursa Major Group and the Scorpio-Centaurus
Association, the search for additional members to nearby
clusters such as the Hyades in Taurus and Praesepe in
Cancer, the discovery of expanding star groups such as the
Zeta Persei Association, the discovery of unseen companions
by means of periodic irregularities in the proper motion,
and the identification of high-velocity stars which is of
importance in the study of stellar populations. (See also
Kapteyn's Star in Pictor, Groombridge 1830 in Ursa Major,
and 61 Cygni)

M9   (NGC 6333) Position 17162s1828. A small bright
              globular star cluster located about 3.5° SE
from Eta Ophiuchi. It was discovered by Messier in May 1764
and the clusters M10, M12, and M14 were all found as well
within the next few nights. M9 is the smallest cluster of
the four, with an apparent diameter variously given as 3',
4', or 6'; numerous outliers extent to possibly 8'. To
Messier the group appeared as a faint nebula without stars,
round and some 3' across. Sir William Herschel resolved it
into a swarm of stars with his 20-foot (focal length)
telescope in 1784, and Lord Rosse (1851) thought it "not
round; on the south side is an outlying portion separated
from the chief portion by a dark passage". Admiral Smyth
called it "a fine object composed of a myriad of minute
stars, clustering to a blaze in the centre and wonderfully
aggregated with numerous outliers seen by glimpses".
         M9 is evidently one of the nearer globular clusters
to the nucleus of our Galaxy, with a computed distance of
about 7500 light years from the Galactic Center. The distance from the Solar System is thought to be about 26,000
light years, which gives an extreme diameter of about 60
light years and a total luminosity of about 60,000 times
that of the Sun, assuming that about half the light has
been absorbed in space. Heavy absorption to the north and
west suggests that the cluster may be dimmed by at least a
magnitude. Our Sun, as a standard of comparison, would
appear as a star of magnitude 19.3 at the distance of M9.
         According to H.B.Sawyer (1963) the total integrated
photographic magnitude of the cluster is 8.92, and the
integrated spectral type is F2. Thirteen variable stars

GLOBULAR STAR CLUSTERS IN OPHIUCHUS. Top: M9 photographed
with the 13-inch telescope at Lowell Observatory. Below:
The cluster M10, photographed with the Lowell 42-inch
reflector.

GLOBULAR STAR CLUSTERS IN OPHIUCHUS. Top: The relatively loose cluster M12, photographed at Lick Observatory. Below: The rich cluster M14, photographed at David Dunlap Observatory in Canada.

DESCRIPTIVE NOTES   (Cont'd)

have been identified in the cluster, an increase of one
dozen over the single one known in 1947. Like most of the
globular clusters, M9 has a very large radial velocity,
134 miles per second in recession.

A smaller globular cluster, NGC 6342, lies about
1.2° to the SE, while another, NGC 6356, will be found at
approximately the same distance to the NE.

M10   (NGC 6254) Position 16545s0402. Globular
           star cluster, forming with its neighbor M12
an interesting pair, presenting a fine study in structural
contrast. The two clusters lie some 3.4° apart, well north
of the main mass of the star clouds of the Ophiuchus Milky
Way. Both are bright and easily located in good binoculars
and both were discovered by Messier in May 1764.

M10 is a rich cluster of magnitude 7 with an
extreme diameter of about 12' and a bright compressed
center. Partial resolution may be achieved with a good 6-
inch or 8-inch telescope, although Messier saw it as a
"nebula without stars" and J.E.Bode in 1774 called it a
"nebulous patch without stars; very pale". William Herschel first resolved it into a "beautiful cluster of extremely
compressed stars" and Admiral Smyth found it " easily
resolvable by moderate means; a rich globular cluster of
compressed stars of a lucid white tint, somewhat attenuated
at the margin and clustering to a blaze in the centre.."
Lord Rosse in 1851 reported a dark lane passing across the
cluster above the center, and thought that the upper one-
sixth of the cluster was noticeably fainter than the rest.

H.Shapley (1933) originally derived a distance of
about 33,000 light years for M10, but modern studies seem
to indicate a smaller distance. T.D.Kinman finds that the
group is probably closer than the great M13, and gives it
a distance of about 16,000 light years; H.B.Sawyer (1963)
has 7 kiloparsecs or about 22,000 light years. The total
integrated photographic magnitude of the group is 7.64;
the integrated spectral type is G0. Only three variable
stars have been reported in this cluster. M10 also has a
fairly low radial velocity when compared to some of the
velocities measured for other globulars; it is receding at
about 43 miles per second.

M12   (NGC 6218) Position 16446s0152. A bright globular star cluster, located about 3.4° NW of M10 and discovered by Messier in May 1764. M12 is a slightly larger but somewhat dimmer group than M10, and has a looser structure showing a relatively slight central condensation. As in the case of most globulars, Messier and Bode both found it unresolvable, and the group was first identified as a cluster by Sir William Herschel in 1783. Smyth found it a "fine rich globular cluster with a cortege of bright stars and many minute straggling outliers". The resolvable mass, Smyth reported, contains several bright spots, evidently clumps of stars, toward the center. K.G. Jones in his book "Messier's Nebulae and Star Clusters" (1968) mentions that the cluster "seems entangled in a long skein of stars lying roughly E-W" and suggests that this is probably Smyth's "cortege". Lord Rosse commented on the outer streams of stars and thought them to show a hint of a spiral arrangement. M12 is not one of the more concentrated globular clusters, but the relatively loose structure does permit more easy resolution in fairly small telescopes, and it is generally a pleasing object in amateur instruments.

According to Sawyer's "Bibliography of Individual Globular Star Clusters" (First Supplement, 1963) M12 has a total integrated photographic magnitude of 7.95 and an integrated spectral type of F7. Only one variable star has been identified in the cluster. Published values for the distance of M12 vary from 16,000 to about 24,000 light years; there appears to be a general agreement that it is virtually at the same distance as M10. The true separation of the two clusters is then about 2000 light years; each group must appear as a bright naked-eye object of about 2nd magnitude as seen by the hypothetical inhabitants of the other cluster. M12, like its neighbor, has a fairly low radial velocity, about 10 miles per second in approach.

M14   (NGC 6402) Position 17350s0313. Globular star cluster, located in a rather blank area of Ophiuchus, some 16° south of Alpha. It was discovered by Messier in June 1764, a few days after the discovery of M9 and M10. He found it "a nebula without stars, not large, faint, and round". William Herschel in 1783 found it easily resolvable and bright with 300X on his 20-foot telescope.

John Herschel saw it as a "most beautiful and delicate globular cluster, not very bright, but of the finest star dust, all well resolved, and excessively rich.....all the stars equal...15..16 magnitude". Isaac Roberts in 1897 obtained one of the earliest high quality photographs of M14 which showed "curves and lines of stars radiating in all directions outward from the dense centre..."

M14 is one of those globulars which lack a sharp central condensation, the distribution of light being very smooth across the disc, with a gradual tapering off at the outer edges. Some hint of resolution may be achieved in an 8-inch or 10-inch telescope, but large instruments are required to show the countless members appearing as if the whole field had been powdered with luminous dust. This is evidently a more remote cluster than M10 or M12, and the stars begin to appear in vast numbers at about magnitude 15.5. Long exposure photographs show that the group is not precisely spherical, but distinctly elongated E-W.

Again from H.B.Sawyer's catalog, the following data have been obtained: Total integrated photographic magnitude = 9.44; total integrated spectral type = G0; apparent modulus = 16.7 or about 70,000 light years, but undoubtedly requiring a large correction for light loss due to strong absorption in this very dusty region of the Galaxy. A large number of variable stars have been identified in M14, 72 of these stars having been detected up to 1963. This cluster shows a radial velocity of 77 miles per second in approach.

A nova which appeared in M14 in 1938 was detected on a series of plates made with the 74-inch reflector of David Dunlap Observatory between 1932 and 1963. The star, a 16th magnitude object, was apparently recorded well after maximum; the peak brightness may have been as high as 10 or 11. There are only two cases of novae in globular star clusters; the only other known example appeared in the very condensed cluster M80 in Scorpius in 1860.

M19   (NGC 6273) Position 16595s2611. Globular star cluster, discovered by Messier in June 1764, four days after the discovery of M14. The position is near the Ophiuchus-Scorpius border, about 7° due east from the bright star Antares. M19 was first resolved by Sir William

DESCRIPTIVE NOTES (Cont'd)

Herschel in 1784; John Herschel called it "a superb cluster resolvable into countless stars 14, 15, 16 magnitude." The surrounding field is extremely rich; the whole area is literally peppered with innumerable faint stars of the Milky Way. A few degrees to the SE however, lies the huge dark cloud called the "Pipe Nebula", a virtually starless area several degrees in diameter. This may be the "vacuity" referred to by Admiral Smyth in the vicinity of the group, though K.G.Jones suggests that Smyth's description may be confused with the cluster M80 in Scorpius which also lies near a dark starless area called by Herschel "a hole in the heavens". Smyth found the cluster itself "a fine insulated globular cluster of small and very compressed stars of a creamy white tinge and slightly lustrous in the centre". Two brighter field stars are superimposed on the cluster, on the north side.

M19 is one of the most oblate globulars, and the flattening of the outline is noticeable even in very small telescopes. H.Shapley estimated that twice as many stars could be counted along the major axis as compared to the minor. The orientation of the longer diameter is nearly N-S; Shapley gives the PA as about 15°. This is one of the clusters which appears to be embedded in the central hub of the Galaxy; the estimated distance from the Galactic Center is about 3000 light years.

H.B.Sawyer (1963) gives the cluster a total integrated photographic magnitude of 8.29 and an integrated spectral type of F5. The cluster appears to be somewhat more remote than M10 or M12, and is also much more heavily obscured, making an accurate distance determination more difficult. Published values range from 20,000 up to about 30,000 light years. Four variable stars have been identified in this cluster. M19 shows a radial velocity of 60 miles per second in recession.

THE MILKY WAY IN OPHIUCHUS

The most impressive feature of the summer sky is the vast chain of Milky Way star clouds, stretching from Cygnus to Sagittarius and defining the central plane of our Galaxy. In the southern part of Ophiuchus is one of the most wonderful regions of the Milky Way, lying only a few degrees from the computed position of the actual Galactic

DEEP SKY OBJECTS IN OPHIUCHUS. Top left: The oblate globular cluster M19; Top right: The open cluster NGC 6633, both
photographed with the Lowell Observatory 13-inch camera.
Below: Spiral galaxy NGC 6384, Palomar 200-inch photograph.

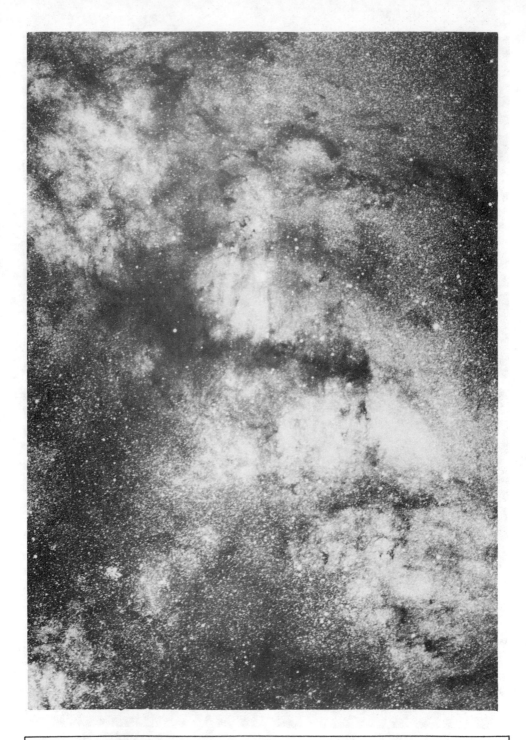

THE MILKY WAY IN OPHIUCHUS. The view is centered on the elongate dark mass of the Pipe Nebula, which appears in an enlarged version on page 1270. Lowell Observatory photo.

DESCRIPTIVE NOTES (Cont'd)

Center. The incredibly rich star fields of this area are
mottled and streaked by a profusion of intricately shaped
dark winding lanes of non-luminous material. E.Barnard
found this portion of the sky to be "fuller of strange and
curious things than any other regions with which my photographs have made me familiar. It is so extremely puzzling
that one attempts a description of it with hesitation. That
most of these dark markings which, in a word, ornament this
portion of the sky are real dark bodies and not open space
can scarcely be questioned".

Lacking our present knowledge of the structure of
our Galaxy, Barnard could not have been aware than in the
Ophiuchus regions we are looking through a vast complex of
dust-streaked spiral arms toward the very nucleus of the
Galactic System. The gigantic dark lane called the "Great
Rift", so conspicuous to the naked eye, is the same huge
equatorial dust band which we see in the edge-on external
galaxies. Beyond this Rift, some 30,000 light years away,
lies the central hub of the Galaxy. Although the statement
has often appeared that the precise position of this hub
is in Ophiuchus, the most accurate modern investigations
place it just across the border in Sagittarius at position
17424s2859. This position coincides with a strong radio
source called "Sagittarius A", believed to be the actual
Galactic Nucleus.

The third magnitude foreground star Theta Ophiuchi
is the observer's guide to the wonders of the Milky Way in
this region. Two degrees to the east and below Theta is the
enormous black cloud called the "Pipe Nebula", clearly
visible to the unaided eye with a total length of about 7°.
The bowl of the Pipe is Barnard's dark cloud B78, about 2°
in diameter, and located just 2° southeast of Theta. The
"stem" is a succession of irregular dark masses extending
westward, and including B59, B65, B66, and B67. The entire
formation is one of the largest of the Milky Way's dark
clouds, outside of those composing the Great Rift. The
distance, although somewhat uncertain, is believed to be
at least several thousand light years; the full extent of
the Pipe must then be several hundred light years.

Near the star Rho Ophiuchi, in the extreme southwest corner of the constellation, is another region of
strangeness and beauty. The star itself is enmeshed in a

DESCRIPTIVE NOTES   (Cont'd)

vast diffuse nebulosity, IC 4604, too faint to be studied
visually. But from this nebulous region a number of dark
narrow lanes flow out toward the east, appearing like black
tentacles obscuring the starry background. These vacant
lanes end near the star Xi Ophiuchi, about 11° distant.
Near Xi also is the large and strangely shaped dark cloud
B63, thickest at its western end where scarcely a single
star shows through over a field half a degree in diameter.
This is an extraordinary region, described by Barnard as
making a picture almost unequalled in interest in the
entire heavens.

A few degrees north of Theta Ophiuchi is a field
where the numerous dark markings assume some of their most
curious shapes. The best known of these is undoubtedly the
dark "S-Nebula", B72, a degree and a half north of Theta
and slightly east. Difficult to detect visually except in
very low power wide-angle instruments, it is revealed on
photographs as a perfect S about 30' in length, winding
snake-like across the massed star clouds; the southernmost
loop is the widest portion but does not appear to be totally obscuring the stars as a few faint stellar points lie
sprinkled across the dark coils. These may, of course, be
foreground objects. Half a degree to the SW is another
smaller dark mass about 8' across which appears totally
blank; it may be seen near the upper right edge of the
photograph on page 1269. This formation is perhaps one of
the dark "globules" which appear in some of the diffuse
nebulae; they are small rounded dense clouds with radii as
small as a few thousand AU, and are often considered to be
new stars in the process of formation.

In her "Bibliography of Individual Globular Star
Clusters" (1957) H.B.Sawyer states that Ophiuchus contains
20 globular star clusters, the greatest number found in any
constellation. Sagittarius and Scorpius are in second and
third place with totals of 17 and 8. Thus more than one
third of the known globulars are contained in three of the
constellations, all located in the same portion of the sky!
The explanation, of course, is that this region marks the
direction of the Galactic Center, toward which these
clusters tend to concentrate. (Refer also to the section
"The Milky Way in Sagittarius".

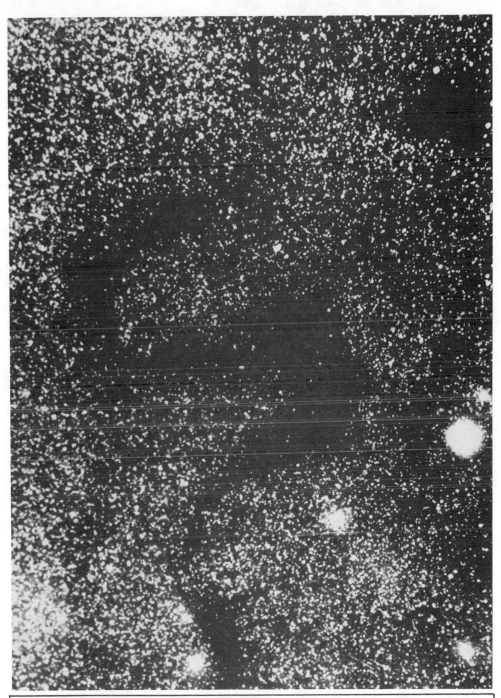

DARK NEBULA B72 in OPHIUCHUS. Barnard's "S-Nebula" in the
Ophiuchus Milky Way. This view was obtained with the 100-
inch reflector ot Mt.Wilson Observatory.

DARK NEBULA B78 in OPHIUCHUS. This dark formation forms the bowl of the "Pipe Nebula", shown in its entirety on page 1266. Photograph by Barnard with the 10-inch Bruce camera.

ORION

LIST OF DOUBLE AND MULTIPLE STARS

NAME	DIST	PA	YR	MAGS	NOTES	RA & DEC
Σ589	4.3	288	55	8 - 8	cpm pair; PA dec, spect G5	04422n0512
A2037	2.9	79	19	7½- 12½	spect A2	04444n0957
A2621	0.2	39	57	8½- 8½	PA inc, spect A0	04470n0207
β883	0.2	38	62	7½- 7½	Binary, 16.3 yrs;	04484n1059
	16.6	162	24	- 14	PA inc, spect dF7; AC PA inc.	
β552	0.8	317	61	7 - 9	binary, 101 yrs;	04490n1334
	44.9	213	06	-12½	PA inc, spect dF6	
Σ609	2.4	72	66	8½- 8½	PA dec, spect F8	04492n0110
Σ612	16.7	198	51	7½- 8	AB cpm, relfix; AC	04516n0718
	46.7	236	27	-13½	optical, dist inc; spect K0	
OΣ90	1.9	341	65	7 - 9	relfix, spect A0	04522n0831
	38.1	95	09	-12		
Σ614	0.2	119	66	9 - 9	(A1019) spect A5	04525s0038
	4.1	68	32	- 9		
Σ620	3.9	236	35	8½- 9½	PA slow inc, spect G0	04555n1353
Σ622	2.4	167	55	8 - 8	(Pi 258) PA dec, spect F5	04555n0136
OΣΣ58	39.4	305	57	5 - 6½	(Sh 49) spect B6,	04562n1428
	54.4	89	57	- 9	B9	
Σ627	21.0	260	32	6½- 7	cpm, relfix, spect both A0	04578n0333
Σ630	14.4	50	32	7 - 8	relfix, spect B8, A	04594n0132
Σ630b	0.5	25	30	8 - 11	(A2630)	
J307	52.2	322	16	7 - 8	spect K0	05012s0236
J307b	0.3	196	10	9 - 9		
14	0.8	71	62	6 - 6½	(OΣ98) binary, about 160 yrs; PA dec, spect Am	05052n0826
Σ643	2.8	300	43	8½- 8½	relfix, spect K2; cpm with 14 Orionis	05052n0820
A2636	0.2	138	63	7 - 7½	binary, 150 yrs; PA inc, spect A0	05063n0309
OΣ100	3.9	252	40	7 - 10	relfix, spect F8	05073n0807
Σ652	1.7	182	53	6½- 7½	relfix, cpm pair; spect F5	05092n0059
Hu 33	0.1	334	37	7½- 8	PA inc, spect B9	05093n0027

LIST OF DOUBLE AND MULTIPLE STARS   (Cont'd)

NAME	DIST	PA	YR	MAGS	NOTES	RA & DEC
A51	1.6	101	57	8½- 9	PA slow inc, spect B9	05100s0306
ρ	7.0	63	38	4½- 8½	(Σ654) relfix, spect K3	05107n0248
0Σ517	0.4	230	62	7 - 7	binary, about 600	05109n0155
	6.9	136	62	- 12	yrs; PA inc, spect A2, G	
Hu1224	1.0	123	56	8 - 9	PA dec, spect A0	05120n1228
β	9.4	202	25	0 - 7	RIGEL (Σ668) (*)	05121s0815
	44.0	1	21	- 14	relfix, spect B8	
Σ667	4.0	312	46	7½- 9	relfix, spect K2	05123s0707
Σ664	5.0	172	49	7½- 8	relfix, spect F0	05124n0823
Σ678	3.6	102	26	8½- 9	slow PA inc,	05150n0438
	75.3	339	08	- 11	spect F0	
τ	36.2	60	19	4 - 11	(β188) (h2259)	05152s0654
τb	3.5	51	22	11- 12	spect B5	
Σ688	10.3	273	50	7 - 7½	relfix, cpm pair; spect F0	05170s1048
Σ692	0.5	343	58	8 - 9	(β190) PA dec,	05180s0805
	35.1	4	16	- 8½	spect F2	
β189	4.2	284	33	7- 11½	relfix, spect B9	05180s0525
Σ693	3.9	9	37	8½- 9	relfix, spect A2	05191s0206
0Σ106	9.3	42	17	7- 10½	relfix, spect F5	05195n0521
A2641	1.0	328	63	8½- 11	binary, 90 yrs; PA inc, spect G5	05200n0234
Hu1225	3.0	18	32	7½- 12	perhaps slight dist dec, spect K0	05201n1419
23	32.0	28	34	5 - 7	(Σ696) (m) relfix, spect B1,A	05202n0330
Σ700	4.7	5	41	8 - 8	relfix, spect A0	05205n0101
Σ697	26.0	285	38	7 - 8	relfix, cpm pair; spect B8, A	05206n1600
Σ701	6.0	141	33	6½- 8½	relfix, cpm pair; spect B8	05209s0828
A847	2.7	160	65	6½- 7½	AB PA dec, dist	05213s0055
A847b	0.3	141	66	8 - 8	inc, spect F5; BC binary, 49 yrs.	
η	1.5	77	66	4 - 5	PA dec, slight dist inc, spect B0 (*)	05220s0226
Σ708	3.0	325	31	8½- 10	dist inc, spect B9	05226n0153

ORION

LIST OF DOUBLE AND MULTIPLE STARS   (Cont'd)

NAME	DIST	PA	YR	MAGS	NOTES	RA & DEC
A848	0.2	138	60	7½- 8	PA inc, spect B9	05230s0035
Σ712	3.2	62	57	7 - 9	PA inc, spect B9	05239n0254
ψ	2.7	324	34	5 - 11	relfix, spect B2	05242n0303
A2643	0.6	13	54	8½- 12½	spect B9	05250n0104
h2268	27.1	300	18	7 - 9½	(Pi 109) spect G5	05263s0825
Da 6	0.6	128	25	7 - 7½	dist dec, PA inc, spect B9	05265s0320
D 8	5.7	49	33	8½- 9	relfix, spect F5	05266s0203
	102	111	03	- 8½		
D 8c	16.9	355	14	8½- 10		
Σ721	24.7	149	21	7 - 9	relfix, spect B3	05269n0307
Σ721b	0.2	174	51	9½- 9½	(β557) PA inc.	
31	12.7	87	33	6 - 11	(Σ725) relfix, cpm, spect K4, F7	05272s0108
32	0.7	54	62	5 - 6½	binary, about 590 yrs; PA dec; spect B5 (Σ728)	05281n0555
Σ726	1.4	262	53	8 - 8½	relfix, spect B9	05281n1013
A2706	3.8	99	21	8 -14	spect F8	05281n0822
33	2.0	26	57	6- 7½	(Σ729) relfix, spect B1	05286n0315
Σ731	4.8	326	21	8½- 9	cpm, PA slight dec, spect A0	05288s0208
δ	52.8	360	32	2 - 6½	(β558) relfix	05294s0020
	32.8	227	22	- 14	spect O9, B2 (*)	
β1048	2.0	357	37	6- 10½	relfix, spect B2	05302s0138
Σ734	1.5	355	67	7 - 8½	relfix, spect B3	05306s0145
	29.7	243	34	- 8½		
Σ734c	0.5	295	43	8½- 9½	(β1049)	
HV 118	27.5	263	01	7- 10½	spect B3	05315s0104
Σ743	2.0	278	54	7 - 8	relfix, cpm pair; spect B8	05322s0425
Σ741	10.2	286	00	7½- 10	spect B5	05324s0009
λ	4.4	44	57	4 - 6	(Σ738) relfix,	05324n0954
	29.1	184	57	- 11	spect O8	
	78.3	271	57	-11½		
OΣ111	2.6	349	43	6 - 10	relfix, cpm pair; spect B8	05325n1013
Σ747	36.0	223	24	5½- 6½	relfix, spect B0, B1; near Iota Ori.	05326s0602

LIST OF DOUBLE AND MULTIPLE STARS (Cont'd)

NAME	DIST	PA	YR	MAGS	NOTES	RA & DEC
θ	135	314	26	5 - 5	Orion Nebula Group	05328s0525
θ^1	13.3	61	25	5 - 6	The Trapezium;	05328s0525
	13.1	311	25	- 7	famous quadruple	
	16.8	342	25	- 8	star in Great Orion	
					Nebula M42 (*)	
θ^2	52.5	92	37	5 - 6½	spect 09	05329s0527
	129	97	26	- 8½		
42	1.6	212	36	5½- 8½	relfix, cpm pair;	05329s0452
					spect B2	
Σ750	4.3	60	40	6 - 8	relfix, spect B2;	05330s0424
					in cluster NGC 1981	
ι	11.4	141	32	3 - 7	(Σ752) relfix, in	05330s0556
	49.7	103	04	- 11	nebula NGC 1980 (*)	
Σ751	15.6	123	31	8 - 9	relfix, spect B9	05332s0101
Σ754	5.3	287	33	6½- 9½	relfix, spect B3	05342s0606
A2651	0.4	172	52	7½- 9	PA inc, spect B9	05346n0444
	8.2	326	21	-12½		
Σ757	1.6	237	43	8 - 8	relfix, spect B9;	05356s0013
	51.8	86	32	- 8½	C= Σ758	
Σ758	11.2	297	34	8½- 9		05356s0013
Σ761	68.3	202	31	8 - 8½	relfix; in field	05361s0236
Σ761b	8.5	268	31	8½- 9	with Sigma Orionis	
σ	0.2	199	63	4 - 6	(β1032) (Σ762)	05362s0238
	12.9	84	34	-7½	fine multiple star	
	11.2	236	36	- 10	spect 09, B3 (*)	
	42.0	61	34	- 6½		
Σ763	5.6	318	26	8 - 8½	relfix, spect F8	05365n1014
0Σ113	10.2	28	25	7- 10½	relfix, spect A0	05371n1300
Σ766	10.0	275	38	7 - 8	cpm pair, relfix,	05374n1520
					spect F0	
ζ	2.6	164	57	2 - 5½	(Σ774) slow PA	05382s0158
	57.6	10	30	- 10	inc, spect 09 (*)	
β1052	0.7	172	43	7 - 8	PA dec, spect F0	05392s0255
A494	0.2	106	62	7 - 7½	binary, 20 yrs; PA	05405s0649
	99.7	233	15	-10	inc, spect F5	
A494c	1.0	199	15	10-13½		
Σ788	7.4	90	24	7½- 9	relfix, spect B9	05421n0349
	36.0	149	05	- 10		
Σ789	16.0	150	58	7 - 8½	(A2655) spect F0	05424n0359
Σ789b	1.1	112	58	8½- 12		

LIST OF DOUBLE AND MULTIPLE STARS (Cont'd)

NAME	DIST	PA	YR	MAGS	NOTES	RA & DEC
Σ 790	7.0	88	33	7 - 9	relfix, spect K1	05436s0417
OΣ 119	0.8	334	54	7½- 8½	PA inc, spect F8	05452n0757
52	1.2	211	66	6 - 6	(Σ 795) slow PA inc, spect A5, F0	05453n0626
A2657	0.1	256	54	8½- 8½	binary, 77 yrs; PA inc, spect F0	05456n0136
J36	1.7	109	52	7 - 9½	PA slow dec, spect F8	05456n0353
Σ 797	7.2	15	31	7 - 10	relfix, spect A0	05458n0441
Σ 798	20.7	181	17	7 - 9	relfix, spect B9	05458s0824
Σ 809	1.1	105	37	8 - 10	(β1188) spect G5	05481s0127
	24.6	97	23	- 8½		
Σ 813	3.0	150	54	8 - 8	relfix, spect A0	05514n1853
OΣ 123	2.1	183	67	7 - 8½	PA inc, spect C5	05514n1014
Ho 20	7.9	276	37	7 - 12	relfix, spect K0	05521n1413
	50.2	287	00	-11½		
Σ 817	18.4	73	35	8 - 8	relfix, spect A5	05522n0702
Σ 816	4.4	289	43	6 - 8½	relfix, spect B9	05522n0551
S503	26.9	326	26	7 - 9	optical triple; AB	05532n1356
	31.5	323	34	- 11	PA dec, AC PA inc, primary spect G5	
β1189	0.2	322	51	8 - 9	PA inc, spect A	05549n0024
	57.6	194	23	-8½		
β1190	1.2	340	45	7½- 10½	spect A0	05549n0001
	6.5	96	45	- 12		
59	36.5	204	25	6 - 9½	(HV 100) relfix; spect A5	05558n0150
OΣ 124	0.3	302	60	6 - 7	PA slight dec, spect gG4, A5	05561n1248
Σ 826	2.0	127	42	8 - 9	PA slow inc, spect B9	05564s0120
OΣ 125	1.6	360	53	7 - 8½	relfix, spect A0	05567n2228
μ	0.2	9	62	4½- 6½	(A2715) (β1056)	05596n0939
	18.2	279	55	- 14	binary, 17½ yrs; PA dec, nearly edge on orbit; spect A2	
Σ 835	2.2	147	51	8 - 9	relfix, spect A3	06020n1819
Σ 838	40.0	327	25	7 - 9	relfix, spect K0	06026n0052
Σ 840	21.7	248	36	7 - 8½	relfix, spect A0	06037n1045
Σ 840b	0.5	156	58	8½- 8½	BC PA dec	

LIST OF DOUBLE AND MULTIPLE STARS (Cont'd)

NAME	DIST	PA	YR	MAGS	NOTES	RA & DEC
A1048	2.8	284	16	8 - 13	spect F2	06039s0057
0Σ133	3.4	35	38	7 - 10	relfix, spect F0	06050n2118
Σ851	3.0	29	42	8 - 8½	relfix, spect A0	06055n0318
Σ848	2.6	110	52	7½- 8	relfix, spect B2;	06057n1359
	14.2	296	12	- 12	in cluster NGC 2169	
	28.4	121	32	- 8		
	43.3	183	10	- 9		
Σ849	1.0	241	53	8½- 9	relfix, cpm pair; spect G0	06058n1725
Σ853	33.2	1	57	8 - 8½	optical, PA & dist inc, spect G5	06064n1140
Σ855	29.4	114	29	6 - 7	relfix, cpm pair; spect A0	06064n0231
J390	1.7	104	60	9- 9½	relfix	06080n1214
Σ867	2.2	155	53	7 - 8½	relfix, spect A0	06087n1723
Σ871	7.3	305	13	8½- 9	relfix, spect A0	06090s0045
Ho 22	0.9	203	66	8- 8	PA slight inc? Spect A2	06107n1016
Kui 24	0.5	139	60	6½- 6½	spect A5	06115n1755
J683	1.2	8	53	8½- 9	spect B9	06117n1726
Σ877	5.6	263	52	7 - 7½	relfix, spect B9	06118n1436
Σ880	5.5	54	38	8 - 8	relfix, spect G5	06127n1036
β193	18.3	90	26	8 - 11	spect B5	06129n0359
	58.3	230	06	- 10½		
RST5225	0.2	284	60	7 - 7	PA inc, spect F5	06133n0111
β894	5.2	138	28	8- 12½	slight PA dec, spect F5	06136n1902
β1019	0.7	282	66	8- 9½	PA inc? spect A0	06136s0252
75	0.1	186	59	6 - 6	(φ331) (β96)	06144n0958
	62.7	258	25	- 10	binary, 8 yrs; PA	
	117	159	25	- 8½	inc, spect A2	
75d	4.9	226	33	8½- 11		
Ho 229	2.1	336	58	6 - 13	PA inc, spect A2	06152n1424
	25.4	186	58	- 12½		
A2809	0.3	196	21	8 - 10	spect A0	06160n1111
Σ891	21.9	293	04	7½- 10½	relfix, spect B8	06169n1219
A2719	0.4	60	51	8 - 8	spect B9	06176n0745
A2666	0.9	304	20	8- 11½	spect M	06184n0236
Ho 233	2.1	36	36	8 - 11	relfix, spect B9	06203n1632
Σ901	20.0	248	13	7½- 9½	relfix, spect B9	06222n1033

ORION

LIST OF VARIABLE STARS

NAME	MagVar	PER	NOTES	RA & DEC
α	0.4--1.3	---	Semi-reg; spect M2 BETELGEUSE (*)	05525n0724
δ	2.4--2.6	5.732	Ecl.Bin; Spect 09 (*)	05294s0020
η	3.2--3.4	7.989	Ecl.Bin; Spect B0 (*)	05220s0226
o'	4.65--4.9	30:	Semi-reg; spect M2	04497n1410
π^5	3.7--3.8	3.700	Ellipsiodal variable; Spect B2	04516n0222
ω	4.4--4.6	Irr	γ Cass type; spect B3	05376n0406
R	9.0--13.5	378	LPV. Spect Ne	04563n0803
S	7.6--13.5	419	LPV. Spect M6e--M8e	05265s0444
T	9.4--12.6	Irr	Erratic, T Tauri type, in Orion Nebula M42, spect B8--A3e (*)	05334s0530
U	5.4--12.5	373	LPV. Spect M6e--M8e (*)	05529n2010
V	8.7--14.5	268	LPV. Spect M3e--M5e	05034n0402
W	6.5--10..	210	Semi-reg; spect N5	05028n0107
X	9.9--14..	422	LPV.	05351s0148
Z	9.7--10.7	5.203	Ecl.Bin; spect B3	05530n1341
RR	9.1--14..	252	LPV. Spect M7e	06001n1623
RS	8.3--9.1	7.567	Cepheid; spect F2--G0	06194n1442
RT	8.0--8.9	320	Semi-reg; spect N3	05305n0707
UX	8.7--12.6	Irr	RW Aurigae type; spect A3e	05020s0351
VV	5.7--6.1	1.485	Ecl.Bin; lyrid, spect B1	05310s0111
BF	9.4--13..	Irr	spect A5e	05348s0637
BK	9.0--14..	335	LPV. Spect M7	05292n0735
BL	6.3--7.0	Irr	Spect N3	06226n1445
BM	8.0--8.7	6.471	Ecl.Bin; in Theta Orionis group in Great Nebula (*)	05328s0525
BN	9.0--12..	Irr	RW Aurigae type, spect A6	05338n0648
BQ	8.7--10..	110	Semi-reg; spect M5e	05541n2250
CK	6.2---6.7	Irr	Spect K2	05277n0410
CO	9.8--12..	Irr	RW Aurigae type; spect Gp	05249n1123
DH	9.4--11.7	165	Semi-reg; spect M0e	05236s0019
DN	9.8--10.5	12.96	Ecl.Bin; spect A2e + gF5	05577n1013
EP	9.5--11..	359	LPV. Spect M10e	04483n0304
ER	9.3--10.0	.4234	Ecl.Bin; W Ursae Majoris type; spect dG1	05088s0837
EU	9.5--11..	328	LPV. Spect M4	05125n0326
EV	9.3--11.1	243	Semi-reg; spect M	05128n0921
EY	9.5--10.4	16.787	Ecl.Bin; spect dA7	05289s0544

LIST OF VARIABLE STARS   (Cont'd)

NAME	MagVar	PER	NOTES	RA & DEC
FO	9.5--10.3	18.80	Ecl.Bin; spect A3	05255n0335
FT	9.1--9.7	3.150	Ecl.Bin; spect A0	06110n2128
FU	9.7--16.	---	Nova-like variable? Spect F2 pec (*)	05426n0903
FX	8.2--10.4	Irr	Spect M3	05391n1449
FZ	9.0--10.0	1.597	Ecl.Bin; spect G0	05388n0235
GI	10.3-11.4	311	Semi-reg; spect M7	06104n1834
GK	9.5--11..	236	Semi-reg; spect N	06150n0834
GQ	8.0--8.8	8.616	Cepheid; spect G0	06085n0938
GW	9.2--10.4	Irr	Spect dK3e	05264n1150
KX	7.1--8..	---	RW Aurigae type, spect B3	05326s0446
LP	8.4---9.3	---	RW Aurigae type, spect B2	05327s0530
NU	6.5---7.6	---	RW Aurigae type, spect B1	05330s0518
NV	9.5--11.3	---	RW Aurigae type, spect F4	05331s0535
V345	8.6--13..	332	LPV.	06134s0103
V351	8.0--11.2	Irr	Spect A7	05417n0008
V352	7.5--8.5	Irr	Spect M7e pec	05593s0221
V359	6.9---9.1	Irr	RW Aurigae type; spect B3	05331s0452
V361	8.1--9.8	Irr	Spect B4- B5	05332s0527
V372	7.4--8.6	Irr	RW Aurigae type, spect A0	05323s0536
V380	9.8--10.5	Irr	T Tauri type? Spect B8--A2e	05340s0645
V429	9.0--9.8	.5017	Cl.Var?; spect G4	04537s0336
V430	8.4--9.2	104	Semi-reg; spect M3--M5	05040n0029
V431	9.3--11.1	122	Spect N	05132n1155
V451	8.5--9.5	Irr	Spect B9e	05287n1059
V530	9.8--10.5	6.111	Spect G0	06031s0311
V535	9.7--10.2	Irr	Spect M5	05209s0437
V586	9.7--11.2	Irr	Spect A0	05346s0611
V642	8.7--9.1	2.757	Spect A0	05567n0914
V643	9.7--10.4	52.42	Spect G--K	06045s0254
V1004	5.9 ±0.01	.054	δ Scuti type; spect Am	05568n0150

LIST OF STAR CLUSTERS, NEBULAE, AND GALAXIES

NGC	OTH	TYPE	SUMMARY DESCRIPTION	RA & DEC
----	J320	◎	Diam 10"; mag 13; vF,S,R; with 14^m central star	05027n1039
1788	32^5	□	B,L, Diam 8' x 5' with 10^m B9 star inv.	05045s0324
----	I.423	□	vF, oval dust cloud 4.5' x 2.5'; reflection nebula	05309s0040
1976	M42	□	!!! Great Orion Nebula + Multiple Star Theta Orionis; Mag 5, diam 65'; eL,eB, Irr; Magnificent object (*)	05329s0525
1977	30^5	□	! 42 and 45 Orionis, with B, vL neby; diam 40' x 25' (*)	05330s0452
1980	31^5	□	F,vL, Diam 14'; surrounds 09 star Iota Orionis (*)	05330s0556
1981		⋯	B,L, scattered group of 10 stars mags 8....10; incl double star Σ750	05330s0424
1982	M43	□	vB,vL,R, with central 8^m star spect 07; detached portion of Orion Nebula M42 (*)	05331s0518
1990	34^5	□	eL,vF,E, Diam 50'; surrounds Epsilon Orionis (*)	05337s0114
1999	33^4	□	10^m star in B neby Diam 16' X 12'	05341s0645
----	I.426	□	9^m star in L,F, neby, Diam about 5'; spect B9	05343s0016
----	I.430	□	vL,vF,E neby about 10' long; 8' np 49 Orionis	05362s0706
----	I.432	□	7^m star in F neby 4' x 5'; 27' north of Zeta Orionis	05385s0131
----	I.434	□	eF,vvL,vmE; 1° long with Zeta Orionis at north end. Incl Dark "Horsehead" Neb B33 (*)	05386s0226
----	B33	■	!! Dark Nebula in I.434 "Horsehead Nebula" (*)	05387s0232
2023	24^4	□	8^m star in L,F,neby; Diam 10'; B33 is 14' sp	05392s0215
2022	34^4	◎	Mag 12, Diam 25"; pB,S,v1E; with 14^m central star, spect continuous	05393n0903

LIST OF STAR CLUSTERS, NEBULAE, AND GALAXIES

NGC	OTH	TYPE	SUMMARY DESCRIPTION	RA & DEC
2024	28^5	▢	! B,vL, Diam 20', with much complex interior detail; 15' nf Zeta Orionis	05394s0152
----	I.435	▢	9^m star with F neby; diam about 3'; spect B8	05405s0220
2068	M78	▢	B,L, diffuse & irregular; Mag 8, diam 8' x 6' with 3 stars invl (*)	05442n0002
2071		▢	10^m star with F neby, diam 4' x 3'; about 15' NE from M78	05446n0017
2141		⭕	L,vRi,F, Diam 8'; about 100 stars mags 15.... Class F	06003n1026
2169	24^8	⭕	S,pmC, diam 4'; about 20 stars mags 8.... Class D; Incl multiple star Σ848	06057n1358
2174-2175		▢	8^m star in L,F, neby Diam 25'; Spect Oe	06067n2031
2186	25^7	⭕	pL,pRi, Diam 5'; about 30 stars, Class E	06094n0527
2194	5^6	⭕	L,Ri, Mag 9; Diam 8'; about 100 stars; Class E (*)	06110n1250

DESCRIPTIVE NOTES

ALPHA   Name- BETELCEUSE or BETELGEUX. Magnitude 0.70 (variable); Spectrum M2 Ia; position 05525n 0724. The name of the star is usually translated "The Armpit of the Giant" or the "Arm of the Central One"; it has also been called "The Martial Star", evidently from its ruddy color. The tint is usually described as golden orange or deep topaz; although this is one of the most famous of the red giant stars, the color is actually more orange than truly "red". The exotic word *padparadaschah*, used in India to designate the rare orange sapphire, might be an appropriate name for Betelgeuse; the approximate meaning is something like *"The Royal Jewel of the Lotus Blossom"*. The star should be compared with the white blaze of Rigel in order to appreciate the color contrast. Betelgeuse is the 11th brightest star in the sky. Opposition date (midnight culmination) coincides closely with the date of the Solstice on December 21.

The "Giant" referred to in the name is of course the Great Hunter or Celestial Warrior *Orion*, most brilliant of the constellations, and visible from every inhabited part of the Earth. As is also the case with Hercules, the figure of Orion has been associated in virtually all ancient cultures with great national heroes, warriors, or demigods. The origin of the name Orion is obscure, though some classical scholars have suggested a connection with the Greek *Arion*, or more probably the Greek word 'Ωαριων which means simply "warrior". Pindar calls the constellation *Oarion*. The British orientalist Robert Brown, Jr., thought to find its origin in the Akkadian *Uru-Anna*, the "Light of Heaven", though the title is also used in early Babylonian writings as a name of the Sun-god. According to R.H.Allen, the constellation was the "Armed King" *Caomai* to the early Irish, *Orwandil* to the Norsemen, and *Ebuorung* to the ancient Saxon tribes. Ovid and Hyginus refer to it as *Urion* which might suggest that the mighty Hunter was regarded also as some variety of star-god. That foremost authority on Hobbit-lore, J.R.R.Tolkien, tells us that the constellation was known in the Third Age of Middle-earth as *Menelvagor*, "The Swordsman of the Sky".

In contrast to Hercules, who has a very definite personality and is credited with a detailed series of exploits, Orion seems to us a vague and shadowy figure. In

"Over the edge of the world now comes forth
Great Orion....Hunter of the Stars....
Behold the gleaming star-fire of his sword!"

Greek myth he was simply a great and powerful hunter, a
son of Neptune, and possibly of a somewhat boastful dispo-
sition, as he is said to have claimed dominion over every
living creature. Homer refers to him as *the tallest and
most beautiful of men.* He had come to love the divine hunt-
ress Diana, but was unintentionally slain by an arrow from
her bow, at the instigation of her brother Apollo. In
another version of the story, Orion was killed by the sting
of the deadly scorpion sent by Juno to punish him for his
arrogant pride, nevertheless he was honored by a place in
the heavens, and the fatal scorpion was placed in the exact
opposite part of the sky so that it could never threaten
him again. Orion is the giant who is said to have pursued
the Pleiades, particularly Merope, and was consequently
blinded by the angry Oenopion, King of Chios. On the advice
of Vulcan, however, Orion climbed to the top of a great
mountain on or near the isle of Lemnos, where, as he faced
the rising sun, his sight was restored. The Greek histor-
ian Diodorus, writing in the days of Julius Caesar, tells
us that Orion was credited with having built the great
harbor-dam at Messana, and the Promontory of Pelorum, in
Sicily.

According to Thomas Hyde in the 17th century, a very
popular Arabian name for Orion was *Al Babadur,* "The Strong
One"; better known is the Arabian name *Al Jabbar,* "The
Giant", obviously derived from the older Syriac *Gabbara,*
the Jewish *Gibbor.* These names, in turn, are the source of
another name, which appears in Longfellow's *"Occultation
of Orion":*

> *"Begirt with many a blazing star,*
> *Stood the great giant Algebar,*
> *Orion, hunter of the beast!*
> *His sword hung gleaming by his side,*
> *And on his arm, the lion's hide*
> *Scattered across the midnight air*
> *The golden radiance of its hair.."*

Equally well known is Tennyson's allusion to the
glittering constellation in the opening passages of the
prophetic *Locksley Hall,* a poem famous also for containing
one of the most exquisite tributes to the Pleiades to be
found in English literature:

THE CONSTELLATION OF ORION as photographed with a wide-angle Tessar lens of 10-inch focus. Rigel and the stars of the Belt appear prominently. Mt.Wilson Observatory.

DESCRIPTIVE NOTES (Cont'd)

"Many a night from yonder ivied casement,
ere I went to rest,
Did I look on great Orion, sloping slowly
to the west.
Many a night I saw the Pleiads,
rising thro' the mellow shade,
Glitter like a swarm of fireflies
tangled in a silver braid."

Orion in classical times was associated with wintry
storms, a tradition mentioned in Babylonian myths, and in
still older Hindu legend. Polybius, in the 2nd century BC,
attributes the destruction of the Roman squadron during
the First Punic War to the fact that the fleet unwisely
sailed with the rising of Orion. Hesiod, in the same vein,
speaks of the appearance of the Great Hunter as a turbulent
time *when the winds battle with thunderous sound....and the*
sullen sea lies hidden in sable cloud... Virgil, Pliny,
and Horace all speak of Orion with adjectives that may be
translated "The Bringer of Clouds", the "Stormy One", or
"He Who Brings Peril to the Seas". Even Milton, in the 17th
century, continues the ancient tradition when he writes in
Paradise Lost of the time
"when with fierce winds Orion arm'd
Hath vexed the Red-sea coast, whose waves
o'erthrew Busiris and his Memphian chivalry..."

The familiar reference to Orion and the Pleiades in
the *Book of Job* (Ch. XXXVIII) has generated much debate,
and it is not at all certain that the popular King James
translation is the correct one:

31. *Canst thou bind the sweet influence of the*
    Pleiades, or loose the bands of Orion?
32. *Canst thou bring forth Mazzaroth in his*
    season? Or canst thou guide Arcturus with his
    sons?

In the *New English Bible,* the "Bands of Orion" appear
simply as "Orion's Belt", the enigmatic *Mazzaroth* is now
interpreted as a reference to the whole cycle of the
Zodiac, while the translation "Arcturus with his sons" may
be an outright error, and is replaced by "Aldebaran and its

train". Other authorities have suggested that the passage actually refers to the Great Bear, which in some ancient writings is regarded as being under the guardianship of Arcturus.

Orion, under the name of *Sahu,* was one of the most important sky figures to the ancient Egyptians, and was regarded as the soul or incarnation of the great god of the afterworld, *Osiris.* On wall reliefs at the Temple of Denderah, he is shown journeying through the heavens in his celestial boat, followed by Sothis (Sirius) who is identified as the Soul of Isis, and is shown as a kneeling cow with a star between her horns. In some of the oldest writings which have come down to us from ancient Egypt, the *Pyramid Texts* of the late 5th Dynasty, the King is promised a celestial journey to the realms of Orion:

"The Great One has fallen....His head is taken by Re, his head is lifted up....Behold, he has come as Orion..... Behold, Osiris has come as Orion, Lord of Wine in the festival....he who the sky conceived and the dawn-light bore. O King, the sky conceives you with Orion; the dawn-light bears you with Orion....by the command of the gods do you live....with Orion you shall ascend from the eastern region of the sky; with Orion you shall descend into the western region of the sky.....Sothis, pure of thrones, shall guide you on the goodly roads which are in the sky in the Field of Rushes....."

References to Orion occur also in the famous *Book of the Coming Forth by Day,* or *Book of the Dead,* which dates back to the very earliest period of Egyptian history. In a late version, the *Papyrus of Ani,* which is at least 3000 years old, the text promises the scribe Ani that he shall enter the heavenly regions, and become one with Orion. Sir W.M.Flinders Petrie, in his contribution to that monumental archeological compendium *Wonders of the Past,* refers to Orion (or Sahu) as "a star god of the myth-strewn firmament"; the accompanying photograph shows Orion as a tall and stately figure whose headdress features a five-pointed star. In the Denderah reliefs, Orion more closely resembles the traditional portrayals of Osiris.

In the sky, Orion appears as the legendary Warrior or Hunter, with his shoulders marked by Alpha and Gamma, his knees by Beta (Rigel) and Kappa, and his great jeweled

ORION and SIRIUS appear (top) on the ancient Egyptian wall
reliefs at Denderah. Below: A section of the Vth Dynasty
Pyramid Texts from the tomb of King Unas at Sakkara.

BETELGEUSE. A "close-up" photograph of one of the largest known stars. The original plate, in red light, was made with the 13-inch camera at Lowell Observatory.

DESCRIPTIVE NOTES (Cont'd)

Belt by the striking row of three bright stars Delta,
Epsilon, and Zeta. Below the Belt, the nebulous gleam of
Theta and Iota Orionis outline the great Sword of Orion.
Beneath his feet crouches Lepus, the Hare; following him
are his two dogs, Canis Major and Canis Minor. Orion, like
Hercules, is usually depicted with a lion-skin shield, held
high on one arm as Orion faces the thundering charge of
Taurus, the Bull, whose baleful red eye, Aldebaran, glares
down from the V-shaped Hyades group, the Bull's Head. The
significance of the Orion-Taurus combat does not appear to
be made clear in any ancient legend; the purely symbolic
interpretation is the obvious one, that the two constellations represent the eternal conflict of good and evil. But
this interpretation, aside from its triteness, does not
seem supported by any of the ancient myths; the Bull was
never regarded as a symbol of evil anywhere in the classical world; on the contrary it was venerated as the embodi-
ment of strength, power, and virility, and was an object of
worship in more than one ancient cult. Picasso uses the
figure of the Bull in a painting which has become perhaps
one of the best known classics of modern art, *Guernica*, but
even here the Bull appears as the incarnation of blind,
destructive, crushing force, rather than deliberate evil.

The name *Betelgeuse* is from the Arabic *Beit Algueze*
or *Bed Elgueze*, very corrupted forms of *Iht al Jauzah*, the
"Armpit of the Giant" or "Arm of the Central One". R.H.
Allen lists a variety of other Arabic names as *Al Mankib*,
"The Shoulder", *Al Dhira*, the "Arm", and *Al Yad al Yamna*,
"The Right Hand". Students of Arabic, however, believe
that the original name, *Al Jauzah*, or *Yad al Jawza*, did not
refer to a "giant" but was a term used for a sheep marked
with a central spot or belt; a fair translation might then
be "The Front Leg of the White-belted Sheep". The present
author believes, however, that star-watchers will continue
to see a celestial giant in the great figure of Orion; the
sheep unfortunately lacks heroic grandeur. The *Alfonsine
Tables* call the star Beldengenze, while Riccioli has it
labeled *Bectelgeuze*. The Sanskrit name, *Bahu*, simply means
"The Arm", and the Coptic *Klaria* signifies "The Armlet".
The Hobbits of Middle-earth, according to Tolkien, knew it
as "Borgil" from an elvish word meaning "Fire-star".

DESCRIPTIVE NOTES   (Cont'd)

The star has also found a place in the modern mythology of
H.P.Lovecraft, for in the tales of his *Cthulhu Mythos*,
Betelgeuse is the home of the "Elder Gods", the infinitely
wise beings of cosmic power who once dwelt on Earth, ages
before the coming of Man. Betelgeuse was also the subject
of a musical tone-poem by Gustav Holst, better known as the
creator of the superbly evocative symphonic suite called
The Planets.

Betelgeuse is the only marked variable among the 1st
magnitude stars. The light changes were probably first
noticed by Sir John Herschel in 1836. In his *Outlines of
Astronomy*, published in 1849, he wrote "The variations of
Alpha Orionis, which were most striking and unequivocal in
the years 1836--1840, within the years since elapsed became
much less conspicuous..." In 1849 the variations again
began to increase in amplitude, and in December 1852 it was
thought by Herschel to be "actually the largest star in
the northern hemisphere". In 1894 it was again near its
peak brilliance. During the 20th century the star has shown
particularly high maxima in 1925, 1930, 1933, 1942, and
1947, while during the decade 1957- 1967 only slight and
uncertain variations were detected. The main period appears
to be about 5.7 years, but there may be shorter superimposed periods which vary between 150 and 300 days. When at
maximum the star sometimes rises to magnitude 0.4 when it
appears not greatly inferior to Rigel; in 1839 and again in
1852 it was thought by some observers to be nearly the
equal of Capella. Observations by the AAVSO indicate that
Betelgeuse probably reached magnitude 0.2 in 1933 and again
in 1942.

At minimum brightness, as in 1927 and 1941, the magnitude may drop below 1.2, a change of light intensity of
about 2 times. Betelgeuse is an irregularly pulsating red
supergiant; studies with the spectroscope reveal that the
diameter of the star may vary by about 60% during the whole
cycle, a difference considerably larger than the radius of
the Earth's orbit! The star is one of the largest known,
and among the naked-eye stars probably holds first place.
It is believed to be at least the size of the orbit of Mars
and at maximum diameter may possibly equal the orbit of
Jupiter. The formula for calculating the diameter appears

in many standard texts; to obtain the result in miles the
equation is given in the following form:

$$D = \frac{d \times 93,000,000}{\pi}$$

D here is the diameter in miles; d= the apparent angular
size in seconds of arc; and π = the annual parallax in
seconds of arc. Since both d and π are subject to a
certain margin of error it is evident that the calculated
diameter will be only reasonably accurate but not absolutely exact. The apparent angular size varies from 0.034" to
0.054" according to the best interferometer measurements,
and the parallax is about 0.006". Inserting these values in
the equation gives a diameter ranging between 480 and 800
million miles, or from 550 to 920 times the size of the
Sun. These figures are considerably larger than previous
estimates which have appeared in many texts; the difference
is due to a recent revision of the distance of Betelgeuse

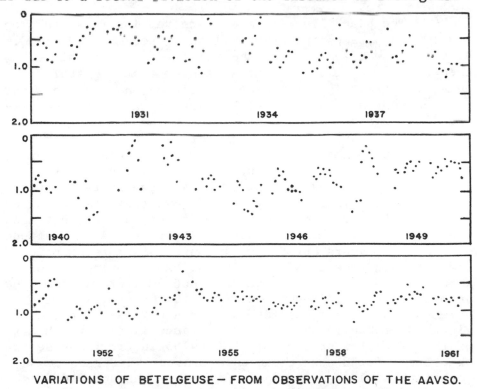

VARIATIONS OF BETELGEUSE – FROM OBSERVATIONS OF THE AAVSO.

DESCRIPTIVE NOTES (Cont'd)

which is now estimated to be about 520 light years. This
star was also the first to have its diameter measured by
means of the beam interferometer invented by A.A.Michelson
and first tested in 1920. The theory and use of this
instrument is explained in optical texts; basically it
consists of a system of mirrors which can be moved to the
opposite ends of a long beam mounted across the mouth of
the telescope. The end mirrors both receive the light of
the star and direct in into the main tube, thus increasing
the effective aperture of the telescope by 8 times or more.
When the images from the two end mirrors are superimposed
in the field of the telescope, a pattern of interference
fringes is created. As the mirrors are moved alternately
toward and away from each other the pattern is seen to
change and vanishes entirely at one point. The separation
of the mirrors required to cancel out the interference
fringes is carefully measured, and an optical formula then
gives the diameter of the light source, in fractions of a
second of arc.

The beam interferometer received its first test on
the 100-inch telescope at Mt.Wilson on December 13, 1920.
Betelgeuse was selected as the first test object since
theoretical calculations had suggested that the star was
unusually great in size. The experiment was a success and
the apparent angular size of Betelgeuse was found to average about 0.044".

It is important to remember that direct interferometer measurements can be made only in the case of stars
that are unusually large, and at the same time not too
distant. For the vast majority of average stars we must
rely upon more indirect methods of computing stellar sizes.

Betelgeuse is not only among the largest, but also
one of the most luminous stars of its class. At a distance
of 520 light years our Sun would appear as a telescopic
star of magnitude 10.8. Betelgeuse has a luminosity of
about 14,000 suns at maximum and 7600 Suns at minimum. The
peak absolute magnitude is about -5.6. Only a few such
objects have been identified in our own Galaxy. A red
giant of probably comparable brilliance lies in the heart
of the Jewel Box star cluster NGC 4755 in Crux, and several
others are found near the Double Cluster in Perseus. Mu
Cephei possibly has a comparable energy output, but lies in

DESCRIPTIVE NOTES (Cont'd)

a heavily obscured region of the sky, and the exact luminosity is still uncertain. Betelgeuse shows an annual
proper motion of 0.03" and a radial velocity of 12.5 miles
per second in recession.

The surface temperature of Betelgeuse is that of a
typical M-type red supergiant, about 3100°K. Only about 13%
of the radiant energy is emitted in the form of visible
light, and Betelgeuse would therefore appear as the brightest star in the sky if our eyes were sensitive to radiation
at all wavelengths. The apparent radiometric magnitude is
about -2.8 at maximum, which is the highest known for any
star.

In volume, Betelgeuse exceeds the Sun by a factor
of at least 160 million even at minumum. Yet the actual
mass of the star is probably no more than about 20 solar
masses, which means that the average density must be in
the range of 0.00000002 to about 0.00000009 the density of
our Sun.   Such star material has a density of less than one
ten-thousandth the density of ordinary air.   A star of
such tenuous nature has often been called a "red-hot
vacuum".

Betelgeuse is one of the few stars in the sky whose
actual disc is theoretically within range of detection by
large modern reflectors; the apparent angular diameter is
about 0.05" whereas the theoretical resolving power of the
200-inch reflector is about 0.02". Owing to the incurable
unsteadiness of the Earth's atmosphere, however, the performance of large reflectors never reaches the theoretical
limit, and it is doubtful that the actual disc of any star
will ever be observed visually by Earth-based telescopes.
On the other hand, through a combination of special photographic techniques and computer-enhancement of images, some
actual detail on the disc of Betelgeuse has been made visible for the first time, at the Kitt Peak National Observatory, in 1975. Using the 158-inch Mayall reflector with an
image-intensifier, astronomers obtained photographs which
were reduced to magnetic data and fed into the Interactive
Picture Processing System (IPPS) which can be adjusted to
enhance any feature present in the images. The mottling of
the disc and the large dusky areas revealed by this method
are evidently true features on the star; they represent

areas of different temperature and light intensity, comparable to the bright flares and dark spots seen on our own
Sun. Betelgeuse is also a source of weak radio radiation,
first detected with the 140-foot "dish" at the National
Radio Astronomy Observatory in 1966, and since confirmed
elsewhere; the red giant Antares has also been shown to be
a source of radio energy.
        Betelgeuse has been listed as a double or multiple
star in both the Aitken and Lick Observatory catalogues;
but it is not certain that any of the faint companions are
more than field stars:

Mag 14.5	Distance 39.8"	PA 109°
14.2	62.0	290
13.5	76.0	347
11.0	174.0	153

The bright primary itself also appears as a suspected spectroscopic binary in some star catalogues, but the estimated
period of about 6 years is close to the long cycle of the
star's light variations; the changes in radial velocity may
thus be a result of the star's pulsation, rather than binary motion.

ENERGY PRODUCTION AND EVOLUTION OF THE RED GIANT STARS.
The extremely tenuous nature of Betelgeuse is one of the
outstanding characteristics of the red giant stars. It may
at first seem incredible that such a body can be at a high
enough temperature to be self-luminous. Yet it is apparent
that the density and temperature must slowly increase
toward the center of the star. Is the central temperature
and pressure sufficient to maintain the cyclic hydrogen-
helium reaction which powers the main sequence stars such
as our Sun? Or do the red giants operate on other nuclear
reactions which lead to a high energy production under very
different conditions? Our problem here is to devise a
workable theoretical model which will solve the mystery of
energy production in the red giants, and which will also
provide a place for these stars in the general picture of
stellar evolution.
        In the past it had generally been believed that the
red giants were stars in the earliest stage of development.
Present knowledge has forced a nearly complete reversal of
this view, and it now seems quite clear that most of the

DESCRIPTIVE NOTES   (Cont'd)

stars of the type are very old stars which have already
passed through the main sequence state, and are now enter-
ing on the final stage of their evolution. The reasons for
accepting this view are partly observational and partly
theoretical. Among observational reasons we may mention the
fact that red giants are common in globular clusters and in
elliptical galaxies, systems which are known to be of great
age, which contain virtually no dust or gas, and in which
star formation has ceased countless ages ago. Conversely,
red giants do not appear in greater numbers in the nebulous
regions of the Galaxy, as they would certainly do if they
had been formed recently from the great gas and dust clouds
of space. Modern theory and observation agree in placing
the red giant stage after the main sequence stage.   The
general picture of this development as we see it today may
be described about as follows:
       Let us assume that a large number of stars have
been formed at about the same time, and that after a relatively short period of gravitational contraction, each star
has achieved a central temperature sufficient to maintain
the hydrogen-helium reaction which produces energy. If we
now plot all these stars on the standard H-R diagram or
Color-Magnitude Diagram (page 84) we will find that they
are all distributed in a rather narrow band running across
the diagram from upper left to lower right. In other words
the stars form a natural series in which the stars, depend-
ing upon the mass of each, range from bright blue hot stars
down to faint red cool ones. This band, is of course, the
Main Sequence. A star near the top of the main sequence
has a mass of 50 or 60 times the solar mass, and a million
times the luminosity of a typical red dwarf.
       This comparison immediately tells us something
about the relative "life expectancies" of the stars. A
very brilliant B-type star is consuming its fuel at such a
rate that its main sequence life cannot last for more than
a few million years. By contrast, our Sun, which is at
least 5 billion years old, is still a main sequence star
and will probably remain so for several billion years more.
The very faint stars, of course, have even greater life
expectancies, and will continue their main sequence exist-
ences for many billion of years.

DESCRIPTIVE NOTES   (Cont'd)

The nature of the evolution of a star after the main sequence stage has been the subject of many studies. The general evolutionary course of a star of about the solar mass is indicated on the diagram opposite. This type of evolution, which is expected to be a stage in the future history of our own Sun, proceeds in the following manner:

The hydrogen reaction operates most efficiently at the center of a star, where the temperature and pressure are greatest. Eventually, the hydrogen supply becomes depleted in this region, and the star develops a "core" of helium. The hydrogen reaction must then operate in a zone surrounding the core. As the core grows in mass, the hydrogen zone slowly increases in radius. This evolution is slow at first, but proceeds more and more rapidly as time goes on. When about 10% of the star's mass has gone into the helium core, the star has evolved from the original point "A" on the diagram to the point "B".

By the time the point "C" is reached, the core contains 20% of the mass of the star. The increase in size and luminosity has become noticeable, and the star can now be classified as a subgiant. As more and more helium is added to the core, the outer regions continue to expand and the star follows the evolutionary path to point "D" where approximately 30% of the mass is contained in the core. Now, although the expansion lowers the star's surface temperature, the increase in size more than compensates for this, and the total radiation therefore increases vastly as the evolution proceeds. When the star arrives at point "E" the diameter has increased by a factor of several hundred, and the luminosity is about 1000 times the original value. The star is now a typical red giant. If we compare our theoretical evolutionary track with the plotted positions of known red giants, and with the actual H-R Diagram of the evolved globular star clusters (pages 990-992) the agreement appears to be very good.

Now at point "E", the star has reached a critical phase in its history. Approximately 40% of the mass is now contained in the core. Since no nuclear energy has been produced in this region, the core has been slowly shrinking with a consequent rise in temperature. Eventually the temperature becomes sufficient for the helium itself to be used as a fuel in nuclear reactions which produce heavier

DESCRIPTIVE NOTES   (Cont'd)

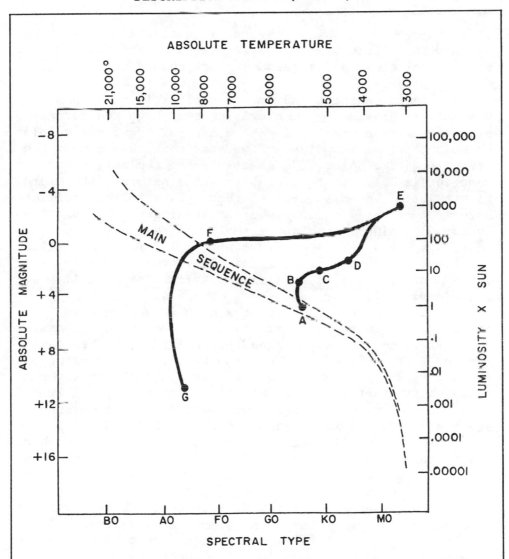

ABSOLUTE TEMPERATURE

THEORETICAL EVOLUTIONARY TRACK FOR A STAR OF ONE SOLAR
MASS.   Point "A" represents the original position on the
Main Sequence.   At "B" the star has evolved to the subgiant stage, while the points C-E represent the evolution
to the true red giant state.   Stars of greater mass have
evolutionary tracks correspondingly higher on the H-R
Diagram.   (Refer to text, pages 1295--1298.

elements such as neon and oxygen. This occurs at the point "E" and the effect is to reverse the expansion phase of the star's evolution, and to send it back toward the left on the H-R Diagram.

An interesting and important link in our chain of evidence is provided by the existence of huge red variable stars, pulsating giants like Mira and Chi Cygni. If we plot these objects on our diagram, we find that they occupy the area around the point "E", exactly where the helium-burning reaction is believed to begin. It seems natural, therefore, to attribute the pulsations of these stars to the conflict between two energy sources: helium burning in the core and hydrogen-burning in the outer shell. If the pulsations do indeed represent the star's attempt to initiate the helium reaction, it may be that some similar mechanism causes the various other types of stellar pulsation, for example the cepheid variable stars and the RV Tauri stars.

The evolution of the star beyond the red giant stage seems fairly clear. Briefly, the star moves toward the left in the H-R Diagram and gradually approaches the region "F". If the mass is not too large, the shrinking and fading carry it eventually to the realm of the super-dense white dwarf stars at "G". Nova outbursts possibly result from instability which precedes the white dwarf state and which characterizes the collapsing stars lying from "F" to "G". Stars of great mass cannot reach the white dwarf state, and it is likely that the necessary reduction in mass is accomplished through the violent outbursts that we call novae and supernovae.

Betelgeuse itself may be expected to have a very spectacular future history. Much more massive than the Sun, its original main sequence position must have been several magnitudes above point "A", and the whole of its evolutionary track is correspondingly higher on the H-R Diagram. Mass loss from the star is already a well observed fact; the star is surrounded by an extensive "corona" extending to several hundred radii from the surface, and the velocity of the gas at the outer edge definitely exceeds the escape velocity. The present mass loss, if it remained constant, would imply a loss of about one solar mass in only a few hundred thousand years. The evolution of such a star proceeds with great rapidity on the astronomic

DESCRIPTIVE NOTES   (Cont'd)

time scale; studies by C.Hayashi and R.C.Cameron (1962)
have shown that a very massive star may evolve from its
main sequence state to the red giant stage in something
like 20,000 years. The bright supergiants Deneb and Rigel
may be at the beginning of this process; perhaps future
inhabitants of the Earth will see them transformed into red
supergiants resembling Betelgeuse. (Refer also to Antares
in Scorpius, Alpha Herculis, Mira in Cetus, Mu Cephei, Zeta
Aurigae, VV Cephei, Beta Pegasi, and Chi Cygni)

BETA   Name- RIGEL, from the Arabic *Rijl Jauzah al
Yusra, the "Left Leg of the Giant". The older
names *Algobar*, *Regel*, and *Riglon* are virtually obsolete.
Rigel is the 7th brightest star in the sky. A magnitude of
0.34 is given in many standard star lists, but recent photoelectric measurements give a revised value of 0.14. There
is some evidnece for a slight variability, which might explain the discrepancies in standard catalogues. The total
range, however, does not appear to exceed 0.1 magnitude.
Spectral type B8 Ia, color bluish-white, position 05121s
0815. Opposition date (midnight culmination) occurs about
December 12. Although labeled *Beta*, Rigel is virtually
always the brightest star in Orion.

Rigel is well past the limiting distance for accurate
parallax measurements, and the exact distance is still uncertain. The star cannot be much closer than 540 light
years, a figure given in many texts, but may easily be
considerably more remote. The presently accepted distance
is about 900 light years. Rigel shows an annual proper
motion estimated to be about 0.001", one of the smallest
values among all the 1st magnitude stars. The radial velocity is about 12 miles per second in recession, with some
variations.

Rigel is a true supergiant, a blazing white-hot star
of intense brilliance and dazzling beauty. Its surface temperature is about 12,000°K and its energy output exceeds
that of our sun by a factor of many thousands; the computed
luminosity is something like 57,000 times the brightness of
our Sun. Rigel is thus one of the most luminous objects
known in our Galaxy, with an absolute visual magnitude of
about -7.1. If such a star was as near to us as Sirius,
it would have an apparent magnitude of about -10 and give

DESCRIPTIVE NOTES   (Cont'd)

us about a fifth the light of a full moon! On the basis
of the spectral type and total luminosity, the computed
diameter of the star is about 50 times the solar diameter,
or approximately 40 million miles. From the mass-luminosity relation, the star is estimated to have a mass of 50
solar masses. The average density of Rigel is about 0.0004
the density of the Sun.

Spectra of Rigel show radial velocity shifts in a
semi-regular period of about 21.9 days; this does not seem
to be attributable to binary motion, but is probably caused
by atmospheric pulsations. Other B and A type giants show
a similar effect, which may be compared with the mysterious
pulsations of the Beta Canis Majoris stars.

THE COMPANION TO RIGEL. A good 6-inch telescope will show
that Rigel is accompanied by a small bluish companion star
of magnitude 6.7, approximately 9" distant in PA 202°. The
small star is not always an easy object for the amateur
telescope, though it may sometimes be glimpsed with a 3-
inch glass under good conditions. The two stars have shown
virtually no change in separation or angle since the early
measurements of F.G.W.Struve in 1831. The pair is generally
thought to be physical one, however, and the radial velocities are very similar. The actual separation is very
great, not less than 2600 AU. Rigel B is itself a close
binary detected by the spectroscope; the period is 9.860
days and the components appear to be about equally bright,
comparable to Regulus in size and luminosity. The total
light of the pair is about 150 times the light of the Sun,
and the combined spectral type is B9. The BC pair has a
computed orbital eccentricity of 0.1, and the primary is
about 2 million miles from the center of gravity of the
system.

In addition to its spectroscopic duplicity, the
companion has been reported as a visual double by a number
of observers. S.W.Burnham measured a distinct elongation
of the image with a 6-inch refractor in 1871, and appeared
to verify the observation with the 18.5-inch telescope at
Dearborn Observatory in 1878. R.G.Aitken estimated a separation of 0.10" in 1903, and obtained the same results in
1911. Barnard, Hussey, and Van den Bos are among the
observers who have recorded a duplicity or elongation of

DESCRIPTIVE NOTES (Cont'd)

the image. However, the observers have, in many cases, been unable to obtain the same measurements again, and the star has often appeared single in the greatest telescopes under very good conditions. Van Biesbroeck has not verified the duplicity with either the Yerkes 40-inch refractor or the McDonald 82-inch reflector. And interferometer observations by Finsen have shown no evidence that the star is a double. The case of Rigel B is thus a curious one. If the star is actually a visual double, the period may be short and the motion quite rapid; the greatest separation being only about 0.10". The star is thus a difficult object in even the greatest telescopes. Rather oddly, a similar uncertainty exists with regard to the reported duplicity of the famous companion to Sirius.

Rigel is the brightest member of a large group of B-type giant stars which include most of the more prominent stars that bejewel the familiar figure of Orion. The region of the "Orion Association", as it is called, is remarkable not only for the abundance of brilliant B-type stars, but also for the vast clouds of bright and dark nebulosity which surround these supergiant suns. The heart of the association is apparently the Great Orion Nebula itself, with its central knot of giant stars. The Orion region is of greatest interest to astrophysicists since it is one of the areas of the heavens where star birth is in progress at the present time. Rigel itself, with its enormous radiating power, must be a very young star; at its present rate of fuel consumption such an object cannot maintain itself for more than a few million years. The presence of numerous variables of the T Tauri class is also considered strong evidence for extreme youth. (Refer also to the section on the Great Orion Nebula M42)

GAMMA   Name- BELLATRIX, the "Amazon Star." Magnitude 1.64; spectrum B2 III. Position 05224n0618. The computed distance of Gamma Orionis is about 470 light years, and the actual luminosity about 4000 times that of our Sun. The annual proper motion is 0.01"; the radial velocity is 11 miles per second in recession.

A faint diffuse nebulosity surrounding the star, shown on the Skalnate Pleso Atlas, is merely an illuminated portion of the general nebulous haze which envelops much of Orion, and which naturally becomes visible in the vicinity of any highly luminous star. Similar masses of nebulous haze are scattered across the entire region to the south of Bellatrix, culminating in the splendor of the Great Nebula itself.

DELTA   Name- MINTAKA, from *Al Mintakah*, the "Belt"; The westernmost star of the three forming the "Belt of Orion". Delta is magnitude 2.20; Spectrum 09 II; Position 05294s0020. This is a giant member of the Orion Association, probably about 1500 light years distant and 20,000 times the luminosity of the Sun. The star is also an eclipsing binary of small range, with a period of 5.7325 days; the light variation is just under 0.2 magnitude. The components of the pair seem to be nearly equal in size and type, and the eccentricity of the spectroscopic orbit is 0.08, with the primary star about 5 million miles from the center of gravity of the system.
Delta Orionis shows an annual proper motion of 0.01"; the radial velocity is 12.5 miles per second in recession.
For the small telescope, the star has a bluish companion of magnitude 6.7, located 52.8" directly north of the primary; the spectral type of the companion is B2 V. No relative change has been seen in over a century and the two stars probably form a physical pair with a projected separation of about 27,000 AU, or nearly half a light year.
Delta Orionis was the first star discovered to show stationary spectral lines (of calcium) which thus proved the existence of interstellar matter; the discovery was made at Potsdam Observatory in 1904.

EPSILON   Name- ALNILAM, "A Belt of Pearls"- The central star in the Belt of Orion. Magnitude 1.70; spectrum B0 Ia; position 05337s0114. Epsilon Orionis is a supergiant star, and a prominent member of the great Orion Association. The estimated distance is about 1600 light years, and the true luminosity about 40,000 times that of the Sun (absolute magnitude about -6.8). The star shows no detectable proper motion; the radial velocity is about 15.5 miles per second in recession.

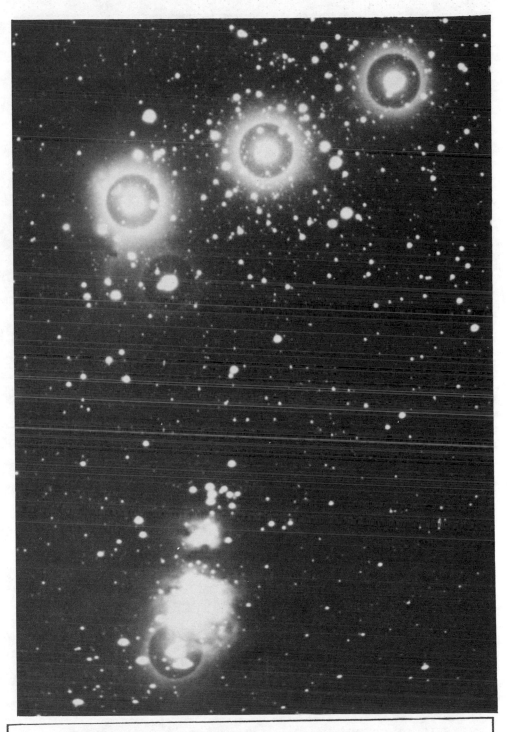

"...Those three stars of the airy Giant's zone,
That glitter burnished by the frosty dark..."

TENNYSON

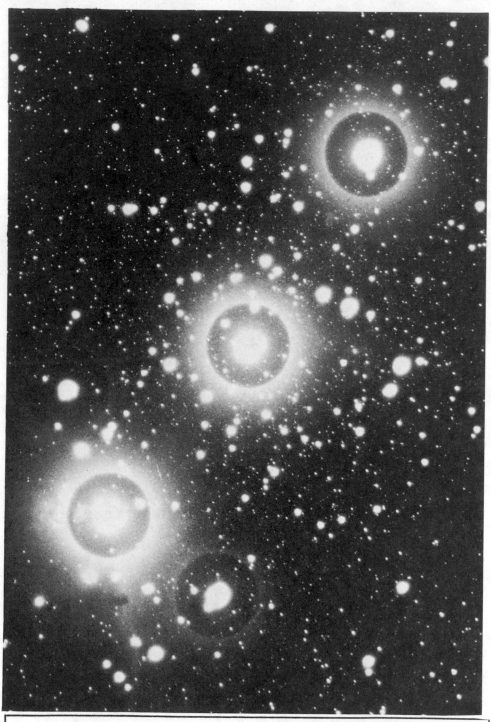

THE BELT OF ORION. Epsilon is at center, Delta at upper
right, and Zeta at lower left. The Horsehead Nebula shows
faintly below Zeta.   Lowell Observatory photograph.

DESCRIPTIVE NOTES (Cont'd)

The star is surrounded by the faint and formless nebulosity NGC 1990, another portion of the great nebulous haze which envelopes the entire constellation and which naturally becomes visible when near so bright a star.

ZETA   Name- ALNITAK, from *Al Nitak*, "The Girdle". The eastern star of the three forming the "Belt of Orion". Magnitude 1.79; spectrum 09 or B0 Ib; position 05382s0158. In distance and luminosity this star is nearly a twin of Epsilon; some 1600 light years away, the absolute magnitude must be about -6.6 and the total luminosity about 35,000 times our Sun. Zeta Orionis shows an annual proper motion of about 0.005"; the radial velocity is about 11 miles per second in recession.

The star is an interesting triple. The close pair are magnitudes 1.91 and 5.5, separation 2.6". This rather difficult pair was first resolved by the German astronomer Kunowsky in 1819, apparently having been missed by William Herschel. Much disagreement exists as to the colors of the components, which seem to show an odd color contrast in large telescopes. R.H.Allen called them "topaz yellow and light purple"; W.T.Olcott has simply "yellow and blue"; E.J.Hartung refers to them as a "brilliant white pair", while the elder Struve in 1836 coined the ponderous term "olivacea subrubicunda" (slightly reddish-olive) for the fainter star. The spectral type of the companion is about B3.

Very slow orbital motion exists in the pair, but the PA has increased by only 8° in the last century; the period must be at least several thousand years. The present projected separation is about 1280 AU.

A third component, which is probably not a true physical member of the system, is a 10th magnitude object, 57.6" distant. No change in separation or angle has been noted in the AC pair.

Zeta Orionis lies in a remarkable field of bright nebulosity which is excited to luminosity by the star. About 15' to the east and slightly north will be found the large nebulous patch NGC 2024, a glowing mass of bright and dark areas with much intricate interior detail, and split by a wide dark lane running from north to south. Directly south of the star a faint reef of nebulosity, IC 434 runs

southward for over a degree, sharply bordered on the east
rim and fading away into a diffuse luminous haze toward the
west. Midway along this nebulous bar, on the eastern edge,
protrudes the great dark mass of the Horsehead Nebula B33,
probably the most famous of the dark nebulae. (Refer to
page 1339)

ETA   Magnitude 3.32; spectrum B0 or B1 V; position
05220s0226. One of the notable members of the
Orion Aggregation. The distance is estimated to be about
940 light years, and the actual luminosity about 4000 times
that of the Sun. Eta Orionis shows an annual proper motion
of less than 0.01"; the radial velocity is about 12 miles
per second in recession.
        The star is a close and rather difficult double for
the amateur telescope. According to S.W.Burnham, the
duplicity was discovered by Dawes with a 4.5" telescope in
1848. Individual magnitudes are 3.6 and 5.0; both stars are
brilliantly white though T.W.Webb thought the companion to
be slightly purplish. A decrease in the PA of about 7° in
the last century indicates a very long orbital period; the
separation of 1.4" has remained nearly constant. For these
stars the projected separation works out to about 400 AU.
There is a third faint star of the 10th magnitude at 115"
in PA 51°, undoubtedly a field star only.
        Eta A is a rapid spectroscopic binary with an
eclipse-type light curve of the lyrid class; the orbital
period is 7.98922 days and the range is only 0.2 magnitude.
Variations in the radial velocity in a period of 9.2 years
also indicate the existence of a fourth star in the system.

THETA   Famous multiple star called the "Trapezium"
in the heart of the Great Nebula M42. For
descriptive data refer to page 1317.

IOTA   Magnitude 2.76; spectrum O9 III; position
05330s0556. This star was known to the Arabs
as "Na'ir al Saif", the "Bright One of the Sword". It lies
about half a degree south of the Orion Nebula and marks the
southern tip of the Sword of Orion. The distance of Iota
Orionis is estimated to be about 2000 light years; the
spectral characteristics are those of a giant of absolute

REGION OF ZETA ORIONIS. The bright nebulosity at top left is NGC 2024, the dark mass of the Horsehead Nebula appears near center.   Lowell Observatory photograph.

DESCRIPTIVE NOTES   (Cont'd)

magnitude about -6, or 20,000 times the luminosity of our
Sun. The star shows an annual proper motion of less than
0.01"; the radial velocity is 13 miles per second in
recession, with definite variations. A large faint nebulous
cloud covers most of the field; this is NGC 1980.

Iota Orionis is an easy and attractive triple star
for the small telescope. The closer companion, 11" distant,
is a bluish star of class B9, apparent magnitude 6.9.   At
50" is a second companion of the 11th magnitude. R.H.Allen
calls the colors of this group "white, pale blue, and grape
red".   No relative change has occurred between the three
stars since the early measurements of F.G.W.Struve in 1831.
The AB pair has a projected separation of about 6500 AU.

In addition to the visual triple, Iota A is itself
a spectroscopic binary with a period of 29.136 days; the
orbit has the rather high eccentricity of 0.76 according
to computations by J.Pearce in 1953.

The easy double star Σ747 lies in the same field
with Iota, about 8' to the SW; this is a fine white pair
with a fixed separation of 36".

KAPPA   Name- SAIPH. Magnitude 2.06; spectrum B0 or
BI; supergiant of luminosity class I with
emission lines in the spectrum. Position 05454s0941. This
is another bright member of the Orion Aggregation, though
the estimated distance of 2100 light years places it somewhat beyond many of the other members. The spectral features suggest an absolute magnitude of about -6.9, or close
to 50,000 times the solar luminosity. Kappa Orionis shows
an annual proper motion of only 0.006"; the radial velocity
is 12.5 miles per second in recession.

LAMBDA   Magnitude 3.40; spectrum 08. Position 05324n
0954. The star marks the Head of Orion. The
computed distance is about 1800 light years; the actual
luminosity is then about 9,000 times that of the Sun. An
annual proper motion of only 0.006" has been measured; the
radial velocity is 20 miles per second in recession.

Lambda Orionis is an elegant double star for the
small telescope, first measured by F.G.W.Struve in 1830.
Many and curious are the reported discrepancies in the
colors of this pair, all the more strange since both stars

are type 0 and should appear simply white. T.W.Webb calls
them "yellowish and purple" while W.T.Olcott found them
yellow and red; to the author of this book the pair has
always appeared sparkling white with just a hint of a light
amber tint. E.J.Hartung calls the star "a brilliant and
easy white pair"; R.H.Allen, perhaps quoting some other
observer, calls them "pale white and violet".

The separation of the stars has remained at 4.4"
since the days of Struve, corresponding to a projected
separation of about 2400 AU. A variable radial velocity is
shown by the fainter component and the star is probably a
spectroscopic binary; the spectrum is also peculiar for the
presence of emission lines. Long exposure photographs show
an extensive but faint diffuse nebulosity over the whole
field.

Lambda Orionis forms a small triangle with the two
fourth magnitude stars Phi-1 and Phi-2. R.H.Allen calls
attention to the fact that the disc of the Moon could be
inserted in this triangle which measures 27' along the west
side and 33' across the south side; although no one would
accept this without verifying it by measurement as the
apparent size of the Moon's disc is grossly exaggerated.
It is similarly difficult to convince many persons that the
Moon, if placed in front of the Pleiades cluster, could not
possibly hide the entire group at once.

PI-3　　　Magnitude 3.19; spectrum F6 V; position 04471n
　　　　　0653.   This is undoubtedly the closest bright
star in Orion, at a distance of 26 light years. A normal
main sequence star only slightly larger and hotter than our
Sun, it has an absolute magnitude of about +3.6, and a true
luminosity of 3 times that of the Sun. The annual proper
motion is 0.47" in PA 88°; the radial velocity is 14 miles
per second in recession.

SIGMA　　　Magnitude 3.73; spectrum 09 V; position 05362s
　　　　　0238. One of the notable members of the Orion
Aggregation, easily found about a degree southwest of Zeta
Orionis. The computed distance is about 1400 light years;
the total luminosity about 5000 times that of the Sun. The
star shows an annual proper motion of about 0.004"; the
radial velocity is 17.5 miles per second in recession.

DESCRIPTIVE NOTES   (Cont'd)

Sigma Orionis is a remarkable multiple star system containing five visible components. The close pair, A&B are about 0.25" apart, a difficult object first resolved by S.W.Burnham in 1888. A 170-year period has been computed for the pair with periastron in 1849; W.D.Heintz (1974) has obtained a semi-major axis of 0.25" with an eccentricity of 0.07 and an inclination of about 165°. The pair has a total mass of about 35 suns, one of the most massive visual binaries known, if these elements are approximately correct. The secondary has a spectral type of. about B3.

At a distance of 11.2" in PA 236° is the 10th magnitude star called "C", and 12.9" away in PA 84° is the 7.5 magnitude star "D". The spectral types of these are A2 and B2 respectively. The most distant component, E, is a 6th magnitude star 42" away in PA 61°; the spectrum of this star resembles that of A, but is remarkable for the great intensity of the helium lines. The luminosity of this star is about 600 times that of the Sun. This star also has a companion of its own, an 8th magnitude star at 30" in PA 231°. In addition, the faint triple star Σ761 lies in the same field, at a distance of 3.5'.

Only in the case of A and B has any relative motion been detected, but there is no doubt that all the components of Sigma Orionis form a vast physical system. The true separation between A and E is at least 20,000 AU, or about a third of a light year! Variations in the radial velocity of the primary suggests that at least one additional unseen component is present in the system.

U   Variable. Position 05529n2010. A bright long-period red variable star, located in the north segment of Orion near the Taurus border, less than 6° from the Crab Nebula. Discovered by J.Gore in 1885, U Orionis is one of the most suitable stars of the type for amateur observations. It has a good light range of about 7 magnitudes, a strong reddish color, and is well placed in the field of the bright stars 54 and 57 Orionis (magnitudes 4.4 and 5.9) which greatly aid the identification.

At maximum the star often outshines 57 Orionis; the greatest recorded brightness is about 5.3. At minimum it may sink to about 12.5. The average period is 373 days, with a change of spectral class from M6 to M8 during the

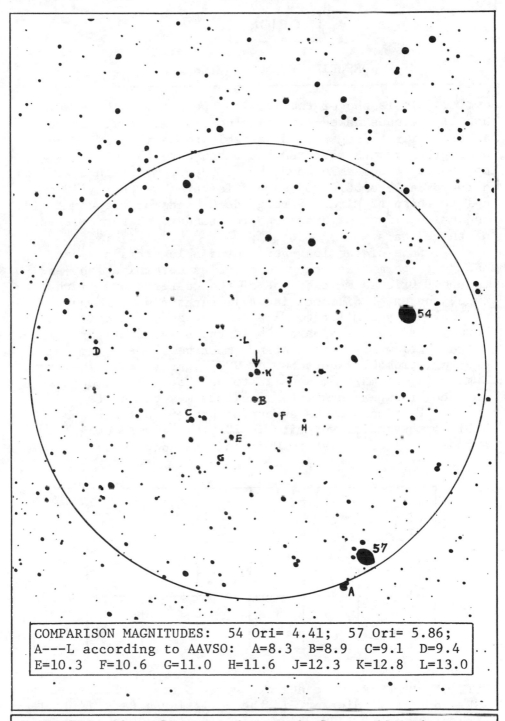

COMPARISON MAGNITUDES: 54 Ori= 4.41;   57 Ori= 5.86;
A---L according to AAVSO: A=8.3   B=8.9   C=9.1   D=9.4
E=10.3   F=10.6   G=11.0   H=11.6   J=12.3   K=12.8   L=13.0

U ORIONIS. Identification chart made from a 13-inch camera
plate at Lowell Observatory. The circle is 1° in diameter
with north at the top.   Stars to about 15th magnitude are
shown.

DESCRIPTIVE NOTES (Cont'd)

cycle. This is one of the few long-period variables which usually reaches naked-eye brightness at maximum. It was also the first long-period variable to be identified by a photograph of its spectrum. At discovery, the star was considered a possible nova, but a spectrum obtained at Harvard in December 1885 showed features virtually identical to those of Mira. Strong dark bands of titanium oxide and bright lines of hydrogen are standard characteristics of these stars.

None of the long-period variable stars are near enough to permit an accurate parallax determination, and though U Orionis must be one of the nearer stars of the type, the exact distance is still uncertain. The star is probably very similar to Mira in size and luminosity, however; since its apparent brightness is about three magnitudes below Mira the distance must be about 4 times greater, probably about 800 or 900 light years. The true luminosity at maximum appears to be about 250 times that of our Sun; the peak absolute magnitude may be about -1.3.

U Orionis shows an annual proper motion of about 0.018"; the radial velocity is 12.5 miles per second in approach. (Refer also to Omicron Ceti, page 631)

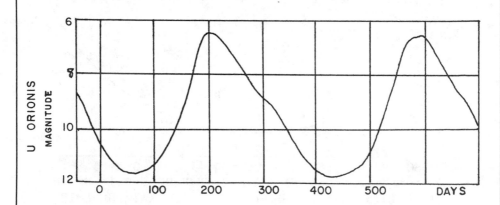

NOTE ON THE CHART— The nearest star to the variable on the ESE, about 0.7' distant, is the eclipsing binary UW Orionis with a period of 1.00805 day and a photographic range of 11.1 to 11.6. It appears to be a star of the Lyrid class with a B-type spectrum.

U ORIONIS. One of the brightest of the long period variable
stars. These photographs were made with the 13-inch camera
at Lowell Observatory.

DESCRIPTIVE NOTES   (Cont'd)

FU   Variable. Position 05426n0903, about 3° NW
from Betelgeuse, and 0.8° east of the small
planetary nebula NGC 2022. FU Orionis is a peculiar star
considered by some authorities as an exceptionally slow
nova, and by others as an odd nebular variable. Before
1937 it was a 16th magnitude object, but in that year it
slowly rose to magnitude 9.7 over an interval of about 3
months. If actually a nova, this star was probably the
slowest one on record. It remained at maximum for 2 years,
and was still relatively bright more than 20 years later.
The magnitude in 1960 was about 11, and a general slow
fading appears to be continuing at the present time.
          FU Orionis is a yellowish star variously classed
as type F, G, or F peculiar. It is almost centrally located in the dark nebula B35, which appears as the almost
totally starless area in the central portion of the chart
on page 1315. The nebulosity was observed to brighten at
the time of the light increase in 1937, and studies by O.
Struve (1939) showed that the nebula was shining by the
reflected light of the star. During the maximum the star's
spectrum was type F, but gradually changed from F5 to G0
and then to G3. The absence of any typical nova spectrum
makes it uncertain whether this star should actually be
classed with the true novae. In any case, the actual luminosity was apparently much inferior to that of a normal nova
since the computed distance would otherwise be improbably
large. The clear association with nebulosity in the Orion
region implies a distance of not over a few thousand light
years, and a maximum luminosity of about 100 times that of
the Sun. On the other hand, C.P.Gaposchkin (1957) finds
that the spectrum resembles the supergiants Rho Cassiopeia
and Gamma Cygni, suggesting an absolute magnitude of −4 to
−6. Perhaps these results can be reconciled by assuming
that the star is being observed through 5 or 6 magnitudes
of obscuration. A.A.Wachmann has suggested that the light
changes are caused by the star's interaction with dense
nebulosity; another intriguing suggestion is the possibility that FU Orionis is actually a newly formed star which is
just reaching the self-luminous state. The surrounding
dark cloud might then be comparable to one of the "globules"
seen in some of the diffuse nebulae and often regarded as
possible proto-stars. (Refer also to page 1199)

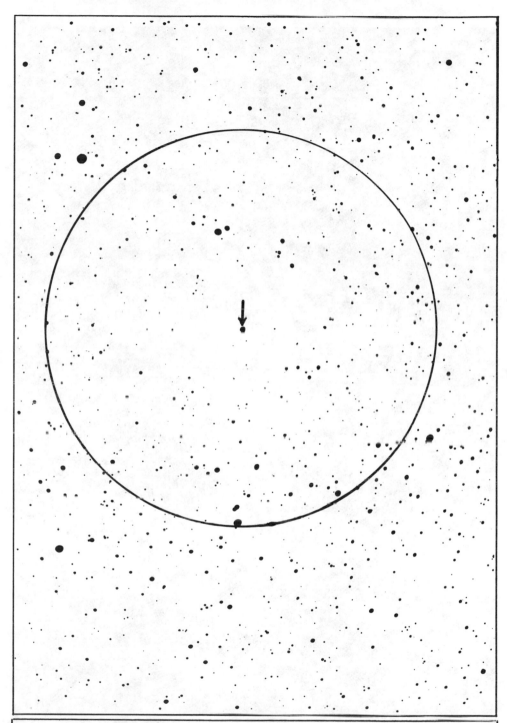

FIELD OF FU ORIONIS from a 13-inch camera plate obtained at Lowell Observatory. The circle is 1° in diameter with north at the top; stars to about 16th magnitude are shown. The starless area near the center is the dark nebula B35.

THE GREAT NEBULA IN ORION (M42) as it appears on a plate
made with the 36-inch Crossley reflector at Lick Observatory. The bright star near the top of the field is Iota
Orionis; nebula NGC 1977 is at the bottom of the print.

M42   (NGC 1976) Position 05329s0525. This is the
"Great Nebula in Orion", generally considered
the finest example of a diffuse nebula in the sky, and one
of the most wonderfully beautiful objects in the heavens.
     It may be seen in a pair of field glasses as a
faint haze spreading out from the famous quadruple star
Theta in the middle of the "Sword of Orion". In the small
telescope it appears as a bright greenish mist enveloping
the star. In a moderately large telescope its appearance is
impressive beyond words, and draws exclamations of delight
and astonishment from all who view it. The great glowing
irregular cloud, shining by the gleaming light of the
diamond-like stars entangled in it, makes a marvelous
spectacle which is unequalled anywhere else in the sky.
Barnard found it resembling a great ghostly bat as it came
drifting into the field of the Yerkes telescope, and spoke
of a feeling of awe and surprise each time he saw it. To
many others it creates, as does no other vista of the
heavens, the single overpowering impression of primeval
chaos, and transports the imaginative observer back to the
days of creation. This irresistible impression is more than
a poetic fancy, as modern astrophysics now confirms, for
the Orion Nebula is undoubtedly one of the regions in space
where star formation is presently underway.
     For the amateur, there is no other object in the
heavens so perfectly suited for observation by low power
wide-angle telescopes. The night should be dark and clear,
of course, and the eyes well dark-adapted. A dramatic way
to appreciate the extent of the nebulosity is to set the
telescope just ahead of the nebula and allow it to drift
slowly across the field. With this method, the glow of the
nebulosity may be traced out far beyond the usual limits.
"A glorious and wonderful sight in a large telescope'.' says
William T.Olcott (1929) "Words fail utterly to describe its
beauty". J.E.Bode referred to it as "the most remarkable
nebula in the heavens....Theta shines four-fold through a
good telescope; there are three very small stars next to
it.....There are altogether seven stars to be seen....all
involved in a lively nebula or light-shimmer..." William
Herschel's great reflector showed the nebula as "an unform-
ed fiery mist, the chaotic material of future suns". K.G.
Jones (1968) has called attention to this passage as an

THE GREAT NEBULA IN ORION. "A single misty star" to the unaided eye, M42 shows impressive detail on this 200-inch
telescope photograph.   Palomar Observatory

DESCRIPTIVE NOTES (Cont'd)

example of Herschel's almost prophetic vision. T.W.Webb
found M42 to be "one of the most wonderful objects in the
heavens; readily visible to the naked eye, yet strangely
missed, as Humboldt says, by Galileo, who paid great attention to Orion...The telescope shows an irregular branching
mass of greenish haze, in some directions moderately well
defined where the dark sky penetrates it in deep openings;
in others melting imperceptibly away...." In their popular
New Handbook of the Heavens, H.J.Bernhard, D.A.Bennett, and
H.S.Rice (1941) refer to the Great Nebula as "visible to
the naked eye and easily observed. Here is found the fascinating colored multiple star, Theta-one Orionis, known as
the trapezium. The nebula is greenish and of irregular
form. Much detail with branches, rifts, and bays; and the
entire nebula with its stars repays long observation." In
western literature, possibly the finest tribute to this
marvel of the skies occurs in Tennyson's *Merlin and
Vivien:*

> "........*A single misty star*
> *Which is the second in a line of stars*
> *That seem a sword beneath a belt of three,*
> *I never gazed upon it but I dreamt*
> *Of some vast charm concluded in that star*
> *To make fame nothing.."*

Tennyson, often called "The Poet of Science", also seems to
be echoing Herschel's vision of the birth of future suns
when he speaks of

> "*.....regions of lucid matter taking form,*
> *Brushes of fire, hazy gleams,*
> *Clusters and beds of worlds, and bee-like swarms*
> *Of suns and starry streams...*"

Here are found, in the words of G.P.Serviss, "*......stars
apparently completed, shining like gems just dropped from
the hand of the polisher, and around them are masses, and
eddies, currents and swirls of nebulous matter yet to be
condensed, compacted, and constructed into suns...*" In
describing one of the first successful photographs of the
Orion Nebula, made by Henry Draper in 1881, R.A.Proctor
found it difficult to find the words to express *"the
thought that seemed so impressive, so thrilling, as to*

*surpass even the feeling of awe with which in the solemn
darkness of night we see some mighty group of suns sweep
into the field of view of the telescope.............that
here on this tiny square inch of shoreline, with its thin
film of chemical sands, had been received the impress of
waves which for years had been traversing the solemn depths
of space.....Here we have mirrored by Nature herself 'that
marvellous round of milky light below Orion'..."*

     "For who would acquire a knowledge of the heavens,"
writes C.E. Barns, "let him give up his days and nights to
the marvels of Orion. Here may be found every conceivable
variation of celestial phenomena: stars, giants and dwarfs;
variables, doubles.. triples.. multiples; binaries visual
and spectroscopic; clusters wide and condensed; mysterious
rayless rifts and nebulae in boundless variety, with the
supreme wonder of all supernal wonders at its heart - the
Great Nebula - before which the learned and the laymen
alike have stood silent in awe and reverence since the first
lens unfolded to man's gaze its true vastness and intricacy,
and which offers abundant field for all the geniuses of
science, with their super-refinements of means and methods,
for generations to come..."

     Mary Proctor, continuing in the tradition of her
famous astronomer-father, wrote in *Evenings With the Stars*
that the Trapezium group yields to a telescope of three
inch aperture: "This is an irregular square with a star on
each corner, and is situated in the midst of a dark gap in
the nebula, within which the four stars gleam brightly...
The radiant mist surrounding them has a greenish tinge,
revealing the vast stellar cloud known as the "Nebula in
Orion" with its *Isles of light and silvery streams, and
gloomy gulfs of mystic shade...*"

     Although indistinctly visible to the naked eye under
the best conditions, the Orion Nebula is not mentioned in
any known ancient or medieval records; it is strange also
that Galileo apparently never noticed this object. The
discovery of the nebula is credited to Nicholas Peiresc in
1611. Naturalist, archeologist, patron of the sciences,
Peiresc is also said to have been the first to verify by
experiment Harvey's discovery of the circulation of the
blood. The next recorded observation of the nebula was made
in 1618 by Cysatus, a Swiss Jesuit. He seems to have made

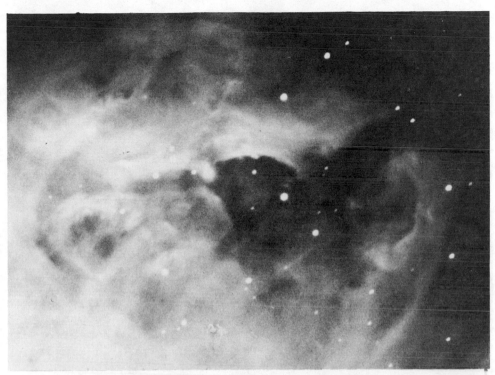

DETAILS IN THE ORION NEBULA. Top: A region of intricate filaments on the west side of the nebula. Below: The most prominent filament, on the east edge. Lowell Observatory

DESCRIPTIVE NOTES   (Cont'd)

mention of it chiefly as a term of comparison for a comet
visible at the time. The nebula remained comparatively
little-known until 1656 when Christian Huyghens published
the first drawing and description of the object, calling
attention to the remarkable multiplicity of the bright star
Theta, in the heart of the luminous cloud. It is recorded
that Sir William Herschel began his observing career by
viewing the Orion Nebula with a reflecting telescope of his
own construction in 1774. The appearance of the nebula so
amazed him that the observation of such objects became a
life-long study. His son, John Herschel, began his own
observing program in 1825; his description of the intricate
details of the Orion Nebula is still quoted today. Of the
bright central area surrounding Theta itself, he wrote:
"I know not how to describe it better than by comparing it
to a curdling liquid, or to the breaking up of a mackerel
sky when the clouds of which it consists begin to assume a
cirrus appearance.."
        M42 was the first nebula to be successfully photo-
graphed, by Henry Draper in 1880. His instrument was an
11-inch refractor, and the exposure was 51 minutes.

        At the time of Herschel, the term "nebula" was used
to designate any object in the sky which appeared as a bit
of cloud or mist. Some astronomers regarded all such objects as merely unresolved star clusters; others thought
that many of them were actually masses of luminous gases
floating in space. In 1864, William Huggins began the
spectroscopic observations of the nebulae, beginning with
the planetary NGC 6543 in Draco. He concluded that the
light of this object, and of the Orion Nebula, came from a
rarified gas excited to luminescence. Eventually it was
learned that the same fact held true for all the diffuse
nebulae and all the planetary nebula. The "spiral" nebulae,
however, were found to be a very different class of object,
and are now recognized as external galaxies.
        The Great Orion Nebula, then, is a vast cloud of
glowing gases, its immensity beyond comprehension, its
physical conditions almost unimaginable. Although among
the nearest objects of the type, the exact distance is not
quite certain, but is almost certainly greater than the 900
light year figure given in many of the older observing

guides. From color and magnitude measurements of the involved group of stars, the minimum distance appears to be
about 1300 light years, while proper motion and radial
velocity measurements by K.A.Strand (1957) indicated a
distance of about 1900 light years. Another study, made at
Palomar Observatory in 1969-1970 gave a modulus of about
8.5 magnitudes, or about 1600 light years. If a value in
the general range of 1600- 1900 light years is accepted,
the diameter of the nebula, as it appears on typical photographs, is about 30 light years, or more than 20,000 times
the diameter of the entire Solar System. The bright central
region is 5 or 6 light years across, but the best photographs show that fainter extensions reach out to cover
nearly the entire constellation of Orion. Although the
density of all this material is less than a millionth the
density of a good laboratory vacuum, yet the total mass
represents enough material to form about 10,000 stars like
our Sun.
        The light of the Orion Nebula is largely fluores-
cense, produced by the strong ultraviolet radiation from
the high-temperature components of Theta Orionis. In the
early days of spectroscopy, a number of unidentified lines
were seen to exist in the spectrum of the nebulae; these
were at first ascribed to a hypothetical element called
"nebulium" but were eventually found to be produced by such
well known substances as oxygen, neon, helium, and nitrogen
under extraordinary conditions never encountered on the
Earth. The greenish color of the nebula is chiefly due to
two strong emission lines of doubly ionized oxygen at 5007
and 4959 angstroms; this is the "forbidden radiation"
resulting from electron transitions from the metastable
levels in the atoms. Unknown on Earth, such transitions
are possible only under the extremely rarified conditions
of the gases in a nebula. In addition to the prominent
greenish tinge, other faint colors are known to be present
but are not detected by the eye. A dramatic color photograph of the Orion Nebula, made by W.C.Miller at Palomar,
shows spectacular hues of red, blue, and purple; these
tints are not seen visually with any telescope because of
the low intensity of the light.
        As in all the nebulae, hydrogen appears to be the
most plentiful constituent. The calculated composition of

DESCRIPTIVE NOTES   (Cont'd)

a cubic foot of the Orion Nebula is about as follows:

Hydrogen-	25,000,000	atoms	Sulphur-	900	atoms
Helium-	2,500,000	"	Neon-	250	"
Carbon	15,000	"	Chlorine-	50	"
Oxygen-	6,250	"	Argon-	38	"
Nitrogen-	5,000	"	Fluorine-	3	"

Much of the intricate structure which appears on photographs may be seen visually in an 8-inch or larger telescope; indeed the eye has an advantage over the camera in the bright central region which invariably appears completely "burned out" on photographs due to strong over-exposure. This central area, known as the "Huyghenian Region" shows remarkable detail and will stand considerable magnification since the surface brightness is high. The intricate curdling of the luminous masses resembles a "mackerel sky" or the patterns traced by frost-work. Just north of this bright region, a dark projection often called the "fish-mouth" separates the main portion of the nebula from the detached mass numbered M43, illuminated by its own central 8th magnitude star. On the west side of the main nebula there is a wonderful region of delicate streamers and slender filaments which can be traced out for immense distances. On the east side there are a few thicker and brighter filaments which curve southward in the direction of the bright star Iota Orionis; the most prominent of these is very clear even in a 6-inch glass. A huge arc of faint nebulosity continues outward from this area, turning back to the westward as it passes Iota, then curving north to join the main nebula again on the west side. This structure is thus a vast loop some 30 light years wide, but too faint to be studied visually.

The Orion Nebula is the brightest feature of a chain of objects called the "Sword of Orion" (photograph on page 1325). Half a degree north of the Great Nebula is the star 42 Orionis (magnitude 4.65, spectrum B3) surrounded by the nebulosity NGC 1977; noticeably fainter than M42 it shows very little detail in small instruments. Another 25' to the north is the scattered grouping of 8th to 10th magnitude stars designated NGC 1981. On the other side of the Nebula, half a degree south, is the bright star Iota Orionis (magnitude 2.76, spectrum O9) with two easy

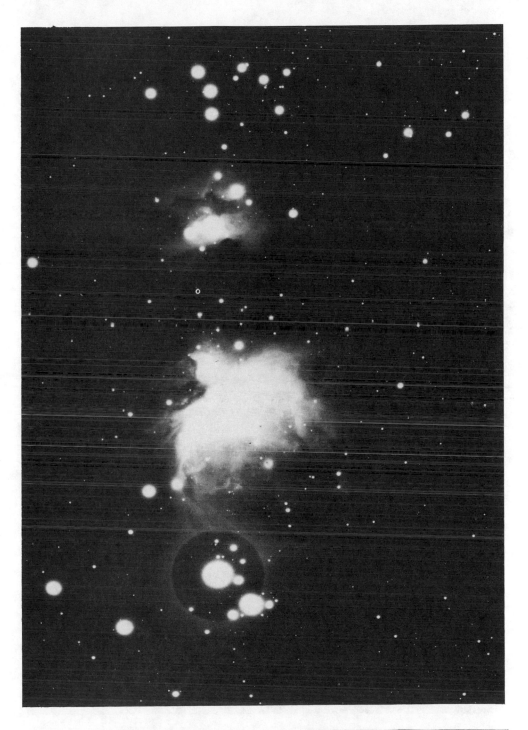

THE SWORD OF ORION. M42 is the bright central mass; above
it are the loose cluster NGC 1981 and the nebulosity NGC
1977. Below M42 is the bright star Iota Orionis. This
photograph was made with the 13-inch camera at Lowell.

CENTRAL REGION OF THE ORION NEBULA M42, showing the entire Huyghenian Region and the multiple star Theta Orionis. The photograph was made with the 120-inch reflector at Lick Observatory.

1326

companions for the small telescope. (See page 1306). And
finally, in the heart of the nebula itself is the noted
quadruple, or multiple star, Theta Orionis.
        This is probably the best known multiple star in
the sky, and one of the most interesting for the small
telescope. The four brighter components form a little
quadrangle called the "Trapezium" and the object is a
favorite of all observers. In many double star catalogs it
has become customary to designate the four stars A,B,C, and
D, but in order of right ascension rather than in the usual
order of brightness. The star called "C" is thus the true
primary of the group with a visual magnitude of about 5.4;
the spectral type is near 06. Star "D" is second in brightness at 6.3 and "A" is third at about 6.8. The spectra of
these two are B0 and A7 respectively. The faintest star,
"B" is an eclipsing binary with a period of 6.471 days, and
is also known under its variable star designation which is
"BM Orionis". Star "A" was in 1975 identified as an eclipsing binary also, with a period of 65.432 days and a visual
range of 6.7 to 7.7; primary eclipse lasts about 20 hours
with a 2½ hour constant minimum at mid-eclipse.
        A fifth star, called "E" was discovered by F.G.W.
Struve in 1826; it is an 11th magnitude object located

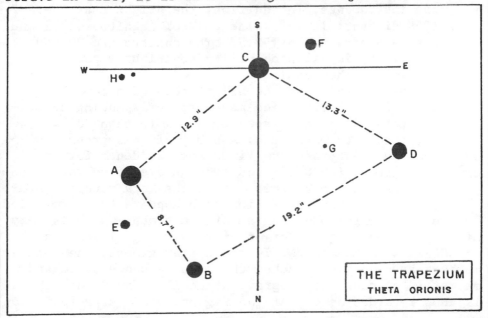

THE TRAPEZIUM
THETA ORIONIS

about 4" from star A, nearly due north. A sixth star, "F",
lies about 4" from C; it is also an 11th magnitude object
and was discovered by John Herschel in 1830. It is a very
curious fact that these two stars were not detected by Sir
William Herschel with his giant telescopes, although both
are visible in a 6-inch telescope today if atmospheric
conditions are good, and have been detected with apertures
of under 3 inches. T.W.Webb suggested that star E "is
believed to have become visible only of late years; perhaps
it may be brightening, as it has been seen with a 3.8 inch
..... Bond's 15-inch achromat has shown it in full daylight
........E. of Rosse sees it very red..." Admiral Smyth
thought the four bright Trapezium stars had the colors
"pale white, faint lilac, garnet, and reddish", but his
impression of reddish tints in any of these stars must be
attributed to the effect of contrast with the greenish
background of the nebulosity.

A very faint star designated "G" was discovered
within the Trapezium by A.G.Clark in 1888 with the 36-inch
refractor at Lick Observatory. It is a 16th magnitude
object. Later in the same year Barnard detected a faint
double star "H" with a separation of 1.3", both components
being about 16th magnitude. These objects can be seen only
in large instruments. In addition, the bright star Theta-2
lies 135" distant; it is a wide pair of magnitudes 5.2 and
6.5, with a separation of 52.5". The spectra are O9 and B1
and the primary in all probability contributes to the
illumination of the nebula.

The Trapezium is the bright core of a compact
cluster of faint stars which may form an expanding association. Within 5' of the Trapezium are more than 300 stars
brighter than 17th magnitude. A study of this group has
been made by K.A.Strand; he finds some evidence for an
expansion rate which would indicate an age of no more than
300,000 years, and make this one of the very youngest star
clusters known. This conclusion also appears to be verified
by the color-magnitude diagram of the group, which is very
similar to the diagrams of other very young clusters such
as NGC 2264 in Monoceros. In these star groups, the fainter
members have not yet reached the main sequence state and
may still be contracting gravitationally. The Orion Nebula
is thus regarded as one of the regions where star formation

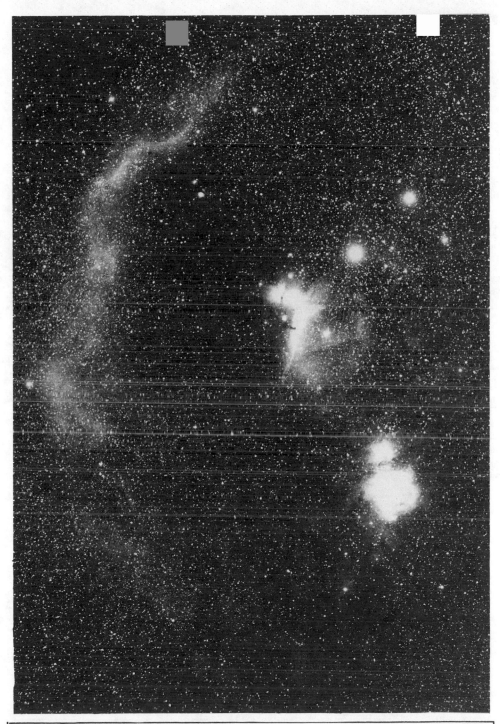

ENVIRONS OF THE ORION NEBULA. M42 is the brightest mass at the right, the three Belt stars appear above. The huge arc at the left is Barnard's "Orion Loop", possibly an ancient supernova remnant.   Photograph by Alan McClure.

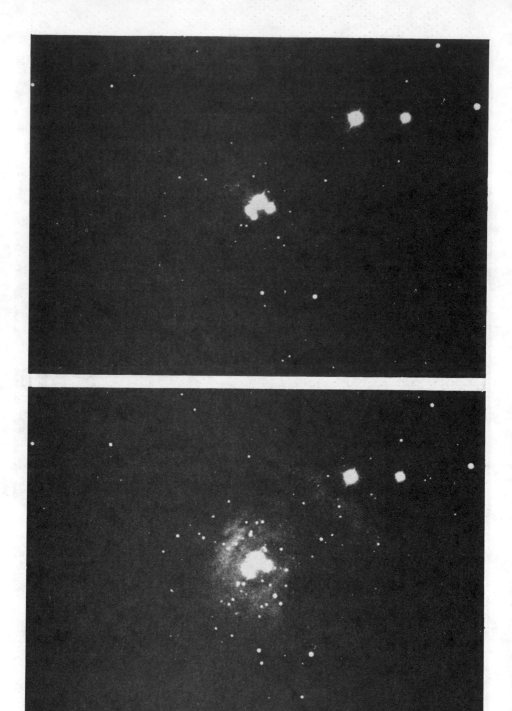

THE TRAPEZIUM GROUP in yellow-green light (top) and infrared (below) showing the cluster of faint red stars which
surrounds the bright quadruple.   Mt.Wilson Observatory

DESCRIPTIVE NOTES   (Cont'd)

is still occurring at the present time. Photographs made
with the 36-inch telescope at Lick Observatory may have
actually recorded a stage in the process; plates obtained
in 1947 and 1954 show conspicuous progressive changes in
several of the small nebulous "knots" in the Orion Cloud.
These small bright condensations, called "Herbig-Haro
Objects" have been suspected to be stars in their early
formative stages. Rather curiously though, none of the
famous dark "globules" seem to exist in M42; these are
present in considerable numbers in the Nebula M8 in Sagittarius and the Rosette (NGC 2237) in Monoceros, and have
been tentatively identified as proto-stars.

    Of great interest is the discovery of a large num-
ber of faint reddish stars, including many erratic variables, scattered throughout the region of the Orion Nebula.
The majority of these appear to be nebular variables of the
T Tauri class, usually considered to be very young stars,
while others are flare stars of the UV Ceti type. The best
known variable in the region is the erratic T Orionis,
discovered by Bond in 1863 during a systematic survey of
the Nebula and its stars. It is located on the outer edge
of the thickest filament on the east side of the Nebula,
about 5.2' south of the Trapezium and 0.6m following in
right ascension. At maximum this star may be about 9th
magnitude, and sometimes remains nearly constant for three
or four months. Then it will display sudden periods of
rapid and irregular fluctuation. Its changes are completely
unpredictable. The spectral type is about early A, but may
vary from B8 to about A3, with emission lines. A similar
star is V361 Orionis, located 2.2' east of Theta-2; when
quite active it may reach maximum every few days. V361 has
a spectral type of about B5.

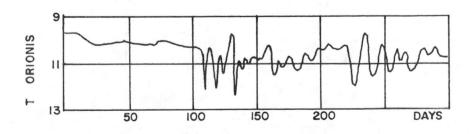

DESCRIPTIVE NOTES   (Cont'd)

Within the bright glow of the Orion Nebula, more
than 50 variable stars have been identified which attain
14th magnitude or brighter at maximum. Most of these have
never been systematically studied, but the majority seem
to show erratic variations which on occasion may be surprisingly rapid.   A typical light range is about 1.5 magnitudes. Stars which attain 12th magnitude or brighter are
included in the accompanying charts and list. These stars
present a rich field for study by the amateur observer.
Because of the rapid nature of the variations, magnitude
estimates should be made every night when possible.   For
comparison purposes, magnitudes of field stars are shown
on the charts, but with decimal points omitted to avoid
confusion with star images. Thus "107" indicates a magnitude of 10.7

Star	Magnitude range	Star	Magnitude range
EZ	11.2-----12.6	IU	8.8-----10.0
V372	7.4----- 8.6	KS	9.9-----10.9
LL	10.9-----12.5	LP	8.4----- 9.3
LQ	11.8-----13.0	LR	11.9-----13.2
LU	12.0-----13.4	LX	11.9-----13.1
BM	8.0-----8.7	MR	10.3-----12.0
MV	11.7-----13.2	TU	11.6-----14..
MX	9.6-----10.5	V358	11.9-----12.4
NP	11.5-----12.6	NQ	11.1-----12.4
AI	12.0-----14.2	AK	11.3-----14.0
NU	6.5----- 7.6	V361	8.1----- 9.8
NV	9.5-----11.3	NZ	11.9-----14.2
AN	10.5-----12.1	T	9.4-----12.6

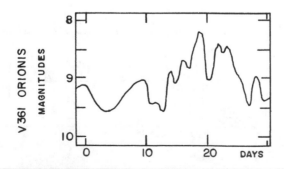

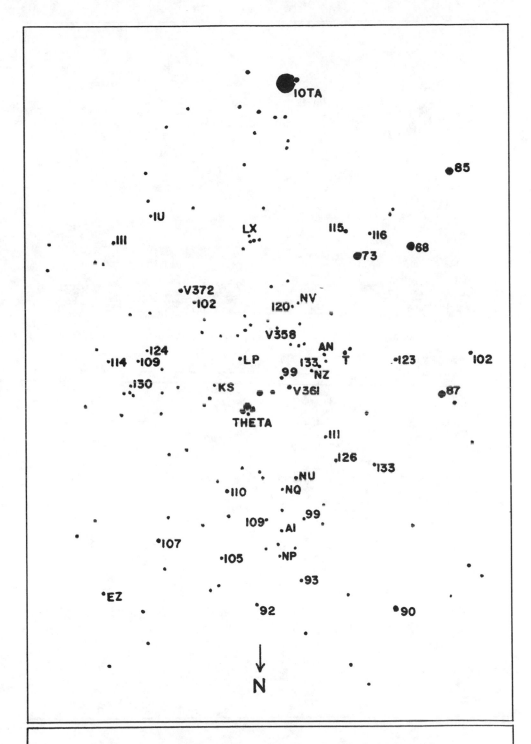

CHART OF VARIABLES and COMPARISON STARS
IN THE ORION NEBULA

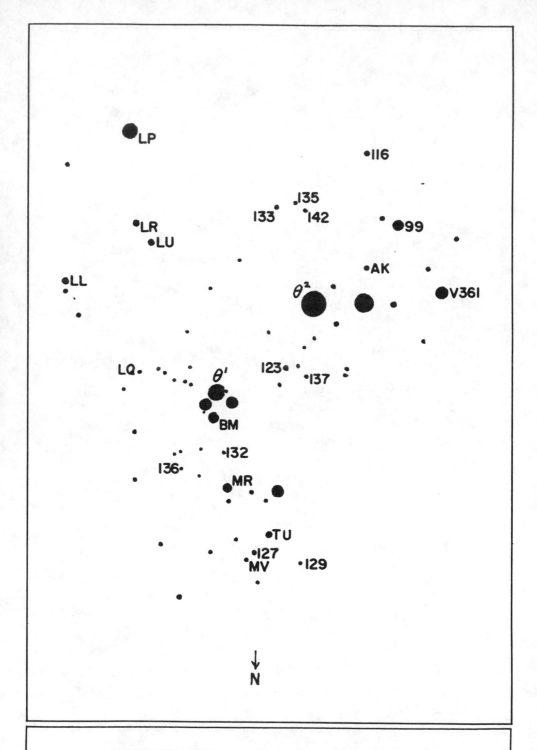

VARIABLE STARS NEAR THE TRAPEZIUM IN
THE ORION NEBULA

DESCRIPTIVE NOTES   (Cont'd)

At Mt.Wilson and Palomar Observatories in 1970, accurate colors and magnitudes were measured for 53 stars fainter than the 12th magnitude within 0.25° of Theta Orionis. This study gave a distance modulus of 8.5 magnitudes, and showed that stars fainter than absolute magnitude +0.5 lie above the main sequence and are apparently still in a stage of gravitational contraction. The survey also identified a number of extremely red stars whose very strong color indices appear to be the result of absorption in the range of 5 to 10 magnitudes, visual. An extreme case of this sort is the strong infrared point source called Becklin's Star, now believed to be a normal main sequence star observed through about 80 magnitudes of absorption!

The visual appearance of the Great Nebula gives an impression of chaotic turbulence. This impression is borne out by spectrum analysis, for it has been discovered that there are large differences in the measured radial velocity at various points within the cloud. While the Trapezium group itself shows a recession velocity of about 20 miles per second, some portions of the Nebula are approaching and some receding with respect to this mean. The largest velocities measured in the Nebula appear to be about 15 miles per second, again taking the mean as a standard. The thick filament on the east side of the Nebula, and the isolated portion M43 both have higher recession velocities than the Trapezium, while the highest approach velocities are found in a region several minutes of arc west of the Trapezium.

In addition to this irregular pattern of turbulence some evidence exists for a general expansion of the entire Nebula; studies of the spectral features of doubly ionized oxygen yield an expansion rate of about 6 miles per second in the central region, decreasing to about 4 miles per second in the outer portions. This expansion is believed to be the result of the radiation of the hot stars of the central cluster; studies of these motions seem to indicate that these stars may be younger than any other known. According to P.Vandervoort (1970) the theoretical model which gives the best representation of the present state of the nebula implies an age of about 23,000 years since the Trapezium stars began to radiate. These results, although still somewhat uncertain, suggest that new stars may be forming in the nebula even at the present time. It would also appear

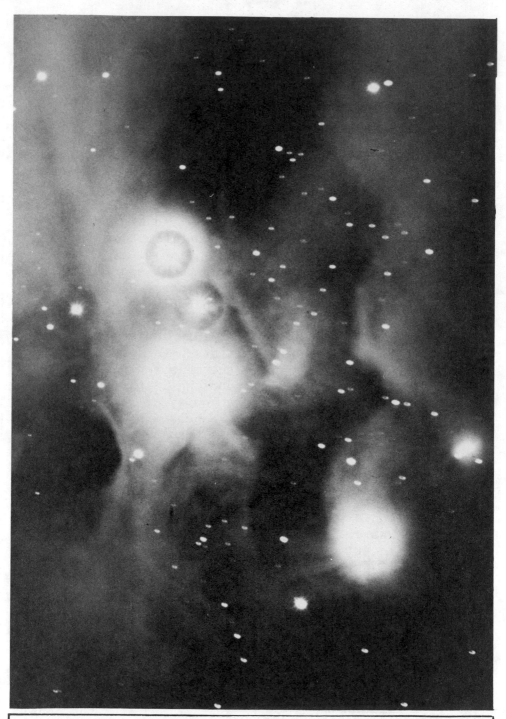

NGC 1977. This nebulosity surrounds the star 42 Orionis, about ½° north of the main mass of the Orion Nebula. Lowell Observatory 42-inch reflector photograph.

that such a region, rich in rapidly evolving and very
massive stars, might show an abnormal frequency of supernovae. Although no outbursts of either novae or supernovae
have been recorded in the 3½ centuries that the nebula has
been under surveillance, there is good evidence that several cataclysms of this type have occurred there in the
past. One probable supernova remnant is the huge arc of
faint nebulosity called "Barnard's Loop", shown on page
1329; it encloses the stars of both the Belt and the Sword,
and in its entirety would cover an area over 10° in diameter. This feature resembles the much smaller Cygnus Loop
(Veil Nebula) but is evidently a much more ancient object.
The eastern arc of Barnard's Loop is fairly well defined,
but only scattered nebulous patches mark the western half;
among these is the oddly shaped "Witch Head Nebula" IC 2118
in Eridanus, which seems to receive its faint illumination
from Rigel in Orion, 2½° away.

There has been much speculation concerning the three
O and B type "Runaway Stars" AE Aurigae, 53 Arietis, and
Mu Columbae, which show high space velocities outward from
the Orion Nebula region. From the present trajectories, it
seems that the three stars were expelled from the Orion
region from about 2.0 to 5.0 million years ago, possibly
through some process connected with supernova explosions.
It is theorized that each of these stars was possibly a
member of a close binary system with high orbital velocities; the explosion of the companion would then release the
other star, which would continue to move out into space at
the same high velocity. The plotted paths of the three
"runaways" are shown on page 288.

The possible presence of a "black hole" in the Orion
Complex remains speculative. The X-ray source in question,
called 2U0525-06, lies close to the position of Theta-2
Orionis, the closest bright star to the Trapezium, at 135"
distance, but the identification, in 1976, had not been
established with certainty. Theta-2 is a known spectroscopic binary with a period of 21.0315 days, spectral type 09
and computed mass of 15 to 20 suns. If the X-ray source is
actually the companion to this star, the system would bear
a close resemblance to the fairly convincing black hole
candidate Cygnus X-1. (Refer also to pages 413 and 793)

THE HORSEHEAD NEBULA. The strange formation is probably
the best known example of a dark nebula.
          Mt.Wilson Observatory, 100-inch telescope plate.

DESCRIPTIVE NOTES (Cont'd)

M78   (NGC 2068) Position 05442n0002. Bright diffuse nebula located about 2.3° NE from Zeta Orionis and virtually on the celestial equator; it was discovered in 1780 by P.Mechain who described it as two fairly bright nuclei surrounded by nebulosity. Messier, who observed it later the same year, thought it "a cluster of stars with much nebulosity". Admiral Smyth saw "two stars in a very wispy nebula", in which Lord Rosse thought to find some indication of spiral structure. According to Lick Observatory Publications Volume XIII the object is a "mass of rather irregular fairly bright diffuse nebulosity whose brighter portion is 6' x 4', involving two 10th magnitude stars...two fainter patches lie 6' W, apparently separated from the main mass by a wide lane of dark matter; the south one of these is NGC 2064 and the northern one NGC 2067".

    M78 is one of the brighter portions of a vast nebulosity which covers much of Orion and which becomes visible in the presence of hot early-type stars. Easily located in small telescopes, the object shows very little detail to the visual observer. The two brighter stars enclosed within the cloud are about 53" apart in PA 18° and the nebulosity itself is rather sharply bordered on the NW side, melting away into the sky background toward the SE. For a space of about half a degree around the nebula, the sky is heavily obscured by an absorbing cloud through which scarcely a star shows.

    The distance is believed to be about the same as Zeta Orionis, or about 1600 light years; the nebulosity then is 2 to 3 light years in diameter. Both of the embedded stars are B-type giants with absolute magnitudes of about −1.5.

B33   Position 05387s0232. This is the famed dark "Horsehead Nebula", undoubtedly the best known example of a dark nebula in the entire heavens. However, it is almost completely invisible to the eye at the telescope and requires long exposure photographs to reveal its strange and spectacular details.

    It is located about half a degree south of the bright star Zeta Orionis, in the long stretch of nebulosity IC 434 which extends for about a degree south of the star. This nebulosity was probably first detected by E.Pickering

THE HORSEHEAD NEBULA. A "close-up" of the famous dark
nebula obtained with the 200-inch reflector at Palomar
Observatory.

on photographs made in 1889, and the dark Horsehead itself
shows clearly on a plate made in 1900 and published in the
Astrophysical Journal in 1903. The significance of the
object was not immediately recognized and the early descriptions refer to it as a "bay" or gap in the nebulosity.
E.Barnard seems to have been the first to recognize it as
a great obscuring mass of some sort, seen against a bright
region of nebulosity.
          Much of the illumination of this region must be due
to the radiation of Zeta Orionis, but it has also been
suggested that IC 434 is actually a collision zone between
two cosmic dust clouds, made visible by the energy of the
impact. The area thus affected measures some 18 light years
in extent, from north to south.
          About halfway along the eastern edge of this mist
cloud is the dark projection of the Horsehead, sharply outlined against the bright background, and irresistibly
conjuring up surrealistic visions of a cosmic chess game.
The horsehead itself is a great obscuring cloud of dark
matter, dust and non-luminous masses of gas in the form of
frozen crystals according to some theorists; perhaps even
large solid particles are involved . The total diameter
a little more than one light year,or about 70,000 times the
distance between the Earth and the Sun. But in addition to
the Horsehead itself, a huge dark cloud appears to border
the bright nebulosity along its entire eastern rim. This
black haze evidently extends for a great distance toward
the east, as shown by the thin star fields in that direction when compared with the much richer fields to the west.
          The distance of the Horsehead is estimated to be
about 1200 light years. The dark cloud may, of course, be
nearer than the bright nebulosity IC 434, but how much
nearer is not definitely known. If IC 434 is actually being
illuminated by Zeta Orionis, its distance must be closely
comparable, about 1600 light years, and this figure quite
obviously sets the maximum possible distance for the
Horsehead.
          For the visual observer, the Horsehead must be
classed as one of the most difficult objects in the sky.
Barnard could find no definite sign of it with the 40-inch
Yerkes refractor in 1913, and observations with other large
telescopes have been equally unsuccessful. With a really

DESCRIPTIVE NOTES   (Cont'd)

excellent sky, a good 8-inch or 10-inch rich-field telescope used with a wide angle eyepiece will sometimes show
a faint hint of the outline.   It is definitely a more
difficult object than the Veil Nebula in Cygnus, and probably somewhat more uncertain than the Merope Nebula in the
Pleiades. Those who have detected the Merope Nebula, however, may expect to eventually succeed with the Horsehead.
Walter S.Houston has detected it with a 5-inch Moonwatch
apogee telescope, and Leslie C.Peltier has succeeded with
a 6-inch refractor and a low-power ocular. One of the
difficulties is the presence of a 2nd magnitude star only
half a degree away; Zeta Orionis must be kept outside the
field or occulted with some sort of temporary obscuring
bar placed in the focal plane of the eyepiece. Only the
darkest and cleanest of skies offer any hope of success.
Finally, in the matter of dark adaption, the hopeful observer could do no better than to employ the method used by
John Herschel, who kept his eyes in total darkness for 20
to 30 minutes when preparing to observe an especially faint
and difficult object.   (Refer also to NGC 6188 in Ara, B86
in Sagittarius, M16 in Serpens, B72 in Ophiuchus, and the
Cone Nebula in NGC 2264 in Monoceros)

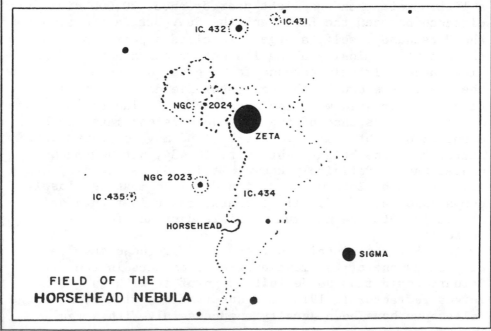

FIELD OF THE
HORSEHEAD NEBULA

REGION OF THE HORSEHEAD NEBULA. A photograph in red light made with a 5.5-inch Zeiss triplet by Alan McClure. For identification of details refer to the chart on page 1342.

THE HORSEHEAD NEBULA. This evocative portrayal hints at the difficulty of visual observations of the Horsehead. The photograph was made with an 8-inch Celestron by David Healy of Manhasset, New York.

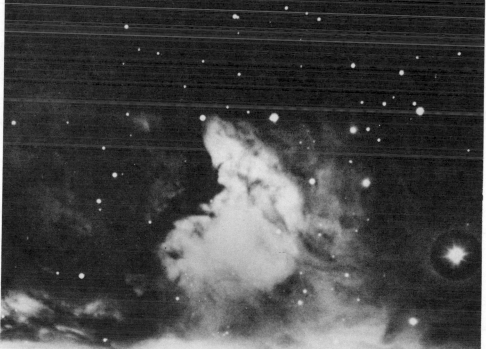

DEEP SKY OBJECTS IN ORION. Top: Star Cluster NGC 2194.
Below: The nebula M43, a detached portion of the Great Nebula M42.   Lowell Observatory photographs

CELESTIAL HANDBOOK- VOLUME TWO

CONSTELLATION INDEX AND STAR ATLAS REFERENCE

CONSTELLATION	PAGE	NORTON'S ATLAS CHART	SKALNATE-PLESO ATLAS CHART
CHAMAELEON	655	16	XVI
CIRCINUS	656	16	XVI XIV
COLUMBA	658	6	XII XIII
COMA BERENICES	661	9	IV IX VIII III
CORONA AUSTRALIS	693	14	XV XIV
CORONA BOREALIS	697	11	IV
CORVUS	716	10	IX XIV VIII
CRATER	723	10	VIII XIII IX
CRUX	726	16	XIV XVI XIII
CYGNUS	735	13	V I IV
DELPHINUS	817	13	XI X V
DORADO	833	15, 16	XVI XII XIII
DRACO	853	1, 2	I IV V III
EQUULEUS	875	13	XI
ERIDANUS	878	6	VI VII XII XVI
FORNAX	895	6	XII VI VII
GEMINI	905	7	III VII VIII
GRUS	945	4	XV XVI XII
HERCULES	950	11	IV X V IX
HOROLOGIUM	997	6, 15	XII XVI
HYDRA	999	8, 10	VIII XIII XIV IX
HYDRUS	1033	15	XVI XII XV
INDUS	1035	14, 15	XV XVI

CONSTELLATION INDEX AND STAR ATLAS REFERENCE

CONSTELLATION	PAGE	NORTON'S ATLAS CHART	SKALNATE-PLESO ATLAS CHART
LACERTA	1039	3	V I II
LEO	1047	7, 9	VIII III IX IV
LEO MINOR	1083	9	III VIII
LEPUS	1087	6	VII XII XIII
LIBRA	1101	12	IX XIV X
LUPUS	1111	12	XIV XVI
LYNX	1123	1, 7	III II I
LYRA	1131	13	V IV
MENSA	1179	15, 16	XVI
MICROSCOPIUM	1180	14	XV
MONOCEROS	1182	7, 8	VII VIII
MUSCA	1215	16	XVI XIV XIII
NORMA	1218	12	XIV XV XVI
OCTANS	1221	15, 16	XVI
OPHIUCHUS	1223	11, 12	X XIV IX
ORION	1271	5, 6	VII III II